Academic Press Rapid Manuscript Reproduction

Proceedings of the International Conference
on Colloids and Surfaces–50th Colloid and Surface Science Symposium,
held in San Juan, Puerto Rico
on June 21–25, 1976

Colloid and Interface Science

VOL. IV

Hydrosols and Rheology

Colloid and Interface Science

VOL. IV

Hydrosols and Rheology

EDITED BY

MILTON KERKER

Clarkson College of Technology
Potsdam, New York

Academic Press Inc.

New York San Francisco London 1976

A Subsidiary of Harcourt Brace Jovanovich, Publishers

ACADEMIC PRESS, INC.
111 Fifth Avenue, New York, New York 10003

United Kingdom Edition published by
ACADEMIC PRESS, INC. (LONDON) LTD.
24/28 Oval Road, London NW1

Library of Congress Cataloging in Publication Data

INternational Conference on Colloids and Surfaces,
50th, San Juan, P.R., 1976.
Colloid and interface science.

CONTENTS: v. 2. Aerosols, emulsions, and surfactants.–v. 3. Adsorption, catalysis, solid surfaces, wetting, surface tension, and water.–v. 4. Hydrosols and rheology. [etc.]
1. Colloids–Congresses. 2. Surface chemistry –Congresses. I. Kerker, Milton, II. Title.
QD549.I6 1976 541′.345 76-47668
ISBN 0–12–404504–9 (v. 4)

PRINTED IN THE UNITED STATES OF AMERICA

Contents

CONTENTS

List of Contributors

G.E. Adams, Department of Applied Mathematics, Research School of Physical Sciences, Institute of Advanced Studies, The Australian National University, Canberra, ACT. 2600, Australia

John L. Anderson, School of Chemical Engineering and Center for Applied Mathematics, Cornell University, Ithaca, New York 14853

M.A. Anderson, Water Chemistry Program, University of Wisconsin, Madison, Wisconsin 53706

C.P. Bean, Physical Science Branch, Energy Science and Engineering, General Electric Corporate Research and Development, P.O. Box 8, Schenectady, New York 12301

P.D. Bhatnagar, National Council of Educational Research, Ministry of Education, Government of India, Regional College, Ajmer, India

R.L. Blokhra, Chemistry Department, Himachal Pradesh University, The Manse, Simla–171001, India

L. Blum, Physics Department, University of Puerto Rico, Rio Piedras, Puerto Rico 00931

N. Bonacci, "Ruder Boskovic" Institute, 41001 Zagreb, Croatia, Yugoslavia

E.B. Bradford, Central Research, Physical Research Laboratory, Dow Chemical USA, Midland, Michigan 48640

D.E. Brooks, Department of Pathology, University of British Columbia, Vancouver 8, Canada

J.C. Brown, Department of Physical Chemistry, School of Chemistry, The University, Cantock's Close, Bristol BS8 1TS, England

William C. Browning, SCT Associates, 610 Whitewing Lane, Houston, Texas 77024

P.E. Cladis, Bell Laboratories, 600 Mountain Avenue, Murray Hill, New Jersey 07974

Felix Dalang, Swiss Federal Institute of Technology, EAWAG, Zurich, Switzerland

R.W. DeBlois, General Electric Corporate Research and Development, P.O. Box 8, Schenectady, New York 12301

R. Despotović, Laboratory of Colloid Chemistry, "Ruder Boskovic" Institute, 41001 Zagreb, P.O. Box 1016, Croatia, Yugoslavia

Victor K. Dunn, Xerox Corporation, 1341 West Mockingbird Lane, Dallas, Texas 75247

A.R. Eggert, The Institute of Paper Chemistry, Appleton, Wisconsin 54911

Birgit Eppler, University of Karlsruhe, Institut für Siedlungswasserwirtschaft, 75 Karlsruhe 1, Postfach 6380, Kaiserstrasse 12, West Germany

Gérard R. Feat, Department of Mathematics, University of Manchester, Manchester M13 9PL, England

N. Filipović-Vinceković, Laboratory of Colloid Chemistry, "Ruder Boskovic" Institute, 41001 Zagreb, P.O. Box 1016, Croatia, Yugoslavia

Shoji Fukushima, Shiseido Laboratories, 1050 Nippacho, Kohokuku, Yokohama, Japan

John W. Gardner, Department of Chemical Engineering, University of Texas at Austin, Austin, Texas 78712

J. Calvin Giddings, Department of Chemistry, University of Utah, Salt Lake City, Utah 84112

Thomas Gillespie, Division of Engineering and Technology, Saginaw Valley State College, University Center, Michigan 48710

P.J. Goetz, Pen Kem Company, P.O. Box 364, Croton-on-Hudson, New York 10520

J.W. Goodwin, Department of Physical Chemistry, University of Bristol, Bristol BS8 1TS, England

John Gregory, Department of Civil & Municipal Engineering, University College London, Gower Street, London WCIE 6BT, England

James N. Groves, Diagnostics Projects, Life Sciences Branch, General Electric, Corporate Research and Development, P.O. Box 8, Schenectady, New York 12301

Hermann H. Hahn, University of Karlsruhe, Institute für Siedlungswasserwirtschaft, 75 Karlsruhe 1, Postfach 6380, Kaiserstrasse 12, West Germany

Christer Heinegård, Swedish Forest Products Research Laboratory, Box 5604, S-114 86 Stockholm, Sweden

A.H. Herz, Research Laboratories, Eastman Kodak Company, Rochester, New York 14650

A. Homola, Department of Physical Chemistry, University of Melbourne, Parkville, Victoria 3052, Australia

D.B. Hough, School of Chemistry, University of Bristol, Bristol BS8 1TS, England

C.P. Huang, College of Engineering, Department of Civil Engineering, 130 Dupont Hall, University of Delaware, Newark, Delaware 19711

Robert J. Hunter, University of Sydney, Sydney 2006, New South Wales, Australia

J.N. Israelachvili, Department of Applied Mathematics, Research School of Physical Sciences, Institute of Advanced Studies, The Australian National University, Canberra, ACT. 2600, Australia

Wanda Y. Jackson, Martin Marietta Laboratories, 1450 South Rolling Road, Baltimore, Maryland 21227

R.O. James, Department of Physical Chemistry, University of Melbourne, Parkville, Victoria 3052, Australia

M. Kahlweit, Max-Planck-Institut für Biophysikalische Chemie, D-34 Göttingen, West Germany

A.M. Khidher, Department of Physical Chemistry, University of Bristol, Bristol BS8 1TS, England

Rudolf Klute, Institut für Siedlungswasserwirtschaft, Universität Karlsruhe, West Germany

J.L. Kolbe, Martin Marietta Laboratories, 1450 South Rolling Road, Baltimore, Maryland 21227

P.H. Krumrine, Center for Surface and Coatings Research, Lehigh University, Bethlehem, Pennsylvania 18015

Shigenori Kumagai, Shiseido Laboratories, 1050 Nippacho, Kohokuku, Yokohama, Japan

Janusz Laskowski, Laboratory of Mineral Processing, Department of Inorganic Chemistry and Metallurgy of Rare Elements, Wroclaw Technical University, 50-370 Wroclaw, Poland

Janusz Lekki, Laboratory of Mineral Processing, Department of Inorganic Chemistry and Metallurgy of Rare Elements, Wroclaw Technical University, 50-370 Wroclaw, Poland

Samuel Levine, Department of Mathematics, University of Manchester, Manchester M13 9PL, England

Tom Lindström, Swedish Forest Products Research Laboratory, Box 5604 S-114 86 Stockholm, Sweden

Kenneth J. Lissant, Petrolite Corporation, Tretolite Division, 369 Marshall Avenue, St Louis, Missouri 63119

James McHardy, Department of Mining and Metallurgical Engineering, McGill University, P.O. Box 6070, Station 'A', Montreal, Quebec, Canada H3C 3GI

A.J. McHugh, Emulsion Polymers Institute, Lehigh University, Bethlehem, Pennsylvania 18015

D.T. Malotky, Water Chemistry Program, University of Wisconsin, Madison, Wisconsin 53706

Richard J. Mannheimer, Southwest Research Institute, 8500 Culebra Road, P.O. Drawer 28510, San Antonio, Texas 78284

F.J. Micale, Center for Surface and Coatings Research, Lehigh University, Bethlehem, Pennsylvania 18015

I. Michaeli, Weizmann Institute of Science, Rehovot, Israel

T.S. Mika, Joseph C. Wilson Center for Technology, Xerox Corporation, Rochester, New York 14644

Michel Mille, School of Pharmacy, University of Wisconsin, Madison, Wisconsin 53706

W.J. Miller, Research Laboratories, Eastman Kodak Company, Rochester, New York 14650

V. Mohan, Department of Chemical Engineering, Illinois Institute of Technology, Chicago, Illinois 60616

L.E. Morford, Central Research, Physical Research Laboratory, Dow Chemical USA, Midland, Michigan 48640

Noboru Moriyama, Research Laboratories, Kao Soap Co., Wakayama-shi 640-91, Japan

Marcus N. Myers, Department of Chemistry, University of Utah, Salt Lake City, Utah 84112

Uwe Neis, Institut für Siedlungswasserwirtschaft, Universität Karlsruhe, West Germany

H. Opperhauser, Martin Marietta Laboratories, 1450 South Rolling Road, Baltimore, Maryland 21227

R.H. Ottewill, School of Chemistry, University of Bristol, Bristol BS8 1TS, England

M.L. Parmar, Chemistry Department, Himachal Pradesh University, The Manse, Simla - 171001, India

J.G. Penniman, Pen Kem Company, P.O. Box 364, Croton-on-Hudson, New York 10520

V.A. Phillips, Martin Marietta Laboratories, 1450 South Rolling Road, Baltimore, Maryland 21227

G.W. Poehlein, Emulsion Polymers Institute, Lehigh University, Bethlehem, Pennsylvania 18015

V. Pravdić, "Ruder Boskovic" Institute, 41001 Zagreb, Croatia, Yugoslavia

Dennis C. Prieve, Faculty of Engineering and Applied Science, State University of New York at Buffalo, Buffalo, New York 14214

P.N. Pusey, Department of Physical Chemistry, School of Chemistry, The University, Cantock's Close, Bristol BS8 1TS, England

C. Christopher Reed, School of Chemical Engineering and Center for Applied Mathematics, Cornell University, Ithaca, New York 14853

Eli Ruckenstein, Faculty of Engineering and Applied Science, State University of New York at Buffalo, Buffalo, New York 14214

T. Salman, Department of Mining and Metallurgical Engineering, McGill University, P.O. Box 6070, Station 'A', Montreal, Quebec, Canada H3C 3GI

R.S. Schechter, Department of Petroleum Engineering, University of Texas at Austin, Austin, Texas 78712

G.V.F. Seaman, University of Oregon Medical School, Portland, Oregon 97201

Alan R. Sears, Diagnostics Projects, Life Sciences Branch, General Electric, Corporate Research and Development, P.O. Box 8, Schenectady, New York 12301

V.P. Sharma, Chemistry Department, Himachal Pradesh University, The Manse, Simla–171001, India

C. Silebi, Emulsion Polymers Institute, Lehigh University, Bethlehem, Pennsylvania 18015

Jan Skalny, Martin Marietta Laboratories, 1450 South Rolling Road, Baltimore, Maryland 21227

John C. Slattery, Northwestern University, Department of Chemical Engineering, Evanston, Illinois 60201

La Rell K. Smith, Department of Chemistry, University of Utah, Salt Lake City, Utah 84112

Ian Snook, Royal Melbourne Institute of Technology, 124 La Trobe Street, Melbourne, Victoria, 3000, Australia

Christer Söremark, Swedish Forest Products Research Laboratory, Box 5604, S–114 86 Stockholm, Sweden

G. Stell, Department of Mechanical Engineering, State University of New York, Stony Brook, New York 11794

Melvin D. Sterman, Research Laboratories, Eastman Kodak Company, Rochester, New York 14650

Dirk Stigter, Western Regional Research Center, Agricultural Research Service, U.S. Department of Agriculture, Albany, California 94710

R.A. Stratton, The Institute of Paper Chemistry, Appleton, Wisconsin 54911

Werner Stumm, Swiss Federal Institute of Technology, EQWAG-ETH, CH 8600 Dubendorf, Zurich, Switzerland

B. Subotić, Laboratory of Colloid Chemistry, "Ruder Boskovic" Institute, 41001 Zagreb, P.O. Box 1016, Croatia, Yugoslavia

Maher E. Tadros, Martin Marietta Laboratories, 1450 South Rolling Road, Baltimore, Maryland 21227

T.F. Tadros, I.C.I. Plant Protection Division, Jealott's Hill Research Station, Brackness RG12 6EY, England

Takeuchi Takashi, Research Laboratories, Kao Soap Co., Wakayama-shi 640-91, Japan

E.D. Tarapore, Department of Materials Science and Engineering, University of California, Berkeley, California 94720

S. Torza, Bell Laboratories, 600 Mountain Avenue, Murray Hill, New Jersey 07974

J.W. Vanderhoff, Center for Surface and Coatings Research, Lehigh University, Bethlehem, Pennsylvania 18015

Garret Vanderkooi, Department of Chemistry, Northern Illinois University, DeKalb, Illinois 60115

William van Megen, Royal Melbourne Institute of Technology, 124 La Trobe Street, Melbourne, Victoria 3000, Australia

A. Vrij, Rijksuniversiteit Utrecht, Van't Hoff Laboratorium voor Fysische-en Colloidchemi, Padualaan 8, Utrecht 2506, The Netherlands

H. Walter, Veterans Administration Hospital, Long Beach, California

D.T. Wasan, Department of Chemical Engineering, Illinois Institute of Technology, Chicago, Illinois 60616

Lun-yan Wei, E. I. du Pont de Nemours, Wilmington, Delaware 19898

R.K.A. Wesley, John L. Smith Memorial for Cancer Research, Pfizer, Inc., Maywood, New Jersey 07607

Claude Wolff, Université de Bretagne Occidentale, Laboratoire d'Hydrodynamique Moléculaire, 6 Avenue Le Gorgeu, 29283 Brest Cedex, France

Frank J.F. Yang, Department of Chemistry, University of Utah, Salt Lake City, Utah 84112

Preface

This is the fourth volume of papers presented at the International Conference on Colloids and Surfaces which was held in San Juan, Puerto Rico, June 21–25, 1976.

The morning sessions consisted of ten plenary lectures and thirty-four invited lectures on the following topics: rheology of disperse systems, surface thermodynamics, catalysis, aerosols, water at interfaces, stability and instability, solid surfaces, membranes, liquid crystals, and forces at interfaces. These papers appear in the first volume of the proceedings along with a general overview by A. M. Schwartz.

The afternoon sessions were devoted to 221 contributed papers. This volume includes contributed papers on the subjects of hydrosols and rheology. Three additional volumes include contributed papers on aerosols, emulsions, surfactants, adsorption, catalysis, solid surfaces, wetting, surface tension, water, biocolloids, polymers, monolayers, membranes, and general subjects.

The Conference was sponsored jointly by the Division of Colloid and Surface Chemistry of the American Chemical Society and the International Union of Pure and Applied Chemistry in celebration of the 50th Anniversary of the Division and the 50th Colloid and Surface Science Symposium.

The National Colloid Symposium originated at the University of Wisconsin in 1923 on the occasion of the presence there of The Svedberg as a Visiting Professor (see the interesting remarks of J. H. Mathews at the opening of the 40th National Colloid Symposium and those of Lloyd H. Ryerson in the Journal of Colloid and Interface Science 22, 409, 412 (1966)). It was during his stay at Wisconsin that Svedberg developed the ultracentrifuge, and he also made progress on moving boundary electrophoresis, which his student Tiselius brought to fruition.

The National Colloid Symposium is the oldest such divisional symposium within the American Chemical Society. There were no meetings in 1933 and during the war years 1943–1945, and this lapse accounts for the 50th National Colloid Symposium occurring on the 53rd anniversary.

However, these circumstances brought the numerical rank of the Symposium into phase with the age of the Division of Colloid and Surface Chemistry. The Division was established in 1926, partly as an outcome of the Symposium. Professor Mathews gives an amusing account of this in the article cited above.

The 50th anniversary meeting is also the first one bearing the new name

Colloid and Surface Science Symposium to reflect the breadth of interest and participation.

There were 476 participants including many from abroad.

This program could not have been organized without the assistance of a large number of persons and I do hope that they will not be offended if all of their names are not acknowledged. Still, the Organizing Committee should be mentioned: Milton Kerker, Chairman, Paul Becher, Tomlinson Fort, Jr., Howard Klevens, Henry Leidheiser, Jr., Egon Matijevic, Robert A. Pierotti, Robert L. Rowell, Anthony M. Schwartz, Gabor A. Somorjai, William A. Steele, Hendrick Van Olphen, and Albert C. Zettlemoyer.

Special appreciation is due to Robert L. Rowell and Albert C. Zettlemoyer. They served with me as an executive committee which made many of the difficult decisions. In addition Dr. Rowell handled publicity and announcements while Dr. Zettlemoyer worked zealously to raise funds among corporate donors to provide travel grants for some of the participants.

Teresa Adelmann worked hard and most effectively both prior to the meeting and at the meeting as secretary, executive directress, editress, and general overseer. She made the meeting and these Proceedings possible. We are indebted to her.

Milton Kerker

MACHINE SIMULATION OF COLLOIDAL DISPERSIONS

Ian Snook and William van Megen
Royal Melbourne Institute of Technology

I. INTRODUCTION

The fact that colloidal dispersions are many-body systems must be explicitly taken into account in attempting to describe their properties. For instance, the process of coagulation, the osmotic pressure (1,2) and the occurrence of phase separation (3) can only be correctly explained in terms of the motions of all the particles in the system coupled by the forces acting between them. Although the expression for the force given by the DLVO (4) theory may be quite a reasonable asymptotic form to describe the interaction between two colloidal particles, its direct use to calculate the gross behaviour (3) of a large assembly of particles is quite invalid.

However, it must be appreciated that many-body problems are, in general, very difficult to solve, so that the introduction of models is necessary.

The model (5) we introduce for colloidal dispersions is one of structureless particles moving in a continuum background characterized by its dielectric constant, ε, and electrolyte concentration, n. The particles are assumed to interact with one another via a pairwise additive potential U composed of screened electrostatic repulsion and van der

Waals attraction terms. We treat this system as thermodynamically stable and determine its macroscopic properties by Monte Carlo averaging over the likely configurations of the particles (5).

To enable this model to be dealt with on a computer it is also necessary to replace the actual system with a very large number of particles by a system containing a much smaller number of particles with appropriate boundary conditions. These boundary conditions are chosen so as to approximate the real system as closely as possible. In this study we simulate a very large number of particles in a large container by using periodic boundary conditions. This introduces no significant error provided the potential is relatively short ranged (6). The simulation of, for example, a small floc particle or a system in a very small container, e.g. a capillary tube, would require different boundary conditions (7).

Direct simulation studies are somewhat tedious and require large amounts of computer time. So several approximate methods are also discussed and their results compared with those from exact Monte Carlo work.

To indicate the main features predicted by these theoretical models we show the results for the osmotic or excess pressure and the conditions under which ordering occurs in a typical monodisperse system. We also exhibit a direct comparison of the theoretical predictions and the limited experimental data on well characterized systems.

II. A MODEL SYSTEM

To show the capabilities of the model we have applied it to a monodisperse system of spherical particles (radius a = 4×10^{-7} m, and surface charge $\sigma_o = .02\ C/m^2$) in an aqueous (relative permittivity $\varepsilon_r = 80$) 1:1 electrolyte of concentration n moles/m^3 at room temperature ($T = 300^oK$). The interaction is described by the simple pair potential (4).

$$U(r_{ij}) = U_R(r_{ij}) + U_A(r_{ij}) \qquad [1]$$

where

$$U_A(r_{ij}) = -\frac{A}{3}\left\{\frac{1}{u_{ij}^2 + 4u_{ij}} + \frac{1}{u_{ij}^2 + 4u_{ij} + 4} + \frac{1}{2}\,\ell n\left(\frac{u_{ij}^2 + 4u_{ij}}{u_{ij}^2 + 4u_{ij} + 4}\right)\right\} \qquad [2]$$

and

$$U_A(r_{ij}) = 2\pi a\varepsilon_o\varepsilon\,\psi_o^2\,\ell n(1 + e^{-\kappa a u_{ij}}) \qquad [3]$$

where r_{ij} is the center to center distance between particles i and j, $u_{ij} = \frac{r_{ij}}{a} - 2$, A is the Hamaker constant to which we assigned the value 9×10^{-21}J, ε_o is the permittivity of free space and κ is given by

$$\kappa^2 = \frac{2\,N_A\,ne^2}{\varepsilon_o\varepsilon_r\,kT}, \qquad [4]$$

where N_A is Avogadro's number, e the electronic charge and k the Boltzmann constant. The surface potential ψ_o determined from the surface charge using the relation (4)

$$\psi_o = \frac{2kT}{e}\sinh^{-1}\left\{\frac{\sigma_o}{\sqrt{8\,N_A\,n\,\varepsilon_o\varepsilon_r\,kT}}\right\} \qquad [5]$$

The properties calculated by the usual Monte Carlo averaging process (5) were the excess or osmotic pressure, P, and the radial distribution function, $g(r)$. Figure 1 shows

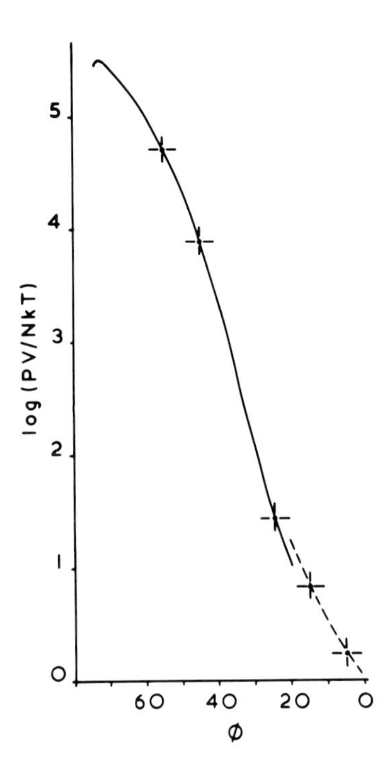

Fig. 1. The logarithm of the reduced pressure, PV/NkT, as a function of the volume fraction, ϕ, for the model system, indicating ——— for the cell model, - - - for first order perturbation theory and –¦– for Monte Carlo results.

the osmotic pressure of this system as a function of the volume fraction ϕ. The concentration of the bulk electrolyte for this system is .1 moles/m^3. The two branches of the curve in this figure represent the pressures of the disordered phase (low volume fractions) and the ordered phase (higher volume fractions). For the ordered phase we used the cell model in which each particle is constrained to a potential cell formed by its nearest neighbours, whilst for the disordered phase we used a hard sphere perturbation model developed for liquids (8). Note the excellent agreement between the results of these approximate methods and those from the Monte Carlo method. At very high volume fractions the pressure reaches a maximum and then drops sharply indicating the onset of coagulation. It is perhaps important to point out that the disorder - to order transition is, like the hard sphere transition, a geometrical effect due to the packing of the particles, i.e. it is a result of the repulsive and not the attractive forces between the particles (9,10). Both the occurrence of ordering (11) and the maximum in the osmotic pressure have been observed experimentally (12).

In Fig. 2 we show the maximum pressure (or coagulation pressure), as a function of the bulk electrolyte concentration assuming a constant surface charge of .02 C/m^2 on the particles. As observed experimentally (12), this pressure first rises with increasing n and then drops sharply. This qualitative behaviour occurs even if constant surface potential rather than constant surface charge is assumed.

Finally, a plot of which phase is stable at a particular volume fraction as a function of n is presented in Fig. 3. This shows the same qualitative features as those observed experimentally by Hachisu et al (11) on a monodisperse system of polystyrene spheres in an aqueous background.

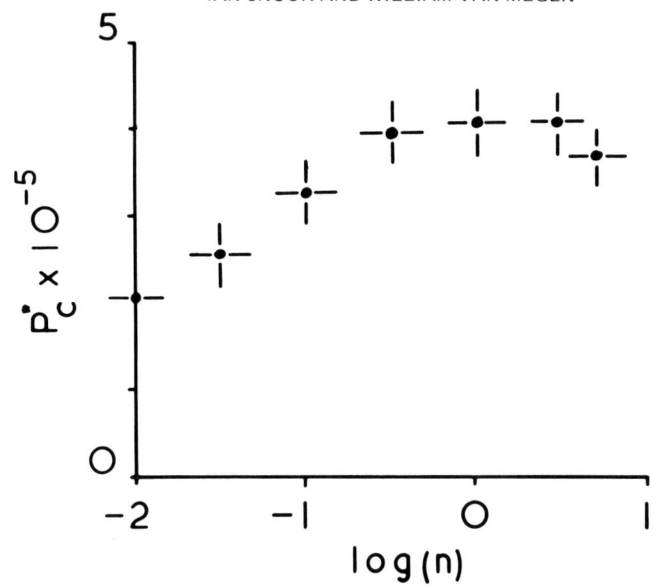

Fig. 2. The coagulation pressure $P_c^* = PV/NkT$ versus the logarithm of the electrolyte concentration (moles/m^3) for the model system.

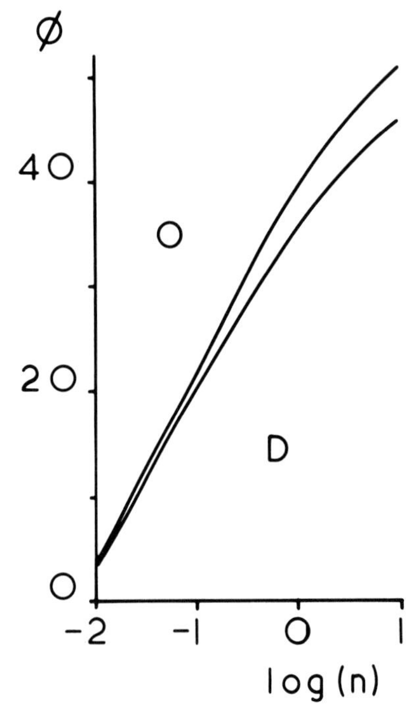

Fig. 3. The regions in which the model system was found to be disordered (D) and ordered (O). The area between the two curves represents the region of coexistence of ordered and disordered dispersions.

III. COMPARISON OF THE MODEL WITH EXPERIMENT

The results presented in the last section show that the model is capable of qualitatively reproducing all the observed features of a stable suspension. However, in order to fully test the model it is necessary to compare its predictions directly with those obtained experimentally on well characterized suspensions. Fortunately, a few such detailed and complete data are available, viz., the pressure cell work of Homola and Robertson (2) and the light scattering experiment of Brown et al (13).

Of course a major problem arises when one wishes to compare directly with experiment, is that the interparticle potential is not known very accurately. In particular, the models available for the repulsive force, which is the dominant component of the total interaction in this study, are rather simplistic in that they are based on the idea of a structureless particle and a structureless background separated by a sharp interface (4).

A further problem is that the many (> 2) particle potentials are not known at all. These are certainly expected to have a significant effect for the situation where the double layers strongly overlap.

Firstly, we consider the results of Homola and Robertson who compressed a polystyrene latex composed of particles of radius 2.95×10^{-7} m with a surface charge of 1.14×10^{-2} C/m^2 in a .1 mole/m^3 aqueous solution of NaCl and observed the osmotic pressure. Figure 4 shows their results as well as those of cell model and Monte Carlo calculations using the pair potential given by eqs. [2-5].

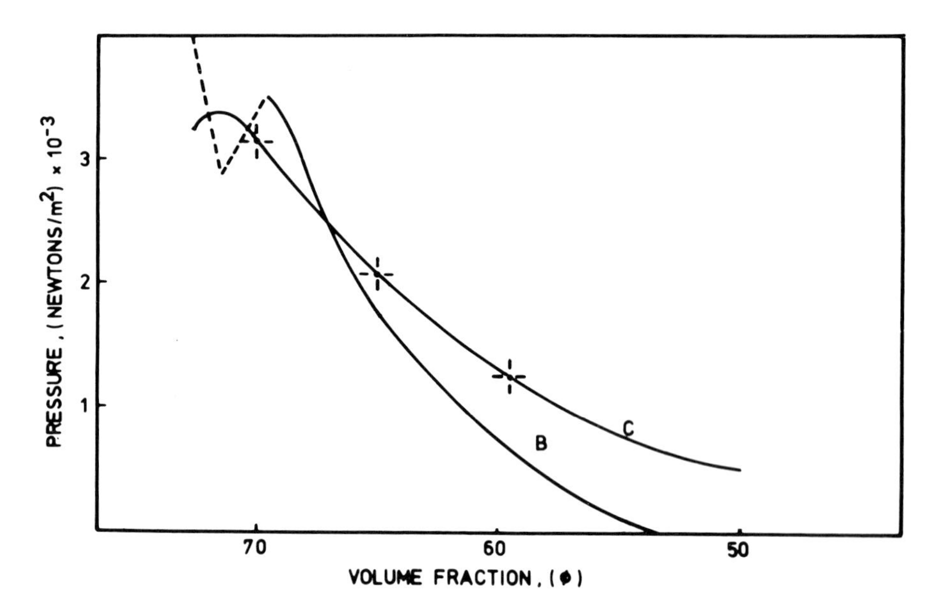

Fig. 4. The experimental (B) and calculated (C) osmotic pressures. Curve C was produced from cell model calculations and the points —¦— *represents corresponding Monte Carlo results.*

From the close agreement one can conclude that the potential given by eqs. [2-5] is at least a reasonable effective pair potential under the conditions of this experiment.

By means of laser light scattering, Brown et al (13) determined the radial distribution functions, $g(r)$ of very dilute (ϕ < .044%) monodisperse systems of small spherical polystyrene particles ($a = 2.3 \times 10^{-8}$ m) in a very dilute monovalent aqueous electrolyte ($n = 10^{-3}$moles/m^3). Their estimate of the particles' surface potential, ψ_0, was in the range of 150-270 mV.

For this system $\kappa a << 1$, and the following form for the double layer repulsion is expected to be more valid (4):

$$U_R(r_{ij}) = \frac{4\pi\, \varepsilon_o \varepsilon_r \psi_o^2\, e^{-\kappa a u_{ij}}}{r_{ij}} \qquad [6]$$

We found the lower limit of the estimated surface potential, viz., ψ_o = 150 mV, to yield the closest agreement between the experiments and results from Monte Carlo calculations.

The position of the first peak in $g(r)$ represents the most probable separation between pairs of particles and Table 1 presents a comparison between the experimental and

TABLE 1
The position, r_{max}, of the first peak in the experimental and calculated $g(r)$ function for various values of ϕ.

ϕ	$r_{max}(cm) \times 10^5$	
	Experimental	*Calculated*
.009	8.2	8.5
.015	6.85	7.1
.030	5.6	5.7
.044	4.8	5.0

calculated values of this distance. The agreement is within a few percent. Figure 5 shows the calculated $g(r)$ for ϕ = 0.044% and in accord with observation shows no long ranged order*. However, at slightly higher volume fractions,

**This was subsequently substantiated by means of a Monte Carlo run with 256 particles which showed $g(r)$ asymptote to unity more clearly.*

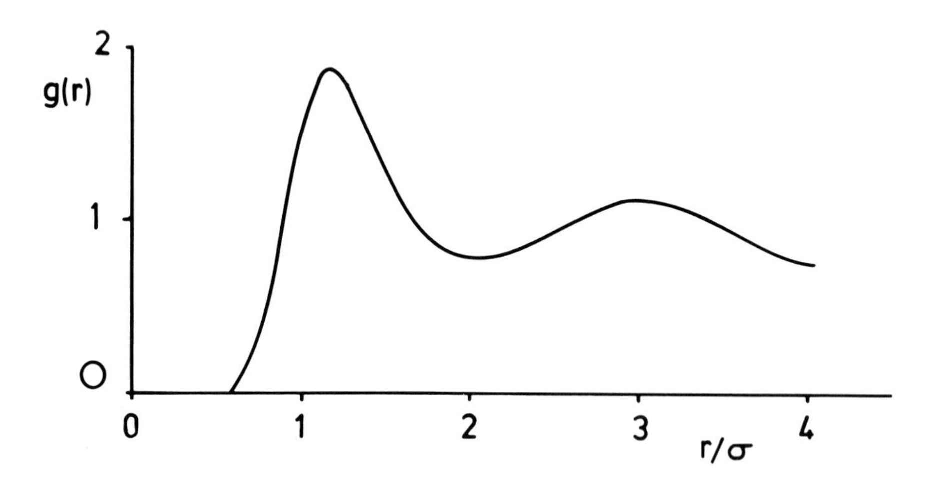

Fig. 5. The radial distribution function, g(r) as a function of the particle separation in units of the particle diameter, σ, for ϕ = .044%.

fig. 6, the calculated radial distribution functions show quite clearly the effects of second and third nearest neighbours, i.e. for $\phi > .1\%$ this system tends to order itself in a crystal like array with mean particle separations approximately eight particle diameters.

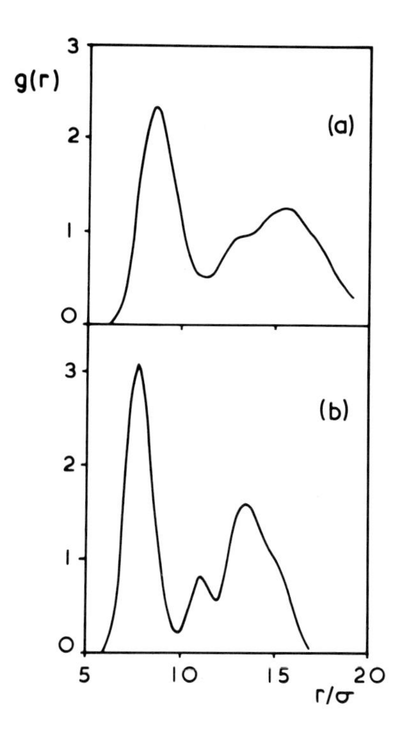

Fig. 6. The radial distribution function for (a) ϕ = .1% and (b) ϕ = .15%.

IV. CONCLUSIONS

We have shown that not only can one obtain all the qualitative features of an electrostatically stabilized colloidal dispersion of spherical particles, but that with reasonable choices of the pair interaction potential, one can reproduce available experimental results reasonably accurately. Furthermore, approximate methods can be developed to describe the dispersion which are much less costly (in computer time) and which aid in the physical understanding of the system, i.e., the cell and hard sphere perturbation models. It is hoped that in future similar models can be developed for the nonequilibrium properties, e.g. diffusion, rheology and kinetics of coagulation of colloidal dispersions; this is currently under investigation.

Finally, this work points to two related areas which need more investigation:

(a) the measurement of properties of well characterized lattices

(b) more detailed and accurate modelling of electrical double layers and their interactions.

V. REFERENCES

1. Barclay, L., Harrington, A and Ottewill, R.H., Kolloid-Z.u.Z. Polymere, 250, 655 (1972).

2. Homola, A. and Robertson, A.A., J. Colloid Interface Sci., 54, 286 (1976).

3. Efromov, I.F., in "Surface and Colloid Science", Vol. 8, (ed. E. Matijevic), Wiley-Interscience, New York, 1976.

4. Verwey, E.J.W. and Overbeek, J.Th.G., "Theory of Stability of Lyophobic Colloids", Elsevier, Amsterdam, 1948.

5. Van Megen, W. and Snook, I., J. Colloid Interface Sci., 53, 172 (1975).

6. Ree, R.H., in "Physical Chemistry, An Advanced Treatise", (ed. D. Henderson), Vol. VIIIA, Chap. 3, Academic Press, New York, 1971.

7. For example; Lee, J.K., Barker, J.A. and Abraham, F.F., J. Chem. Phys., 58, 3166 (1973).
McGinty, D.J., ibid, 58, 4733 (1973).

8. Van Megen, W. and Snook, I., J. Colloid Interface Sci., in press.

9. Rowlinson, J.S. in "Essays in Chemistry" (eds. J.N. Bradley, R.D. Gillard and R.F. Hudson), Vol. 1, Academic Press, New York, 1970.

10. Van Megen, W. and Snook, I., Chem. Phys. Letters 35, 399 (1975).

11. Hachisu, S., Kobayashi, Y. and Kose, A., J. Colloid Interface Sci., 42, 342 (1973).

12. Homola, A.M. and Robertson, A.A., Can. J. Chem. Eng., 53, 389 (1975).

13. Brown, J.C., Pusey, P.N., Goodwin, J.W. and Ottewill, R.H., J. Phys. A., 8, 664 (1975).

ION-PAIR CORRELATION FUNCTIONS IN ELECTRIC DOUBLE LAYER THEORY

Samuel Levine and Gérard R. Féat, Department of
Mathematics, University of Manchester.

An integral equation of the Kirkwood type can be derived for the pair correlation function between two ions situated in a planar electric double layer by charging one of the ions. The use of this function in a modified Poisson-Boltzmann equation for the double layer potential is discussed. The correlation function should be symmetrical in the two ions when these are equidistant from the interface. This symmetry is lost when a closure procedure is introduced to solve the integral equation, which implies linear superposition of the self-atmospheres of the two ions. Two methods of restoring this symmetry are proposed. The relevance of the pair-correlation function to the discreteness-of-charge effect for adsorbed ions in the Stern inner region is explained. For two oppositely-charged species on the adsorption plane, the ion-pairing theory of Bjerrum and Fuoss, as extended by Poirier and DeLap, is applied in determining the discreteness-of-charge potential.

I INTRODUCTION

The well-known inadequacies of the Poisson-Boltzmann (P.B.) equation have influenced the development of strong electrolyte theory in directions which differ markedly from the classical approach of Debye and Huckel. In recent years, however, Outhwaite (1,2,3) has constructed a method of successive approximations for dealing with the so-called fluctuation corrections to the P.B. equation, which promises to revive the use of a modified form of this equation in electrolyte theory. When volume exclusion terms associated with ion size are included, Outhwaite's method of extending the Debye-Huckel theory yields thermodynamic results for symmetrical electrolytes comparable with those based on more recent statistical mechanical theories (3,4) which are often regarded as having superseded the Debye-Huckel theory. Being a field equation, the P.B. equation is highly suited to describing the electrical properties of inhomogeneous interfaces and consequently has been one of the cornerstones of electric double layer theory in colloid science. The application to the electric double layer of Outhwaite's modification to the P.B. equation, which is an extension of earlier work (5,6,7) should be of some importance to future developments in this field. Perhaps the most significant correction to the P.B. equation centers around an adequate "ion-pair" correlation function. The present paper is concerned with a study of this function, in particular its symmetry properties, for ions situated both in the diffuse region and in the Stern inner region of the electric double layer.

II ION-PAIR CORRELATION FUNCTION IN THE DIFFUSE LAYER

This correlation function appears in the modified P.B. equation for the diffuse layer in the following way. Consider a plane interface between a non-aqueous medium and an aqueous electrolyte solution at which an electric double layer is formed. The bulk aqueous phase contains per unit volume n_i ions of species i, charge e_i, where $i=1,\ldots,s$. For simplicity, the diffuse layer is given a uniform, dielectric permittivity ε. Introducing the Guntelberg charging process, we imagine that an ion i, situated in the diffuse layer at position $\underset{\sim}{r}_1$, distant x_1 from the plane interface, carries a fraction ξ_1 of its full charge e_i. Let $\Psi_1^{(i)}(2,\xi_1)$ be the mean potential at position $\underset{\sim}{r}_2$, distant x_2 from the interface, in the presence of the above ion i. Then the potential of the mean electrostatic force acting on the ion i is

$$W_i(x_1)=e_i\int_0^1 d\xi_1 \lim_{\underset{\sim}{r}_2\to\underset{\sim}{r}_1} [\Psi_1^{(i)}(2,\xi_1)-\{\Psi_1^{(i)}(2,\xi_1)\}_{x_1\to\infty}] \cdot \quad (2.1)$$

This is the difference in the work of charging ion i at distances x_1 and $x_1=\infty$ (the electrolyte interior). Suppose that the ion is regarded as a point charge at the center of a sphere of dielectric permittivity ε_0 which is much smaller than the dielectric permittivity of the aqueous medium. Then $\Psi_1^{(i)}(2,\xi_1)$ will tend to infinity as $\xi_1 e_i/(\varepsilon_0|\underset{\sim}{r}_2-\underset{\sim}{r}_1|)$ as $\underset{\sim}{r}_2\to\underset{\sim}{r}_1$. This infinity in the integrand of (2.1) is removed when the corresponding potential at $x_1=\infty$ is subtracted away. The mean density of ions i at $x=x_1$ is

$$n_i(x_1)=n_i\zeta_i(x_1)\exp[-W_i(x_1)/\mathrm{k}T], \quad (2.2)$$

where k is Boltzmann's constant, T the absolute temperature and $\zeta_i(x_1)$ is the product of an exclusion volume term due to ion size and a term arising from the polarization energy of the ion i in the electric field of the diffuse layer. With the ion model described above, such polarization energy is present when the ion is completely uncharged). The diffuse layer mean potential $\psi(x_1)$ at x_1 satisfies Poisson's equation

$$\frac{d^2\psi(x_1)}{dx_1^2} = -\frac{4\pi}{\varepsilon}\sum_{i=1}^{s} e_i n_i(x_1) \,. \tag{2.3}$$

The part of $\Psi_1^{(i)}(2,\xi_1)$ which is the self-atmosphere or fluctuation potential at position $\underset{\sim}{r}_2$ due to ion i, charge $\xi_1 e_i$ situated at $\underset{\sim}{r}_1$, is defined as

$$\phi_1^{(i)}(2,\xi_1) = \Psi_1^{(i)}(2,\xi_1) - \psi(x_2), \tag{2.4}$$

which we shall call a "first-order" fluctuation potential. This satisfies the equation

$$\nabla^2\phi_1^{(i)}(2,\xi_1) = -\frac{4\pi}{\varepsilon}\sum_{j=1}^{s} e_j n_j(x_2)[g^{(ij)}(12,\xi_1)-1]. \tag{2.5}$$

The Laplace operator ∇^2 refers to position $\underset{\sim}{r}_2$ in the diffuse layer and $g^{(ij)}(12,\xi_2)$ is the pair correlation function between ions i and j, charges $\xi_1 e_i$ and e_j, situated at positions $\underset{\sim}{r}_1$ and $\underset{\sim}{r}_2$ respectively. The two differential equations (2.3) and (2.5) for the potentials $\psi(x_2)$ and $\phi_1^{(i)}(2,\xi_1)$ must be solved simultaneously, but this is only possible if $g^{(ij)}(12,\xi_1)$ can be expressed in terms of the fluctuation potentials. If these are assumed to vanish, then $\psi(x_1)$ satisfies the classical P.B. equation. Omission of the charging parameter ξ_1 will signify that $\xi_1=1$.

In strong electrolyte theory $g^{(ij)}(12)$ is symmetrical with respect to interchange of ions i and j and, for two ions of different species, this property continues to hold in a planar electric double layer provided that the two ions are equidistant from the interface. An integral equation of the Kirkwood type from the theory of liquids can be set up for $g^{(ij)}(12)$ by the usual charging process on one of the ions i and j. In the subsequent approximation inherent in the closure procedures required to obtain a solution of the integral equation, the above symmetry property can be lost. For 1-1 electrolytes the results of Outhwaite indicate that this lack of symmetry is not crucial but for unsymmetrical electrolytes and electric double layers it is certainly more important. There are a number of ways of retaining the symmetry, two of which will be briefly described. Assuming pair-wise additivity the electrostatic part of the interaction energy between two ions, charges e_i and e_j, at positions r_1 and r_2 in the double layer, is written as $e_i e_j v(12)$ where the function $v(12)$ can account for electrostatic imaging due to variations in the dielectric permittivity at an interface. From statistical mechanics it can be shown that (7,8)

$$-kT \ln [g^{(ij)}(12)/g^{(ij)}(12,\xi_1=0)] = e_i e_j v(12) \tag{2.6}$$

$$+e_i\int_0^1 d\xi_1 \sum_{k=1}^{s} e_k \int_V v(13)[n^{[ijk]}(123,\xi_1) - n^{[ik]}(13,\xi_1)]dV_3$$

Here $g^{(ij)}(12,\xi_1=0)$ is the pair correlation function when ion i has been discharged. $n^{[ik]}(13,\xi_1)$ is the conditional volume density of ions of species k, charge e_k at some position r_3 in the double layer, given that an ion i of species i, charge $\xi_1 e_i$, is at position r_1. The corresponding volume density when two ions i and j, charges $\xi_1 e_i$ and e_j, are situated at r_1 and r_2 respectively is $n^{[ijk]}$. The volume V may include an inner Stern region occupied by absorbed ions. Note the unequal treatment of ions i and j in regard to the charging process.

Following Outhwaite (3,8) and the present authors (9) we transform the right-hand side of (2.6) as follows. In the definition (2.4) of the self-atmosphere potential due to an ion of charge $\xi_1 e_i$ at r_1, let the general position r_2 become r_3, distant x_3 from the interface. We can define similarly the first-order self-atmosphere potential at r_3 due to an ion of charge $\xi_2 e_j$ at r_2 by

$$\phi_2^{(j)}(3,\xi_2) = \Psi_2^{(j)}(3,\xi_2) - \psi(x_3), \tag{2.7}$$

where $\Psi_2^{(j)}(3,\xi_2)$ is the corresponding total potential at r_3. A 'second order' fluctuation potential $\phi_{12}^{(ij)}(3,\xi_1)$ at position r_3 in the presence of two ions i and j of charges ξe_i and e_j, situated at r_1 and r_2 respectively, is defined by writing the corresponding total potential at r_3 in the form

$$\Psi_{12}^{(ij)}(3,\xi_1) = \psi(x_3) + \phi_1^{(i)}(3,\xi_1) + \phi_2^{(j)}(3) + \phi_{12}^{(ij)}(3,\xi_1), \tag{2.8}$$

where the omission of ξ_2 in $\phi_2^{(j)}(3)$ means that $\xi_2=1$. Then (2.6) can also be expressed as (8,9,10).

$$-kT \ln[g^{(ij)}(12)/g^{(ij)}(12,\xi_1=0)] = e_i\phi_2^{(j)}(1)$$

$$+e_i\int_0^1 d\xi_1 \lim_{\underset{\sim}{r}_3\to\underset{\sim}{r}_1} \phi_{12}^{(ij)}(3,\xi_1)\ . \tag{2.9}$$

$\phi_2^{(j)}(1)$ is the potential at $\underset{\sim}{r}_1$ due to ion j at $\underset{\sim}{r}_2$ in the absence of ion i and $e_i\phi_2^{(j)}(1)$ is the electrostatic energy of ion i in this potential field. A first approximation to $g^{(ij)}(12)$ is obtained by omitting the integral term on the right-hand side of (2.9) and putting $g^{(ij)}(12,\xi_1{=}0){=}1$. This implies linear superposition of the self-atmospheres of the two ions i and j, but symmetry in the correlation function is lost. The roles of ions i and j can be interchanged and a fraction of charging ξ_2 attached to ion j. In place of (2.8) and (2.9) we obtain

$$\Psi_{12}^{(ij)}(3,\xi_2)=\psi(x_3)+\phi_1^{(i)}(3)+\phi_2^{(j)}(3,\xi_2)+\phi_{12}^{(ij)}(3,\xi_2), \tag{2.10}$$

and

$$-kT\ \ln[g^{(ij)}(12)/g^{(ij)}(12,\xi_2{=}0)]=$$

$$e_j\phi_1^{(i)}(2)+e_j\int_0^1 d\xi_2 \lim_{\underset{\sim}{r}_3\to\underset{\sim}{r}_2} \phi_{12}^{(ij)}(3,\xi_2), \tag{2.11}$$

where, as before, absence of the symbol ξ_1 implies $\xi_1{=}1$. If ion i carries a charge $\xi_1 e_i$, then (2.11) is replaced by the equation defining $g^{(ij)}(12,\xi_1)$ and the first term on the right-hand side of (2.11) becomes $e_j\phi_1^{(i)}(2,\xi_1)$. On approximating $g^{(ij)}(12,\xi_1)$ by $\exp[-e_j\phi_1^{(i)}(2,\xi_1)/kT]$ and linearising with respect to $\phi_1^{(i)}(2,\xi_1)$, (2.5) simplifies to Loeb's equation (5). By using approximate solutions of Loeb's equation, modified forms of the P.B. equation are obtained (6,11,12). The most satisfactory form at present is due to Outhwaite (11) and Bell and Rangecroft (12), who derive a differential-difference equation for $\psi(x)$ which is a generalisation of that obtained by Stillinger and Kirkwood (13). This modification of the P.B. equation predicts that $\psi(x)$ becomes oscillatory with respect to x above some critical electrolyte concentration, indicating that the classical P.B. equation has become inadequate.

The formulae (2.9) and (2.11) are derived from classical statistical mechanics and still apply when the dielectric permittivity is not uniform. In particular, they hold for the ionic model of a point charge at the center of a sphere of dielectric permittivity ε_0, since the interaction energy of two such ions i and j takes the form $e_ie_jv(12)$, which is assumed here. Although not explicitly evident $g^{(ij)}(12)$, defined by (2.9) or (2.11) should have the required symmetry property with respect to interchange of ions i and j, when these are equidistant from the interface. A first approximation to $g^{(ij)}(12)$ which retains symmetry and is based on omitting the integral term on the right-hand sides of (2.9) and 2.11) is obtained by adding (2.9) and (2.11). This yields

$$-kT\ \ln[g^{(ij)}(12)/\{g^{(ij)}(12,\xi_2=0)g^{(ij)}(12,\xi_1=0)\}^{\frac{1}{2}}]$$

$$= \tfrac{1}{2}[e_i\phi_2^{(j)}(1)+e_j\phi_1^{(i)}(2)]+\tfrac{1}{2}e_i\int_0^1 d\xi_1 \lim_{\underset{\sim}{r}_3\to\underset{\sim}{r}_1}\phi_{12}^{(ij)}(3,\xi_1)$$

$$+\tfrac{1}{2}e_j\int_0^1 d\xi_2 \lim_{\underset{\sim}{r}_3\to\underset{\sim}{r}_1}\phi_{12}^{(ij)}(3,\xi_2). \tag{2.12}$$

An alternative method of introducing explicit symmetry into $g^{(ij)}(12)$ is to imagine that ions i and j are charged simultaneously at the same rate ξ. This gives

$$-kT\ \ln[g^{(ij)}(12)/g^{(ij)}(12,\xi=0)]=\int_0^1 d\xi[e_j\phi_1^{(i)}(2,\xi)$$

$$+e_i\phi_2^{(j)}(1,\xi)]+\int_0^1 d\xi[e_i\lim_{\underset{\sim}{r}_3\to\underset{\sim}{r}_1}\phi_{12}^{(ij)}(3,\xi)+e_j\lim_{\underset{\sim}{r}_3\to\underset{\sim}{r}_2}\phi_{12}^{(ij)}(3,\xi)]. \tag{2.13}$$

In principle, all four relations (2.9), (2.11)-(2.13) give the same correlation function. In (2.9), ion j has its full charge e_j and ion i is being charged and in (2.11) the two ions are interchanged. $\phi_2^{(j)}(1)$ in (2.9) is identical with $\phi_2^{(j)}(1,\xi=1)$ in (2.11) and similarly $\phi_1^{(i)}(2)$ in (2.11) is identical with $\phi_1^{(i)}(2,\xi=1)$ in (2.13). If $\phi_1^{(i)}(2,\xi)$ and $\phi_2^{(j)}(1,\xi)$ are proportional to the charges on the respective ions i and j, so that they are both proportional to ξ, then the first integral on the right-hand side of (2.13) is identical with the first term on the right-hand side of (2.12). It can be shown that apart from a small correction associated with ion size, neglecting the "second-order" fluctuation potential $\phi_{12}^{(ij)}$ implies linear superposition of the self-atmosphere charge densities of the two ions (9). In general, such superposition will yield a correlation function $g^{(ij)}(12)$ which depends on the path taken in charging the two ions i and j. We show as follows that the first terms on the right-hand sides of (2.12) and (2.13) are not identical even though they are both symmetrical with respect to ions i and j. The self-atmosphere potential $\phi_1^{(i)}(2,\xi_1)$ defined in (2.4) represents the change in the potential at a general point $\underset{\sim}{r}_2$ in the diffuse layer due to fixing an ion i, charge $\xi_1 e_i$ at $\underset{\sim}{r}_1$. Placing this ion at $\underset{\sim}{r}_1$ requires the removal of the diffuse layer charge which would be occupying the ion's exclusion volume. The amount of charge removed is independent of $\xi_1 e_i$ and this is manifested as a contribution to $\phi_1^{(i)}(2,\xi_1)$, called the cavity potential, which is independent of ξ_1. Consequently the contribution to the first term in (2.13) from the cavity potentials of ions i and j apparently equals twice the corresponding contribution in (2.12). In (2.13), ions i and j are charged simultaneously, whereas in (2.12) they are charged consecutively.

The equality of the 'first-order' terms on the right-hand sides of (2.12) and (2.13) requires that $\phi_1^{(i)}(2,\xi)$ and $\phi_2^{(j)}(1,\xi)$ be both proportional to ξ. The corresponding equality between (2.9) and (2.11) implies that $e_i\phi_2^{(j)}(1) = e_j\phi_1^{(i)}(2)$, which is satisfied in the linear Debye-Huckel theory of electrolytes. Although the cavity potential terms do not satisfy these conditions, they form a minor part of the first-order fluctuation potential provided the total diffuse layer charge contained in the exclusion volume of an ion is much less in magnitude than the charge of the ion itself. Nevertheless it is desirable to find a reason for the difference by a factor of two between cavity potential terms in (2.12) and (2.13). This apparent discrepancy is due to the approximations which yield the first order term only. In addition to assuming that the self-atmosphere charge distributions of ions i and j are additive outside their exclusion volumes, an ion size is given to the 'central' ion only in (2.9) and (2.11) (the i ion with (2.9) and the j ion with (2.11)) whereas all the other ions are treated as point charges. (This concept of assigning a size to the 'central' ion only is implied in the Debye-Huckel theory of electrolytes). Now (2.12) represents simply an average of (2.9) and (2.11) and so the model in which one ion only has a size is retained. In contrast the first order terms on the right-hand side of (2.13) are obtained by giving sizes to both ions i and j, all other ions being treated as point charges. We must assume exclusion volumes from which diffuse layer charge is removed to be greater in (2.12) than in (2.13), indeed by an amount which yields the same cavity potentials.

Outhwaite (1,2,3) considered the charging of one ion only and therefore his starting point was an equation equivalent to (2.9). He introduced a method of successive approximations which in the first stage beyond the P.B. equation signifies the omission of the second order potential $\phi_{12}^{(ij)}$. In the second stage he used the equivalent of (2.12), but because he was charging one ion, the last integral on the right-hand side of (2.12) was omitted. This means that his closure procedure is not symmetrical in i and j. He set up a system of equations to determine the second order potential $\phi_{12}^{(ij)}$ for a bulk symmetrical electrolyte. However, this approach has not been applied to the electric double layer.

III. APPLICATION TO TWO ADSORBED IONS

Consider ion i, charge e_i, adsorbed on a plane (the inner Helmholtz plane or i.h.p.) in the Stern inner region, at the origin of co-ordinates. The change in potential $\phi^{(i)}(\rho)$ at distance ρ on the i.h.p. due to placing the ion at the origin $\rho = 0$, can be expressed as the sum of three terms

$$\phi^{(i)}(\rho) = e_i\phi_o(\rho) + \phi_a^{(i)}(\rho) + \phi_d^{(i)}(\rho). \tag{3.1}$$

$\phi_o(\rho)$ is the potential change at distance ρ due to a unit point charge at the origin $\rho = 0$, $\phi_a^{(i)}(\rho)$, the so-called disc potential, is the change in potential at distance ρ, due to the removal of the charge of mean density σ from the exclusion area on the i.h.p. surrounding the ion i at the origin. This is the two-dimensional counterpart of the cavity potential described in section II. The third term $\phi_d^{(i)}(\rho)$ in (3.1) is the potential change at ρ due to the redistribution of surface charge beyond the exclusion area on the i.h.p. when ion i is placed at the origin. The potential due to ion j, charge e_j, which corresponds to (3.1) is

$$\phi^{(j)}(\rho) = e_j\phi_o(\rho) + \phi_a^{(j)}(\rho) + \phi_d^{(j)}(\rho). \tag{3.2}$$

Suppose now that ions i and j are situated on the i.h.p. at separation ρ. The formulae (2.9) to (2.13) still apply because we have not specified the dependence of the basic interaction potential $e_ie_jv(12)$ on the inhomogeneous dielectric permittivity at the interface and therefore the two ions may be situated anywhere in the electric double layer. We shall only consider the first-order terms on the right-hand sides of (2.9)-(2.13). Then (2.9) and (2.10) yield respectively the following pair correlation functions $g^{(ij)}(\rho)$ for ions i and j on the i.h.p.,

$$-kT \ln [g^{(ij)}(\rho)/g^{(ij)}(\rho,\xi_1=0)] = e_j\phi^{(i)}(\rho), \tag{3.3}$$

$$-kT \ln [g^{(ij)}(\rho)/g^{(ij)}(\rho,\xi_2=0)] = e_i\phi^{(j)}(\rho). \tag{3.4}$$

A symmetrical form is obtained by adding (3.3) and (3.4) to obtain

$$-kT \ln [g^{(ij)}(\rho)/\{g^{(ij)}(\rho,\xi_1=0)\ g^{(ij)}(\rho,\xi_2=0)\}^{\frac{1}{2}}]$$

$$=e_i e_j\ \phi_0(\rho)+\tfrac{1}{2}[e_i\phi_a^{(j)}(\rho)+e_j\phi_a^{(i)}(\rho)] \qquad (3.5)$$

$$+\tfrac{1}{2}[e_i\phi_d^{(j)}(\rho)\ +\ e_j\phi_a^{(i)}(\rho)],$$

making use of (3.1) and (3.2). The difference between (3.5) and the unsymmetrical forms (3.3) and (3.4) becomes quite evident when the exclusion areas of the two ions i and j are equal and their charges e_i and e_j are equal in magnitude but opposite in sign. Then the disc potential terms in (3.3) and (3.4) are also equal in magnitude and opposite in sign, resulting in their mutual cancellation in (3.5). Alternatively, on charging both ions at the same rate ξ, (2.12) yields in place of (3.5),

$$-kT \ln[g^{(ij)}(\rho)/g^{(ij)}(\rho,\xi=0)] = e_i e_j\phi_0(\rho) \qquad (3.6)$$

$$+[e_i\phi_a^{(j)}(\rho)+e_j\phi_a^{(i)}(\rho)]' \ +$$

$$+\int_0^1 [e_i\phi_d^{(j)}(\rho,\xi)\ +\ e_j\phi_d^{(i)}(\rho,\xi)]'d\xi,$$

where the prime indicates that the potential functions differ from those in (3.5), because of the difference between the exclusion areas in (3.5) and (3.6), similar to the difference between the exclusion volumes in (2.12) and (2.13).

The so-called discreteness-of-charge effect for an adsorbed ion i involves determining the potentials $\phi_a^{(i)}(0)$ and $\phi_d^{(i)}(0)$ at the center of the ion (14). $\phi_a^{(i)}(0)$ is expressed as an integral which can be found numerically but the evaluation of $\phi_d^{(i)}(0)$ is much more difficult. We outline here an approximate method of obtaining $\phi_d^{(i)}(0)$ for the case where two oppositely charged, univalent ionic species i and j are situated on the i.h.p. We make use of the symmetry property of the pair correlation function between i and j and adapt to two dimensions the ion-pairing theory of Bjerrum and Fuoss (15), as modified by Poirier and DeLap (16). The 'unlike-partners only' definition of an ion-pair, as defined by Poirier and DeLap, is employed. The probability of finding an ion j at a distance ρ from an ion i, in a circular ring of thickness $d\rho$, under the condition that neither ion has another unlike partner at any distance less than ρ, is given by

$$G^{(ij)}(\rho)d\rho=2\pi\rho\nu^{(i)}g^{(ij)}(\rho)P^{(i)}(\rho)\ P^{(j)}(\rho)d\rho. \qquad (3.7)$$

Here $\nu^{(i)}$ and $\nu^{(j)}$ are the mean number densities of ions i and j on the i.h.p., $P^{(i)}(\rho)$ is the probability that ion i does not have another unlike partner on the i.h.p. within a radius ρ, and $P^{(j)}(\rho)$ is defined similarly. $P^{(i)}(\rho)$ and $P^{(j)}(\rho)$ are readily expressed in terms of $G^{(ij)}(\rho)$ and $G^{(ji)}(\rho)$ respectively. Making use of the symmetry relation

$$g^{(ij)}(\rho) = g^{(ji)}(\rho), \tag{3.8}$$

it is possible to show that (17)

$$G^{(ij)}(\rho) = 2\pi\rho\, g^{(ij)}(\rho) \frac{\nu^{(j)}(\nu^{(j)}-\nu^{(i)})^2 \exp(-B)}{(\nu^{(j)} - \nu^{(i)}\exp(-B))^2}, \tag{3.9}$$

where, if a_i is the radius of the exclusion area of ion i,

$$B = 2\pi(\nu^{(j)}-\nu^{(i)}) \int_{a_i}^{\rho} \rho' g^{(ij)}(\rho')d\rho' . \tag{3.10}$$

Ion-pairing theory is used in the following way to determine the potential $\phi_d^{(i)}(0)$. When ion i is fixed at the origin $\rho=0$, the mean surface charge density σ on the i.h.p. is assumed unchanged beyond a certain distance $\rho=b$, to be chosen below. We imagine the mean charge density σ to be removed from the disc of radius b, center at origin, causing a change in potential at the center of ion i which is given by

$$\phi_b^{(i)}(0) = -2\pi\sigma \int_0^b \rho\phi_o(\rho)d\rho . \tag{3.11}$$

The probability of finding ion j at a distance ρ from ion i in the range (a_i, b), such that neither ion i nor ion j has a second unlike partner at any distance less than ρ equals

$$P^{(ij)}(\rho) = \int_{a_i}^{\rho} G^{(ij)}(\rho')d\rho' ,$$

where a_i is equated to the distance of nearest approach between ions i and j. We now imagine a 'fraction' of a j ion, $P^{(ij)}(b)$, brought back into the area between $\rho = a_i$ and $\rho = b$ to form an ion-pair with ion i situated at $\rho = 0$. This process yields

$$\phi_d^{(i)}(0)=\phi_b^{(i)}(0) - \phi_a^{(i)}(0) + \phi_p^{(i)}(0), \tag{3.13}$$

where $\phi_b^{(i)}(0)$ is given by (3.11),

$$\phi_a^{(i)}(0) = -2\pi\sigma \int_0^{a_i} \rho\phi_o(\rho)d\rho, \tag{3.14}$$

the disc potential at the center of ion i, and

$$\phi_p^{(i)}(0) = e_j \int_{a_i}^{b} G^{(ij)}(\rho)\phi_o(\rho)d\rho \;. \tag{3.15}$$

The Stern inner region consists of an 'inner zone' and an 'outer zone', of thicknesses β and γ respectively. The inner zone is situated between the non-aqueous phase of the interface and the i.h.p. The outer zone lies between the i.h.p. and the outer Helmholtz plane (o.h.p.) the boundary between the Stern region and the aqueous diffuse layer. The dielectric permittivities of the non-aqueous phase, the inner and outer zones are assumed to be ε_a, ε_1 and ε_2 respectively. The 'basic' potential function $\phi_o(\rho)$ which depends on β and γ, the dielectric permittivities of the different parts of the interface, the mean potential ψ_d at the i.h.p. and the electrolyte concentration c in the aqueous medium, is obtained as an integral (14,17) which is determined numerically. In the Bjerrum-Fuoss theory of strong electrolytes the ion-pair correlation function is given by its form at infinite dilution of electrolytes, i.e. the unscreened Coulomb interaction energy of an ion-pair is used. Here the analogous approximation is

$$g^{(ij)}(\rho) = \exp\,[-e_i e_j \phi_o(\rho)/kT]. \tag{3.16}$$

This means that on the right-hand side of (3.5) (or(3.6)), we retain only the leading term, which is negative. The second term, due to the disc potential, will vanish for equal exclusion areas of ions i and j. To determine the sign of the third term on the right-hand side of (3.5) (or(3.6)), suppose that the surface charge density on the i.h.p. is negative,

$$\sigma = e_j\nu^{(j)} + e_i\nu^{(i)} = e_o(\nu^{(i)} - \nu^{(j)}) < 0 \;, \tag{3.17}$$

where ions i and j denote the anion and cation respectively and e_o is the proton charge. Then $\phi_d^{(j)}(\rho)$ is due to the reduction in $|\sigma|$ because of Coulomb repulsion beyond a distance a_j from the center of ion j, where a_j is the radius of the ion's exclusion area. This means that $\phi_d^{(j)}(\rho)$ is positive and by a similar argument, $\phi_d^{(i)}(\rho)$ is negative. The two members of the third term are both positive and therefore opposite in sign to the leading term on the right hand side of (3.5), a result which holds regardless of the sign of σ. Thus (3.16) over-estimates $g^{(ij)}(\rho)$, although this effect will be slightly offset by the neglected exclusion volume corrections on the left-hand side of (3.5). By (3.10) B is also over-estimated. However, the quantity required is the probability function $G^{(ij)}(\rho)$ and this has $g^{(ij)}(\rho)$ as a factor but diminishes as B increases. In view of the uncertainty in the behaviour of $G^{(ij)}(\rho)$ and of the approximations inherent in the ion-pairing method used here, the simple form (3.16) should suffice.

As an example of a charged interface with two oppositely charged ions in the Stern region, consider a completely ionized anionic monolayer at the air/water interface. Let j denote the head-group anion and i the adsorbed counter-cation, both of which are assumed to be situated on the i.h.p. The model of the Stern region will be that desribed above. We wish to calculate the discreteness-of-charge potentials, as given by (3.11) and (3.13)-(3.15) for the counter-ion i. The areas per head-group and counter ions are respectively $A^{(j)}=1/\nu^{(j)}$ and $A^{(i)}=1/\nu^{(i)}$ completely ionized. The relation between these areas would be determined by establishing an adsorption isotherm for the counter-ion but this is not attempted here. We express $A^{(i)}$ in terms of $A^{(j)}$ by assigning a fixed potential ψ_d at the o.h.p. For a 1-1 electrolyte at concentration c in mol./1. use of the Gouy-Chapman theory for the diffuse layer yields the simple relation

$$\sigma = e_o \left[\frac{1}{A^{(j)}} - \frac{1}{A^{(i)}}\right] = \left[\frac{cN\varepsilon kT}{1000\pi}\right]^{\frac{1}{2}} \sinh\left[\frac{e_o\psi_d}{2kT}\right] \qquad (3.18)$$

where N is Avogadro's number and $A^{(j)}<A^{(i)}$. In the Bjerrum-Fuoss theory of strong electrolytes a minimum is sought in the function which corresponds to our $G^{(ij)}(\rho)$ and this would fix the distance $\rho=b$. In the present application to the air/water interface the form of $\phi_o(\rho)$ is such that $G^{(ij)}(\rho)$ decreases rapidly with increase in ρ but has no minimum. We have chosen the condition that the basic potential interaction energy $e_o^2\phi_o(b)$ equals kT. In the table, $A^{(j)}$ is varied in the range $100(100)1000$ A^2, and $A^{(i)}$ is given by (3.18) for T=10°C, c=0.0316 mol./1., ε = 84.15 and $e_o\psi_d$/kT = -1. Other parameters required are $\varepsilon_a=1$, $\varepsilon_1=20$, $\varepsilon_2=40$, $\beta=3A$, $\gamma=2A$ and $a_i=5A$. The function $G^{(ij)}_{a}{}^{1}(\rho)$ is so defined that only one j ion is allowed in the region $a_i<\rho<b$ around an i ion on the i.h.p. To test whether this condition is acceptable, we have compared b with the nearest neighbour distance d between j ions for the different areas $A^{(j)}$, assuming a regular two-dimensional hexagonal close packing of these ions. At given c, ψ_d and T, σ = -1.10μC cm^{-2} and (3.14) yields $e_o\phi_a^{(i)}(0)$= 0.34kT for the disc potential

$\phi_a^{(i)}(0)$. The function $\phi_o(\rho)$ is also specified and in particular $e_o^2\phi(a_i)=2.5kT$. The condition $e_o^2\phi_o(b)=kT$ yields b=9.5A where, by (3.11), $e_o\phi_b^{(i)}(0)=0.55kT$. For $b/d_m<0.5$ the ion-pairing method used here seems reasonable. The fraction of ion-pairing is seen to exceed 0.6. Somewhat larger fractions, together with the greater values of b and of $\phi_p^{(i)}(0)$ are obtained with the condition $e_o^2\phi_o(b)=\frac{1}{2}kT$.

TABLE 1.
Discreteness-of-charge potential at adsorbed counter-ion i on i.h.p. in presence of head-group ion j, at fixed charge density on i.h.p.

b=9.5A, $e_o\phi_d^{(i)}(0)=e_o\phi_p^{(i)}(0)-0.11kT$, $e_o^2\phi_o(b)=kT$.

$A^{(j)}(A)$	$A^{(i)}(A)$	$p^{(ij)}(b)$	$e_o\phi_p^{(i)}(0)/kT$	b/d_m
100	107	0.94	-2.10	0.88
200	232	0.88	-1.87	0.63
300	378	0.83	-1.70	0.51
400	551	0.78	-1.56	0.44
500	760	0.74	-1.45	0.40
600	1019	0.71	-1.36	0.36
700	1345	0.67	-1.27	0.33
800	1770	0.64	-1.20	0.31
900	2347	0.62	-1.14	0.30
1000	3175	0.60	-1.08	0.28

REFERENCES

1. Outhwaite, C.W., J. Chem. Phys. 50,2277(1969).
2. Outhwaite, C.W., Molecular Phys. 20,705(1971).
3. Outhwaite, C.W., in "Specialist Periodical Reports, Statistical Mechanics" Vol.2,p.188. The Chemical Society, London, 1975.
4. Burley, D.M., Hutson, V.C.L. and Outhwaite, C.W., Molecular Phys. 27,225(1974).
5. Loeb, A.L., J. Colloid. Sci. 6,75(1951).
6. Bell, G.M. and Levine,S., in "Chemical Physics of Ionic Solutions" (B.E. Conway and R.G. Barradas,Eds.) p.409,

Wiley, New York, 1966.
7. Bell, G.M. and Levine, S., in "Statistical Mechanics, Foundations and Applications, Proc. I.U.P.A.P. meeting, Copenhagen, 1966 (T.A. Bak, Ed.) p.258, Benjamin, New York, 1967.
8. Outhwaite, C.W., Molecular Phys. 27,561(1974).
9. Féat, G.R. and Levine, S. (submitted for publication).
10. Outhwaite, C.W., and Thomlinson, M.M., Chem. Phys. Letters. 25,375(1974).
11. Outhwaite, C.W., Chem. Phys. Letters, 7,636(1976).
12. Bell, G.M. and Rangecroft, P.D., Molecular Phys. 24,255(1972).
13. Stillinger, F.H. and Kirkwood, J.G, J. Chem. Phys, 33,1282(1960).
14. Levine,S., Robinson, K., Bell, G.M. and Mingins,J., J. Electroanal. Chem. 38,253(1972).
15. Fowler, R.H. and Guggenheim, F.A, "Statistical Thermodynamics", p.409, Camb. Univ. Press, 1949.
16. Poirier, J.C. and DeLap, J.H, J. Chem. Phys. 35,213(1961).
17. Levine, S. and Féat, G.R. (submitted for publication).

ACKNOWLEDGMENT: We are indebted to the Science Resarch Council of the United Kingdom for a postdoctoral research assistantship to G.R.F.

THE ELECTRICAL DOUBLE LAYER OF γ-Al_2O_3 - ELECTROLYTE SYSTEM

C. P. Huang
Department of Civil Engineering
University of Delaware

Abstract

A modified Grahame's method was used to compute the distribution of ions at the hydrous γ-Al_2O_3 electrolyte interface. The characteristics of the double layer at a reversible oxide-electrolyte interface as such differ significantly from those of a polarized mercury-solution and reversible AgI-, Ag_2S-solution systems. Presumably, the electrostatic field strength plays an important part in determining the electrical double layer structure. The validness of the Gouy-Stern version of the electrical double layer theory was verified by direct electrolyte adsorption measurement. It appears that the model holds satisfactorily at dilute concentration of indifferent electrolyte, viz. NaCl.

I. INTRODUCTION

Studies of the nature of electrical double layer have been prompted mostly by the investigations made with polarized mercury-solution interface. Extensive data obtained with this system have been reported by numerous investigators (1, 2). Generally, this information primarily relates the differential capacity of the mercury electrode to certain variables: polarized potential, the concentration, and the nature of electrolyte present. Therefore, an accurate determination of the double layer capacity is always a necessity.

More quantitative analysis of the electrical double layer to differentiate the contributory components of double layer capacity was first undertaken by Grahame and Soderberg (3). They computed the excess of cations and anions in the double layer and subdivided the anions into those in the diffuse part and those in the inner part of the double layer of mercury-solution system in particular.

The applications of electrical double layer theory obtained from the studies with mercury electrode to a reversible solid-solution system were further demonstrated by several authors with silver iodide (4-6), silver sulfide (7), hematite (8), rutile (9), and zinc oxide (10). The results generally showed that the accommodation of ions at the reversible electrode-solution interface differs significantly from that at the mercury-solution interface and that silver iodide and silver sulfide also behave differently from hydrous oxide regardless of the fact that both are reversible solid-solution systems.

Information on the ionic components of charge in a reversible hydrous oxide-solution system is rather scanty. In this present work, the Grahame method (3) modified by Lyklema (5) was used to analyze the distribution of ions in the hydrous γ-$A\ell_2O_3$ - solution interface. The importance of electrostatic field strength to the properties of double layer was illustrated by comparing the capacity of various solids. The validness of the Gouy-Stern model of electrical double layer was also tested by direct measurement of electrolyte ($NaC\ell$) adsorption. All experiments were done at 25 $\pm$ 1 C.

II. MATERIALS AND METHOD

A. Materials

A commercial product "Alon" was used. It is a fine powder of approximately spherical shape and has an average diameter of 0.03 μm. The specific surface area was found 117 m^2g^{-1} (11). From x-ray diffraction analysis, the oxide consists of 90% Al_2O_3 in the gamma form.

Prior to use, Alon was pretreated with dilute strong base (10^{-2}M NaOH) followed by elutriation with triple-distilled water till a constant conductivity was reached in the supernatant. Details of the treatment procedure were reported in an earlier publication (12).

A stock solution of sodium chloride was prepared by dissolving Mallinckrodt AR grade salt into triple-distilled water without receiving further purification.

Standard acid (HCl) and base (NaOH) were purchased from Anachemica. Triple-distilled water with specific conductivity less than 2.6 x 10^{-6} mho cm^{-1} was used.

B. Alkalimetric Titration

To determine the surface charge density, alkalimetric titration technique was followed. The overall system for titration set-up can be described by the following cell:

$$(+)\,Pt,H_2(g)/HCl(m_1),NaOH(m_2),NaCl(m_3),\text{hydrous } Al_2O_3$$

$$H_2O/AgCl,Ag(-) \qquad [1]$$

A Kiethley model 660 differential voltmeter was used to record the EMF readings (or E) from which the corresponding values of pH were computed (13). The titration cell was a flat-bottomed 300 ml Teflon beaker with a rubber stopper through which passed the electrodes, and the microburette. Detail of the titration experiment has also been reported earlier (12).

C. Na^+ and Cl^- Adsorption Experiment

The procedures for suspension preparation were essentially the same as for the alkalimetric titration experiment except that the pH values were recorded with a Beckman Model G pH meter. After pH adjustment, 40 ml of the suspension was pipetted from the titration vessel for assay. Solution was extracted from the suspension by a number of different methods: (1) centrifuging at high speed (19,600 g) for 20 minutes; (2) centrifuging at low speed (9,800 g) for 40 minutes; (3) passing through 100 mμ millipore filter pad. All the means of separation gave results equal within the errors of experiment. The supernatants were examined for escaping colloidal particles by Tyndall-light; results were negative.

The amounts of adsorption were obtained by comparing the concentrations of sodium or chloride originally added and finally remained in the supernatant. Sodium was analyzed with an atomic absorption spectrophotometer, Perkin-Elmer, model 303; chloride was determined by mercuric ion titration modified from Kolthoff and Stenger (14).

III. DATA EVALUATION AND RESULTS

A. Surface Charge

The surface charge density of hydrous oxide can be readily determined by measuring the amounts of proton or hydroxide ion consumption by the solid phase from alkalimetric titration. Figure 1 gives the surface charge density as function of solution pH at various levels of ionic strength. The point at which all curves meet, pH 8.5 ± 0.2, is set as the zero point charge (15).

In this context, it is assumed that H^+ and OH^- ions are the sole potential determining species and that the Nernst equation, $\psi_0 = -(2.303\ RT/F)(pH-pH_{zpc})$ is followed. Regardless of the reservation expressed by some researchers (16, 17), the Nerstian relationship holds surprisingly well for the γ-$A\ell_2O_3$

B. Differential Capacity

By replacing the potential axis of a polarized interface by pH (or E), the Lippmann-Helmholtz equation of a differential capacity for a reversible electrode can be derived from the surface charge density (Figure 1).

$$C = -\left(\frac{\partial\sigma_o}{\partial\psi_o}\right)_{P,T,\mu_s} = -\frac{F}{RT}\left(\frac{\partial\sigma_o}{\partial pH}\right)_{P,T,\mu_s}$$

$$= -\left(\frac{\partial \sigma_o}{\partial E}\right)_{P,T,\mu_s} \qquad [2]$$

where μ is the chemical potential of the indifferent electrolyte.

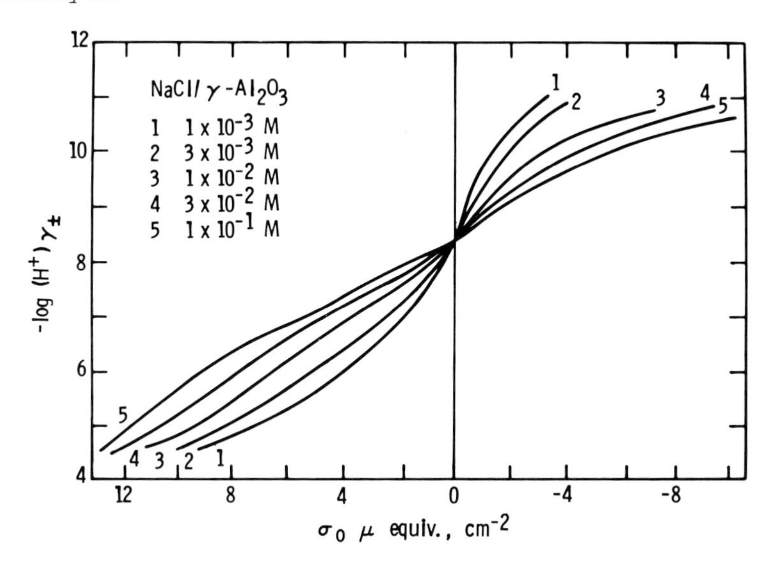

Fig. 1 Charge Distribution on the Solid Side of the Double Layer.

The computed differential capacity for γ-$A\ell_2O_3$ in $NaC\ell$ electrolyte is shown in Figure 2. The steepness of the plots of C vs. ψ_0 or pH, as well as the magnitude of C per sec, gives information on the sorbability of the ionic species. Clearly, Na^+ ions exhibit slightly more affinity toward the hydrous γ-$A\ell_2O_3$ surface than the $C\ell^-$ ions.

The differential capacity of the diffuse part of the electrical double layer is also shown for comparison. Excellent agreement between hydrous γ-$A\ell_2O_3$ and the idealized Gouy-Stern double layer is seen at pH_{zpc} and low concentration of ionic strength.

C. Interfacial Tension

Upon integration of Eq. [2] one obtains

$$\gamma = \left(\frac{RT}{E}\right) \int \sigma_o \, dPH + K = \int \sigma_o \, dE + K \qquad [3]$$

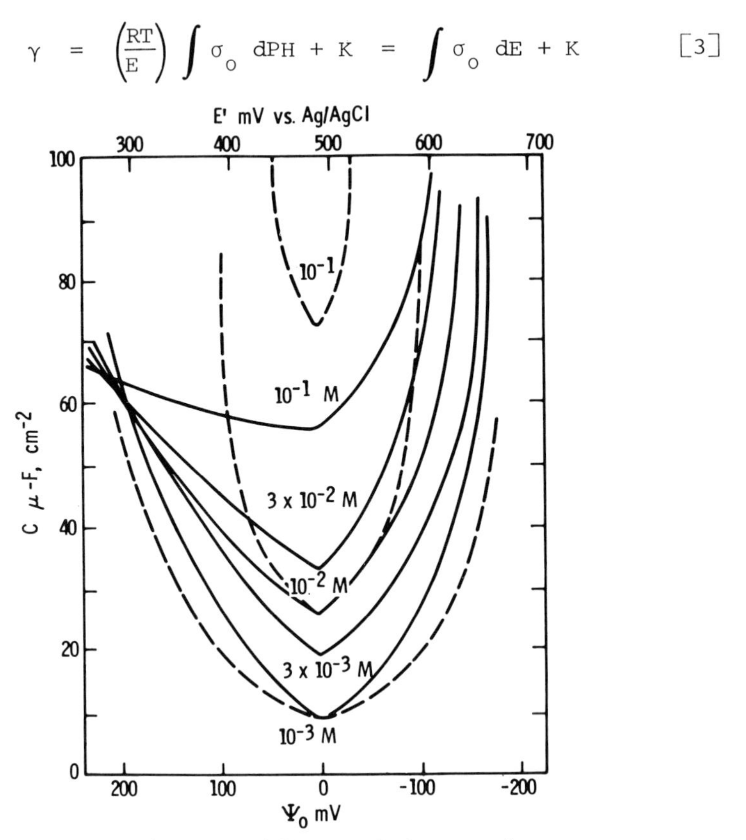

Fig. 2 Differential Capacity as Function of ψ_o. The dashed lines represent the theoretical capacity of the diffuse part of the double layer.

where K is the integration constant, which can be eliminated with the assumption that for uncharged surface the surface excesses due to cations and anions other than the potential determining ions are zero; otherwise, the value of K can be estimated by experimental measurement of γ at the point of zero charge.

Figure 3 shows the γ for γ-$A\ell_2O_3$ in the presence of $NaC\ell$ electrolyte. This does not give any real information in addition to the previous data, since in contrast to the polarized electrode the integration constant is not experimentally known ($K = 400 \pm 300$ dyne cm^{-1}) (18).

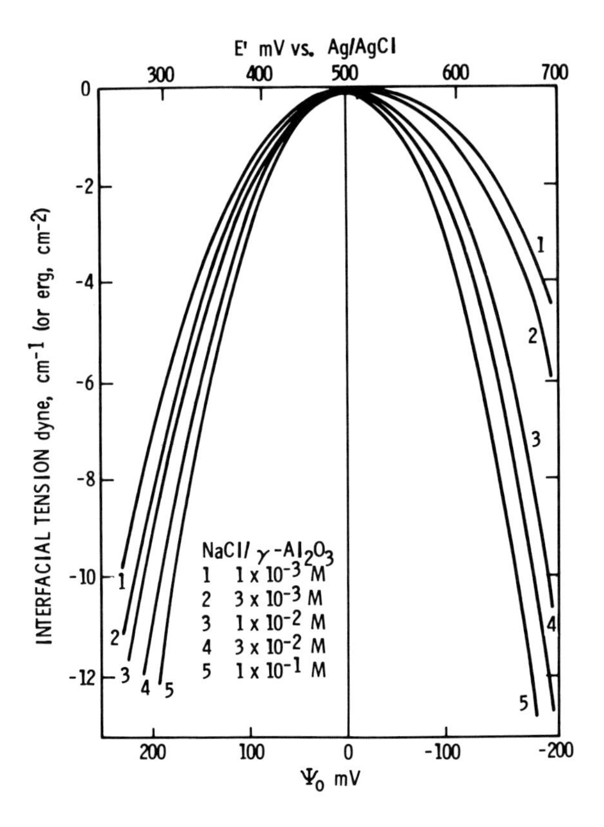

Fig. 3 Variation of Interfacial Tension with ψ_0.

D. <u>Ionic Components of Charge</u>

In view of the experimental results shown in Fig. 1, it is proper to depict the double layer as consisting of a surface charge resulting from adsorbed H^+ and OH^- (and their complexes) and an equal but opposite counter charge, composed of indifferent electrolyte; thus,

$$\sigma_0 = -(\sigma_+ - \sigma_-) \tag{4}$$

where $\sigma_+ \equiv F\Gamma_+$ and $\sigma_- \equiv F\Gamma_-$; Γ_+ and Γ_- are, respectively, the interfacial excesses of the indifferent cation and anion. σ_+ and σ_- can be positive or negative, depending on whether the appropriate ions are attracted to or repelled from the surface.

Application of the simplest Gouy-Stern model, with the only assumption that at pH_{zpc} $\sigma_+ = \sigma_- = o$, or in the absence of specific adsorption of both the Na^+ and $C\ell^-$ ions, yields the expressions

$$\sigma_+ = -F\left(\frac{\partial\gamma}{\partial\mu_s}\right) - \frac{\sigma_o}{2} \qquad [5a]$$

$$\sigma_- = +F\left(\frac{\partial\gamma}{\partial\mu_s}\right)_{c_o} - \frac{\sigma_o}{2} \qquad [5b]$$

Figure 4 shows the calculated σ_+ and σ_- values. It clearly indicates that at pH < pH_{zpc}, Na^+ ions tend to be expelled from the surface, and that at pH > pH_{zpc}, $C\ell^-$ ions are kept away from the surface.

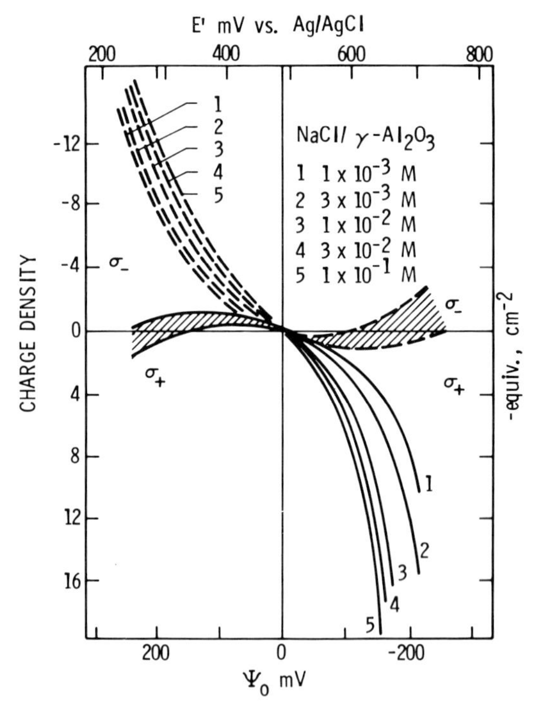

Fig. 4 The Distribution of the Ionic Components of Charge on the Solution Side of the Double Layer.

E. Capacity Components

Moreover, differential capacity can be subdivided into C_+ and C_- by defining

$$C_+ = -\frac{\partial\sigma_+}{\partial\psi_o}\bigg|_{\mu_s} \quad \text{and} \qquad [6a]$$

$$C_- = -\frac{\partial\sigma_-}{\partial\psi_o}\bigg|_{\mu_s} \qquad [6b]$$

where C_+ and C_- are components of differential capacity due to cations and anions, respectively, and $C = C_+ + C_-$. The computed components of capacity for γ-$A\ell_2O_3$ together with that for Hg and AgI is shown in Fig. 5. By virtue of their definitions, C_+ and C_- measure the change of adsorption of cations and anions in the double layer with surface potential, ψ_o. The C_+ and C_- are virtually mirror images of one another, especially for Hg and AgI at low ionic strength. As surface potential becomes progressively negative, C_+ begins to tail off and reaches a constant value for Hg and AgI; however, C_+ continues to increase with more negative potential for γ-$A\ell_2O_3$. The results demonstrate that anions not cations are generally favored by mercury and AgI surface; while both cations and anions are about equally sorbable by γ-$A\ell_2O_3$.

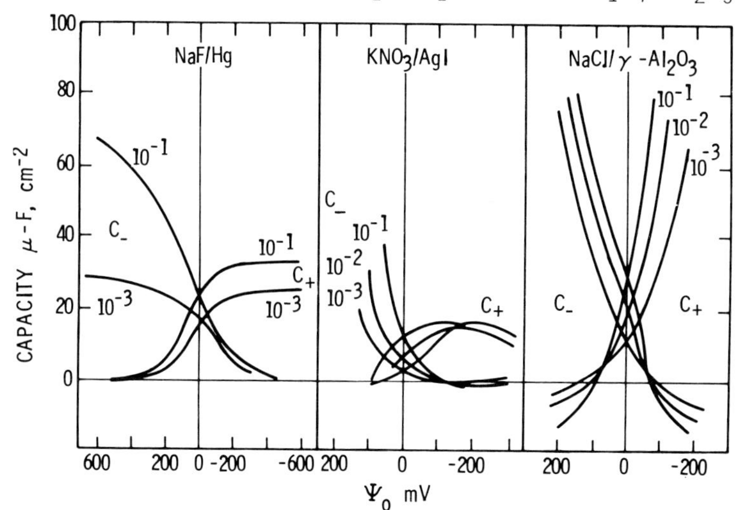

Fig. 5 Comparison of Distribution of Components of Differential Capacity Among Hg (1), AgI (5), and γ-$A\ell_2O_3$.

IV. DISCUSSION

The results obtained from the study have clearly indicated that there is significant difference in the double layer structure of Hg, AgI, Ag_2S, and hydrous oxides.

Many factors have been postulated to account for the difference in the properties of electrical double layer; namely, the intrinsic properties of the solid, the nature of the electrolyte, and the interaction among the solid, the electrolyte and the water phase (19). The structure of electrical double layer is largely determined by the extent of inter-reaction among these three components and, seemingly, no general rule can be established (8).

According to the Gouy-Stern model of double layer structure, the differential capacity may be subdivided into a contribution within the compact layer and a capacity within the diffuse part of the double layer, these contributions being connected in series as

$$\frac{\partial \psi_o}{\partial \sigma_o} = \frac{\partial (\psi_o - \psi_s)}{\partial \sigma_o} + \frac{\partial \psi_s}{\partial \sigma_o} \quad \text{or} \qquad [7a]$$

$$\frac{1}{C} = \frac{1}{C_{comp}} + \frac{1}{C_{diff}} \qquad [7b]$$

where ψ_s is the potential at plane of the closest approach. At high ionic strength and small ψ_s, $1/C_{comp} >> 1/C_{diff}$; while at low ionic strength and large ψ_s, $1/C_{diff} > 1/C_{comp}$. Therefore, the capacity in very dilute solutions and at pH_{zpc} is not expected to be affected by the solid phase.

A comparison of the differential capacity for various solids verifies this argument. Figure 6 shows that in dilute solution ($<10^{-3}$M, 1:1 electrolyte), no significant difference in C can be observed for different solids. This is particularly obvious at pH_{zpc}. With increasing ionic strength, and ψ_o the differential capacity or C_{comp} differs among solids; the capacity increases at a given surface charge as the electrostatic field strength increases. In Fig. 6, the solids have been arranged from left to right in order of increasing field strength, F_v.

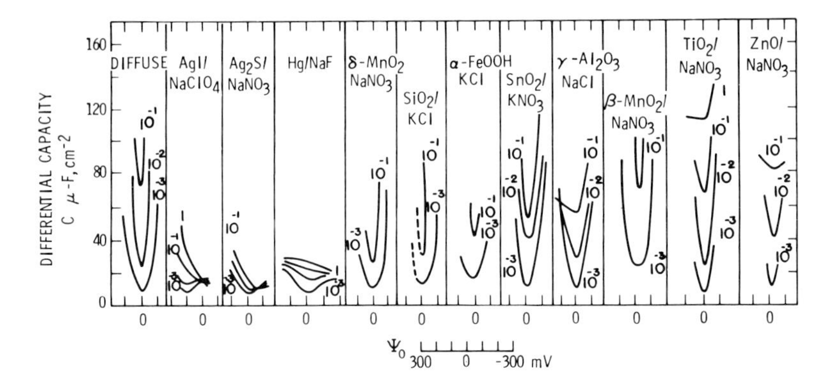

Fig. 6 Comparison of Differential Capacity Among Various Solids. Theoretical values for diffuse double layer have been computed according to the Gouy-Stern theory of single flat double layer. C values plotted for AgI, Ag_2S, Hg, TiO_2, and ZnO are from Lyklema, Iwasaki, and deBruyn (7); Grahame (1); Berube and deBruyn (9); and Block and deBruyn (10), respectively. The solids are listed from left to right in order of increasing electrostatic field strength.

The chemical composition of the surface depends on the type of chemical bonds produced in the subdivision of the solid phase, and thereby determines the polarity of the solid. For instance, if graphite is subdivided, a non-polar hydrophobic surface results because primarily van der Waals bonds are broken in the fracturing process; if the fracture involves rupture of ionic or covalent bonds as in $A\ell_2O_3$, MnO_2, or quartz, a polar and in turn hydrophilic surface will result. The polarity of solid must also affect the extent of the ionic interaction at the surface. This is apparent from the observation that solids of the same chemical composition can have different capacity (Fig. 7). Such an effect must largely be accounted for by the influence which electrostatic field strength has on the structure of the adjacent water layer and on the coulombic interaction of surface ionic groups with cations.

For the system which contains simple indifferent electrolyte, F_V is seen inversely proportional to volume of unit cell per metal (or central atom) V. It is also seen from Fig. 6 that the C values are much larger for oxide interfaces than for Hg, AgI, and Ag_2S interfaces. Among the oxides, the

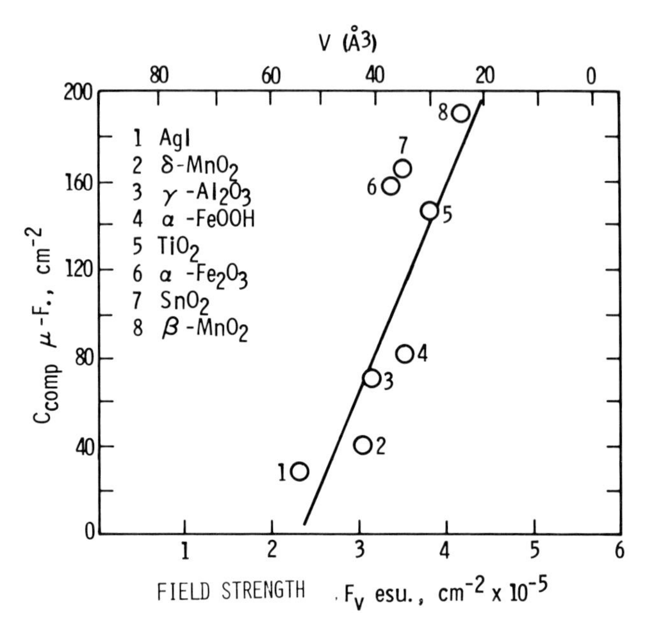

Fig. 7 Effect of Field Strength on Double Layer Capacity. Field strength is calculated by Hückel equation. V is the volume in A^3 occupied per every central metal atom of the crystal.

amorphous modifications (amor SiO_2, δ-MnO_2 and γ-Al_2O_3) have lower C_{comp} values than the more crystalline and more polar solids. Apparently, the water molecules at the very vicinity of a strong-field-strength solid are drastically modified and more counter-ions are thus accommodated at the surface (19). The penetration of ions into the compacted part of the double layer can significantly enhance its capacity. Breeuwsma and Lyklema (8) have recently attributed this phenomenon to the surface porosity of oxide. The surface charge of a porous surface was found greater than those of a non-porous surface. They related that extra ions may be confined within the surface micropores. However, they did not rule out the possibility of ion penetration into some depth of the surface layer. The effect of porosity on the double layer properties of γ-Al_2O_3 is insignificant, since the surface of γ-Al_2O_3 was found rather smooth by t-test (11).

The validness of the Gouy-Stern theory of double layer and the modified Grahame's method were also tested by direct measurement of electrolyte adsorption at concentration of

10^{-3}M (NaCℓ). The results shown in Fig. 8 agree well with what was calculated semitheoretically. However, for high ionic strength, substantial quantities of ion adsorption were detected even when the surface was oppositely charged. It can be concluded that the Gouy-Stern theory is primarily valid for lower ionic strength and that the assumption, $\sigma_+ = \sigma_- = o$, at pH_{zpc} can no longer hold at high ionic strength.

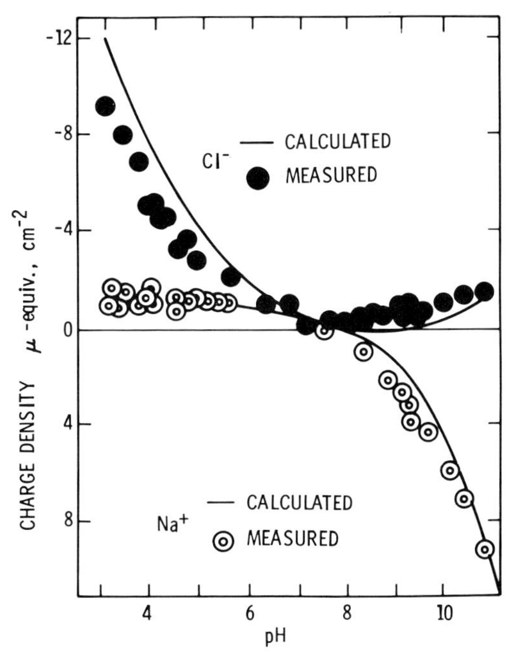

Fig. 8 Charge Distribution in the Solution Side of the Double Layer Calculated by the Modified Grahame's Method and Measured by Direct Electrolyte Adsorption for an Ionic Strength of 10^{-3}

V. CONCLUSION

The characteristics of the double layer at the γ-$A\ell_2O_3$ - electrolyte interface differs significantly from those at a polarized mercury- or reversible AgI- and Ag_2S-electrolyte interface because of the electrostatic field strength of the solids. Ions tend to penetrate to a much larger extent into the compact part of the double layer of the aluminum oxide interface than to the double layer of mercury, AgI, or Ag_2S interface.

The distribution of ions at the γ-$A\ell_2O_3$ - solution interface has been evaluated from the alkalimetric titration of the aqueous dispersions of the oxide, and with the aid of the Gouy-Stern version of the double layer theory with the only assumption that $\sigma_+ = \sigma_- = o$ at pH_{zpc}. The results from model calculations have been verified by adsorption measurement of simple electrolyte, i.e., $NaC\ell$ from dilute solution at a concentration less than 10^{-3}M.

Acknowledgement. Valuable advice given by Werner Stumm is acknowledged. This work was supported in part by a NSF Grant ENG75-07176.

VI. REFERENCES

(1) Grahame, D. C., Chem. Rev. 41, 441 (1947).
(2) Grahame, D. C., and Parson, R., J. Am. Chem. Soc. 83, 1291 (1961).
(3) Grahame, D. C., and Soderberg, B. A., J. Chem. Physics 22, 449 (1954).
(4) Lyklema, J., and Overbeek, J. T., J. Colloid Sci. 16, 595 (1961).
(5) Lyklema, J., Trans. Faraday Soc. 59, 418 (1963).
(6) Lyklema, J., Disc. Faraday Soc. 42, 81 (1966).
(7) Iwasaki, J., and de Bruyn, P. L., J. Am. Chem. Soc. 62, 594 (1958).
(8) Breeuwsma, A., and Lyklema, J., Disc. Faraday Soc. 52, 324 (1971).
(9) Berube, Y. G., and de Bruyn, P. L., J. Colloid Interface Sci. 28, 92 (1968).
(10) Blok, L., and de Bruyn, P. L., J. Colloid Interface Sci. 32, 533 (1970).
(11) Huang, C. P., and Stumm, W., Surface Sci. 32, 287 (1972).
(12) Huang, C. P., and Stumm, W., J. Colloid Interface Sci. 43, 409 (1973).
(13) MacInnes, D. A., "The Principles of Electrochemistry," p. 187, Dover Publications, Inc., New York, 1961.
(14) Kolthoff, I. M., and Stenger, V. A., "Volumetric Analysis," Vol. 2, p. 242, p. 256, Interscience Publications, New York, 1947.
(15) Parks, G. A. and de Bruyn, P. L., J. Phys. Chem. 66, 967 (1962).
(16) Levine, S., and Smith, A. L., Disc. Faraday Soc. 52, 290 (1971).
(17) Wright, H. J. L., and Hunter, R. J., Aust. J. Chem. 26, 1183 (1973).

(18) Schindler, P. W., in "Equilibrium Concept in Natural Waters," (Stumm, W., Ed.), Vol. 67, p. 196, Advances in Chemistry Ser. Am. Chem. Soc., 1967.
(19) Stumm, W., Huang, C. P., and Jenkins, S. R., Croat Chem. Acta. 42, 223 (1970).

THE DIRECT MEASUREMENT OF ELECTROSTATIC SURFACE FORCES

D.B. Hough and R.H. Ottewill
University of Bristol, England

I. SUMMARY

The normal electrostatic pressure between two solid surfaces, with an aqueous solution between them, has been determined in an apparatus which enabled the thickness of the liquid film and the applied pressure to be measured simultaneously. One solid surface was an optically polished glass prism and the other an optically smooth spherical cap of transparent rubber. The electrostatic charge on the surfaces was effected by the adsorption of sodium dodecyl sulphate. The experimental curves obtained were in reasonable agreement with those expected from theory.

II. INTRODUCTION

It is well-known that interfaces can acquire an electrostatic charge which gives rise to an electrical double layer in the neighbourhood of the interface. This applies to both macroscopic and microscopic interfaces. The stability of many colloid dispersions, in fact, frequently has its origin in the increasing magnitude of the electrostatic repulsive force as the particles approach to distances of the order of twice the thickness of the diffuse electrical double layer. The pressure arising from the interaction of two planar diffuse double layers was, in fact, derived by Langmuir(1) in 1938. Subsequently, the development of the theory of stability of lyophobic colloids developed by Derjaguin and Landau(2) and Verwey and Overbeek(3), the so-called DLVO theory, emphasized the importance of both electrostatic repulsion and van der Waals attraction. Although the latter forces have been investigated somewhat extensively using macroscopic surfaces (see for example the review by Israelachvili and Tabor(4)) very few direct measurements have been made of the distance dependence of electrostatic repulsive forces. Derjaguin et al(5) demonstrated directly the existence of repulsive forces and their variation with electrolyte concentration and surface potential using polarized crossed-wires and Peschel, Aldfinger and Schwarz(6) have measured the interaction pressure between two fused silica plates in electrolyte solutions.

Recently, however, one of the major contributions in this area has been the work of Roberts and Tabor(7). They used an optically smooth spherical cap of transparent rubber which was pressed against a flat glass surface in an aqueous electrolyte solution to give a thin liquid film between the glass and the rubber. The rubber deformed easily over local protrusions so that over the compression region the surfaces were essentially parallel. The distance between the surfaces was obtained by reflectance measurements. The apparatus was so designed that a compressive load could be applied, by means of a weight, normal to the parallel surfaces to balance the electrical double-layer force acting normally between the plates(8). A similar apparatus was constructed in our laboratories by Lewis(9) and used to extend the work of Roberts and Tabor. However, it was found that this type of apparatus could not easily be used to measure very low pressures, the region where the theoretical treatments are at their best and van der Waals forces can be neglected, and it could not easily be adapted for film drainage measurements.

A different apparatus has therefore been constructed to enable both low pressure measurements and kinetic measurements to be obtained. The apparatus is described in the present paper. It has been used to study the thin liquid films formed between glass and rubber surfaces in aqueous solutions of sodium dodecyl sulphate (SDS).

III. EXPERIMENTAL

Materials

BDH sodium dodecyl sulphate was used as starting material. It was purified by continuous liquid-liquid extraction of a 10% solution in an ethanol-water mixture (70:30) with 60-80 petroleum ether. The purified material was recrystallized from alcoholic solution at $-10^{\circ}C$. A freshly prepared aqueous solution did not exhibit a minimum in the surface tension against concentration curve.

The sodium chloride was BDH Analar material which was roasted at $800^{\circ}C$ before use. All water was twice-distilled and used direct from the still.

Apparatus for Measuring Repulsive Forces

The central part of the apparatus used for the direct measurement of repulsive forces is shown in Figure 1a. An essential component consisted of a cylinder (A) of synthetic polyisoprene rubber which had one end moulded into the form

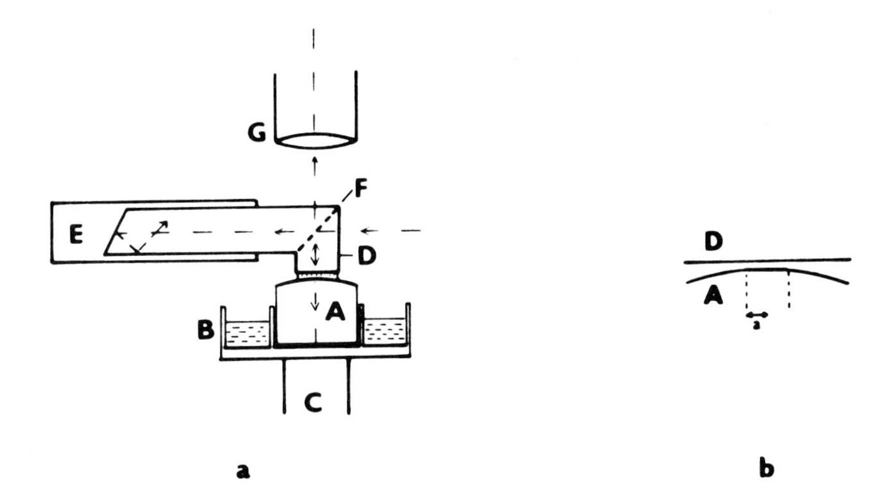

Figure 1: a) Schematic diagram of device for measuring interaction forces. A, polyisoprene with spherical cap; B, support cup for rubber; C, connexion to micrometer drive; D, glass prism; E, prism clamp and Rayleigh horn; F, beam splitter; G, microscope. b) Diagram illustrating flattening of rubber surface under applied pressure.

of an optically smooth spherical cap. The flat end of the cylinder was supported in a blackened metal cup (B). The latter was rigidly mounted on the drive movement (C) of a micrometer unit (The L.S. Starrett Co. Ltd., U.K.). Lockable joints within the drive assembly allowed the spherical surface of the rubber to be both rotated around the drive axis and to be tilted. The micrometer drive also enabled movement of the rubber in the vertical direction to be obtained in a very precisely controlled manner.

The glass prism (D) which had an optically polished lower surface, was mounted rigidly, relative to the micrometer movement, in a blackened recess (E). The latter also acted as a Rayleigh Horn and trapped the light transmitted horizontally through the beam splitting interface (F).

The aqueous solution was placed between the lower surface of the glass prism and the spherical rubber surface. Initially, the two surfaces were maintained at a distance so that no distortion of the rubber occurred, i.e. at zero applied pressure. A thin liquid film was formed by driving the rubber surface upward, via C, so that the rubber became distorted as shown in Figure 1b. Distortion of the rubber was observed through the microscope (G) by changes in the Newton's rings formed by the glass-liquid-rubber interfaces.

The thickness of the thin liquid film at a specified applied pressure was determined from reflectance measurements using as incident radiation a 10 mW Helium-Neon laser (λ_o = 632.8 nm). The reflected light from the film was taken, via the microscope G, to a photomultiplier detector fitted with an aperture system so that only light reflected from the film centre reached the photomultiplier. The signal from the latter was converted into a voltage and taken on to a chart recorder. The latter signal was directly proportional to the reflected intensity.

Preparation of Rubber Cylinders

The rubber used was a synthetic, 95% cis, polyisoprene (kindly provided by Mr. R. Dove of Avon Rubber Co. Ltd., Melksham, U.K.) containing 2.5% w/w dicumyl peroxide. About 4 g of the uncured rubber was cut into the shape of a cone and placed apex down against the surface of a glass lens which was situated at the bottom of a cylindrical stainless steel mould, 21.5 mm in diameter. Both lens and mould had been pre-heated to 150°C in the hot-press. A stainless steel plunger was pressed gently downwards (0.3 atmosphere) over a time period of about 3 min. This moulded the rubber. Curing of the rubber took place in the absence of applied pressure.

The material prepared in this way was optically clear with a refractive index of 1.52 (λ_o = 632.8 nm) and a Young's modulus of ca 2×10^6 Nm^{-2}. Just prior to use the lens was slowly peeled from the rubber under propan-2-ol to minimize the adherence of dust particles to the spherical surface; the surface was then allowed to dry in a dust-free atmosphere.

The Glass Prism

The prism (D) was made from Chance Pilkington Hard Crown Glass, type 519604, and had a refractive index of 1.52 (λ_o = 632.8 nm). The lower face of the prism had dimensions of 7.6 cm by 1.3 cm and had been polished flat to within one fringe at λ_o = 589 nm. All faces of the prism not used for light transmission were blackened. Before mounting the prism in the recess (E) the lower face was cleaned by successive washings in propan-2-ol, 50/50 nitric acid and water. Zero contact angle with a water drop was taken as an indication of a clean surface.

Procedure for Observations on Thin Films

The spherical surface of a rubber cylinder was placed ca 1 mm below the lower face of the prism and adjusted so that the Newton's rings, so formed, were concentric with the centre of the field of view of the microscope (G). The rubber was then slowly advanced towards the glass surface until contact, in the form of a circular area, was made between the two surfaces. Only a very small amount of light was reflected from this intimate contact area and this was recorded as the background intensity.

Approximately 0.5 cm^3 of solution of surfactant was then applied to the space around the rubber-glass contact region where it was held by capillary suction. The rubber surface was then retracted until contact was broken and it returned to its spherical shape. About 15 min was allowed to elapse in order for the surfactant to adsorb at the glass-solution and rubber-solution interfaces. The rubber was then advanced until the inter-surface distance was ca 240 nm, a distance less than the dark fringe at $\lambda_o/2n$ and greater than that of the bright fringe at $\lambda_o/4n$ ($n \triangleq$ average refractive index of the film).

The thin film was formed quickly at this point by rapidly advancing the rubber in order to produce an essentially plane-parallel area with a radius, a, (see Figure 1b) not greater than 7.5×10^{-3} cm, as measured with a fine graticule in the eye-piece of the microscope; larger areas gave dimple formation. The applied pressure was obtained from this area by an independent calibration.

As the film thinned the reflected intensity was recorded as a function of time. After equilibrium had been attained (constant intensity) the pressure on the film was increased and a new set of readings taken. A succession of pressure applications hence gave a curve of applied pressure against equilibrium film thickness. Moreover, from the initial curve of reflected intensity against time (first pressure application) it was possible to obtain pressure against thickness data.

Film thicknesses were calculated from the reflected intensity measurements using the equation given by Caballero (10). The intensity for a particular film thickness was compared to the intensity of the first bright interference fringe. A refractive index sequence, glass-adsorbed surfactant layer-water-adsorbed surfactant layer-rubber, was assumed. A thickness of 1 nm was taken for the adsorbed layers and the refractive index was taken to correspond to dodecane (1.42). The film thicknesses, however, were relatively insensitive to the last two parameters down to 7 nm.

IV. RESULTS

Kinetics of Film Drainage

Experimentally, the thickness of the aqueous core of the film, h, i.e. total thickness between the glass and rubber surfaces minus twice the length of a dodecyl chain, was determined as a function of time. The results obtained using as the solution phase both 6.0×10^{-3} mol dm^{-3} SDS and 6.0×10^{-3} mol dm^{-3} SDS + 1.4×10^{-2} mol dm^{-3} NaCl are shown in Figure 2. The curves, given in the form $1/h^2$ against Δt, where Δt = time interval from the start of observations, are linear at low time intervals and then bend over and become parallel to the time axis.

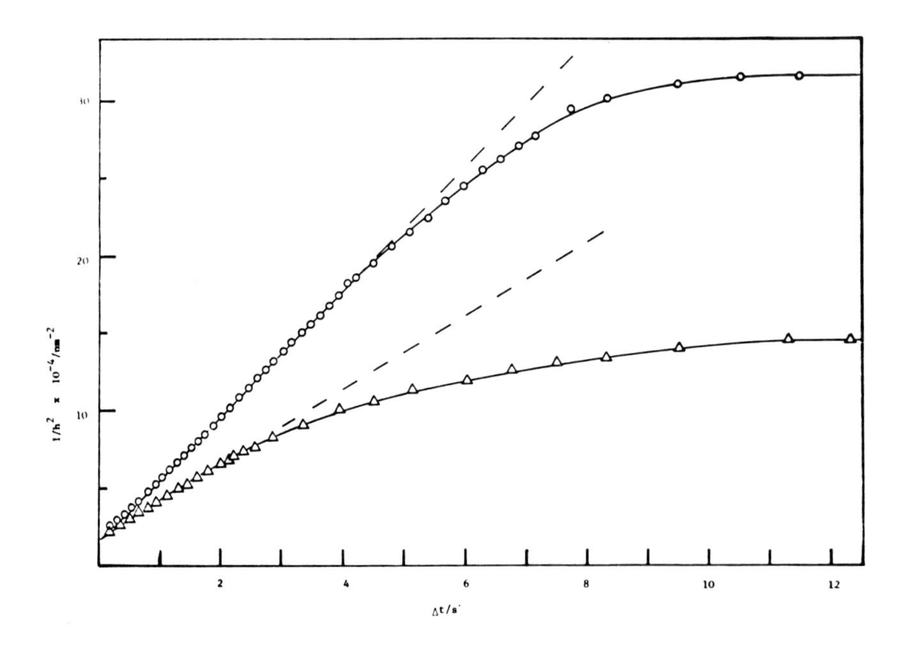

Figure 2: Curves of $1/h^2$ against Δt illustrating film drainage under applied pressure. –△–, 6×10^{-3} mol dm^{-3} SDS; –○–, 6×10^{-3} mol dm^{-3} SDS + 1.4×10^{-2} mol dm^{-3} NaCl.

It was shown by Reynolds(11) that the flow of liquid between two parallel discs of radius, a, under a net applied pressure, p. is given by,

$$\frac{d(1/h^2)}{d(\Delta t)} = \frac{4}{3\eta a^2} p \qquad \ldots (1)$$

The boundary condition taken by Reynolds was that the velocity of liquid flow was zero at both surfaces and this appears to be fulfilled in the present experiments. As mentioned earlier, once a small pressure was applied, the contact area formed immediately, and remained constant in cross-section during the drainage process.

For the thicker films, h > 50 nm, the normal repulsive pressure $p_R(h)$, arising from double-layer compression was zero and consequently equation (1) was obeyed with p = the applied pressure, p_a. This is well illustrated by the data for $1/h^2 < 7 \times 10^4$ nm^{-2} in 6×10^{-3} mol dm^{-3} SDS and for $1/h^2 < 19 \times 10^4$ nm^{-2} in 6×10^{-3} mol dm^{-3} SDS containing 1.4 x 10^{-2} mol dm^{-3} sodium chloride. From these curves in the linear regions the viscosity of the solution, η, was calculated. It was normally found to have a value of approximately 2.4 times that of the bulk solution.

Once the film has thinned to a certain distance the double layer repulsion pressure, $p_R(h)$, acting normally to the surface begins to have an effect and the net pressure in the film becomes,

$$p = p_a - \frac{3\eta a^2}{4} \frac{d(1/h^2)}{d\Delta t} \qquad \ldots (2)$$

and since all the quantities on the right-hand side are known then $p_R(h)$ can be calculated for a particular value of h. In the present work p_a was taken as the average applied pressure and was determined from an independent calibration curve of applied load (mg) against a^3 (measured a). The value of η used was that obtained from the initial region of the $1/h^2$ against Δt curve.

The results obtained using the kinetic approach are shown as open circles in Figures 3 and 4. This provided a very useful means of obtaining data at very low pressures.

Equilibrium Films

It can be seen from equation (2) that as $p_R(h)$ increases with decrease in h, a value of h is reached at which the net pressure = 0, and $p_R(h) = p_a$. This point was taken as the position of equilibrium, i.e. a position of constant thickness with the applied pressure exactly balanced by the repulsive pressure in the film. The film thickness was thus directly obtainable from the horizontal position of the curve of $1/h^2$ against Δt.

With equilibrium films the film was allowed to thin to a constant value and the film thickness obtained from reflectance measurements. The results obtained using this procedure are shown as open triangles in Figures 3 and 4.

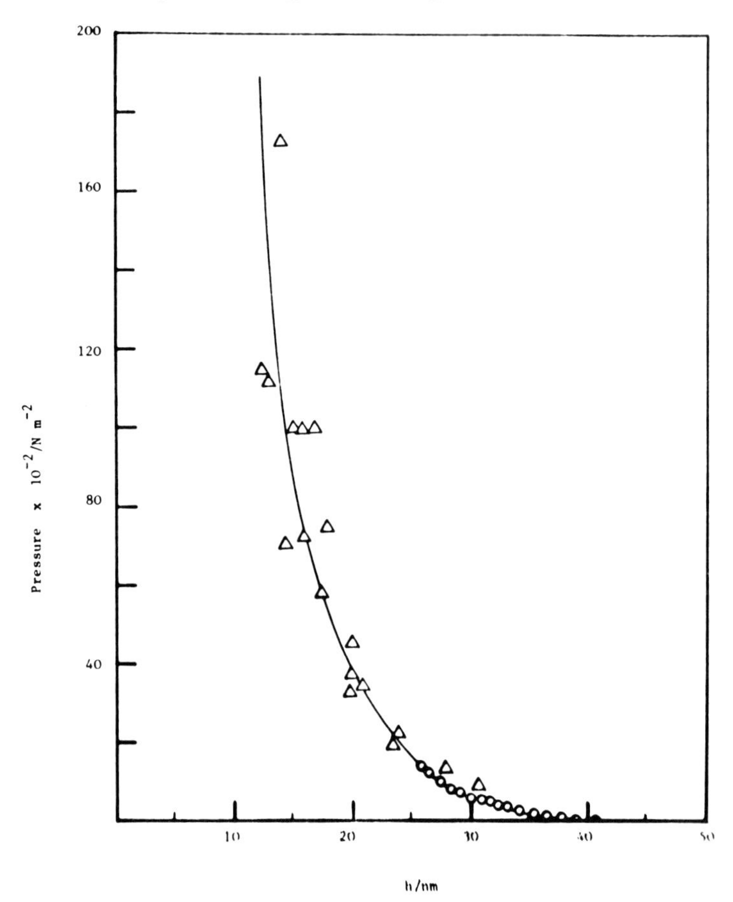

Figure 3: Pressure, $p_R(h)$ against h for 6 x 10^{-3} mol dm^{-3} SDS —o—, kinetic points; —△—, equilibrium points.

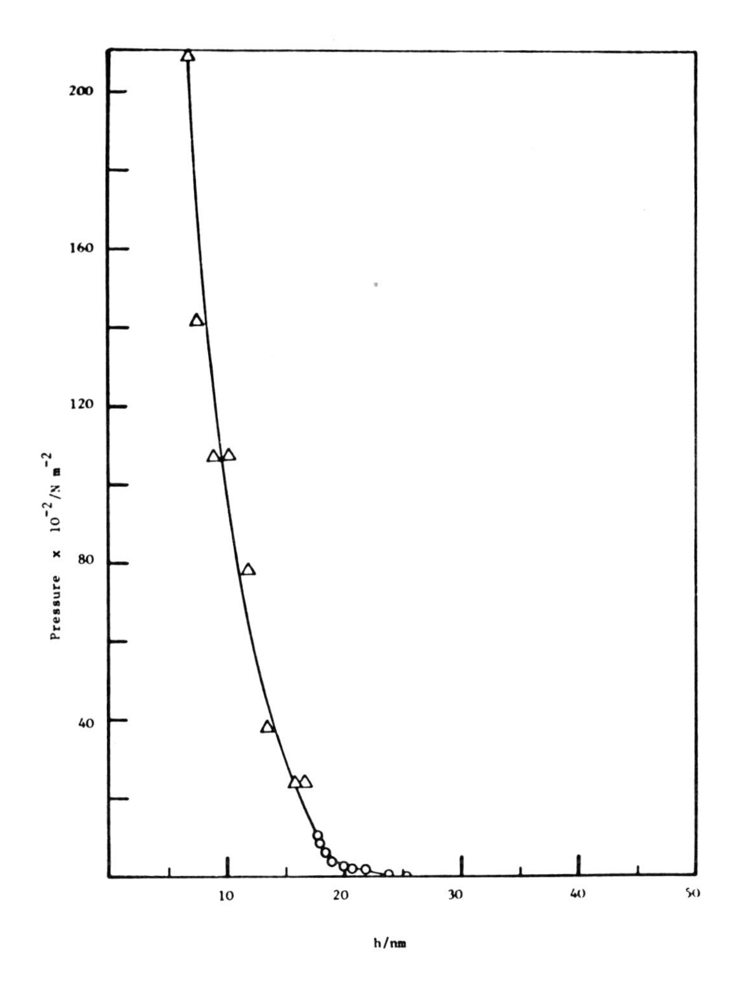

Figure 4: Pressure, $p_R(h)$ against h for 6×10^{-3} mol dm^{-3} SDS + 1.4×10^{-2} mol dm^{-3} NaCl. —○—, kinetic points; —△—, equilibrium points.

V. DISCUSSION

Although the glass surface and the polyisoprene surface were optically smooth neither of the surfaces was characterized in detail chemically. The glass surface was presumed to be composed of siloxane groupings with some isolated silanol groups (ionizable) and the work of Lewis(9) indicated that the polyisoprene surface appeared to be predominantly hydrophobic with some charged sites present, probably arising from the decomposition of dicumyl peroxide. It was clear from the experiments that both surfaces adsorbed dodecyl sulphate ions, and it was the latter in the adsorbed state which conferred a charge on both surfaces. Below concentrations of 2×10^{-3} mol dm^{-3} SDS it was not possible to form stable thin films, an indication of the inherently low charge on the native surfaces.

The deviations from the simple Reynolds' equation observed with the draining films showed that repulsive forces became operative within the films as their thickness decreased. In 6×10^{-3} mol dm^{-3} SDS solutions (Figure 3) the repulsive force became significant at distances of the order of 40 nm. In the higher ionic strength it became operative at 25 nm. These results demonstrate the dependence of electrostatic repulsive forces on the ionic strength of the medium.

Langmuir(1) showed that the electrostatic pressure between two plates, each having a surface potential ψ_o, was given by,

$$\pi_{el} = 2 n_o kT (\cosh u - 1) \qquad \ldots (3)$$

where n_o = the number of ions per cm^3 of each type, k = Boltzmann constant, T = absolute temperature and u is given, for a 1:1 electrolyte, by,

$$u = e \psi_{h/2}/kT \qquad \ldots (4)$$

where e = fundamental units of charge and $\psi_{h/2}$ = the electrostatic potential mid-way between the plates. The relation between h and the surface potential is given by,

$$-\frac{\kappa h}{2} = \int_z^u \frac{dy}{[2(\cosh y - \cosh u)]^{\frac{1}{2}}} \qquad \ldots (5)$$

where $z = e \psi_o/kt$, $y = e\psi/kT$ and κ = the reciprocal of the Debye-Hückel screening distance. Tables are available giving solutions of this integral(3,12) or alternatively it can be integrated numerically to give the potential at the mid-plane as a function of h at constant values of κ and z.

Under conditions where the interaction is weak, that is where $\kappa h > 2$, a useful approximation(3) is given by

$$\pi_{el} = 64 n_o kT \tanh^2 (z/2) \exp(-\kappa h) \qquad \ldots (6)$$

In Figure 5 a comparison is given of the experimental results for applied pressure against h with the theoretical values of π_{el} obtained using the tables of Devereux and de Bruyn(12). The theoretical curves have been obtained using z values of 4 and 10. As can be seen from Figure 5 there appears to be reasonable agreement between theory (z = 10) and experiment at the very low pressures. However, as the pressure increases, the experimental pressure becomes less than the theoretical pressure for z = 4. Some disagreement could undoubtedly be due to experimental errors in both the pressure and aqueous core thickness. Data are not available at present for either the surface potentials or the electrokinetic potentials of the rubber and the glass. An estimate of the surface potentials in the presence of SDS was made, however, using equation (6) since this equation is in good agreement with the exact solutions of Devereux and de Bruyn for the range of κh values covered in the present work. It was possible to draw good straight lines through plots of $\ln p_R(h)$ against h and from the slopes values of κ were obtained. From the intercepts at h = 0 values of ψ_o were then calculated using the values of κ obtained from the slopes. $\psi_o = -34$ mV was obtained in 6×10^{-3} mol dm^{-3} SDS and $\psi_o = -23$ mV was obtained at the higher ionic strength. The values of κ obtained were respectively 1.83×10^8 m^{-1} and 3.45×10^8 m^{-1} compared with the calculated values of 2.55×10^8 m^{-1} and 4.65×10^8 m^{-1}. Although these values of ψ_o are sufficiently high to ensure significant electrostatic repulsion they are nevertheless substantially lower than one would expect from a monolayer of dodecyl sulphate ions. A feasible explanation for these low values of ψ_o would be that as the films thin the surfaces do not remain at constant potential. This could be a consequence of some desorption of the surfactant from the surfaces and/or deionization of some surface ionic groups.

For the film thicknesses investigation in this work van der Waals attractive forces have been neglected. This assumption appears reasonable since at h = 10 nm the calculated attractive force is only ca 4% of the measured $p_R(h)$. Furthermore, the exponential form of the $p_R(h)$ against h curve supports the contention that only forces arising from electrical double layers have been measured.

The apparatus constructed has enabled direct measurements of electrostatic repulsive forces between macroscopic surfaces in an aqueous medium to be made. For the weak interaction forces investigated reasonable agreement was obtained between experiment and theory.

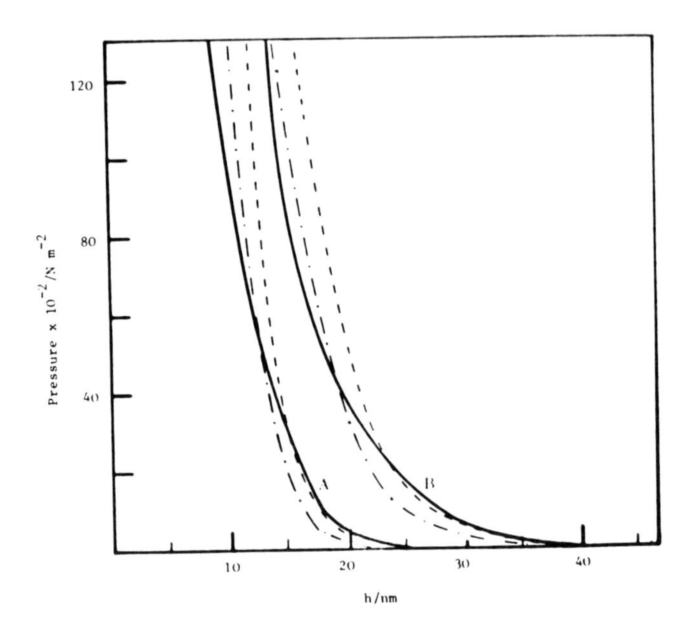

Figure 5: Pressure against h curves. ——— , experimental pressure $p_R(h)$; A in 6×10^{-3} mol dm^{-3} SDS + 1.4×10^{-2} mol dm^{-3} NaCl; B in 6×10^{-3} mol dm^{-3} SDS. Theoretical pressure, π_{el}; - • -, z = 4; - - - z = 10.

VI. ACKNOWLEDGEMENTS

We wish to thank the Scientific Research Council for support of this project.

VII REFERENCES

1. Langmuir, I., J.Chem.Phys., 6, 873 (1938).
2. Derjaguin, B.V. and Landau, L., Acta Physicochim. U.R.S.S. 14, 633 (1941).
3. Verwey, E.J.W. and Overbeek, J.Th.G., "Theory of the Stability of Lyophobic Colloids" Elsevier, Amsterdam, 1948.

5. Derjaguin, B.V., Voropayeva, T.N., Kabanov, B.N. and Titiyevskaya, A.S., J.Colloid Science, 19, 113 (1964).
6. Peschel, G., Aldfinger, K.H. and Schwarz, G., Naturwiss., 61, 215 (1974).
7. Roberts, A.D. and Tabor, D., Proc.Roy.Soc., A325, 323 (1971).
8. Roberts, A.D., J.Phys. (D), 4, 423, 433 (1971).
9. Lewis, P., Ph.D. Thesis, University of Bristol (1972).
10. Caballero, D., J.Opt.Soc.Amer., 37, 176 (1947).
11. Reynolds, O., Phil.Trans.Roy.Soc., 177, 157 (1886).
12. Devereux, O.F. and de Bruyn, P.L., "Interaction of Plane-Parallel Double Layers", M.I.T. Press, Cambridge, Mass., 1963.

THE DETERMINATION OF THE RADIAL DISTRIBUTION FUNCTION FOR INTERACTING LATEX PARTICLES

J.C. Brown,* J.W. Goodwin,† R.H. Ottewill† and P.N. Pusey*

* Royal Radar Establishment, Malvern, England
† University of Bristol, England

I. SUMMARY

The addition of mixed-bed ion-exchange resins to polystyrene latices leaves the particles in a dispersion medium which is mainly composed of counter-ions from the particles themselves. Under these conditions the particles form a geometrically regular array which on examination by visible light shows well-defined diffraction maxima in the angular envelope of scattering intensity. Fourier transformation of the angular intensity data leads to radial distribution curves. The interpretation of the observations is discussed in terms of long-range electrostatic interactions between the particles.

II. INTRODUCTION

The importance of electrostatic repulsive forces in stabilizing colloid systems has been recognized for a considerable period and in the 1940's was treated theoretically in some depth(1,2). The considerable attention paid to the study of coagulation processes, which usually occur in moderately strong electrolytes, has meant that studies in very low electrolyte concentrations have been infrequent. In the latter situation, however, interesting phenomena occur and the long range of the electrostatic forces frequently leads to a "degree of order" in the particle arrangements. In concentrated dispersions studies of the ordered arrangement have been reported by Klug et al(3), Luck et al(4), Hiltner and Krieger(5), Barclay et al(6) and Hachisu et al(7). In more dilute dispersions ordering was reported by Williams and Crandall(8) and in an earlier communication(9) we have reported measurements on very dilute polystyrene latices at very low ionic strengths by conventional light scattering and by photon correlation spectroscopy.

In this communication we show that a dilute polystyrene latex (mean particle radius = 23.1 nm), after treatment with mixed-bed ion-exchange resin for two weeks, exhibits a well-defined diffraction maximum in the angular light scattering intensity envelope indicative of spatial ordering of the particles. Analysis of the angular intensity data leads to values of the structure factor and hence by Fourier inversion to a radial distribution function.

An analysis of the variation of the position of the maximum of the diffraction peak with volume fraction of the latex also provides a simple means of determining particle size.

III. EXPERIMENTAL

Materials

The polystyrene latex (RB66) was prepared using the method described by Ottewill and Shaw(10) using sodium dodecyl sulphate as the surface active agent; the latter material was removed after the preparation by extensive dialysis. The dialysed latex was diluted to a concentration of ca 2×10^{-3} g cm^{-3} with either distilled water or 10^{-3} mol dm^{-3} sodium chloride solution in order to provide a stock dispersion. The particle size distribution was determined by electron microscopy. The number average particle radius, $\langle R \rangle$, based on analysis of 1600 particles was 23.1 nm, with an estimated systematic error of about 5%. The standard deviation of the particle size was 19% of the mean.

The mixed-bed ion-exchange resin was prepared using the method described by Vanderhoff et al(11). A small amount of resin, ca 50 granules, after washing thoroughly with ultra-filtered distilled water was added to the latex in the light scattering cells. The latex was filtered through a well-washed millipore filter (pore size 0.22 μm).

The latex concentration was determined to ± 5% after scattering measurements had been completed. A known weight of sample was dried at 70°C and the residual polymer dissolved in a known weight of dioxan. The optical density of the dioxan solution was determined at a wavelength of 280 nm and converted to concentration using an independently determined calibration curve.

Light Scattering

The light scattering measurements were made using a system 4300 light scattering spectrometer (Precision Devices and Systems, Spring Lane, Malvern). For absolute intensity measurements a CRL argon laser (λ_o = 488 nm) was used; this was monitored continuously so that a correction for drifts in laser intensity could be made. The error of intensity measurements was ca 10%. For some samples the angular positions of the diffraction peaks were determined using a Brice-Phoenix, Series 2000, light scattering photometer.

IV. LIGHT SCATTERING THEORY

Time Average Light Scattering - Non-Interacting Particles

The latex particles used in the present work had a radius of about 1/20th of the wavelength of the incident light used. The particles were also fairly monodisperse. As a first approximation therefore Rayleigh-Gans theory(12) was applied.

For a spherical particle of volume V in 1 cm^3 of medium of refractive index, n_o, illuminated by unpolarised light of wavelength, λ_o (in vacuo), the Rayleigh ratio

$$R_\theta^R = \frac{I_\theta \cdot r_o^2}{I_o} \qquad \ldots (1)$$

is given by,

$$R_\theta^R = \frac{9\,\pi^2\,n_o^4}{\lambda_o^4}\,V^2\left(\frac{m^2-1}{m^2+2}\right)^2(1+\cos^2\theta) \qquad \ldots (2)$$

where I_θ = intensity of the scattered light at an angle θ to the incident direction, I_o = the incident intensity, r_o = distance between scattering particle and point of intensity measurement and m = the ratio of the refractive indices of the particle and the medium.

For the condition that $m \rightarrow 1$, then

$$R_\theta^R = \frac{2\,\pi\,n_o^4}{\lambda_o^4}\,V^2\,(m-1)^2\,(1+\cos^2\theta) \qquad \ldots (3)$$

and in the Rayleigh-Gans region,

$$R_\theta^{R.G.} = R_\theta^R \cdot P(\theta) \qquad \ldots (4)$$

where $P(\theta)$ = the particle scattering form factor.

For vertically polarized incident radiation, i.e. a laser beam, and a system of N particles per cm^3, then

$$R_\theta^{R.G}(\perp) = \frac{2\,\pi^2\,n_o^4}{\lambda_o^4}\,N\,V^2\,(m-1)^2 \cdot P(\theta) \qquad \ldots (5)$$

and the time average intensity of scattering at an angle θ is given as,

$$I_\theta\ (\perp) = \left\{ \frac{I_o\ 2\ \pi^2\ n_o^4\ (m-1)^2}{\lambda_o^4\ r_o^2} \right\} N\ V^2\ P(\theta) \qquad \ldots (6)$$

All the terms in the curly brackets are constant for a given apparatus at a chosen wavelength. Moreover, the volume fraction of particles ϕ is equal to NV and therefore,

$$I_\theta\ (\perp) = \text{Const.}\ \phi V.\ P(\theta) \qquad \ldots (7)$$

and hence for non-interacting particles, the observed intensity $I_\theta\ (\perp)$ divided by $\phi V\ P(\theta)$ should be constant. For spherical particles the particle scattering form factor is given by(12),

$$P(\theta) = 1 - \frac{K^2R^2}{5} \ \ldots\ldots \qquad \ldots (8)$$

where R = the radius of the scattering sphere and K equals the magnitude of the scattering vector and is given by,

$$K = \frac{4\ \pi\ n_o}{\lambda_o} \sin\theta/2 \qquad \ldots (9)$$

<u>Time Average Light Scattering - Interacting Particles</u>

Under conditions where the particles interact, equation (7) requires correction to allow for the phase correlations introduced by the interactions. Thus a structure factor, $S(\theta)$, must be introduced giving for this situation,

$$I_\theta\ (\perp) = \text{Const.}\ \phi\ V\ P(\theta)\ S(\theta) \qquad \ldots (10)$$

Experimentally, the structure factor will be given by,

$$S(\theta) = \frac{I_\theta\ (\perp)}{\text{Const.}\ \phi\ V\ P(\theta)} \qquad \ldots (11)$$

Theoretically, $S(\theta)$ is given by,

$$S(\theta) = 1 + \frac{4\ \pi\ N}{K} \int_0^\infty (g(r) - 1)\ r \sin Kr.\ dr \quad \ldots (12)$$

where r = the centre to centre interparticle separation and g(r) is the particle-pair distribution function. The latter can thus be obtained from the experimental values of $S(\theta)$ by Fourier inversion of equation (13), to give,

$$N(g(r) - 1) = \frac{1}{2\ \pi^2\ r} \int_0^\infty \left[S(\theta) - 1 \right] K.\ \sin Kr.dK \quad \ldots (13)$$

For non-interacting scattering particles $S(\theta) = 1$ and equation

(7) applies.

V. RESULTS

Dependence of Angular Scattering Intensity Envelope on Electrolyte Concentration

In figure 1 the variation of the intensity of the scattered light as a function of angle is shown for a dilute sample of latex RB 66 ($\phi = 3.4 \times 10^{-4}$) determined initially in 10^{-3} mol dm^{-3} sodium chloride solution and then redetermined after the sample had remained for two weeks in contact with a mixed-bed ion-exchange resin. The results obtained in the presence of electrolyte are essentially those expected for a Rayleigh-Gans scattering system of non-interacting particles. After ion-exchange treatment, however, a well-defined peak becomes apparent in the scattering envelope indicating the introduction of some order into the system and the superposition of a structure factor on the scattering.

Figure 1: Scattering intensity, I_θ, in arbitrary units against scattering angle, θ, for latex RB 66 ($\phi = 3.4 \times 10^{-4}$); ---, in 10^{-3} mol dm^{-3} sodium chloride solution; ——, after mixed-bed ion-exchange treatment

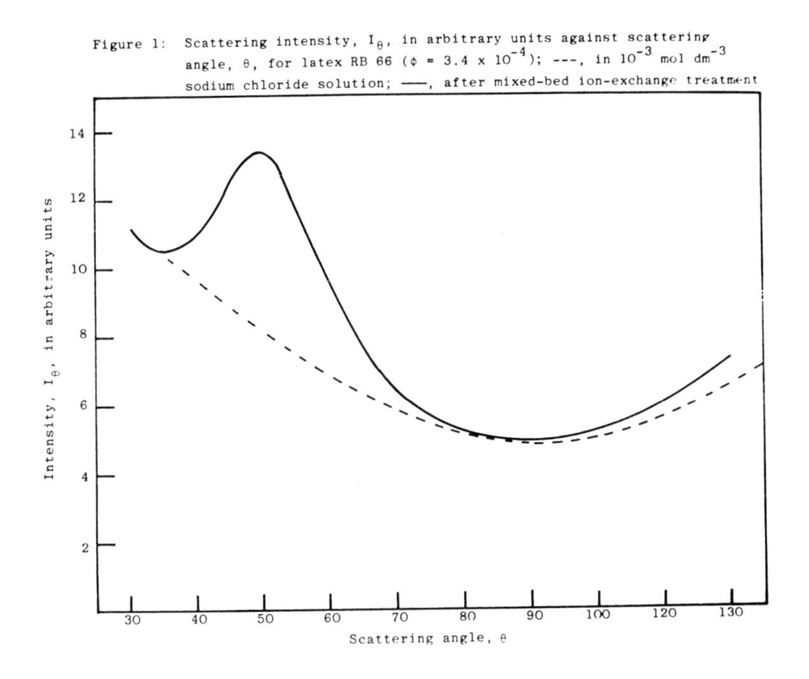

Examination of the scattering envelopes for ion-exchanged systems at different volume fractions indicated that the

angular position of the peak changed with volume fraction. The experimental data are given in Table 1.

Table 1

Variation of Peak Position with Volume Fraction

λ_o = 488 nm $\quad n_o$ = 1.3368

Sample	Volume Fraction	$\underline{N}$	θ_{max}
1	0.746×10^{-4}	1.24×10^{12}	26.9^o
2	1.00×10^{-4}	1.67×10^{12}	30.0^o
3	1.74×10^{-4}	2.90×10^{12}	36.2^o
4	3.42×10^{-4}	5.70×10^{12}	46.2^o
5	5.08×10^{-4}	8.46×10^{12}	53.5^o

Determination of the Structure Factor

In order to obtain the values of $S(\theta)$ the raw intensity data were divided by $P(\theta)$, by volume fraction and by a normalising factor obtained on systems for which $S(\theta) \rightarrow 1$, or on dilute systems for which $S(\theta) \rightarrow 1$ as $K \rightarrow \infty$.

The experimental curve of $S(\theta)$ as a function of K obtained on sample 3 after exposure to ion-exchange resin for a period of two weeks is given in Figure 2. These data were Fourier transformed according to equation (13) to yield the quantity $N(g(r) - 1)$. In order to carry out the transformation smoothed lines were drawn through the $S(\theta)$ against K curves and values of $S(\theta)$ obtained at equal intervals of ΔK, namely 10^4 cm^{-1}, for K values between 0.6×10^5 and 3.3×10^5 cm^{-1}. The transformation was performed numerically with integration in equation (13) replaced by summation. It is probable that the largest sources of error in $N(g(r) - 1)$ are the lack of intensity data at small values of K and possible errors in data normalization(14). Nevertheless the data are sufficiently accurate to show the form of the radial distribution function for the interacting latex system.

The radial distribution function obtained from the measurements made on sample 3 is given in Figure 3.

Simple Treatment of Peak Positions in the Angular Scattering Envelope

The sharp peaks obtained in the angular scattering intensity envelopes imply that considerable spatial ordering of

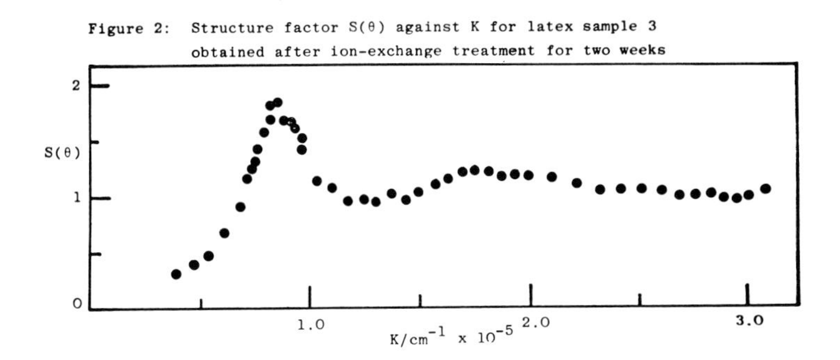

Figure 2: Structure factor S(θ) against K for latex sample 3 obtained after ion-exchange treatment for two weeks

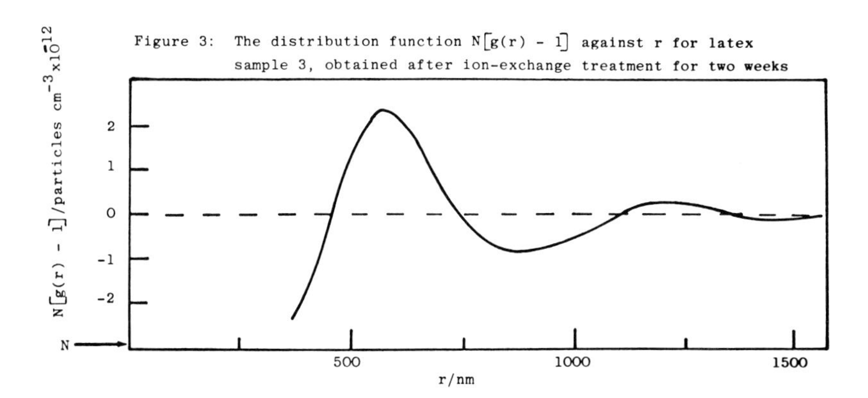

Figure 3: The distribution function N[g(r) - 1] against r for latex sample 3, obtained after ion-exchange treatment for two weeks

the particles had occurred in the ion-exchanged systems. Several features of the ordering can be illustrated using a Bragg diffraction analysis. For a Bragg "reflection"

$$m\lambda_o/n_o = 2\ d_{hk\ell}\ \sin\ \beta \qquad \ldots\ (14)$$

where β = the Bragg angle, m the order of diffraction and $d_{hk\ell}$ the interplanar spacing for a set of $hk\ell$ planes in the lattice. In terms of the scattering angle at the peak position, θ_{max}, then

$$m\lambda_o/n_o = 2\ d_{hk\ell}\ \sin\ (\theta_{max}/2) \qquad \ldots\ (15)$$

If a unit cell is defined in the lattice to have a side length 2R + h, where h = the surface to surface distance between the spheres then the volume fraction will be defined by,

$$\phi = \frac{\text{volume of spheres}}{\text{volume of unit cell}} = \frac{A.\ R^3}{(2R + h)^3} \qquad \ldots (16)$$

where A is a constant depending on the lattice type (see Table 2).

Table 2

Values of A for Different Lattice Types

Lattice Type	A
simple cubic	$4\pi/3$
body-centred cubic	$8\pi/3$
face-centred cubic	$16\pi/3$

The interplanar spacing is given by,

$$d_{hk\ell} = \frac{2R + h}{B} \qquad \ldots (17)$$

where B depends also on the lattice type (see Table 3).

Table 3

Values of B for Different Lattice Types

Lattice Type	d_{100}	d_{110}	d_{111}
simple cubic	1.0	$\sqrt{2}$	$\sqrt{3}$
body-centred cubic	-	$\sqrt{2}$	-
face-centred cubic	-	-	$\sqrt{3}$

Combining equations (16) and (17) gives,

$$d_{hk\ell} = \left(\frac{A}{\phi}\right)^{1/3} \frac{R}{B} \qquad \ldots (18)$$

and the combination of (15) and (18) yields,

$$\lambda = \frac{\lambda_o}{n_o} = \frac{2R}{B} \left(\frac{A}{\phi}\right)^{1/3} \sin(\theta_{max}/2) \qquad \ldots (19)$$

or

$$\lambda^3 = \frac{8R^3\ A}{B^3\ \phi} \sin^3(\theta_{max}/2) \qquad \ldots (19a)$$

Table 1 gives values of θ_{max} as a function of ϕ and these experimental data are shown plotted in Figure 4. For comparison the theoretical positions of θ_{max} expected for reflections, from the 011 planes of a body-centred lattice, the 111 planes of a face-centred lattice and the 100 planes of a simple cubic lattice, are also plotted. The value of R was taken as 23.1 nm. The agreement between the experimental results and the simple diffraction theory is remarkably good.

Figure 4: Angular peak position, θ_{max}, against latex volume fraction, ϕ: —— (110) body-centred lattice; – – – (111) face-centred lattice; --- (100) simple cubic lattice. $R = 23.1$ nm; $\lambda_o = 488$ nm

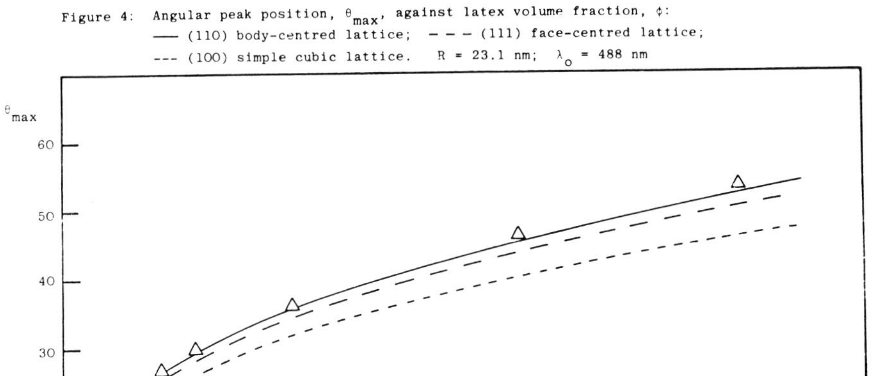

An alternative form of data treatment is suggested by equation (19a) and in Figure 5 the results from Table 1 are plotted in the form of ϕ against $\sin^3(\theta_{max}/2)$. The excellent linear curve obtained appears to confirm the validity of equation (19a) and suggests that at least with respect to peak position the particulate dispersion appears to behave as an ordered lattice.

The slope of the ϕ against $\sin^3(\theta_{max}/2)$ curve is given by,

$$\frac{d\phi}{d\sin^3(\theta_{max}/2)} = \left[\frac{8A}{B^3\lambda^3}\right] R^3 \qquad \ldots (20)$$

Assuming a body-centred lattice and 011 reflections a value of 22.9 nm was obtained for R whilst assuming a face-centred lattice and 111 reflections a value of 22.2 nm was obtained.

Figure 5: Volume fraction, ϕ, against $\sin^3(\theta_{max}/2)$ from data obtained at $\lambda_0 = 488$ nm

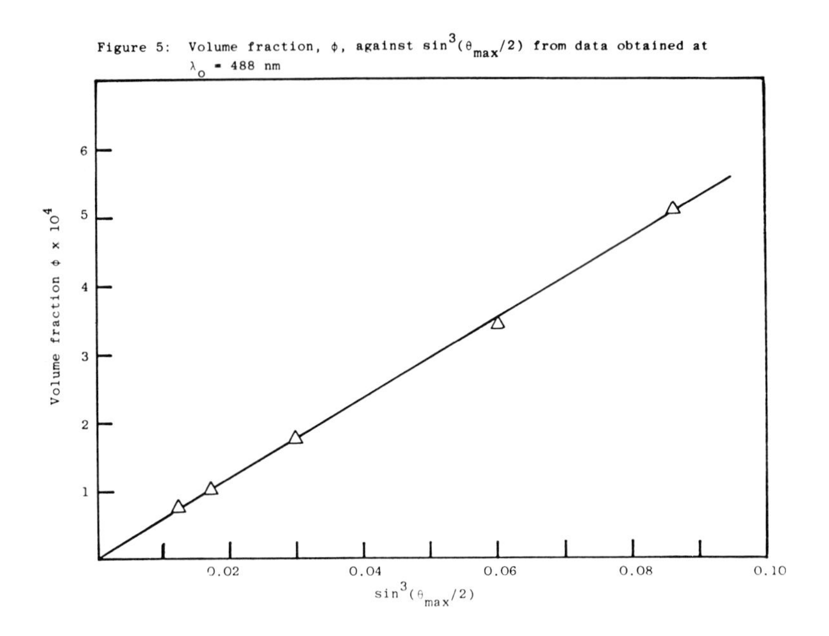

These values can be compared with those obtained from electron microscopy and from photon correlation spectroscopy (9) (see Table 4).

Table 4

Comparison of Particle Radii (RB66) obtained by Various Methods

Method		Radius (nm)
Photon Correlation Spectroscopy (9)		25.0
Electron Microscopy (number average)		23.1
Electron Microscopy (weight average)		25.2
ϕ vs $\sin^3(\theta_{max}/2)$	f.c.c.	22.2
ϕ vs $\sin^3(\theta_{max}/2)$	b.c.c.	22.9

[N.B. The radius obtained by photon correlation spectroscopy is not a number average.]

The measurement of θ_{max} as a function of volume fraction appears to offer a simple means of determining the size of particles in a liquid dispersion provided that ordering can be achieved. An alternative approach is to use a single volume fraction and measure the variation of θ_{max} with λ. Alternatively both ϕ and λ could be varied leading to a plot of $\phi\lambda^3$ against $\sin^3 (\theta_{max}/2)$.

VI. DISCUSSION

The latex used in this work was prepared by persulphate initiation and hence the stabilizing electrical charges on the surface were sulphate groups(15). The surface charge density, as determined by conductometric titration, after extensive dialysis followed by ion-exchange treatment, was 1.2 $\mu C\ cm^{-2}$; this corresponds to one charge per 13.5 nm^2. Since ion-exchange resin was used in the light scattering cells the principal counter-ions were H^+ ions and the concentration of these depended on the volume fraction and degree of ionization of the latex. In the case of sample 3 a total hydrogen ion concentration of 2.66 x 10^{-6} mol dm^{-3} was possible with full ionization. pH measurements on latex samples at similar concentrations to those used gave pH values of about 6, all the evidence implying a bulk concentration of about 10^{-6} mol dm^{-3}. Thus at this concentration the Debye-Huckel screening distance ($1/\kappa$) was 305 nm at 25°.

For $\kappa R \ll 1$ the potential energy of electrostatic repulsion is given(2) by the equation,

$$V_R = \frac{\varepsilon R^2 \psi_o^2}{r} \exp\left\{- \kappa (r - 2R)\right\} \qquad \ldots (21)$$

where ψ_o = the surface potential and ε = the relative permittivity of the dispersion medium. Direct determination of the value of ψ_o has not yet been achieved but using the surface charge density we obtain from the Debye-Hückel approximation for a spherical particle $\psi_o \simeq 370$ mV. However, the Debye-Hückel estimate is poor for high potentials and the planar estimate is probably more reliable. Moreover, potential is likely to be reduced in the low ionic strength environment by suppressed ionization of the surface groups on the polyelectrolyte particle surface (e.g. Tanford(16)). It can only be concluded that ψ_o lies in the range 200 ± 50 mV. Taking the lower estimate, ψ_o = 150 mV, with R = 23.1 nm, $1/\kappa$ = 305 nm at r = 580 nm gives V_R = 7.8 kT. Although the arguments are approximate at this stage, it is clear the appreciable electrostatic interaction exists between the particles at the long distances (ca 25 particle radii) involved in the light scattering experiments. The force increases steeply at shorter distances indicating on theoretical grounds that the particles would be situated in a potential energy well.

The analysis of the variation of θ_{max} with volume fraction ϕ tend to support this suggestion and indicate that the electrostatic repulsive forces between the particles are sufficiently strong that the particles arrange themselves in a loosely ordered array.

The analysis of the angular intensity measurements on the ion-exchanged samples leads to the structure factor $S(\theta)$ and hence to the function, $(N\ g(r) - 1)$, as a function of the interparticle separation r. Figure 3 shows a clear peak corresponding to a definite nearest-neighbour shell together with evidence for a more diffuse next-nearest-neighbour shell. The value of r at the first maximum (r_{max}) is 580 nm, indicating a particle separation of 25 radii. The curve starts from a minimum value for r, at -N (see arrow on Figure 3), of approximately 330 nm. At this point $g(r) = 0$ and consequently the probability of finding a particle as close as this to another is zero. Consequently at this distance the repulsive interaction energy between the particles must be rising very steeply, thus providing experimental confirmation that the particles are located in a potential energy well with relatively steep sides.

Numerical integration of the function $4\pi N \int_0^{r_{min}} r^2\, g(r)\, dr$ for the first peak leads to a coordination number of 7; although only approximate this can be considered not an unreasonable estimate. In the case of a body-centred lattice each particle would be expected to have 8 nearest neighbours disposed towards the corners of a cube. For a volume fraction of 1.74×10^{-4} the side-length of the unit cell $(2R + h)$ for a body-centred lattice is 788 nm, which gives for half the body diagonal, the nearest particle centre to centre distance, a value of 682 nm. This is not very close to r_{max} = 580 nm but is closer to the value of r at which $4\pi r^2 g(r)$ reaches a maximum, i.e. 650 nm.

If each particle is considered to have a sphere of influence of $r_{max}/2$ then the quantity

$N \frac{4}{3} \pi \left(\frac{r_{max}}{2}\right)^3$ can be regarded as a "packing fraction" giving the fraction of the total volume occupied by the spheres of influence. For all the samples, the calculated packing fraction lay in the range 0.5 ± 0.025. However, there is a considerable systematic error on these data which could be as high as 30% and the results do not rule out the packing fraction of 0.68 expected for a body-centred array or a value of 0.64 for random close-packing; the latter value is obtained with some simple liquids(17).

It has been quite clearly established in the present work that at low volume fractions of latex and at very low electrolyte concentrations the particles, under the influence of long-range electrostatic repulsive forces, form a sufficiently ordered array to exhibit well-defined diffraction phenomena with visible light.

VII. REFERENCES

1. Derjaguin, B.V. and Landau, L., Acta Physicochim.U.R.S.S., 14, 633 (1941).
2. Verwey, E.J.W. and Overbeek, J.Th.G., "Theory of the Stability of Lyophobic Colloids" Elsevier, Amsterdam, 1948.
3. Klug, A., Franklin, R.E. and Humphreys-Owen, S.P.F., Biochim.Biophys.Acta, 32, 203 (1959).
4. Luck, W., Klier, M. and Wesslau, H., Die Naturwissenschaffen, 14, 485 (1963); Berichte der Bunsengesellschaft, 67, 75, 85 (1963).
5. Hiltner, P.A. and Krieger, I.M., J.Physical Chem., 73, 2386 (1969).
6. Barclay, L., Harrington, A. and Ottewill, R.H., Kolloid Z.u.Z.Polymere, 250, 655 (1972).
7. Hachisu, S., Kobayashi, Y. and Kose, A., J.Colloid Interface Sci., 42, 342 (1973).
8. Williams, R. and Crandall, R.S., Phys.Lett., 48A, 224, (1974).
9. Brown, J.C., Pusey, P.N., Goodwin, J.W. and Ottewill, R.H. J.Phys.A: Gen.Phys. 8, 664 (1975).
10. Ottewill, R.H. and Shaw, J.N., Kolloid Z.u.Z.Polymere, 215, 161 (1967).
11. Vanderhoff, J.W., van den Hul, H.J., Tausk, R.J.M. and Overbeek, J.Th.G., in "Clean Surfaces: Their Preparation and Characterization for Interfacial Studies" (G. Goldfinger Ed. p.15) Marcel Dekker, New York, 1970.
12. Kerker, M., "The Scattering of Light and other Electromagnetic Radiation", Academic Press, New York, 1969.
13. Chen, S.H., in "Physical Chemistry: Liquid State" (H. Eyring, D. Henderson and W. Jost, Eds.), vol.8A, Academic Press, New York, 1971.
14. Pings, C.J., in "Physics of Simple Liquids" (H.N.V. Temperley, J.S. Rowlinson and G.S. Rushbrooke, Eds.) Amsterdam, North Holland, 1968.
15. Hearn, J.H., Ottewill, R.H. and Shaw, J.N., Br.Polymer J., 2, 116 (1970).
16. Tanford, C., "Physical Chemistry of Macromolecules", Wiley, New York, chap.7, 1961.

17. Bernal, J.D. and King, S.V., in "Physics of Simple Liquids" (H.N.V.Temperley, J.S. Rowlinson and G.S. Rushbrooke Ed.) Amsterdam, North Holland, 1968.

ROLE OF PHYSICAL INTERACTIONS IN THE REVERSIBLE ADSORPTION OF HYDROSOLS OR GLOBULAR PROTEINS: APPLICATIONS TO CHROMATOGRAPHIC SEPARATIONS

Dennis C. Prieve
Carnegie-Mellon University

Eli Ruckenstein
State University of New York at Buffalo

I. ABSTRACT

The profile of total interaction energy (Born, van der Waals and double-layer) may assume several shapes -- some leading to reversible adsorption. Hydrodynamic and hydrophobic chromatographies are each identified with a shape and discussed. A third shape, in which a surmountable maximum occurs, is also identified. Adsorption and desorption rates and equilibrium are computed for this case and are shown to be highly sensitive to particle properties, including radius, surface charge and surface chemistry. Attaching short hydrocarbon chains containing ionizable groups onto the column packing provides flexibility in column design. Chain length affects van der Waals attraction and the number of attached chains affects the surface charge. This new particle separation technique is called potential-barrier chromatography and appears to offer finer discrimination than existing methods.

II. INTRODUCTION

Sol stability has remained the central topic in colloid science for well over half a century. The general acceptance of the DLVO(1,2) theory stems from its ability to explain why rates of flocculation decrease dramatically with decreases in ionic strength, when the ionic strength is lower than some critical value. The emphasis has generally been upon the conditions necessary for a stable sol.

Less attention has been given to the quantitative prediction of flocculation rates. Smoluchowski(3) and Fuchs(4) have established theories for rapid and slow coagulation, respectively. Further, Ruckenstein & Prieve(5) suggest how Fuchs' result may be extended, by inclusion of short-range Born repulsion, to model the kinetics of repeptization. When interaction forces are unimportant (rapid coagulation), experiments generally confirm the theory; however, Reerink & Overbeek(6) and Ottewill & Shaw(7) were unable to obtain quantitative agreement between the theory and their experiments -- regarding the dependence of flocculation rates upon ionic strength and particle size -- under conditions where interparticle forces are important. Because the accurate prediction of rates from Fuchs' theory requires extremely precise values for interaction forces, this lack of quantitative agreement reflects a need to refine existing expressions for these forces.

Physical forces, which form the basis of the DLVO theory, are also important in determining the rate of adsorption of hydrosol particles onto a collector surface. Ruckenstein & Prieve(8) and Spielman & Friedlander(10) computed the rate of deposition of particles by neglecting fluid convection inside a thin region (~100Å thick) near the collector, called the interaction-force boundary-layer, and by neglecting interaction forces outside the region. When a diffusion boundary layer exists, which is thick compared to the interaction-force boundary layer, the resistance to transport offered by the interaction forces can be lumped into a boundary condition on the usual convective-diffusion equation. This condition takes the form of a first-order reaction on the collector surface.

Ruckenstein & Prieve(5) later included Born repulsion among the interaction forces to explain reversibility in particle adsorption. Rates of adsorption and desorption, as well as the number of particles adsorbed at equilibrium, were computed using existing expressions for the interaction forces. Applications to chromatographic separation of hydrosols were suggested, based on these computations.

One goal of the present paper is to extend this analysis for computing the adsorption/desorption rates and adsorption equilibrium to the case of globular proteins. These particles differ from those previously considered in Ref. 5 because they are much smaller and acquire their surface charge by the dissociation of acidic and basic amino acid residues. Conformation of protein molecules is usually such that these hydrophilic residues are exposed to the surrounding water, while more hydrophobic residues form the core of the protein globule. For such particles, simultaneous dissociation equilibria need to be considered when calculating the charge density of the particle at each position relative to the collector. Particles with amphoteric surfaces are interesting because a charge reversal can be induced by the overlap of double-layers. These results may also be applied to other particles whose surfaces bear acidic and basic sites (e.g. metal oxides).

Another goal is to discuss more fully the chromatographic separation of hydrosols or globular proteins. Several methods, which involve colloidal forces between the particle and grains of the column packing -- such as hydrodynamic (11,12,13), hydrophobic(14, 15,16) and potential-barrier chromatographies(5) -- are classified according to the shape of the interaction energy profile. Chemical attachment of short hydrocarbon chains to the column packing can provide flexibility in design of the column. The length of such chains is shown to strongly affect the van der Waals attraction, while the number of chains attached per unit area and dissociation characteristics of any ionizable groups on these chains provide control of the surface charge density. Calculations regarding the effect of chain length and dissociation characteristics are also presented.

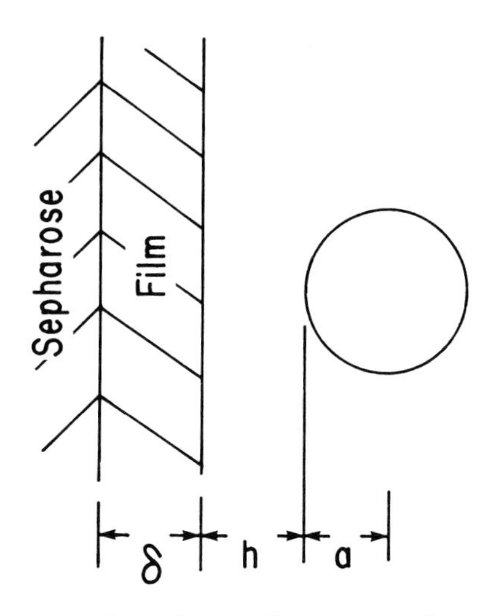

Fig. 1. Schematic of a spherical particle near a much larger grain of Sepharose column packing, where the grain is coated with a thin film of other material.

III. INTERACTION FORCES

The unretarded van der Waals attraction between a sphere and a thick plate, coated with a thin film of different material (see Fig. 1), was estimated by the atomic superposition principle:

$$\phi_{VdW}(h) = A\,f\left(\frac{h+\delta}{a}\right) + A_f\left[f\left(\frac{h}{a}\right) - f\left(\frac{h+\delta}{a}\right)\right]$$
$$f(H) = \frac{1}{6}\ln\left(\frac{H+2}{H}\right) - \frac{1}{3}\,\frac{H+1}{H(H+2)} \qquad [1]$$

where A is Hamaker's constant for the plate-sphere interaction and A_f is Hamaker's constant for the film-sphere interaction. Lifshitz continuum approach could also have been used, but [1] is adequate for the present purpose if A and A_f are treated as measurable properties.

Born repulsion is a short-range atomic interaction which

results from the overlap of electron orbitals. By integrating an inverse twelfth-power interatomic repulsion energy over the sphere and plane in the manner of Eq. [1], one obtains:

$$\phi_{Born}(h) = \frac{(\sigma'/a)^6}{7560}\left\{A\, g\left(\frac{h+\delta}{a}\right) + A_f\left[g\left(\frac{h}{a}\right) - g\left(\frac{h+\delta}{a}\right)\right]\right\}$$

$$g(H) = \frac{H+8}{(H+2)^7} + \frac{6-H}{H^7} \qquad [2]$$

where σ' is the atomic collision diameter.

Double-layer interactions between plane surfaces bearing single(17) and multiple(9,18) ionizable groups have been computed. For multiple groups the nonlinear Poisson-Boltzmann equation is numerically solved subject to boundary conditions of the following type:

$$\left.\frac{d\psi}{dx}\right|_{x=0} = -\frac{4\pi e}{\epsilon}\left[\sum_{j=1}^{m_b}\frac{[H^+]_r\, N_{bj}}{[H^+]_r + K_{bj}\exp(e\psi_o/kT)} - \sum_{i=1}^{m_a}\frac{K_{ai}N_{ai}}{K_{ai} + [H^+]_r\exp(-e\psi_o/kT)}\right] \qquad [3]$$

which arise through consideration of the following dissociation equilibria:

$$G_{ai}H(\text{surf}) \overset{K_{ai}}{\rightleftarrows} G_{ai}^-(\text{surf}) + H^+(\text{aq})$$

for $i=1,\cdots,m_a$

$$G_{bj}H^+(\text{surf}) \overset{K_{bj}}{\rightleftarrows} G_{bj}(\text{surf}) + H^+(\text{aq})$$

for $j=1,\cdots,m_b$

where ψ is the local electrostatic potential, x is the distance, e is the protonic charge, ϵ is the dielectric constant of the fluid, ψ_o is the potential inside the diffuse layer, $[H^+]_r$ is the hydrogen ion concentration far from any surface (where $\psi = 0$), N_{ai} (N_{bj}) is the total number of surface sites of type ai (bj) per unit area, and K_{ai} (K_{bj}) is the corresponding equilibrium constant. Using this method to calculate the double-layer force per unit area $p(\ell')$ between two plates separated by ℓ', the force between a sphere and a plate will be calculated from Derjaguin's approximation(19):

$$\phi_{DL}(h) = 2\pi a \int_h^\infty \int_\ell^\infty p(\ell')d\ell' \, d\ell \qquad [4]$$

Bell, Levine & McCartney(20) have stated that this approximation introduces an error of less than 10% for sphere-sphere interactions, provided $\kappa h \leq 2$ and $\kappa a \geq 5$. This error is expected to be less for sphere-plate geometry.

Proteins may display some of the properties of lyophilic colloids with regard to the importance of solvation effects. These effects are neglected in the present paper, but they may affect the form of the expression for $\phi(h)$, probably decreasing the sensitivity to ionic strength(2).

IV. RATES OF ADSORPTION AND DESORPTION OF HYDROSOLS

By assuming that convection can be neglected inside a very thin ($\sim$100Å) region next to the collector and that interaction forces can be neglected outside of this region, Ruckenstein & Prieve(5) found that the net adsorption rate may be calculated by solving the usual convective-diffusion equation subject to a reversible, first-order reaction-like boundary condition on the collector surface:

$$-J = K_f c_o - K_r n \qquad [5]$$

where $-J$ is the net adsorption rate per unit area, c_o is the volumetric concentration of particles near the surface and n is the number of adsorbed particles per unit area.

When the total potential energy of interaction

$$\phi(h) = \phi_{VdW}(h) + \phi_{Born}(h) + \phi_{DL}(h) \qquad [6]$$

possesses a maximum $\phi_{max} = \phi(h_{max})$ and a primary minimum $\phi_{min} = \phi(h_{min})$, where $h_{min} < h_{max}$, such that $\phi_{max} \gg kT$ and $\phi_{max} - \phi_{min} \gg kT$, they show that the apparent rate constants can be computed from the $\phi(h)$ profile according to:

$$K_f = D(h_{max}) \, (\gamma_{max}/2\pi kT)^{1/2} \, \exp(-\phi_{max}/kT) \qquad [7a]$$

$$K_r = D(h_{max})(\gamma_{max}\gamma_{min})^{1/2} \exp[-(\phi_{max}-\phi_{min})/kT]/2\pi kT \qquad [7b]$$

where $D(h_{max})$ is the local diffusion coefficient evaluated at h_{max}, $\gamma_{max} = -d^2\phi/dh^2|_{h_{max}}$ and $\gamma_{min} = +d^2\phi/dh^2|_{h_{min}}$. Thus ϕ_{max} is the "activation" energy (barrier) for adsorption and $\phi_{max} - \phi_{min}$ is the activation energy for desorption. Calculations in the current paper differ from the previous one in that the double-layer contribution to $\phi(h)$ includes the effect of the equilibria for ionizable surface groups. Solutions to the convective-diffusion equation, subject to boundary condition [5], exist(5,8); however, for simplicity, if the resistance of the interaction-force boundary layer is assumed to be rate-controlling, the rate is given directly by [5] after replacing c_o by c_∞, the bulk particle concentration.

V. RESULTS AND DISCUSSION

A. Classification of Interaction Profiles and Chromatography Methods

Three distinct types of interaction profiles are illustrated in Fig. 2. One contains no extrema. This corresponds to a potential energy of interaction which monotonically increases as the separation is decreased. Here the attraction is overwhelmed by repulsion, which prevents the adsorption of particles. This occurs in "hydrodynamic chromatography" as observed by Small(13). Because particles will be eluted from a nonporous chromatography column before a molecular tracer, it is easy to distinguish this mechanism from others discussed below. Larger particles are eluted first because, due to their finite size, their centers are excluded from a region near the surface of packing grains where the fluid velocity is slowest. Since particles sample all accessible regions by Brownian motion, this exclusion from slow regions gives larger particles a larger average velocity. Although no adsorption occurs, interaction

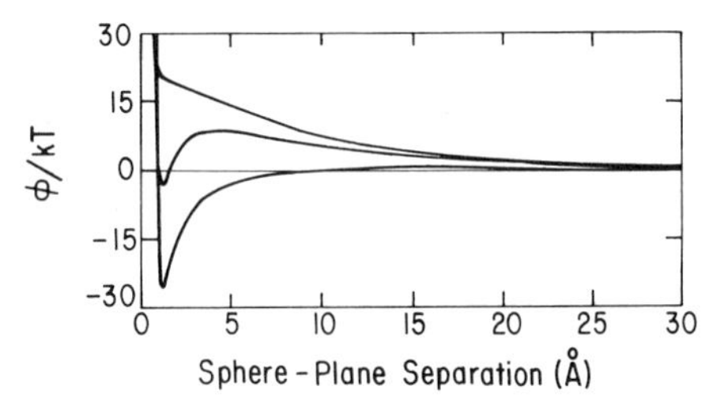

Fig. 2. Classification of chromatographic methods according to the shape of the profile of interaction energy. Upper to lower curves coincide most closely with hydrodynamic, potential-barrier and hydrophobic chromatographies, respectively.

forces can still affect the elution time through their influence on the distribution of particles near the grain surface. To a first approximation, repulsive forces may be considered to decrease the fraction of the total column volume which is accessible to the particles, by preventing particles from getting as close to the grains. Small(13) reports that as the ionic strength of the carrier solution is decreased, the elution time of particles is also decreased. This can be explained by considering the increase in double-layer repulsion caused by a lowering of the ionic strength.

A second situation depicted in Fig. 2 is the case in which the profile has only a primary minimum. Electrostatic interactions must be attractive or only very weakly repulsive. Rapid (limited only by convective-diffusion) adsorption of particles into the energy well occurs. If the well is deep ($\gg kT$), the adsorbed particles can be eluted only by changing the interaction profile. For instance if the electrostatic interaction is attractive, adding salt to the carrier may weaken the attraction sufficiently to raise the well and elute the particles. On the other hand, if the well is shallow ($\leq kT$), particles may rapidly adsorb on a grain as the pulse of particles sweeps by the grain and then desorb more or less slowly by Brownian motion once the pulse has passed. In either case particles can be expected to remain in the column longer than a molecular tracer, in contrast to hydrodynamic chromatography. Particles, differing in size or chemical composition, will possess interaction energy wells of different depths, allowing separation through physical adsorption.

Selectivity in physical adsorption has been used as a mechanism for chromatographic separation. Shaltiel and coworkers(14,15) selectively adsorbed glycogen synthetase and/or glycogen phosphorylase from crude rabbit muscle extract using a Sepharose column, which was modified by chemically binding amino alkyl chains on the packing surface. In experiments where the attached chain is uncharged(14) they found the strength of the attraction between glycogen phosphorylase _b_ and the column increased markedly as the length of the attached chain increased: from no effect to a slight increase in elution time, to an easily reversible physical adsorption, and finally to a much stronger physical adsorption. The overall effect is explained in terms of a hydrophobic (van der Waals) attraction between the protein and the column, which increases as the length of the attached chains is increased. Thus the method is called "hydrophobic chromatography."

Additional control over the interactions between proteins

and the packing can be obtained by attaching side chains to the packing which have charged end-groups. For instance, Shaltiel and Er-el(15) attached diamino-alkyl groups to Sepharose packing and passed crude rabbit muscle extract through the column. Under the conditions of the experiment, the terminal amine group on the alkyl chain is positively charged, whereas the proteins are above their iso-electric points and so are negatively charged. Again, the length of the alkyl chains was varied. For less than four methylene groups in the chain, no protein is adsorbed on the column. With exactly four methylene groups, glycogen synthetase was adsorbed, and could later be desorbed by adding salt, but phosphorylase was not adsorbed. Further increases in the number of methylene groups caused the synthetase to be adsorbed more tightly so that it could not be desorbed, except in denatured form. With five methylene groups, the glycogen phosphorylase is partially adsorbed but desorbes in the presence of a high concentration of salt. These experiments serve to demonstrate that rapid physical adsorption is sufficiently selective to be useful in separating proteins and that also the chain length increases the van der Waals attraction.

If neither van der Waals attraction nor double-layer repulsion dominate, then the third situation shown in Fig. 2 can result. This corresponds to a potential energy profile with a surmountable maximum and a primary minimum of finite depth. Then slow adsorption occurs whose rate can be carefully controlled by adjusting the height of the maximum through the pH and ionic strength. Two proteins which differ slightly in surface charge, size or shape can be expected to have maxima in their interaction profiles of slightly different heights. Because adsorption rates are exponentially sensitive to the height of this barrier(5), the protein with the slightly larger maximum may be eluted while the other is adsorbed. Of course, the column has to be properly "tuned" by adjusting pH, ionic strength, and the length of hydrocarbon chains attached to the column. This is a new separation process which we have named "potential-barrier chromatography." Because both van der Waals and electrostatic interactions are attractive in Shaltiel's experiments, no barrier exists. However, similar experiments have been performed in which the electrostatic interaction is repulsive.

Yon(16) passed a solution of bovine serum albumin through a column packed with Sepharose, which had been modified by the attachment of carboxypropionylaminodecyl chains. The attached chains have terminal carboxylic acid groups which can be expected to be fully dissociated for pH greater than five. The isoelectric point of the protein is 4.7-4.9, so at higher pH both the column and the particle are negatively charged. For pH's above the isoelectric point of the protein, an increase in pH will shift the dissociation equilibria of surface groups (see after Eq. [3]) in favor of fewer associated protons, thus making the charge on the protein's surface more negative. If the chemically-bound carboxylic acid groups on the column grains are already fully dissociated, then an increase in pH causes no further change in the negative charge on that surface. So the overall effect of increasing pH is to increase the double-layer repulsion between the protein and the column grains. When the repulsion is strong enough, an insurmountable barrier prevents adsorption. Indeed, this is what Yon observed: at a pH of 5.0 or 6.5 all the albumin is adsorbed and held on the column, but at 8.0 or 9.5 the fraction adsorbed decreases with pH. The occurrence of similar charges on both particle and column packing offers the following two advantages: 1) it permits the establishment of an energy barrier to adsorption which can discriminate more finely among proteins or hydrosols via potential-barrier chromatography and 2) the resulting repulsion might be increased by flushing the column with a solution of low ionic strength to desorb proteins trapped on the column without denaturing them.

Table 1 summarizes the three chromatographic methods discussed above, where separation is affected by colloidal forces. Also listed for comparison are gel-permeation and affinity chromatographies. The former occurs when the packing is porous, making the interior volume of the grains accessible to particles below a certain size, thus increasing their elution volume. In affinity chromatography, chemical (rather than physical) adsorption occurs. Such covalent bonds are strong and highly specific. Elution of the bound protein is accomplished by the addition of an agent to cleave the bond. Regarding adsorption specificity, potential-barrier chromatography is seen by the authors as the physical analogue to affinity chromatography, with the height of the barrier in the former corresponding to chemical specificity of surface sites on the column packing of the latter.

B. Sensitivity of Adsorption and Desorption Rates in Presence of Energy Barrier

Calculations of the apparent rate constants for adsorption and desorption of hydrosols and proteins were performed for situations where a potential energy barrier ϕ_{max}, greater than a few kT, opposes adsorption or desorption. This corresponds to potential-barrier chromatography. Fig. 3 demonstrates the effect of chemical constitution of the particle surface on K_f and K_r and hence on the adsorption and desorption

TABLE 1

CHROMATOGRAPHIC METHODS FOR HYDROSOLS AND PROTEINS

Method	Mechanism	Particle-Column Interaction
Hydrodynamic Chromatography	Movement of particles into and out of regions having low fluid velocity external to packing grains	Repulsion dominates
Hydrophobic Chromatography	Rapid physical adsorption of particles on packing grains	Attraction dominates so that a minimum is the only extremum
Potential-Barrier Chromatography	Slow adsorption of particles over a potential energy barrier	Potential energy profile has both a maximum and a primary minimum
Gel Permeation Chromatography	Movement of particles into and out of pores in packing grains	——
Affinity Chromatography	Rapid chemical adsorption of particle on packing grain	Short-range interaction leading to covalent bonds

rates. Three particle surfaces are compared: a particle bearing a single type of strong acid group ($pK_a = 0$), a particle with a single type of a weak acid ($pK_a \sim pH$ near surface of particle far from collector), and a particle with equal numbers of an acidic and a basic group near its isoelectric point. In each case the number of surface groups was adjusted to give the same surface potential on the particle when it is located far from the collector ($\psi_o|_{h=\infty} \equiv \zeta = -25\text{mV}$). Differences in the rate constants for the three cases result from differences in changes in the surface potential and charge density which occur as the particle surface is brought nearer the collector. To maintain electroneutrality as two opposing surfaces are brought together, there exists a tendency for the counterion concentration to increase to offset the decrease in volume of solution between the surfaces. The corresponding decrease in pH (H^+ is a counterion) decreases the charge on the particle

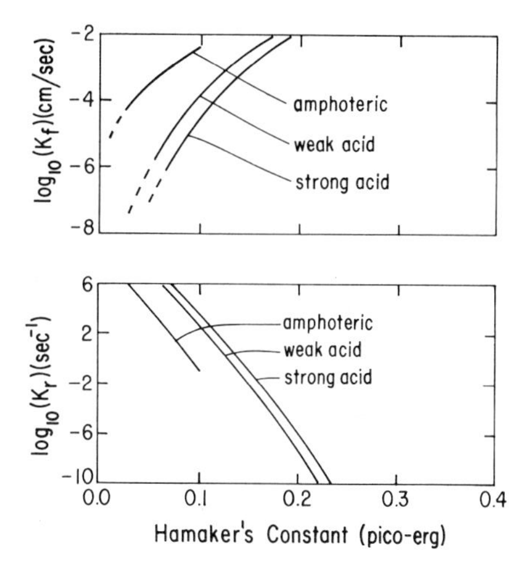

Fig. 3. Affect of chemistry of the particle's surface on rate constants. In each case, the number of surface groups on the particle was adjusted to give ζ = -25mV. The collector was taken to be a strongly acidic surface with ζ = -50mV (N_a = 2.576×10^{13} cm^{-2}, pK_a = 0). Characteristics of the particle's surface are: N_a = N_b = 6.861×10^{17} cm^{-2}, pK_a = 3.127, pK_b = 10 (amphoteric); N_a = 2.311×10^{13} cm^{-2}, N_b = 0, pK_a = 6.58 (weak acid); N_a = 1.147×10^{13} cm^{-2}, N_b = 0, pK_a = 0 (strong acid). Other fixed parameters include: ionic strength = 0.1 mol/ℓ, pH_r = 7, σ' = 2Å, a = 100Å, ϵ = 74.3, δ=0, T = 300°K. Dotted lines do not rigorously meet assumptions in Eq. [7].

bearing the weak acid groups by shifting the dissociation equilibrium, whereas the strongly acid surface remains fully dissociated. This decrease in charge weakens the repulsion and therefore leads to larger values of the adsorption rate constant than the strongly acidic constant-charge case. Likewise, smaller values of the desorption rate constant are obtained for the weak acid compared to the strong one.

The third situation depicted in Fig. 3 is the particle having the amphoteric surface. Because the surface has both acidic and basic groups, not only does the charge density on the particle decrease as it is brought closer to the collector, but the charge can also change sign. This causes the rate constants to be dramatically different from those where the surface charge is constant (strong acid). To better understand the changes which are induced by the overlap of double-layers, Fig. 4 gives the charge density, surface potential and double-layer force as a function of separation distance for the interaction of two plane surfaces where one is amphoteric and the other is a strong acid. The corresponding electrostatic potential contours (solutions to Poisson-Boltzmann equation) for various separations are shown in Fig. 5. Results summarized by Figs. 4 and 5 were used to obtain the rate constants

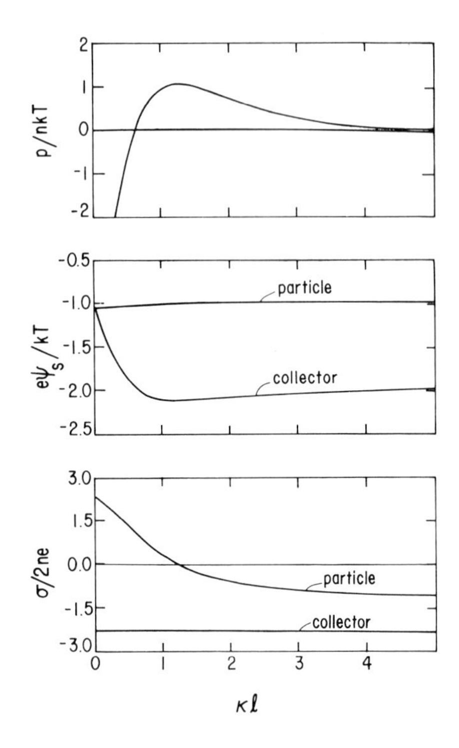

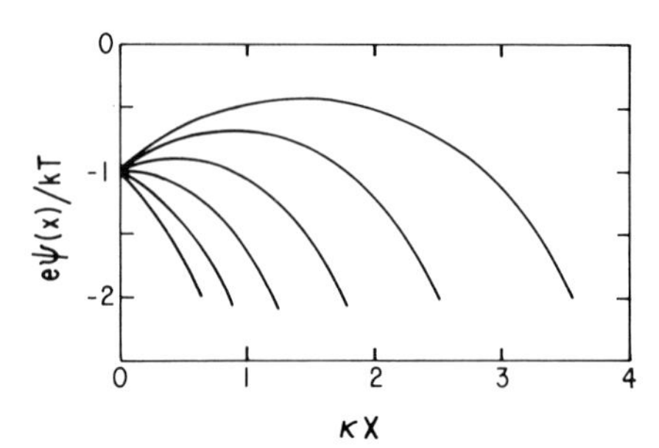

Fig. 5. Electrostatic potential profiles (solutions to the Poisson-Boltzmann eqn.) corresponding to several separations of the plane surfaces in Fig. 4, where κx is the distance measured from the amphoteric surface. The curves terminate at $\kappa x = \kappa \ell$, which is the strong acid surface. Note the sign reversal of the slope at x = 0: this reflects charge reversal on the amphoteric surface.

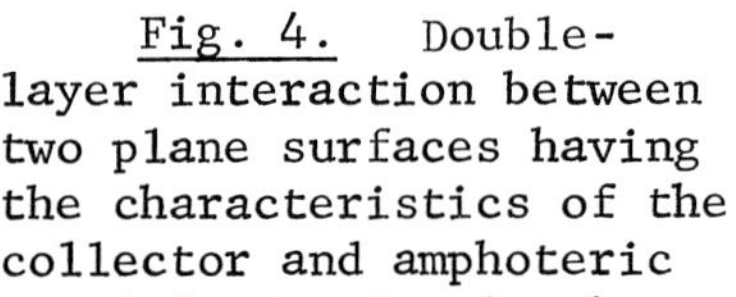

Fig. 4. Double-layer interaction between two plane surfaces having the characteristics of the collector and amphoteric particle in Fig. 3, where $\kappa \ell$ is the plate separation expressed in Debye lengths, p is the force per unit area, n is the ionic strength and σ is the surface charge density. Note sign reversal in force and particle charge occurring at different $\kappa \ell$.

---

in Fig. 3 for the amphoteric particle. Note that the double-layer force is repulsive at large distance, where the surfaces have charges of like sign, but becomes attractive at small distances where the surfaces have opposite sign. This behavior is discussed at greater length by Prieve & Ruckenstein(21): a reversal of charge and force can be induced only if the surfaces are different. Thus this effect may occur in adsorption but not in flocculation. Charge reversal on the amphoteric particle in Fig. 3 dramatically increases K_f and decreases K_r. For remaining cases below, all surfaces are strongly acidic.

Fig. 6 illustrates the effect of a hydrocarbon chain (or layer) attached to the collector surface if the van der Waals

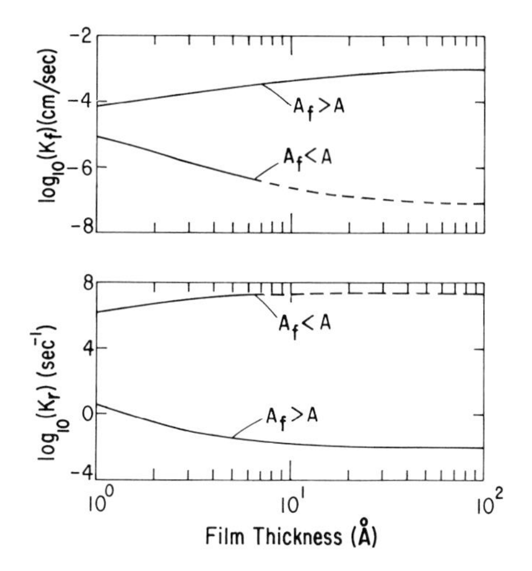

$\log_{10}(K_f)$(cm/sec)

$\log_{10}(K_r)$(sec^{-1})

Particle Radius (Å)

Fig. 6. Effect of length of hydrocarbon chains attached to collector surface. Attached chains are modeled as the continuous film in Fig. 1. Hamaker's constant for the film-particle interaction is 0.5×10^{-13} erg ($A_f < A$) or 1.5×10^{-13} erg ($A_f > A$), while Hamaker's constant for the Sepharose-particle interaction is 1.0×10^{-13} erg. Surface of particle and collector are strongly acidic (constant charge), while other parameters coincide with those of Fig. 3.

Fig. 7. Effect of particle radius on rate constants. Surface of particle and collector are strongly acidic (constant charge), $A = 1 \times 10^{-13}$ erg, $\delta = 0$, while other parameters coincide with those of Fig. 3.

interaction of that layer with the particle is different from that of the remainder of the collector. If the Hamaker constant of the chain (film) is larger than that of the collector, the adsorption rate constant increases appreciably with chain length whereas the desorption rate constant decreases. The effect is noticeable even for chains as short as 1Å. No further changes occur at large chains lengths (> 100Å) because now the entire collector acts as if it has the Hamaker constant of the hydrocarbon chains. However, if the chains are long and flexible enough, entropic repulsion may also occur. Although ignored in the present calculations, this affect may permit potential-barrier chromatography to be used on non-charged systems (e.g. with nonpolar solvents), provided that entropic repulsion and van der Waals attraction generate a barrier to adsorption of several kT.

Chromatographic separation of particles is often based

on differences in size. Fig. 7 shows the effect of particle radius upon the rate constants. Since $\phi(h)$ is roughly proportional to the radius, these curves are almost linear. Note that every 20Å increase in radius causes both K_f and K_r to decrease by an order of magnitude, but the equilibrium constant, K_f/K_r is not appreciably changed. Consequently, if particles of two slightly different radii are present and particles of both sizes reversibly adsorb on the column packing, the larger particles will be retained appreciably longer in the column since adsorption and desorption is slower.

Fig. 8 illustrates the effect of ζ (the surface potential of the particles at $h = \infty$). In this context, ζ reflects the number of acidic groups on the particle's surface: more dissociated acid groups on the surface corresponds to a more negative ζ (according to the Guoy-Chapman equation). Every 5mV increase in $-\zeta$ decreases K_f and increases K_r by about an order of magnitude, thus the equilibrium constant is shifted two orders of magnitude in favor of non-adsorbed particles. Consequently, potential barrier chromatography should also be capable of separating particles of the same size which differ only slightly in surface charge (or potential).

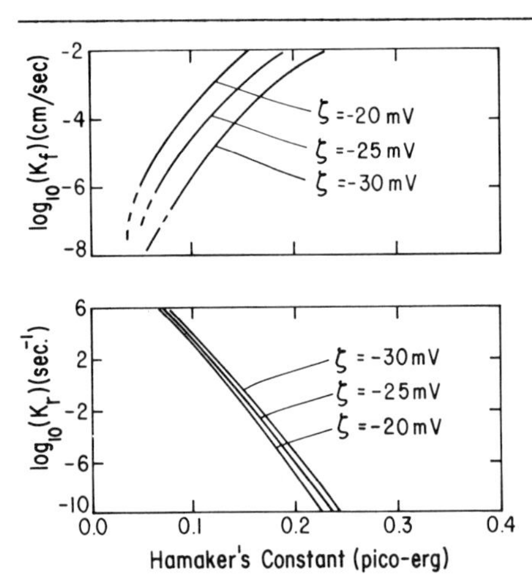

Fig. 8. Effect of charge density on particle's surface, expressed as surface potential when particle is far from collector. Other parameters correspond to the case of a constant-charge particle in Fig. 3.

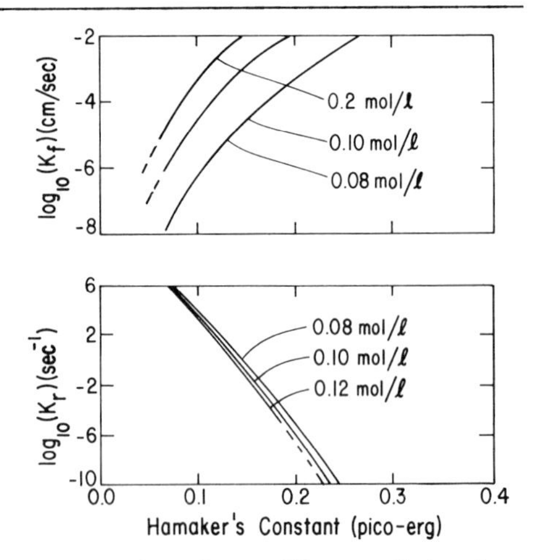

Fig. 9. Effect of ionic strength on rate constants. Other parameters correspond to the case of a constant-charge particle in Fig. 3.

Finally, Fig. 9 shows the sensitivity to ionic strength. A small decrease in ionic strength causes a substantial decrease in K_f and an increase in K_r. As might have been anticipated, this shifts the equilibrium substantially in favor of nonadsorbed particles. Recalling the middle curve of Fig. 2, a large enough decrease in ionic strength can raise the primary minimum to correspond to positive ϕ. Then the primary minimum is no longer the lowest energy state so that, if the desorption barrier $\phi_{max}-\phi_{min}$ is low enough, particles will rapidly desorb. Thus particles which are essentially irreversibly adsorbed from a carrier having one ionic strength may be desorbed by flushing with a solution of lower ionic strength.

VI. SUMMARY AND CONCLUSIONS

Colloidal forces, usually associated with flocculation, cause physical adsorption of hydrosols and proteins onto a collector surface and can also impose a energy barrier whose height greatly affects the adsorption rate. By including Born repulsion, together with van der Waals attraction and double-layer repulsion, the primary minimum in the profile of potential energy of interaction is made to have finite depth. This allows physical adsorption to be reversible.

Hydrodynamic, hydrophobic and potential-barrier chromatographies are three methods which are strongly affected by colloidal forces. They can be distinguished by the shape of the potential energy profile: the profile corresponding to the first method contains no extrema and repulsion dominates. In the second method, the profile contains only a minimum into which rapid adsorption occurs. The third situation, where a maximum in potential energy regulates adsorption at the primary minimum, is suggested here as a new method called potential-barrier chromatography.

Calculation of adsorption and desorption rates were performed under conditions corresponding to potential-barrier chromatography of globular proteins using the method of Ruckenstein & Prieve(5). Double-layer forces were computed by solving the nonlinear Poisson-Boltzmann equation subject to boundary conditions deduced from the equilibrium dissociation of acidic and basic surface groups. These calculations demonstrate an extraordinary sensitivity to particle size, Hamaker constant, surface charge density, as well as to the surface chemistry (i.e. the type of acidic or basic groups yielding a given surface charge density).

Double-layer interactions may be modified by changing the ionic strength or pH. Van der Waals attraction may be modified by chemically binding short hydrocarbon chains to the grains of the column packing. Increasing the number of methylene groups in these alkyl chains increases the attraction. The effect of chain length on adsorption rates was computed by considering a thin continuous film coating the collector, which has a Hamaker constant different from the packing grains. Rates were found to be significantly affected by chain length. When these chains contain ionizable groups the surface density of the attached chains also affects the double-layer repulsion.

To separate a complex mixture of hydrosols existing chromatographic methods may be employed first to obtain a coarse separation into a number of fractions. Highly similar particles in each fraction may then be separated by potential-barrier chromatography. Ionic strength, pH, the number of methylene groups in attached chains and the number of attached chains per area must be carefully adjusted to give a surmountable maximum in the potential energy profile. If appropriate control on these variables can be exercised, potential-barrier chromatography offers the possibility of finer discrimination than existing methods.

VII. ACKNOWLEDGMENT

One of us (DCP) gratefully acknowledges partial support from the Materials Research Laboratory Section, National Science Foundation, through the Center for the Joining of Materials at Carnegie-Mellon University.

VIII. REFERENCES

1. Derjaguin, B.V., and Landau, L.D., Acta Physiochim. 14, 633 (1941).
2. Verwey, E.J.W., and Overbeek, J. Th. G., "Theory of the Stability of Lyophobic Colloids," Elsevier, Amsterdam, 1948.
3. Smoluchowski, M., Physik. Z. 17, 557 (1916); 17, 585 (1916); Z. Physik. Chem. 92, 129 (1917).
4. Fuchs, N., Z. Physik. 89, 736 (1934).
5. Ruckenstein, E., and Prieve, D.C., A.I.Ch.E. J. 22, 276 (1976). [Further discussion on the kinetics of repeptization will be published in a paper currently in press in A.I.Ch.E. J.]
6. Reerink, H., and Overbeek, J.Th.G., Disc. Faraday Soc. 18, 74 (1954).

7. Ottewill, R.H., and Shaw, J.N., Disc. Faraday Soc. 42, 154 (1966).
8. Ruckenstein, E., and Prieve, D.C., J.C.S. Faraday II 69, 1522 (1973).
9. Prieve, D., "Rate of Deposition of Brownian Particles under the Action of London and Double-Layer Forces," Ph.D. Dissertation, Univ. of Delaware, 1974.
10. Spielman, L.A., and Friedlander, S.K., J. Colloid Interface Sci. 44, 22 (1974).
11. DiMarzio, E.A., and Guttman, C.M., Polymer Letters 7, 267 (1969).
12. DiMarzio, E.A., and Guttman, C.M., Macromolecules 3, 131 (1970).
13. Small, H., J. Colloid Interface Sci. 48, 147 (1974).
14. Er-el, Z., Zaidenzaig, Y., and Shaltiel, S., Biochem. Biophys. Res. Comm. 49, 383 (1972).
15. Shaltiel, S., and Er-el, Z., Proc. Nat. Acad. Sci. USA 70, 778 (1973).
16. Yon, R.J., Biochem. J. 126, 765 (1972).
17. Ninham, B.W., and Parsegian, V.A., J. Theor. Biol. 31, 405 (1971).
18. Prieve, D.C., and Ruckenstein, E., J. Theor. Biol. 56, 205 (1976).
19. Hogg,H., Healy, T.W., and Fuerstenau, D.W., Trans. Faraday Soc. 62, 1638 (1966).
20. Bell, G.M., Levine, S., and McCartney, L.N., J. Colloid Interface Sci. 33, 335 (1970).
21. Prieve, D.C., and Ruckenstein, E., to be presented at 69th Annual Meeting of A.I.Ch.E., Chicago, November 28-December 2, 1976.

THE STRUCTURE CHANGE WITH TEMPERATURE IN THE AQUEOUS SOLUTION OF AN ETHOXYLATED SURFACTANT AND THE STABILITY OF CARBON BLACK DISPERSIONS

Shigenori Kumagai and Shoji Fukushima
Department of Colloid Science, Shiseido Laboratories

ABSTRACT

The structure of an ethoxylated surfactant in a dilute aqueous solution and its effect on the dispersion of carbon black was studied. We found that carbon black particles strongly coagulate when a surfactant solution forms liquid crystals. Dilute aqueous solutions of polyoxyethylene n-dodecyl ether with 5 to 5.4 moles of ethylene oxide formed liquid crystals at certain temperatures (mesophase points). At the mesophase point, the pronounced coagulation of carbon black was determined by particle size measurement and confirmed microscopically. The adsorption isotherms of surfactants on carbon black were also measured.

I. INTRODUCTION

The dispersion of carbon black in water with nonionic surfactants is affected by various factors such as HLB, concentration, temperature etc. In addition to these factors, structural change of surfactant molecules in an aqueous solution is expected to affect the stability of disperse systems.

Various mesophases *e. g.* the neat or middle phase have often been observed in nonionic surfactant-water system (1). Even in a dilute aqueous solution, formation of the mesophase was reported by Harusawa *et al.* (2) and others (3, 4).

Although the effect of nonionic surfactants on the dispersion of carbon black in water was reported by Moriyama *et al.* (5), the formation of liquid crystals was not dealt with in their paper.

We studied the effect of a liquid-crystal mesophase on the stability of a carbon black dispersion, using a homologous series of polyoxyethylene n-dodecyl ethers.

II. EXPERIMENTAL

A. Materials

Surfactants used were monodispersed tetra-, penta or hexaoxyethylene n-dodecyl ether ($C_{12}E_4$, $C_{12}E_5$ or $C_{12}E_6$) from Nikko Chemicals Co., Ltd. The purity was confirmed to be above 99% by gas chromatograph. Carbon black (Lamp Black 101 from Degussa Co., Ltd.) was used without further purification. The BET specific surface area was 18.7 m^2/g.

B. Determination of Phase Transition and Separation

About 10 ml of aqueous solutions containing 0.01 mole/l of the surfactant was sealed in a glass tube of 15 mm diameter. Variation of the state of solutions with temperature was visually observed in a thermobath. The phase separation temperature and the phase transition temperature were determined after equilibration at the respective temperatures.

C. Particle Size Distribution

Suspensions were prepared by dispersing 5.0 g of the carbon black into 500 ml of 2 m mole/l surfactant aqueous solutions. After 24 hours, apparent size distributions of the carbon particles or aggregates were determined with Andreasen's pipette at various temperatures.

D. Adsorption Isotherm

One gram of the carbon black was dispersed in 100 ml of the surfactant solutions of known concentrations. After 24 hours, the concentrations of surfactant in supernatant liquids were determined as described elsewhere (6).

E. Microscopic Observation

Suspensions containing a small amount of the carbon black in 5% surfactant solution were observed using a Leitz ORTHOPLAN microscope fitted with a hot stage with the rate of temperature increase of about 2°C/min.

III. RESULTS

Fig. 1 shows cloud points, double cloud points and mesophase forming temperatures (mesophase points) of 0.01 mole/l surfactant solutions against the mole numbers of ethylene oxide of the surfactants. The mesophase appeared when the latter was between 5 and 5.4. However the lower value was not well defined because the transition from mesophase to quasi-mesophase continuously took place. At the quasi-mesophase area the solution exhibited flow birefringence.

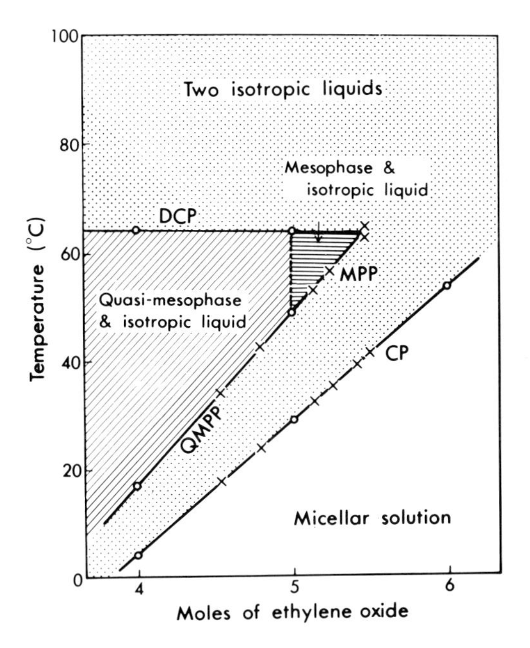

Fig. 1. Cloud point (CP), double cloud point (DCP) and mesophase point (MPP) of 0.01 mole/l $C_{12}E_n$ aqueous solution vs. mole number of ethylene oxide of the surfactants.

o ; The surfactant is not mixed.
x ; The surfactant is mixed.

Fig. 2 shows the adsorption isotherms for $C_{12}E_5$ on carbon black. The isotherms were of the Langmuir type at the temperatures below 29°C. Surface excess, *i. e.* the apparent adsorbed amount, however, abruptly increased above the cloud point. The equilibrium surfactant concentrations of supernatant liquids were 5×10^{-5} mole/l, which almost equaled to the c.m.c. This suggests that above the cloud point the surfactant molecules separated from micellar solutions were concentrated near the surface of carbon black.

The apparent particle size distributions of carbon black in the surfactant solution at various temperatures are shown in Fig. 3. When the apparent particle size was larger than 50 μm, it was not measurable because of rapid sedimentation. It was noteworthy that the distribution suddenly shifted to larger particle size at the temperature near 50°C with $C_{12}E_5$, which corresponds with the mesophase point. To establish these temperatures more precisely, suspensions were observed under the microscope as a function of rising temperature.

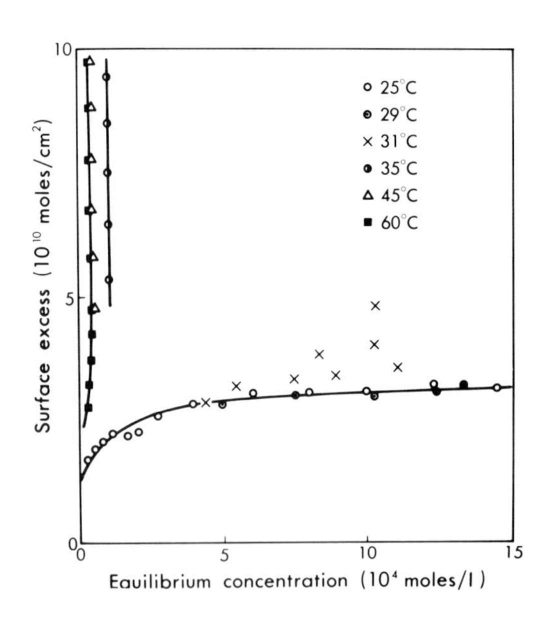

Fig. 2. Apparent adsorption isotherms of $C_{12}E_5$ on the carbon black.

Fig. 4 shows that the carbon particles are in a coagulated state at 49°C, *i. e.* the mesophase point. This was confirmed also with other surfactants forming the mesophase.

(a)

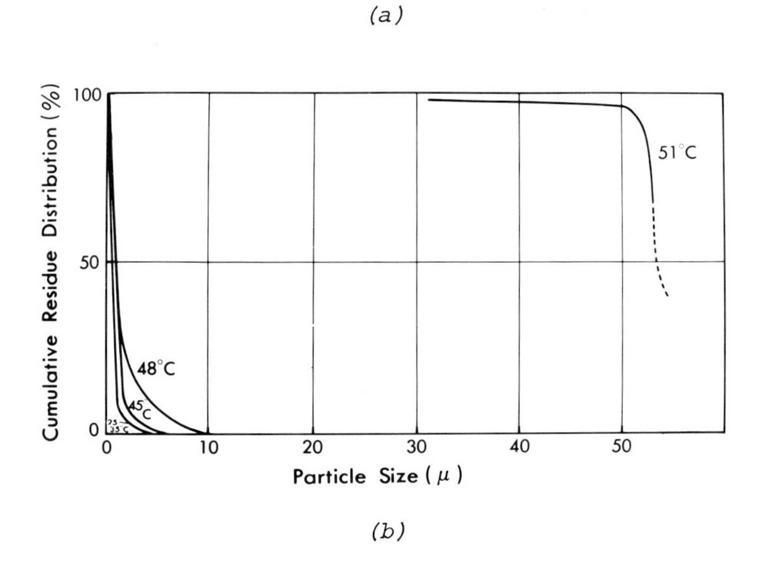

(b)

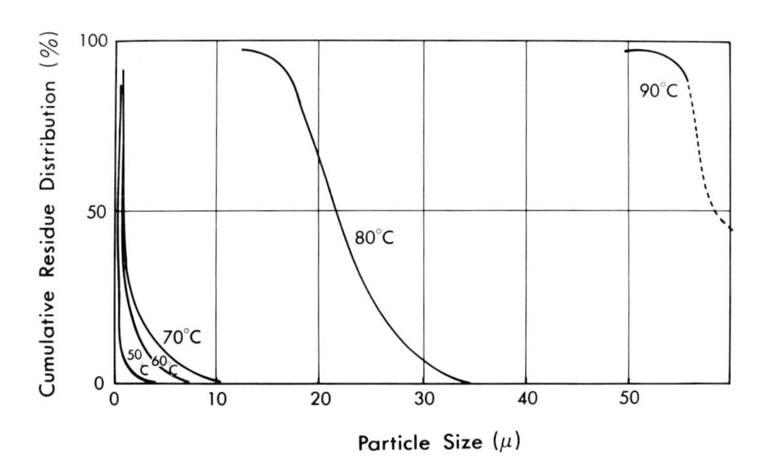

Fig. 3. Particle size distribution of carbon black in $C_{12}E_5$ (a) and $C_{12}E_6$ (b) aqueous solutions.

(a)

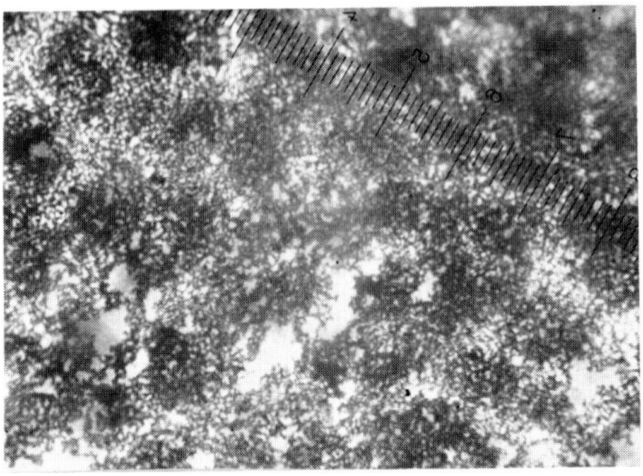

(b)

(c)

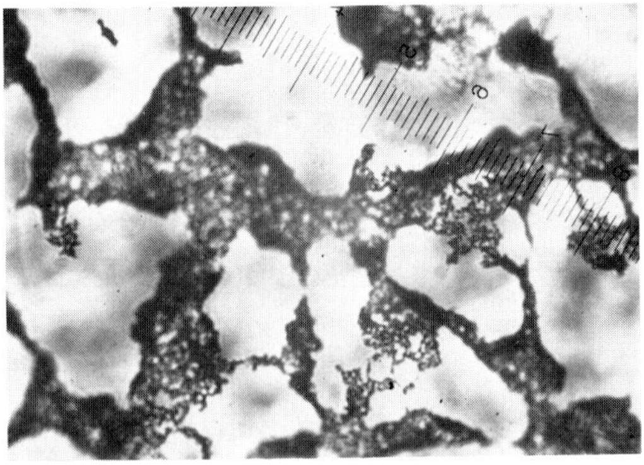

Fig. 4. Microphotographs of carbon particles in the surfactant concentrated phase of $C_{12}E_5$ at 48°C (a), at 49°C (b) and at 50°C (c).

Fig. 5 shows that the average particle size of carbon black increases with increasing temperature. The most abrupt change was observed with surfactants having between 5 and 5.4 mole ethylene oxide.

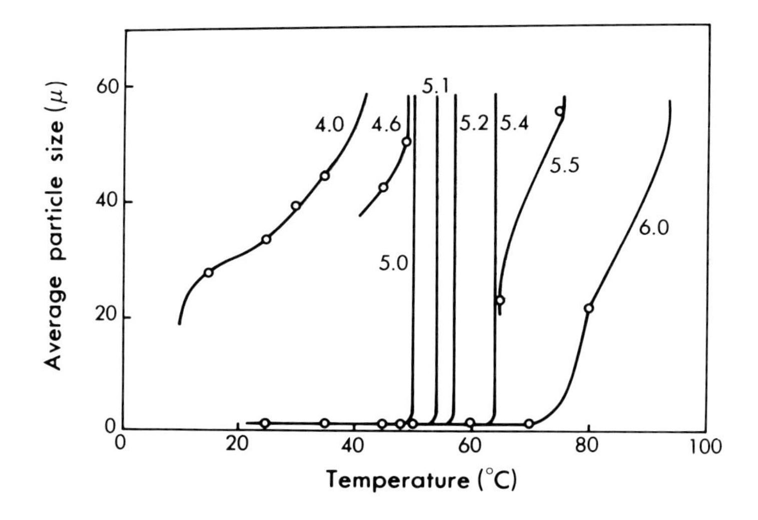

Fig. 5. Average particle size of carbon black plotted against temperature. The numbers indicate the moles of ethylene oxide.

IV. DISCUSSION

It is known that due to the hydrophobic surface carbon black does not disperse in pure water. The dispersibility is remarkably improved with the aid of surfactants, which adsorb at the solid/liquid interface.

Essentially, the HLB of nonionic surfactants depends on temperature (7, 8). Moriyama *et al.* have reported in an analogous experiment that the region of stable carbon black suspensions or polyethylene suspensions extends towards higher temperature with an increase in ethylene oxide chain length of the surfactants (5, 9).

In the present experiment, the carbon particles strongly coagulated at the temperature at which the surfactant solution forms a mesophase. This phenomenon is illustrated by the contour lines for the average particle size as shown in Fig.6.

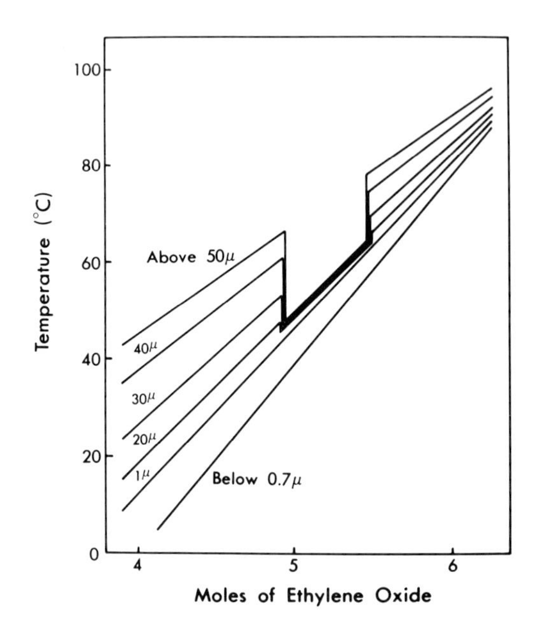

Fig. 6. Contour lines for the particle size of carbon black in $C_{12}E_n$ aqueous solution at various temperatures.

These lines for particles above 20 μm are discontinuous in the mole range of ethylene oxide between 5 and 5.4 on account of the abrupt coagulation of carbon black at the mesophase points.

Fig. 2. shows that the adsorption isotherm was of a Langmuir type below the cloud point. The extreme increase in the surface excess above the cloud point means that the carbon particles adsorb the surfactant being separated from the solution. The adsorption above the cloud point, however, is an apparent state, since the separated surfactant phase may simply concentrate on the surface of carbon particles. Corkill *et al.* (10) and Kravchenko (11) reported essentially the same results, although the surfactant concentration at which the adsorption curve suddenly rose was higher than found in this work. The difference of the concentration may be attributed to that of the solubility of the surfactant.

The increase in the particle size of carbon black beyond the cloud point was not as drastic as that beyond the mesophase point. The gradual increase of the particle size with temperature in other ranges of ethylene oxide moles than those of mesophase has been studied by Moriyama *et al.* (5). The drastic increase above the mesophase point is, therefore, a specific phenomenon.

When the surfactant molecules are orderly oriented to form liquid crystalline structure, the carbon particles tend to be excluded from the surfactant phase. This may be responsible for the remarkable coagulation of carbon particles as the mesophase point.

V. ACKNOWLEDGMENT

The authors wish to thank Dr. K. Meguro, Dr. E. Matijević and Dr. P. Becher for discussions and suggestions regarding this work. The authors also wish to thank Drs. S. Ohta and T. Mitsui for their support and advices.

VI. REFERENCES

1. Balmbra, R. R., Clunie, J. S., and Corkill, J. M., Trans. Faraday Soc. 58, 1661 (1962).
2. Harusawa, F., Nakamura, S., and Mitsui, T., Colloid & Polymer Sci. 252, 613 (1974).
3. Kenjo, K., Bulletin Chem. Soc. Japan 39, 685 (1966).
4. Saito, H., J. Chem. Soc. Japan 92, 223 (1971).
5. Moriyama, N., Hattori, K., and Shinoda, K., J. Chem. Soc. Japan 90, 35 (1969).
6. Fukushima, S., and Kumagai, S., J. Colloid Interface Sci. 42, 539 (1973).
7. Shinoda, K., J. Colloid Interface Sci. 24, 4 (1967).
8. Shinoda, K., and Saito, H., J. Colloid Interface Sci. 26, 70 (1968).
9. Hattori, K., Moriyama, N., and Shinoda, K., J. Chem. Soc. Japan 88, 1030 (1967).
10. Corkill, J. M., Goodman, J. F., and Tate, J. R., Trans. Faraday Soc. 62, 979 (1966).
11. Kravchenko, I. I., Colloid J. U. S. S. R. 33, 312 (1971)

DISPERSIBILITY AND STABILITY OF CARBON BLACK IN WATER

Victor K. Dunn
Xerox Corporation

Abstract

A centrifugation method to measure the effect of surfactants and dispersion equipment on the dispersibility and stability of carbon black in water is presented. This method is based on centrifuging dispersed suspensions of a constant pigment loading at a given set of conditions and comparing the weight concentrations of pigment remaining in the centrifuged fluid. Applications of this method to evaluate dispersion equipment and surfactants are demonstrated.

I. INTRODUCTION

This paper will discuss a centrifugation method of evaluating the dispersibility and stability of carbon black pigments in water. The method will be applied to investigate the influence of two types of dispersion equipment and various commercial surfactants on dispersibility and stability of a channel black pigment.

Dispersibility and stability of carbon black in water are considered on the following basis: in dry powder form, carbon black exists mostly as particles that are interlinked to form three-dimensional, grape-like network structures (1). In thermal blacks, the structures consist of well defined spherical particles that are interlinked. In furnace and channel black, the structures consist of primary aggregates and secondary agglomerates. The primary aggregates are clusters of particles that are fused together and are usually non-dispersible. The secondary agglomerates are loosely bound agglomerates formed by the primary aggregates. To disperse the powder in water, water must displace a powder/air interface with a powder/water interface. Subsequently, shearing energy such as from a ball mill or ultrasonic device is required to break the agglomerates into small, individual units of colloidal dimension. The extent to which the powders can be broken down into colloids is defined as dispersibility. The ability of the resulting colloids to resist agglomeration to form larger particles as the dispersion ages is defined as stability.

Dispersibility and stability are two important factors in the preparation of carbon black hydrosols. These factors are controlled by the type of carbon black, dispersion equipment, and surfactant used. Therefore, experiments are required to determine the best combination of materials and equipment to prepare a stable and well dispersed carbon black hydrosol.

II. METHOD TO MEASURE DISPERSIBILITY AND STABILITY

The method consists of determining the weight concentrations of pigment in a dispersion before and after centrifuging at a given set of conditions as shown in Figure 1. A yield, ϕ, is computed as follows:

$$\phi = \frac{P}{P_o} \times 100 \tag{1}$$

where P_o and P are the concentrations of the suspended pigment in weight percent before and after centrifugation.

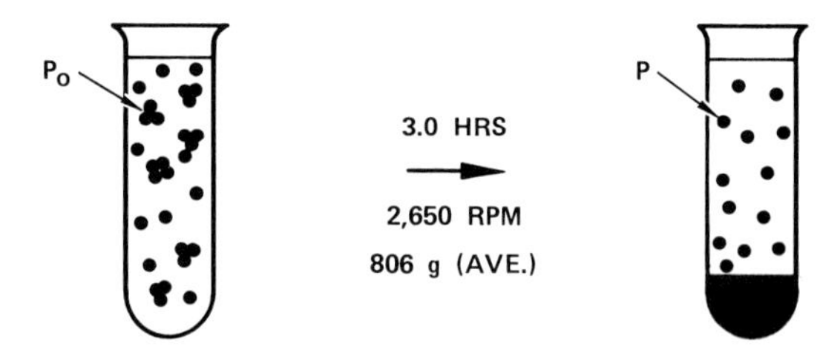

Fig. 1. The yield is computed from the concentration of carbon black in suspension before (P_o) and after centrifugation (P).

The yield, ϕ, is a relative measure of the degree of dispersion achieved (dispersibility). It varies from 0 to 100%. Poorly dispersed samples contain a high proportion of large agglomerates. These agglomerates will be centrifuged down to the bottom of the centrifuge tube as they will experience a greater centrifugal force than the dispersed colloids. Therefore, a low yield value will be obtained. For highly dispersed samples, a high proportion of pigment exists as colloids. More pigment will remain in suspension after centrifugation and a high yield value will be obtained.

The stability of a dispersion is measured by the change in ϕ as a function of the age of the dispersion. Colloids in a stable dispersion will not agglomerate as the dispersion ages and the yield, ϕ, will remain constant. For unstable dispersions, the yield will decrease as the dispersion ages. The rate of decrease is a measure of instability.

One should note that the above discussion of yield in relationship to dispersibility and stability is a simplified picture. A wide range of particle (floc) size probably exists before and even after centrifugation. The amount of sediment (and amount suspended) depends on a number of variables such as floc structure and density, initial particle size distribution, interparticle distances, and attractive and repulsive forces between particles.

The advantage of the centrifugation method is that it can be used on concentrated dispersions as the results of this paper show. Samples for analysis can be determined without dilutions. Method of determination is simple and expensive equipment is not required. Other methods (2) of measuring the degree of dispersion and stability, such as by disc centrifugation, coulter counter, microscopic counter (3), and optical density (4) are often applicable only to dilute systems. Concentrated dispersions will have to be diluted for such a measurement and the change in environment may cause a change in particle size.

The disadvantage of the present method is that it is yet a relative measurement of particle size and stability. Considerably more information could be obtained if the particle size remaining in suspension after centrifugation under a given set of conditions were determined experimentally and theoretically for concentrated dispersions. However, such experimental determinations have not yet been performed, and theoretical determinations are complicated by interparticle interactions and heterodispersity of the particles. Theoretical analysis of centrifugal sedimentation of particles is simplified if the particle concentration is very dilute and Stokes' sedimentation equation is applicable (5). Such an analysis, using our present centrifugation conditions, shows that the maximum radius of particle remaining in suspension after centrifugation is 0.072 microns. It is uncertain how this value actually relates to our high particle concentration dispersions.

In each of the actual yield determinations, a 10.0 ml aliquot of the dispersion was placed in a heavy-walled glass centrifuge tube (14mm internal diameter) and centrifuged at 2650 rpm, with an average acceleration of 806g, for 3.0 hours in an

International Clinical Centrifuge, Model CL. At the end of centrifugation, the top 4.0 ml layer was withdrawn carefully with a pipet and placed into a weighed aluminum weighing dish (diameter 57mm, height 20mm). The weight of the liquid was determined using a semi-microbalance. The pan was then dried for 5.0 hours at 150°C in an oven, cooled for 30 minutes in a desiccator, and reweighed. Another 4.0 ml aliquot was withdrawn from the original uncentrifuged dispersion and weighed and dried using the same procedure. The terms P_o and P were determined from the weights of the dispersion in the pan prior to and after drying.

For such a gravimetric determination, non-volatile, soluble solids other than the pigment will be included in P and P_o. The non-pigment residue comes from surfactants and bases such as sodium hydroxide and may account for 0 to 1.0% of P and P_o. The non-pigment residues will not affect the changes in yields of a dispersion since the amount is relatively constant. The yields so computed will, however, slightly deviate from the actual yield based on pigment alone.

The application of this method to carbon black hydrosols will be illustrated in the next three sections. In Section III, this method will be used to evaluate the effectiveness of two types of laboratory dispersion equipment in dispersing carbon black. In Section IV, the application of this method to measure stability will be demonstrated. In Section V, the method will be used to determine an effective surfactant dispersant for carbon black.

III. EVALUATION OF DISPERSION EQUIPMENT

The purpose of this experiment is to find an effective laboratory dispersion equipment for carbon black. Two relatively new types of commercial dispersion equipment were evaluated on their efficiency in dispersing carbon black in water using the centrifugation method. One was an ultrasonic unit, Branson Sonifier Model W185 (manufactured by Heat Systems, Ultrasonics, Inc., Plainview, New York). The other was a rotor-stator unit, Super Dispax System Model SD-45N (manufactured by Tekmar Company, Concinnati, Ohio).

Both systems were described as being highly effective in dispersing pigments. The ultrasonic unit was used with the standard 1/2" disruptor horn generally from 0 to 150 watts of sonic energy at 20 kHz. The rotor-stator unit is propelled by a universal motor with 390 watts maximum output at 10,000 rpm. The specific rotor-stator head used was G-456 which consists of a rotor and stator containing concentric circular layers of vertical, square-toothed blades. The blade layers of the rotor and stator alternate to provide the shearing and grinding action to disperse pigments.

A. Materials

All materials were used directly from suppliers' bottles. The carbon black used was Neo Spectra Mark II, a channel black supplied by City Service Company. The company listed the surface area determined by N_2 adsorption as 906 m^2/g and the arithmetic mean particle diameter as 13 mμ. Dry sodium hydroxide pellets (ACS certified) were supplied by Fischer Scientific Company. Freshly distilled water was used.

B. Procedure

Two identical 500.00 gm suspensions of carbon black in a basic solution of an anionic surfactant in water were prepared. The weight percent of each component in the suspension was: Neo Spectra Mark II, 5.800%; anionic surfactant, 0.505%; sodium hydroxide, 0.202%; and distilled water, 93.493%. Surfactant was added to help disperse and stabilize the pigments. Base was added because information from the supplier suggests that the surfactant is more effective in a basic medium. The suspensions were prepared by making a batch solution of all the components except the pigment and adding 29.00 gms of carbon black to 471.00 gms of the solution in a 500 ml wide-mouth polypropylene bottle. The suspensions were then stirred with a magnetic stirrer for 10.0 minutes and left to equilibrate at room temperature (23° ± 1°C) for 20.0 hours. One bottle was then dispersed with the ultrasonic unit at 112 watts with stirring, and 10.0 ml aliquots were withdrawn from the bottle after 0.0, 15.0, 30.0, and 45.0 minutes of irradiation for centrifugation analysis as described in the method section.

The suspension in the other bottle was transferred to a glass vessel (500 ml capacity) with an integral flow-thru cooling jacket and dispersed with the rotor-stator unit at setting #5 (middle setting) while being cooled with water. Aliquots of 10.00 ml were withdrawn after 0.0, 5.0, 10.0 and 25.0 minutes for centrifugation analysis. Control samples of 4.0 ml aliquots were withdrawn from both units after dispersion was completed to determine the total weight percent of non-volatile solids in the dispersion before centrifugation (P_o) and were heated at the same time with the centrifuged samples.

C. Results and Discussion

The ultrasonic unit was more effective in dispersing carbon black than the rotor-stator unit under the conditions described previously. During ultrasonic dispersion, the suspension remained as fluid as water; however, excessive heat was generated and the process was stopped several times to let the suspension cool down. Heat was also generated during dispersion by the rotor-stator unit. The suspension increased in viscosity as time of dispersion increased until a thick paste was formed.

Figure 2 is a plot of yields (ϕ) as a function of time of dispersion. The results appear to show that the ultrasonic unit is far more efficient in dispersing carbon black in water than the rotor-stator unit. Yields for the ultrasonic unit were 82.4% after 15 minutes and up to 96.6% after 45 minutes of irradiation which indicates that almost all the pigments were well dispersed after 45 minutes. In comparison, the rotor-stator unit had yields of 17.5% and 20.5% after 10 and 25 minutes of dispersion, respectively.

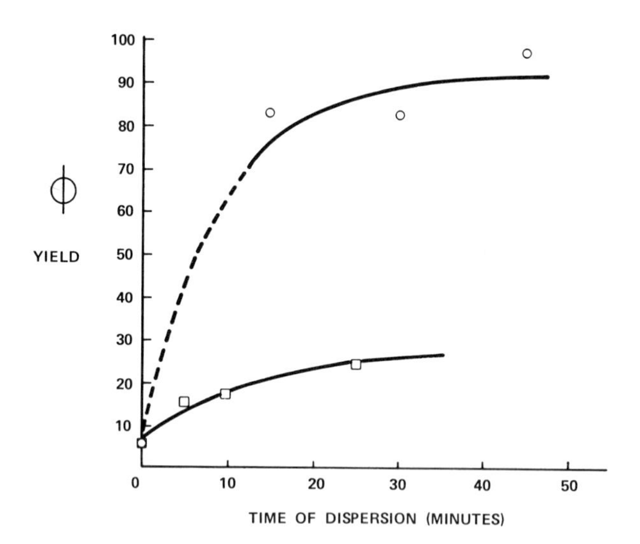

Fig. 2. A comparison of the effectiveness of two types of dispersion equipment on dispersing carbon black in water. ○, *ultrasonic equipment;* □, *rotor-stator equipment.*

The ultrasonic unit also is better in terms of simplicity of operation and equipment cleanup. Dispersion only requires the tip of the probe to be immersed in the suspension. Contamination is minimized because of the small surface area of the probe which is in contact with the suspension. Cleanup involves dipping the probe in clean water and removing the adhering pigment ultrasonically.

IV. STABILITY MEASUREMENTS

The centrifugation method was used to determine the stability of the dispersion that was prepared using the ultrasonic unit. Aliquots of 10.0 ml were withdrawn from the original dispersion at various times after it was prepared and subjected to analysis as described in the method section.

The stability data is shown in Table 1 and Figure 3. Values of the concentration of percent solid in suspension after centrifugation (P) and the respective yields (ϕ) are given as a function of the age of the dispersion. The yield value computed for each day deviates about ± 1 unit from the average value based on 2 to 4 determinations (at 23°±1°C).

Figure 3 illustrates yield as a function of the age of dispersion. The plot shows that the dispersion was unstable although it was initially well dispersed with the yield decreasing significantly from 96.5% to 24.1% over 15 days. There appears to be two stages of flocculation. The first stage was more rapid with the yield decreasing from 96.5% to 34.8% in the first 6 days. The second stage was slower and the yield only decreased from 34.8% to 24.1% from the 6th day to the 15th day.

These results demonstrate that the centrifugation method is applicable to measuring the stability of carbon black pigments in water. The stability of various dispersions can be compared by measuring the rate at which the yield decreases as a function of time. A stable system will have a constant yield with a zero slope.

V. EFFECT OF SURFACTANTS ON DISPERSIBILITY

The purpose of this experiment was to find an effective dispersant for carbon black. The surfactants selected for this experiment are some of the typical dispersing agents used in dispersions or emulsions. In this experiment, 100.0 gm samples of the suspension consisted of 1.0% surfactant, 6.0% Neo Spectra Mark II channel black, and 93.0% distilled water by weight were prepared in 125 ml wide-mouth glass bottles. A solution of the surfactant and water was first made and carbon black was added last. The suspensions were equilibrated in an Eberbach 6000 shaker for 20.0 hours at 23° ± 1°C and 140 cycles/minute. Each of the bottles were then dispersed by the ultrasonic unit at 110 watts for 5.0 minutes. Aliquots of 10.0 ml were withdrawn from the bottles for centrifugation to determine the yield.

TABLE 1
Stability Data on a Carbon Black Hydrosol[a]

Time (Days)	*P(Weight %)*	*Yield*[c] *(%)*
0.0	6.709[b]	96.6
	6.664	
	6.754	
4.0	4.554[b]	65.5
	4.656	67.0
	4.452	64.1
6.0	2.419[b]	34.8
	2.492	35.9
	2.429	
	2.435	
	2.318	33.4
11.0	2.148[b]	30.9
	2.203	31.7
	2.124	
	2.097	30.2
	2.167	
15.0	1.677[b]	24.2
	1.664	24.0
	1.683	
	1.686	24.3
	1.674	

a. *The hydrosol was prepared by ultrasonic dispersion. The composition is described in Section IIIC.*
b. *Average value.*
c. *Yield is equal to* $(\frac{P}{P_o}) \times 100$. *The average value of* P_o *was 6.945%.*

Dispersion yield is very sensitive to the type of agent as shown in Table 2. Values were found to range from 0.0% for a poorly dispersed sample, to 89.9% for a well dispersed sample. As all the non-ionic surfactants examined gave low yields, it seems to indicate that non-ionic surfactants alone are not effective in dispersing the carbon black. The only cationic agent tested, Aerosol C-61, was not effective. Of the anionic surfactants evaluated, two had low yields and three had relatively high yields.

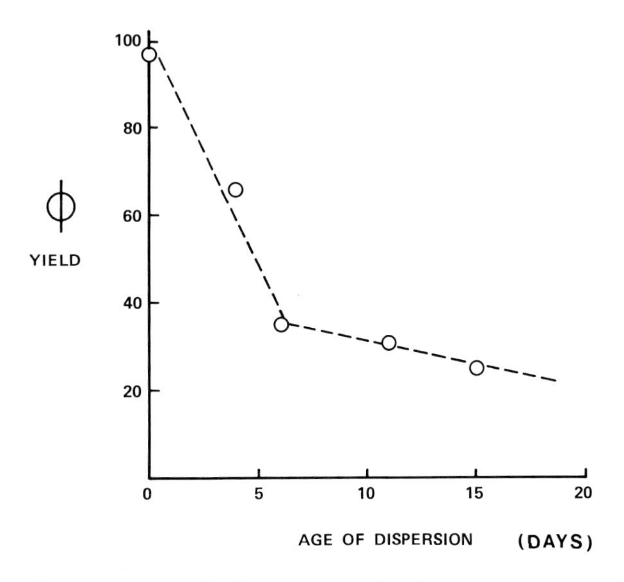

Fig. 3. Stability plot of the carbon black hydrosol prepared by the ultrasonic unit.

The extent to which the carbon black was dispersed was also reflected by the viscosities of the samples. Although the weight percent of added carbon black was the same on all dispersions, highly dispersed formulations such as the one containing surfactant C, had viscosities of about 2 centipoise as measured by a Brookfield viscometer. Poorly dispersed samples such as the sample without surfactant had viscosities in the range of 100 centipoise. These observations agree well with the experimental results using the rotor-stator unit. The sample dispersed by that unit was also poorly dispersed and was paste-like. Qualitative increases in viscosity were also observed during stability measurements. As a dispersion ages and yield decreases, the dispersion gradually becomes more viscous due to an increase in particle size.

VI. CONCLUSIONS

The results of this investigation indicate that the centrifugation method is a simple and practical comparative method to study the effect of dispersion composition and conditions on dispersibility and stability of pigments in a liquid. The

TABLE 2
The Effect of Surfactants on the Dispersibility of Carbon Black in Water

Trade Name	*Chemical Description (6)*	*Type*	*Yield(%)*
Surfactant A		A	89.8
Surfactant B		A	75.8[a]
Surfactant C		A	72.1[b]
Aerosol OT	Dioctyl ester of sodium sulfosuccinic acid	A	4.3
BRIJ 35	Polyoxyethylene lauryl ether	N	2.3
Emulplor EL-620	Polyoxyethylated vegetable oil	N	1.7
Diazopon SS-837	Polyoxyethylated alkylphenol	N	0.8
Triton X-100	Octylphenoxy polyethoxy-ethanol	N	0.4
Priminox R-15	$RNH(CH_2CH_2O)_{15}H$	N	0.1
Aerosol C-61	Ethanolated allyl guanidine amine complex	C	0.0
Daxad 30	Carboxylated polyelectrolyte	A	0.0
No Surfactant			0.0

a. Average value from three separate runs, 74.2, 76.6 and 76.6%

b. Average value from two separate runs, 72.9 and 71.2%

results also show that an ultrasonic dispersion unit is a more effective laboratory device for dispersing carbon black in water than the rotor-stator; and that some anionic surfactants are apparently very effective carbon black dispersants.

Acknowledgement

The author is grateful to Dr. Linda T. Creagh for her helpful comments and suggestions.

References

1. Marsh, G. E., "Properties of Carbon Black which Influence Dispersion in Printing Inks," Special Blacks Technical Service Report S-13, Cabot Corporation, Massachusetts.

2. Vincent, B., Advances in Colloid and Interface Science, 4, 193 (1974).

3. Parfitt, G.D., and Lawrence, S.G., J. Colloid Interface Sci. 35, 675 (1971).

4. Dunn, V.K., and Vold, R.D., J. Colloid Interface Sci. 54, 22 (1976).

5. Mysels, K.J., "Introduction to Colloid Chemistry," p. 43, Interscience Publishers, New York, 1959.

6. Brown, H., Christenson, R., Judge, M., and Kessler, A., "Detergents and Emulsifiers," North American Edition, Manufacturing Confectioner Publishing Co., New Jersey, 1975.

STABILITY OF COLLOIDAL KAOLINITE SUSPENSIONS IN THE PRESENCE OF SOLUBLE ORGANIC COMPOUNDS

Rudolf Klute and Uwe Neis
Institut für Siedlungswasserwirtschaft
Universität Karlsruhe

SUMMARY

Colloidal kaolinite suspensions were flocculated with a cationic polyelectrolyte in the presence of varying amounts of benzoic and humic acid. Using liquid szintillation and particle counting techniques, polymer adsorption on the clay surface and changes of the relative particle number were measured. In addition, kinetic experiments on adsorption and flocculation were performed.

The rate of polymer adsorption and flocculation was found to be influenced significantly by the amount of humic acid present in solution, whereas in the case of benzoic acid no effect could be detected.

The results indicate that depending on the polyelectrolyte dosage organic compounds may increase or decrease the stability of clay suspensions.

I. INTRODUCTION

The adsorption-desorption behaviour of polymeric molecules on surfaces in solid-liquid systems has gained increasing interest in recent years. Jenckel and Rumbach (1) were the first to propose a model for polymer adsorption. Ruehrwein and Ward (2) by employing the Jenckel and Rumbach model of polymer adsorption to dispersed systems proposed a bridging mechanism for flocculation of highly concentrated clay suspensions. A mathematical model was developed by La Mer and Healy (3).

Based on these early investigations a number of authors have studied the adsorption of polymers and the subsequent flocculation in dispersed systems. Recent studies of Fleer and Lyklema (4) confirm the bridging model, whereas Kasper (5) and

Gregory (6) postulate an uneven charge distribution on the particle surface caused by adsorbed cationic polymer molecules to be responsible for flocculation. Black et al. (7) showed that the adsorption of hydrolized polyacrylamide onto kaolinite particles is influenced by the presence of inorganic salts which may be explained by changes of polymer chain conformation and configuration (8).

Stenius et al. (9) studied the adsorption of negatively charged lignosulfonates on kaolinite particles, presuming complex formation between hydroxyl groups of the polymer and aluminium atoms on the clay edges as adsorption mechanism. Huang and Liao (1o) studying the adsorption of pesticides on clays interprete their results by postulating hydrogen bonding as the main adsorption mechanism. Chen et al. (11) found a significant influence of cationic polyelectrolytes and polymeric humic acid on the adsorption of phosphate on kaolinite. The results are explained in terms of charge reversal by polyelectrolyte adsorption respectively complex formation of humic acid at the kaolinite surface. This leads to an increase respectively decrease in phosphate adsorption. Narkis et al. (12) studying the influence of humic and fulvic acid on the flocculation of clays observed only a stabilizing effect when synthetic polymers were used as flocculants. However, very high concentrations of organic acids and an unusual analytic technique do not allow a satisfactory explanation of the phenomena observed.

In this situation the question arises whether the different observed phenomena can be interrelated. Therefore, it is the purpose of this study to evaluate the influence of selected organic molecules (cationic polyelectrolyte, benzoic acid, humic acid) on the mode of adsorption and flocculation in dilute clay-polymer systems.

II. EXPERIMENTAL

Materials

The clay used in this investigation was Kaolinite # 3 as provided from Ward's Natural Sci. Est. Physico-chemical details of this material are described in (13). An aqueous suspension of the material was prepared by micro-mesh-sieving, treatment in H_2O_2-solution, monoionisation, repeated settling and resuspension. This was used as stock suspension of Na-kaolinite with a particle size distribution between 1.5 µm and 1o µm approximately. A predetermined amount of this stock solution was diluted with deionized water to give an initial clay concentration of 85 mg/l. The pH was held constant at 6.5. Addition of polyelectrolyte and organic acids did not cause significant change of the pH. The surface of kaolinite in suspension was determined to 12.9 m^2/g (BET method).

Benzoic acid was Merck reagent grade. A stock solution of 25o mg/l humic acid (Fa. Roth, Karlsruhe, No. 2-7821) was filtered twice before application (o.45 µm and o.o1 µm Sartorius membrane filters).

The cationic polyelectrolyte used was Praestol 434 K (Chemische Fabrik Stockhausen, Krefeld, W.-Germany), a copolymer of acrylamide and trimethylaminoethyl methacrylate. The molecular weight, given by the manufacturer, is $4x1o^6$, the chain containing 7o % functional groups. The polylectrolyte used for adsorption measurements was labeled with H^3 (specific activity 2.o µ Ci/mg) following the Wilzbach method (14).

Methods

Particle Counting

Flocculation experiments were performed with 1oo ml suspensions stirred with 6o rpm for 6o minutes in 15o ml beakers. After 3o minutes settling the particle size distribution was measured with a particle analyzer (Coulter Counter ZB, Coulter Electronics, Hialeah, Flo.,USA). Details of this method are described in (15). The method was improved by installing a multi-channel-analyzer (ND 1oo, Nuclear Data Co., Schaumburg, Ill., USA) which allowed immediate registration of the whole size distribution.

Adsorption Measurements

All equipment surfaces in contact with polyelectrolyte were coated with Siliclad (product of Clay-Adams Inc., New York, N.Y., USA) in order to prevent adsorption. The amount of polyelectrolyte adsorbed on kaolinite was considered to be the difference between the initial and the remaining polyelectrolyte concentration in the supernatant after flocculation and 3o minutes sedimentation. Sedimentation was considered to have fewer disadvantages on liquid-solid separation than other methods. A liquid szintillation spectrometer (Packard Model 3375, Packard Instruments Co., Inc., Downers Grove, Ill., USA) was used to determine the polyelectrolyte concentration in the supernatant by counting a mixture of 1o ml polyelectrolyte solution and 1o ml of a premixed szintillator cocktail (Instagel, Packard Instruments Co., Inc., Downers Grove, Ill., USA) at fixed instrument settings for 5o minutes at 8^0 C. The measurements with benzoic acid (C^{14} -labeled; specific activity o.86 μ Ci/mg) followed the same procedure.

III. RESULTS

In a first series of experiments the adsorption of the cationic polyelectrolyte Praestol 434 K onto kaolinite was studied. In order to investigate probable mutual interactions between this polyelectrolyte and organic compounds further adsorption experiments were performed with addition of benzoic and humic acid (Fig. 1 and 2).

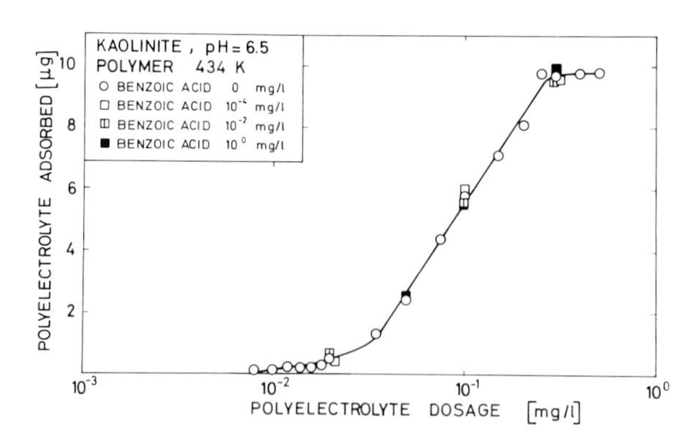

Fig. 1. Adsorption of the cationic polyelectrolyte Praestol 434 K on kaolinite clay versus polyelectrolyte dosage without and with benzoic acid.

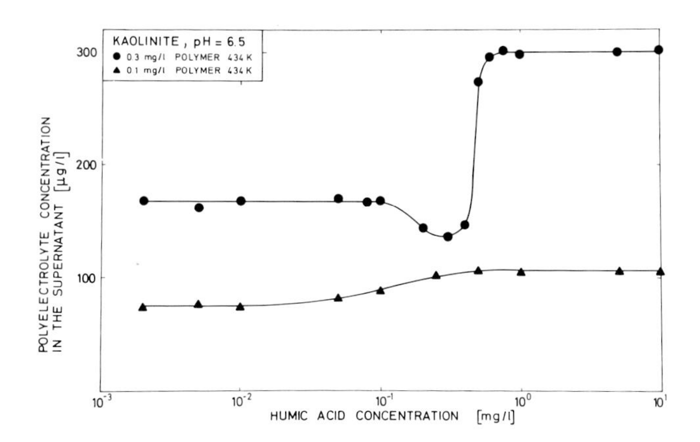

Fig. 2. Concentration of the cationic polyelectrolyte Praestol 434 K in the supernatant of flocculated and settled kaolinite suspensions versus humic acid concentration in solution.

The adsorption of the polyelectrolyte is a function of the initial concentration. In the low concentration range only small amounts are adsorbed, whereas at high initial concentrations saturation is observed. The intermediate range is characterized by a steep and linear increase in adsorption. No influence of benzoic acid on polyelectrolyte adsorption is observed.

However, in the presence of humic acid a quite different adsorption behavior is evoked (see Fig. 2). Depending on the initial polyelectrolyte dose and on the humic acid concentration, adsorption may be favoured or reduced. Complete inhibition of polyelectrolyte adsorption occurs at higher humic acid concentrations than o.7 mg/l.

Next, flocculation experiments with identical reaction conditions with respect to the solid and solute phase were performed (Fig. 3 and 4).

The flocculation of kaolinite in pure polyelectrolyte solutions is again a function of the initial concentration (Fig. 3). In low concentration ranges no or only weak flocculation is detected, whereas at high initial polyelectrolyte concentrations complete restabilization occurs. The inter-

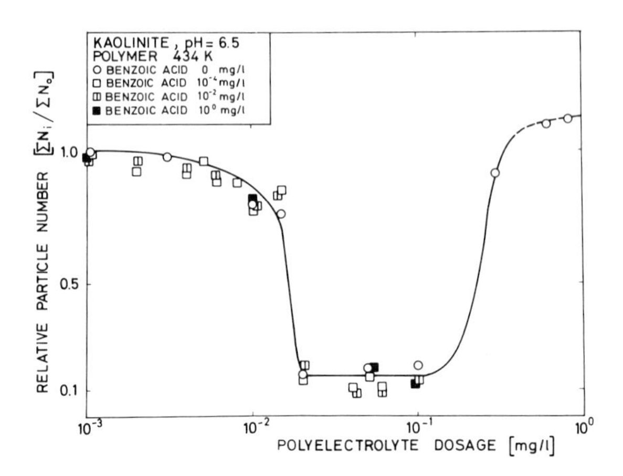

Fig. 3. Flocculation of kaolinite suspensions versus cationic polyelectrolyte dosage. Flocculation rates are presented as relation of the total particle number ΣN_i *in the supernatant after a flocculation time of 60 min. and 30 min. sedimentation to the total initial particle number* ΣN_o*. The results of additional dosage of benzoic acid are included.*

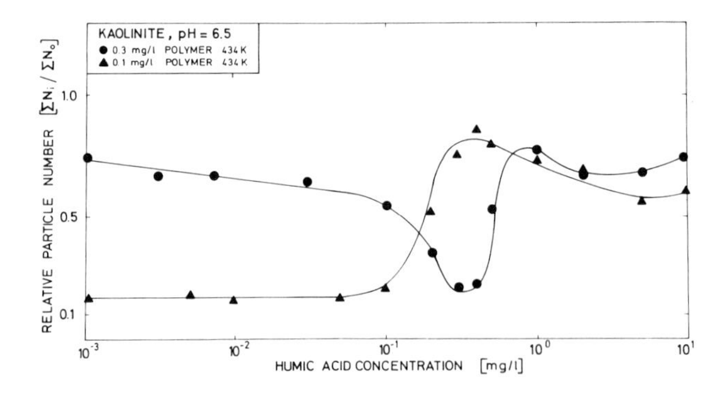

Fig. 4. Flocculation rates of kaolinite suspensions (ΣN_i *= total particle number after a flocculation time of 60 min. and 30 min. sedimentation,* ΣN_o *= total initial particle number) versus humic acid concentration in solution. Cationic polyelectrolyte (Praestol 434 K) concentration is 0.1 and 0.3 mg/l.*

mediate range can be characterized as optimal flocculation range indicated by a minimum in particle numbers in the supernatant. The transition to this optimal range is characterized by a steep increase resp. decrease in flocculation rates. Within the usual experimental error no influence of benzoic acid on flocculation rates can be detected.

On the basis of these data and assuming polyelectrolyte concentrations of o.1 mg/l and o.3 mg/l to be optimal resp. to restabilize the system the influence of various amounts of humic acid on flocculation was examined. Again, in a distinct humic acid concentration range C_d (mg/l) $o.1 < C_d < 1.o$ flocculation may be favoured or reduced. Humic acid concentrations > o.7 mg/l provoke high stability of the clay suspension, irrespectively of the polyelectrolyte dosage (see Fig. 4).

Finally, kinetic investigations on polymer adsorption and flocculation were performed. It was the purpose of these experiments to evaluate the influence of humic acid during the initial phase of the adsorption-flocculation reaction. First kinetic data under reaction conditions described above revealed no additional information. Therefore, simultaneous dosage of humic acid and polyelectrolyte was applied (Fig. 5).

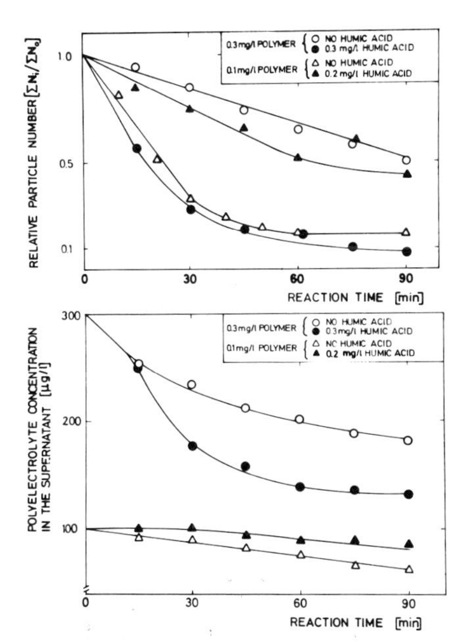

Fig. 5. Effect of humic acid on the flocculation and adsorption kinetics of kaolinite.

During the initial reaction phase (≤ 3o minutes) there seems to be an apparent influence of humic acid on adsorption. In addition, flocculation appears to be enhanced or reduced for o.3 resp. o.1 mg/l polyelectrolyte. Longer reaction periods confirm the results shown in the preceeding figures.

IV. DISCUSSION

It is evident from Fig. 1 that there is a significant effect of initial polyelectrolyte concentration on the adsorption process. Between adsorption and flocculation a close correlation can be depicted by comparison of Fig. 1 and 3. In both curves three distinct regions are noticable at corresponding polyelectrolyte concentrations. A more detailed explanation may be derived from the data given in Table 1.

TABLE 1

Comparison between Polyelectrolyte adsorbed and Flocculation Efficiency in Kaolinite Suspensions

Initial Polyelectrolyte Dosage	Polyeletrolyte adsorbed		Flocculation Efficiency
mg/l	µg/l	% of initial	$\Sigma N_i/\Sigma N_0$
0.01	1	10	0.86
0.02	5	25	0.15
0.035	13	37	0.15
0.05	25	50	0.15
0.075	44	59	0.15
0.1	56	56	0.15
0.15	71	47	0.19
0.2	85	42.5	0.35
0.3	97	32	0.90
0.5	98	20	> 1.00

As can be seen from Table 1, optimum flocculation efficiency is obtained already at 25 % polyelectrolyte adsorption relative to initial. The assumption of a specific charge density on the negative clay surface of o.31 μ C cm^{-2} (calculated from cation exchange capacity) yields a total amount of $2.1x1o^{6}$ unit charges per 1 l suspension.

At 25 % polycation adsorption the amount of positive charges adsorbed on the clay surface can be calculated to $1.0x10^{16}$ unit charges, approximately. A comparison of both values leads to the conclusion that charge neutralization occurs within this concentration range (exactly at 0.03 mg/l initial polyelectrolyte concentration). This result is confirmed by studies of several other authors (7, 12, 16). They found that at optimum flocculation concentrations for clay particles the electrophoretic mobility was still negative.

Increasing the initial polyelectrolyte dosage provokes still increasing adsorption leading to a relative maximum. Finally at saturation an adsorbed amount of $19.4x10^{16}$ in terms of unit charges can be calculated. This value is about ten times that of the kaolinite surface charge, thus resulting in complete restabilization of the system (see Fig. 3).

These results may lead to the following destabilization mechanism in clay-cationic polyelectrolyte systems. A critical flocculant concentration is necessary to destabilize the system. This destabilization corresponds to a maximum reduction of the total particle number. Increasing the flocculant concentration leads to further polyelectrolyte adsorption. There appears to be no effect of additional polyelectrolyte adsorption on the flocculation reaction until the clay particles are covered with too many polycations. Complete coverage of all surface adsorption sites results in total inhibition of the flocculation reaction (see Fig. 3), which may be explained in terms of the bridging model (3) or by steric effects (17).

From the results in Fig. 1 and 3 it appears that the effect of benzoic acid on polyelectrolyte adsorption onto the clay surface or on flocculation is negligible. Apparently, no reactions between benzoic acid and the polyelectrolyte molecules occur that influence both processes. To conclude that low molecular, weakly dissociated organic compounds in general behave in the same manner, would not be more than guesswork. Additional influence, e.g. of pH-variation and of clay surface modification must be taken into account.

When humic acid is present in solution, probable interactions between humic acid and polyelectrolyte may serve to explain the observed results. No polyelectrolyte molecule present in solution is able to adsorb in high humic acid concentration, hence no flocculation is observed (Fig. 2 and 4). Lowering the humic acid concentration provokes a certain amount of free polycations to be available for adsorption, thus flocculation is induced. The extent of adsorption and flocculation depends on the initial polyelectrolyte dosage. At very low acid concentrations decrease in adsorption of polyelectrolyte seems to occur paralleled by an increase in particle number. One might consider this as stabilization of the clay suspension. No similar effect is observed using o.1 mg/l polyelectrolyte. In this case the flocculation curve stays at a minimum level indicating complete destabilization.

Assuming humic acid-polycation interactions to be responsible for these oberservations may not explain all facts described above. However, interactions between the dissolved organic species seem to take place in first order. Therefore, reactions on the clay surface are more or less inhibited, depending on the concentrations of solute and cosolute.

This explanation would confirm the results of Narkis et al. (12). Further interactions of dissolved macromolecules, e.g. non-specific association of cationic and anionic polymers are described comprehensively by Morawetz (18).

The kinetic results as depicted from Fig. 5 indicate that the presence of humic acid apparently effects adsorption and flocculation during the initial phase of reaction (≤ 3o min.) since at longer periods corresponding curves can be considered to be roughly parallel. It is of interest that this behaviour holds true for favouring as well as inhibiting effects of humic acid. So, increase or decrease in adsorption or flocculation might be provoked by the same mechanism. These first data do not supply more details about the process involved. Formation of humic acid-polycation complexes as well as electrostatic interactions between both reactants may be proposed.

V. REFERENCES

1. Jenckel, E., and Rumbach, B., 2. Elektrochem. 55, 612 (1951).
2. Ruehrwein, R.A., and Ward, D.W., Soil Sci. 73, 485 (1952).
3. La Mer, V.K., and Healy, T.W., Rev. Pure and Appl. Chem. 13, 112 (1963).
4. Fleer, G.J., and Lyklema, J., J. Colloid Interface Sci. 46, 1 (1974).
5. Kasper, D.R., "Ph.D. Thesis", California Institute of Technology (1971).
6. Gregory, J., J. Colloid Interface Sci. 42, 448 (1973).
7. Black, A.P., Birkner, F.B., and Morgan, J.J., J. Amer. Water Works Ass. 57, 1547 (1965).
8. "Water, a Comprehensive Treatise". Vol. 4. Aqueous Solutions of Amphiphiles and Macromolecules. (F. Franks,Ed.). Plenum Press New York (1975).
9. Stenius, P., Le Bell, J., Bergroth, B., and Stenlund, B., Paperija Puu-Papper och Trä 56, 463 (1974).
1o. Huang, J.-Ch., Liao, Ch.-S., Proc ASCE, J. San. Eng. Div., SA 5, 1o57 (197o).
11. Chen, Y.R., Butler, J.N., and Stumm, W., J. Colloid Interface Sci. 43, 421 (1973).
12. Narkis, N., Rebhuhn, M., and Sperber, H., Isreal J. Chem. 6, 295 (1968).
13. American Petroleum Institute, Project 49: "Clay Mineral Standards", Columbia University, New York (195o).
14. Wilzbach, K.E., "Advances in Tracer Methodology", Vol. 1. Plenum Press, New York (1963).
15. Neis, U., Eppler, B., and Hahn, H., in "Water Resources Instrumentation" (R.J. Krizek, E.F. Mosonyi, Eds.), Vol. 1, p. 134. Ann Arbor Sci. Publ. Inc. (1974).
16. Black, A.P., and Vilaret, M.R., J. Amer. Water Works Ass. 61, 2o9 (1969).
17. Heller, W., and Pugh, T., J. Polymer Sci. 48, 2o3 (196o).
18. Morawetz, H., "Macromolecules in Solution", Wiley Interscience, New York (1975).

DIFFERENCES IN COLLOID STABILITY OF BACTERIA, ALGAE AND CLAYS AT IDENTICAL AND DEFINED SOLUTION CONDITIONS

Hermann H. Hahn and Birgit Eppler
University of Karlsruhe / West Germany

Previous investigations in these authors' laboratory have identified the behavior of clays and metal oxides in simple electrolytes with respect to destabilization and aggregation. In general one observes increasing destabilization with increasing concentration of mono- and divalent counterions, such as might be encountered in natural waters. - The aim of this present investigation was to study the relative colloid stability of various organic colloids, i.e. bacteria and algae, under comparable conditions.

Colloid stability, defined as relative rate of aggregation or as collision efficiency was determined by following changes in turbidity of the suspension with time. Concurrent measurements with the Coulter Counter (for bacterial cells) served for calibration and interpretation. Bacteria (E. coli and Bac. cereus) and algae (Chlorella vulgaris), reproducibly cultured in specially designed chemostats, were suspended in ionic media containing varying amounts of NaCl and $CaCl_2$ (10^{-3} to 10^{-1} M/l) at controlled pH values ranging from 4 to 9. Hydrodynamic conditions affecting mostly the transport of the destabilized colloids were kept constant and quantifyable.

The observed results show that for both types of bacteria (with a pH_{iso} = 2 to 3) studied, aggregation was optimal at concentrations at or below 10^{-3} M/l and pH values smaller than 4 to 5 (that is colloid stability was lowest at these conditions). Increase in salt concentration and pH brought about increasing colloid stability. - The algae investigated even showed noticeable aggregation in distilled water with steadily decreasing collision efficiency (increasing colloid stability) at higher concentrations. The pH range investigated showed a slight optimum for algal aggregation around neutral values. - These findings run contrary to the above mentioned results on the stability of clays and metal oxides in aqueous systems. Microscopic investigations of destabilized and restabilized bacteria indicate the exudance of cell material, possibly affecting changes in the stability. The behavior of algal cells, differing from that of bacterial cells, might still ask for other explanations.

Quantitative determination of colloid stability and aggregation behavior of bacterial and algal cells together with other relevant physiological information will help to predict transport and distribution of such cells from points of discharge and areas of algal bloom.

I. INTRODUCTION

The properties of natural waters that characterize them as ecological systems as well as sources of water supply are determined to a significant extent by dissolved and suspended solids. The suspended phase in particular affects directly and, as a matrix for various surface chemical reactions, indirectly a large number of water quality parameters. Aggregation and destruction of aggregates of suspended particles play a significant role in the description of this solid phase. - Ettner et al. (1) concluded that the distribution of deposited clays in an investigated natural system could not be explained by hydraulic phenomena alone. Weiss (2) indicated that inorganic and organic suspensa may react in a similar fashion or with each other in natural systems. Hahn et al. (3) analyzed existing data on clay stability as well as on clay aggregation and sedimentation and argued that aggregation can explain certain processes of clay deposition. - In all studies it is directly or indirectly stated that further investigations on the stability of inorganic and organic suspensoids in natural waters are needed.

Comparing prominent characteristics of suspended solids in natural systems with those of well defined colloids (e.g. size and shape, electrophoretic mobility, surface chemistry) and taking into consideration that the dissolved phase of natural water contains inorganic and organic substances that have shown to affect colloid stability in various ways one might expect that all destabilization and transport mechanisms known (compare for instance O'Melia (4)) can be invoked. Thus, detailed studies have been made to explain and possibly predict stability of various clays and metal oxides using the Gouy Chapman theory on double layer compaction (3). In continuation of this work and in appreciation of the limited applicability of the concept of double layer compaction a number of laboratory studies have recently tried to determine the stability of various clays and metal oxides in solutions resembling natural waters or consisting of natural water samples (for instance (5) and (6)). It was found that postulates derived from double layer considerations could explain qualitatively the effect of pH, concentration and charge of "inert" counter ions and mixtures thereof upon the stability of different clays and metal oxides. Quantitative confirmations appeared problematic due to a number of parameters that have to be estimated independently. In addition the effect of even low concentrations of organic substances as found in polluted systems has been studied in a global manner ((5) and (7)). Under the conditions investigated, an increase in colloid stability was noted which could not be explained conclusively and quantitatively by existing theories on colloid stability. - Similar investigations on the stability of organic colloids, foremost microorganisms, in natural waters were not reported in this context.

Aggregation of microorganisms, especially bacteria has been investigated in connection with water purification, biotechnological processes, as well as colloid chemical studies. Studies on the formation and destruction of bacterial flocs in the so-called activated sludge process have focussed on the relationship between culture conditions (nutrients, microorganisms, etc.) and aggregation tendencies (for instance Tenney et al. (8)). Subsequently it was postulated and shown valid that aggregation of microbial cells under specified conditions is affected by an interaction of polymers excreted by these cells (9). Attempts were made to define the conditions under which such cell material would be activated and excreted (10), (11), (12). In particular, McGregor (13) showed these phenomena by illucidating the effect of temperature and shear stress upon the release of microbial polymers from E. coli and Pseudomonas suspensions. Furthermore, he demonstrated that these polymers may enhance flocculation in one instance and hinder it in other instances. - Parallel to these findings a series of investigations was focussed on the exploration of the validity of colloid chemical theories concerning the aggregation of mostly hydrophobic colloids. Santoro et al. (14) studied the effect of the charge of various cations as well as the influence of solution pH upon aggregation of various bacterial and yeast cells and compared it with the coagulation of different clays. Rubin et al.((15) and (16)) concentrated on the coagulation of E. coli by mono-, di-, tri- and tetravalent cations and concluded that the Schulze-Hardy rule was obeyed, with E. coli being similar in this instance to hydrophobic colloids. Bradley et al. (12) could show that the effect of shear flow on aggregation and aggregate destruction of activated sludge organisms was comparable to that of inorganic particles. - Coagulation or flocculation of algae is an important step in purifying water for supply purposes. Ives (18) pointed out very early that charge density, algae size and electrokinetics affecting substances determine the amount of chemical coagulant for algae aggregation. McGarry (19) extended these results by including (cationic) polyelectrolytes as flocculant aides.

Summarizing the various questions underlying the above described studies and abstracting from the published results one can conclude that aggregation of microorganisms, in particular bacteria has been investigated with emphasis on the substances inducing aggregation, the conditions under which such cell material is set free and the global effect of hydraulic parameters. The process of algal aggregation appears less researched but the effect of metal hydroxo complexes, precipitating metal hydroxides and cationic polymers on algae coagulation and/or flocculation has been reported. - From a viewpoint of colloid stability of microorganisms in natural systems the effect of varying concentrations of such substances that are found in natural waters, in particular mono- and divalent cations, is not documented satisfactorily. For a description of the fate of microorganisms in natural systems shape and viability of aggregated cells should be known, too. Thus the here presented study concentrated on a quantitative determination of colloid stability of two representative microorganisms, one bacteria and one algae. Solution conditions varied in the range found in natural systems. Sodium and calcium were used as aggregation causing agents. The global effect of complex organic pollutants was investigated. Colloid stability was measured reproducibly in repeated experiments by more than one analytical technique while hydrodynamic parameters were kept constant and quantifyable.

II. MATERIALS AND METHODS

1. Microorganisms

Bacillus cereus (Botanic Inst. Karlsruhe) and Chlorella vulgaris (211/11 h Goettingen) were used as representative and ubiquitous microorganisms. They were grown in self-built continuous culture apparatus (chemostat) at 37°C (B. cereus in filtered TSB (Difco)) respectively 27°C (Chlorella). Bacteria were harvested after about ten generation times (30 minutes), washed three times in distilled water and resuspended in salt solutions intended for aggregation experiments. Algae were harvested after about 25 generation times, controlled by dark light cycles of altogether 24 hrs and suspended directly in salt solutions prepared for aggregation studies. (Suspension in distilled water induced immediate aggregation.)

2. Chemicals

All coagulants and other chemicals were reagent grade (Merck). Stock solutions of sodium - and calcium chlorides were prepared with double-distilled water which was filtered (Millipore HA 0.10 μ pore size). They were stored in stopped glass bottles at low temperature. Adjustment of pH was made with 10^{-1} molar HCl and NaOH respectively.

3. Aggregation experiments.

Colloid stability was determined from coagulation or flocculation rates. Such aggregation rates were measured as change in particle size distribution recorded by a Coulter Counter Model ZB. Microorganisms were suspended in aliquots of 100 ml containing a predetermined amount of salt and placed on a jar test apparatus after an initial reading was taken. Particle number concentrations were in the range 10^{+7} per ml (B. cereus) respectively 10^{+6} per ml (Chl. vulgaris). Stirring rates, i.e. coagulation transport conditions were kept constant at 40 rpm for bacteria (no sedimentation, less stable aggregates) and 70 rpm for algae (sedimentation, stronger aggregates). All aggregation experiments were done at controlled room temperature of 20°C ± 2°C. After the initial reading the changing particle size distribution was determined at fifteen minutes intervals. Each beaker was used once for a reading in the case of B. cereus (lower floc strength) and in the case of Chlorella twice (increased floc strength). Total aggregation time did not exceed three hours. Parallel to electric particle concentration measurements microscopic inspection and counting (in case of B. cereus electronmicroscopy) was done, as well as turbidity measurements with a Zeiss Elko (480 - 530 μm) in order to interpret aggregation phenomena. Furthermore, bacteria were plated parallel to counting experiments to determine changes in viability.

4. Coulter Counter measurements.

Bacterial suspensions were analyzed with a 100 μ capillary with a measuring range from 2 to 40 μ; individual particles were in the order of size of 1.5 to 5.4 μ. A 50 μ orifice with a measuring range from 1 to 20μ proved adequate for the analysis of algal suspensions, with individual cells ranging from 2 to 5.5 μ. The required electrolyte for this type of analysis was given by solution conditions as set for aggregation studies. Initially, reproducibility was checked for B. cereus through 15 to 18 experiments each for different settings; due to very low scattering reproducibility for algal counting appeared established after about 6 experiments each. Counting of individual cells and aggregates thus seemed feasible for both types of microorganisms. - Routine measurements were performed in duplicate.

III. RESULTS

The here reported data stem from investigations with controlled, unchanged and quantifyable hydrodynamic characteristics. Only very simple solution parameters such as hydronium ion concentration, type and concentration of counter ion, were varied systematically. The results are presented in parallel for both microorganisms investigated in order to facilitate comparison. There are basically three phenomena to be reported:

> Effect of pH changes, consequences of varying salt concentrations and influence of the type of counter ion used. (The term 'counter ion' is applied to denote ions of a charge opposite to the assumed charge of the microorganisms; it is not implied that it acts in a similar way as counter ions known from theories on double layer compaction.)

1. pH effects.

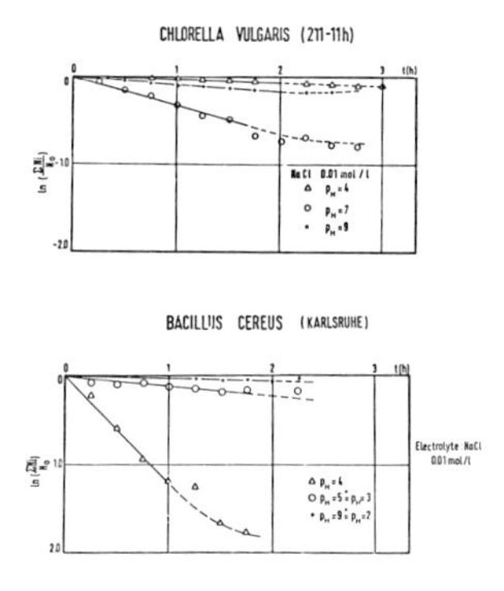

Fig. 1. Aggregation rates for B. cereus and Chl. vulgaris at NaCl = $1o^{-2}$ M/l and varying pH.

Fig. 1 shows aggregation rates for bacteria and algae, depicted as logarithm of the changing particle concentration (dimensionless!) at given NaCl concentration and varying pH. Both types of microorganisms show similar and marked variation in aggregation rate (i.e. colloid stability) with varying solution pH. At low pH values aggregation rates are low (colloid stability high) in both systems. With increasing pH the aggregation rate increases, reaches a maximum at a given $pH_{inst.}$ and decreases again upon further increases in pH. While this effect of changing solution pH upon colloid stability is the same for both microorganisms, the $pH_{inst.}$ at which maximum aggregation rates are found is significantly different: for B. cereus it is between pH = 3 and pH = 5 while for Chl. vulgaris this value lies between 6 and 8.

2. Concentration effects of counter ion.

The change in aggregation rates due to varying concentrations of Na^+ (expressed in the same manner as in Fig. 1) has been shown in Fig. 2 for the bacterial and algal suspensions.

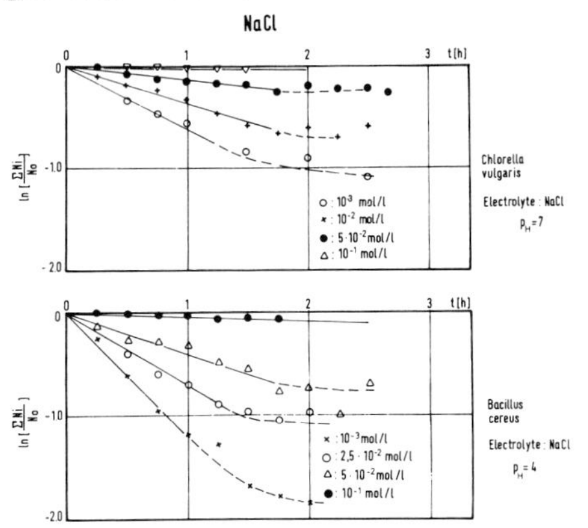

Fig. 2. Aggregation rates for B. cereus and Chl. vulgaris for varying concentrations of NaCl at fixed pH = 4 (B. cereus) resp. pH = 7 (Chl. vulgaris).

Fig. 3 shows the effect of varying concentrations of Ca^{+2}. Both salts were in the chloride form.

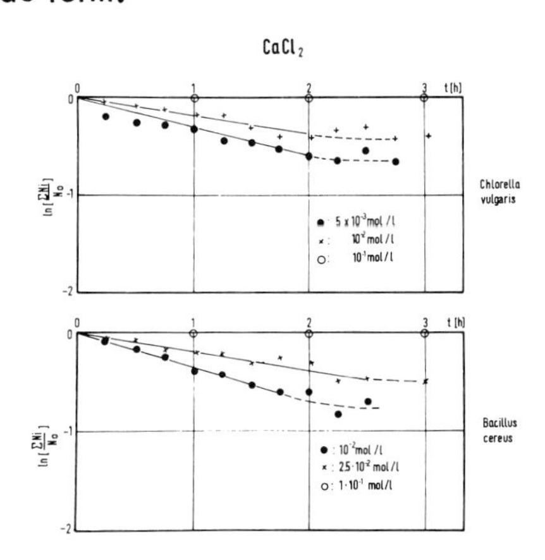

Fig. 3. Aggregation rates as in Fig. 2, for varying concentrations of $CaCl_2$.

The following observations can be made: first, an increase in the concentration of counter ion reduces aggregation rates, i.e. increases colloid stability; second, this reduction in aggregation rate with increasing salt concentration is similar for the bacteria and the algae; third, absolute aggregation rates and therefore absolute colloid stability values are in a similar order of size for both types of microorganisms at comparable salt concentrations (taking differences in number concentration and energy dissipation into account); fourth, the stabilizing effect of increasing salt concentration is the same for both types of counter ions; fifth, at identical concentrations, the aggregation rate is lower for Ca^{+2} than for Na^{+}, for both types of microorganisms; and sixth, at concentrations "extrapolated to zero" (distilled water) algae aggregated immediately as mentioned above. - All experiments with bacteria reported here were done at pH = 4, all experiments with algae at pH = 7.

3. Effect of type of counter ion.

In order to demonstrate clearly the different effects of Na^{+} and Ca^{+2} upon microorganism aggregation, colloid stability values have been calculated from the observed aggregation rates. (Method of calculation and assumptions implied in such calculations have been described by (20).) Fig.4 shows calculated colloid stability values as function of

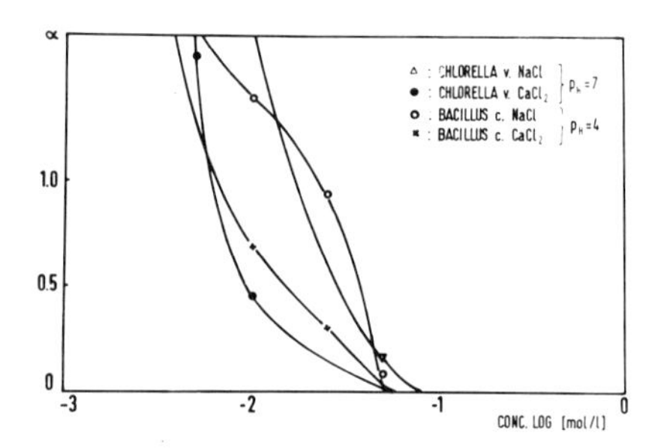

Fig. 4. Calculated colloid stability values for B. cereus (pH = 4) and Chl. vulgaris (pH = 7) for NaCl and $CaCl_2$ at varying concentrations.

concentration of Na^{+} and Ca^{+2} for B. cereus (pH = 4) and for Chl. vulgaris (pH = 7). It is found that the characteristics of the counter ion, dosed to identical concentrations, has a marked effect. Furthermore, one can deduce that Ca^{+2} leads to a higher colloid stability than Na^{+} at identical concentration. Finally it can be stated that this effect is observed for both types of microorganisms. The change in colloid stability when at identical concentrations Na^{+} is substituted for Ca^{+2}, or vice versa, is in the same order of size in bacterial and algal suspension.

IV. DISCUSSION

1. Comparison with existing coagulation theories.

A juxtaposition of colloid stability values, calculated for various pH values at given Na^+ concentration, with electrophoretic mobility values of B. cereus (reported by (21)), Fig.5, shows that stability is significantly determined by charge effects on the cell surface. This same observation is made in the case of Chl. vulgaris suspensions (surface charge values reported by (22)). -

While the pH dependence of microorganism stability seems to be in agreement with known inorganic colloid stability phenomena, the effect of the counter ion concentration is opposite to predictions on the basis of double layer theory (23) and empirical evidence (24): increase of salt concentration leads to an increase of colloid stability within the concentration range investigated (10^{-3} molar to 10^{-1} molar - see Figures 2 and 3). This is observed for both types of microorganisms.

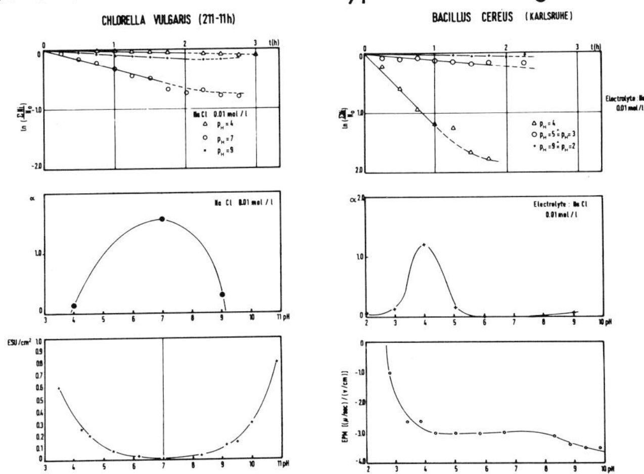

Fig. 5. Comparison of colloid stability and reported EPM, respectively charge density values.

Similarly the different effects of Na^+ and Ca^{+2}, a higher stability when at identical concentrations the (higher charged) Ca^{+2} is used, contradicts known hypotheses on colloid coagulation (see Fig.4). Here too, both types of microorganisms show the same phenomena. The results of Rubin et al. ((15), (16)) could not be confirmed. The obeyance of the Schulze-HardyRule is not given in this instance. Physiological conditions, here set to resemble natural systems (viability was maintained in all aggregation experiments), might have been significantly different in the former study. Conversely Santoro et al. (14) note also a narrower size distribution of bacterial suspensions, i.e. higher aggregation rates, for Na^{+} than for Ca^{+2}.

Summarizing one could state that colloid stability, as defined by Fuchs (25), is for bacteria and algae, as investigated in this study, determined to a significant extent by surface charge phenomena. - On the other hand, cell aggregation appears to be independent of surface charges on the cell. This might be confirmed by testing series of differently charged ions and determining electrophoretic mobility and aggregation rates of physiologically unaltered microorganisms.

2. Mechanisms of microorganismic aggregation.

As suggested by electronmicroscopic pictures of aggregated bacterial suspensions, taken parallel with aggregation experiments, bacteria aggregate through polymers released by the cells. Such explanations have been put forward by Tenney et al. (8), Busch et al. (9), Harris (10), Wuhrmann et al. ((11), (12)), McGregor et al. (13) and Santoro et al. (14), amongst others. In other instances it was also found that such cell material hindered aggregation. It seems reasonable to consider type of bacteria, culture conditions and characteristics of chemicals dosed as decisive parameters in the aggregation of bacteria through cell-generated polymers.

Literature on aggregation of algae is not abundant. Algal flocculation appears to have been investigated with precipitating hydroxides and polymeric flocculation aides (Ives (18), McGarry (19)). The question whether cell-born substances could be released under certain conditions and then react , has not been mentioned in those studies. - In this investigation it was observed that algae aggregated without the addition of any chemicals, i.e. in distilled water. From this phenomenon and the fact, that all effects observed with bacterial suspensions could also be noted in algal suspensions, it could be concluded that algae, too, aggregate and escape aggregation through material generated and released by the cells under specific conditions.

In summary one must note that no conclusive explanation for the mode of aggregation of microorganisms exists. The existing hypothesis of the destabilizing and stabilizing role of polymeric material released by cells could not be proven wrong; rather, the here presented data seem to confirm it in an indirect way. Furthermore, the data suggest that aggregation of algae might occur in a similar fashion: the quantitatively described relationship between aggregation rate and concentration and type (charge) of chemical added can best be explained by such hypotheses. Such hypotheses could be somewhat more confirmed if a quantitative relationship between cell material released and extent of aggregation was shown and if the same material could be proven to aggregate other colloids.

3. Conclusions.

Aggregation of microorganisms is determined by aspects of colloid stability (if stability is defined analogously to Fuchs) and the release of flocculating or coating cell material. Adjustment of cell stability through variation of the surface charge might be a necessary condition for aggregation; the release of floc-forming cell material in specific dosages and modes, however, appears to be a sufficient condition for aggregation of cells that remain physiologically intact.

Comparing values of microbial colloid stability calculated from the observed aggregation rates with those reported for inorganic colloids (26) one finds that they are in a similar order of size. Table (1) furthermore shows that the stability of clays decreases with increasing salt concentration and charge while that one of microorganisms increases.

TABLE 1

Calculated Colloid Stability Values for Organic and Inorganic Colloids at pH ~ 7 and Varying Salt Concentrations

Mol/l pH = 7		Bacillus cer.	Chlorella v.	Kaolin (5)
Na^+	10^{-2}	(0.01)	1,6	0.03
	10^{-3} RHEIN	(0.005)	–	0.015
	10^{-1}	$(< 10^{-5})$	$< 10^{-5}$	0.17
Ca^{++}	$5 \cdot 10^{-3}$	(0.01)	1,5	0.38
	10^{-2}	(0.0028)	0.4	0.38

If inorganic and organic suspensa are removed to similar extents in fresh water systems, the latter one will most likely remain much longer in suspension than clays in brackish or sea water. (The question of mutual aggregation of different colloids has not been considered in this context.) - If not only simple inorganic ions of low charge constitute the ionic medium but also complex organic material as found in 'polluted' systems, then an additional stabilization similarly as in the case of inorganic colloids is observed (Table 1).

V. REFERENCES

(1) Ettner, R.J., Hoyer, R.P., Partheniades, E., and Kennedy, J. F., Am. Soc. Civil Eng. Hydraulics Div. Proc., V 94, no HY6 1439 (1968).
(2) Weiss, C.M., J. of Sewage and Industrial Wastes 23, 227 (1951).
(3) Hahn, H.H. and Stumm, W., Am. J. Science 268, 354 (1970).
(4) O'Melia, C.R., in "Physiochemical Processes" (W.J. Weber Jr, Ed.) p. 61 Wiley Interscience, New York (1972).
(5) Neis, U., "Experimentelle Bestimmung der Stabilität anorganischer Schwebstoffe in natürlichen Gewässern" Dissertation, University of Karlsruhe (1974).
(6) Edzwald, J.K., Upchurch, J.B. and O'Melia, C.R., Environ. Sci. Technology 8, 58 (1974).
(7) Hahn, H.H., Dickgießer, G. and Klute, R., Wasserwirtschaft (sent in for publication).
(8) Tenney, M.W. and Stumm, W., J. WPCF 37/10, 1370 (1965).
(9) Busch, P.L. and Stumm, W., Environ.Sci.Technology 2,49 (1968).
(10) Harris, R.H., "Aggregation of Microorganisms" Dissertation, Harvard University / Cambridge (1972).
(11) Wuhrmann, K., and Fazeli, A. in "Flockung", Veröffentl. Wasserchemie der Universität Karlsruhe (H. Sontheimer, Ed.) Karlsruhe (1967).
(12) Peter, G. and Wuhrmann, K. in "Advances in Water Pollution Research" II-1 (S.H. Jenkins, Ed.), Pergamon Press, Oxford (1971).
(13) McGregor, W.C., and Finn, R.K., Biotechnology and Bioengineering Vol. XI, 127 (1969).
(14) Santoro, T., Stotzky, G., Archives of Biochemistry and Biophysics, 122, 664 (1967).
(15) Rubin, A.J., and Hanna, G.P., Environ. Sci. Technology 2, 358 (1968).
(16) Rubin, A.J., Hayden, P.L., Hanna, G.P., Water Research 3, 843 (1969).
(17) Bradley, R. A., and Krone, R.B., J. San.Eng.Div. 97 (1971).
(18) Ives, K.J., J. of Biochem. Microbiol. Technology and Engineering, 1, 37 (1959).
(19) McGarry, M.G., J. WPCF, 42, 191 (1970).

(20) Hahn, H.H., Stumm, W., J. Colloid Interface Sci. 28, 139 (1968).

(21) Lemp, J.F., Asbury, E.D., and Ridenouv, E.O., Biotechnology and Bioengineering, XIII, 17 (1971).

(22) Ives, K.J., Soc. Water Treatment and Examin., Sutton 5, 41 (1959).

(23) Overbeek, J.T.G. in "Colloid Science" (H.R. Kruyt, Ed.), p. 115, Elsevier, Amsterdam (1952).

(24) Van Olphen, H., "An Introduction to Clay Chemistry", Interscience, New York (1963).

(25) Fuchs, N., Z. Physik, 79, 736 (1934).

(26) Eppler, B., Neis, U. and Hahn, H.H. in "Progress in Water Technology" (S.H. Jenkins, Ed.) p. 207, Pergamon Press, Oxford (1975).

VI. ACKNOWLEDGEMENTS

The authors gratefully acknowledge the financial assistance obtained from the Deutsche Forschungsgemeinschaft (SFB 80, Ha 679/6) and the invaluable analytic help of U. Pagel.

REVERSAL OF CHARGE AND FLOCCULATION OF MICROCRYSTALLINE CELLULOSE WITH CATIONIC DEXTRANS

Christer Heinegård
Tom Lindström
Christer Söremark
Swedish Forest Products Research Laboratory

I. SYNOPSIS

A suspension of microcrystalline cellulose has been used as model for a cellulose surface. The amounts of cationic dextrans of different molecular weights and degrees of substitution required for charge neutralization of cellulose particles and for flocculation have been investigated. The influence of pH, salt concentration and sol concentration on the amount of dextran required for charge neutralization has also been investigated. The results show that the amount required for charge neutralization decreases when the molecular weight or the degree of substitution is increased and when pH, salt concentration or sol concentration is decreased. The results are interpreted in terms of changes in the conformation of the polymer in solution. The results suggest that destabilization of the cellulose sol is accomplished by a charge neutralization mechanism.

II. INTRODUCTION

During recent years there has been a growing interest in cellulose/polymer interactions as a result of the wider use of polymers as wet-end additives in the papermaking process. These additives are commonly used to improve water drainage and retention of fines and fillers during the formation of the fibrous mat. Currently, the use of cationic polymers for this purpose is receiving particular attention.

The mode of action of cationic polymers in negatively charged colloids has been extensively studied, principally for inorganic and latex suspensions. Flocculation in these systems may be accomplished by charge neutralization (1-3), bridge formation (4-5) or a combination of both mechanisms (6-7).

Some work has been done to determine the relative importance of each mechanism for the destabilization of cellulose suspensions (8-9). A new flocculation mechanism for cellulose suspensions under kinetic conditions has also been proposed (10). The term "flocculation" will be employed regardless of the true destabilization mechanism.

In this study a microcrystalline cellulose was chosen as a model for cellulose surfaces (11, 12).

The cationic polyelectrolytes used in this study were two series of modified dextrans: one series with different degrees of substitution (D.S.) and one with different molecular weights (M_W).

The parameters measured were the viscosity of the polyelectrolytes, the amount needed for charge neutralization of the cellulose and the electrophoretic mobility as a function of molecular weight, degree of substitution, pH, salt concentration and finally sol concentration. The adsorption on cellulose of the polyelectrolytes used was also determined. In a separate experiment the extent of flocculation was determined as a function of molecular weight and degree of substitution of the polyelectrolytes.

III. EXPERIMENTAL

A. Materials

The microcrystalline cellulose used in this study was manufactured by Sigma Chemical Company, USA, and delivered as a dry powder labelled Sigmacell type 38. Before use, the microcrystalline cellulose was suspended in water at a concentration of 200 g/l and dialyzed against distilled water for 3 days. The particle size of the suspended particles was approximately 38 μm according to the manufacturer. The specific surface of the dry powder was determined by argon adsorption to be 1.32 m^2/g.

The diethylaminoethylethers of dextran were prepared as described by Painter (13), with dextrans procured from Pharmacia Fine Chemicals, Sweden. The structure of the cationic dextrans is illustrated in Fig. 1. Two series of dextrans were prepared. One with constant degree of substitution (0.10) and varying molecule weight ($2.2 \cdot 10^4$, $6.3 \cdot 10^4$, $1.5 \cdot 10^5$, $2.4 \cdot 10^5$) and the other with constant molecular weight ($1.5 \cdot 10^5$) and varying degree of substitution (0.03, 0.07, 0.10).

$$Cl^- \; H-\overset{C_2H_5}{\underset{C_2H_5}{N^+}}-CH_2-CH_2-O-DEXTRAN$$

Fig. 1. Structure of cationic dextrans.

The molecular weight was reported by the manufacturer and represents the virgin polymer and the degree of substitution was determined by Kjeldahl nitrogen analyses.

All chemicals used were of the highest grade commercially available. In all experiments, distilled and ion-exchanged water was used with a specific conductivity less than $2 \cdot 10^{-6}\ \Omega^{-1},\ cm^{-1}$.

B. Methods

Stable sols of microcrystalline cellulose were prepared according to the following procedure. An aqueous suspension containing 1 g/litre (pre-dialyzed) was mechanically disintegrated (Polytron "Kinematisches Hochfrequenz-Gerät", manufacturered by Kinematica GmbH, Switzerland) for 10 minutes. Subsequent centrifugation at 1,000 g for 10 minutes gave a supernatant liquid that contained a sol that was stable for weeks. Scanning electron micrographs showed the sol particle to be irregular in shape, with diameters ranging from 1 to 5 μm.

The final concentration of the sol prepared in this way was determined from several freeze-drying experiments to be 110 ± 5 mg/l.

Flocculation was considered to have occurred if the sol had settled out within 24 hours. In the experiments showing how different parameters (M_W, D.S., pH, NaCl) affect the amount needed for charge neutrality ("cationic demand") the final concentration of the dispersion was in all cases 10 g/l. These dispersions were not disintegrated prior to use.

Viscosity measurements were performed using Cannon-Ubbelohde Semi-Micro Dilution Viscosimeters (Cannon Instrument Co., USA). No shear rate dependency was found for the highest molecular weight cationic dextran (D.S. = 0.1, $M_W = 2.4 \cdot 10^5$).

The pH-value was not adjusted. It was thus approximately 6, except for one series of experiments, where HCl and NaOH were used.

The electrophoretic mobilities of the particles were determined by using a particle electrophoresis instrument (Rank Brothers, England) equipped with a flat cell. In all experiments the electrophoretic mobility was determined at equilibrium, i.e. 24 hours.

The adsorption experiments were carried out in thermostated (20 ± 1°C) adsorption vessels, equipped with a mechanical stirrer to ensure good mixing. The adsorption values (3h) were found to be unaffected by the stirring rate if the stirring rate was decreased 5 times. The mixture of the predialyzed non-disintegrated microcrystalline cellulose (6 g/l) and polymer was stirred for 3 hours at 20°C. This time was chosen on the basis of test runs showing that after 1 h the adsorption value had reached more than 95 per cent of its final value (24 h). Parallel blank runs with only cellulose and polymer respectively were run. After 3 hours the particles were separated by centrifugation and subsequent filtration of the supernatant on 0.1 μm nuclepore membrane filter (Wallabs, California).

The concentration of the polyelectrolyte determined by

measuring the refractive index with a differential refractometer (Waters Associates, England). The reproducibility of the concentration determination was found to be within ± 1 mg/l of polymer.

IV. RESULTS AND DISCUSSION

A. Reversal of Charge

1. *Molecular Weight*

The efficiencies of cationic dextrans of different molecular weights (M_W) in recharging a dispersion of microcrystalline cellulose are illustrated in Fig. 2. As can be seen the polyelectrolyte with the highest molecular weight is more effective in recharging the particles.

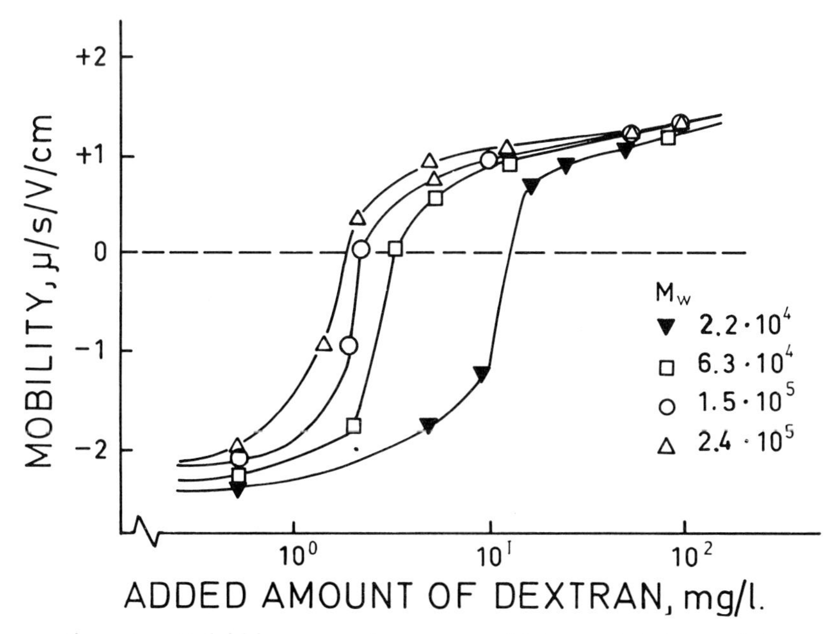

Fig. 2. Mobility of microcrystalline cellulose (10 g/l) versus added amount of cationic dextrans of different molecular weights.

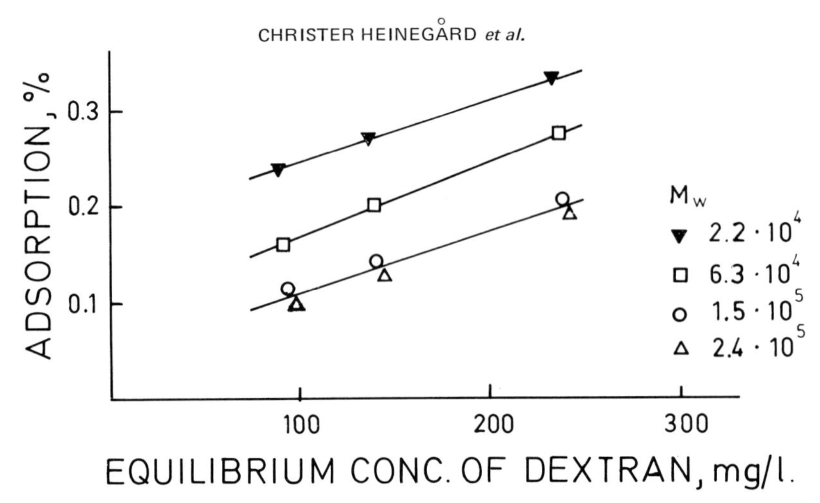

Fig. 3. Adsorption on microcrystalline cellulose (6g/l) versus equilibrium concentration of cationic dextrans of different molecular weight.

Up to the zero point of charge the adsorption of the two lowest molecular weight samples was almost complete.

Thus the decrease with increasing molecular weight in the amount of polyelectrolyte required for charge reversal is accompanied by a corresponding decrease in adsorption. Moreover adsorption experiments (Fig. 3) revealed that an increase in molecular weight led to a decrease in the amount of polymer adsorbed onto the cellulose particles at polymer additions exceeding that required to pass the zero point of charge. This can be interpreted in terms of the repulsive forces between the polymer and the surface of the sol at dosages exceeding the isoelectric point (10).

As the adsorption is almost complete below the zero point of charge, the total ionic exchange capacity of cellulose can be compared with that of the adsorbed amount of cationic polymer.

The total carboxyl content of microcrystalline cellulose was determined to be 0.8 μeqv/100 g (14). If this is compared with the total amount of added cationic groups at the isoelectric point, this will give values ranging from 10.2 μeqv/100 g cellulose ($M_w = 2.4 \cdot 10^5$) up to 74 μeqv/100 g cellulose ($M_w = 2.2 \cdot 10^4$). Thus only a small fraction of the total number of charged sites of the cellulose dispersion needs to be neutralized.

An analogous influence of molecular weight on the cationic demand for cellulose dispersion has also been observed in the case of cationic polyacrylamides (11).

In that case results were interpreted in the light of the shape factor of the polyelectrolyte in solution.

Thus the higher the molecular weight the more expanded is the polymer in solution, so that it can be adsorbed in a more expanded conformation and thus cover more negative sites on the cellulose surface. It has been pointed out elsewhere (15-18) that a qualitative relationship exists between the configuration in solution and in the adsorbed state.

2. *Degree of substitution*

As expected an increase in the degree of substitution (D.S.) reduces the amount of cationic dextran required for reversal of charge (Fig. 4). Other authors have found the same influence of D.S. (10-12, 19-20).

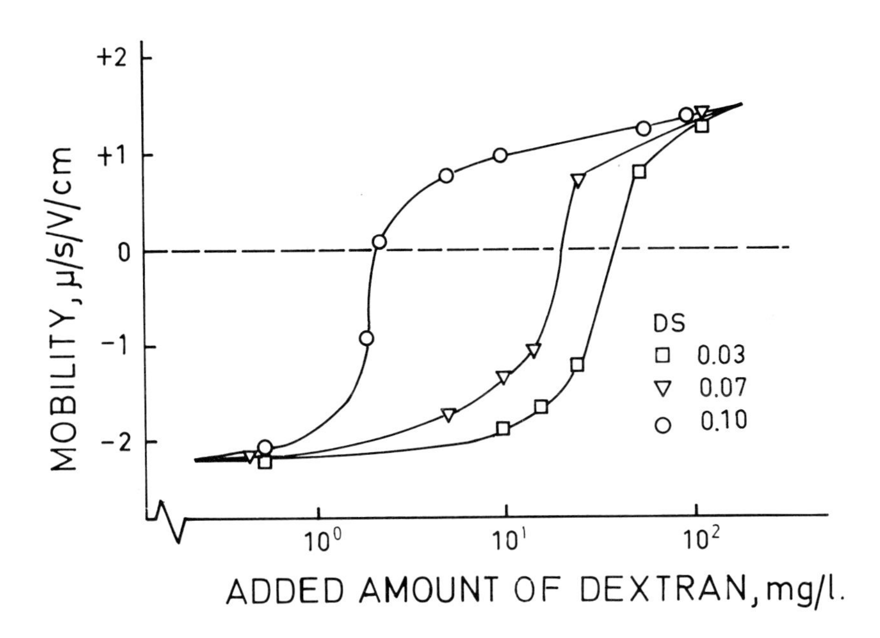

Fig. 4. Mobility of microcrystalline cellulose (10 g/l) versus added amount of cationic dextrans of different degrees of substitution.

The effect of D.S. on the adsorbed amount at high polymer dosages is given in Fig. 5. Thus an increase in charge density led to a decrease in the adsorbed amount of cationic polymer.

It was also noticed from the adsorption experiments that adsorption was almost complete below the zero-point of charge.

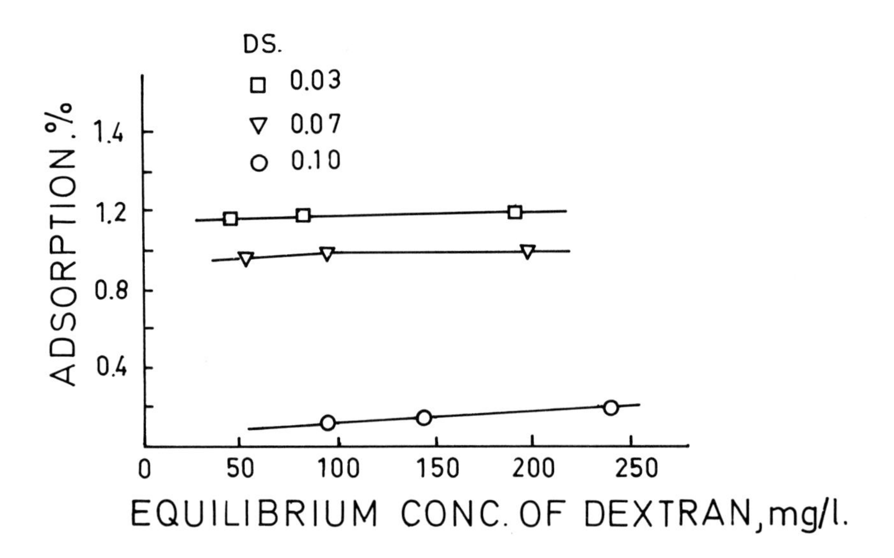

Fig. 5. Adsorption on microcrystalline cellulose (6 g/l) versus equilibrium concentration of cationic dextrans of different degrees of substitution.

No quantitative relationship exists between D.S. and cationic demand and thus the shape factor of the molecule has to be taken into account.

Viscosity data show that the molecule expands with the degree of substitution owing to electrostatic repulsion between the charged amino-groups (Fig. 6).

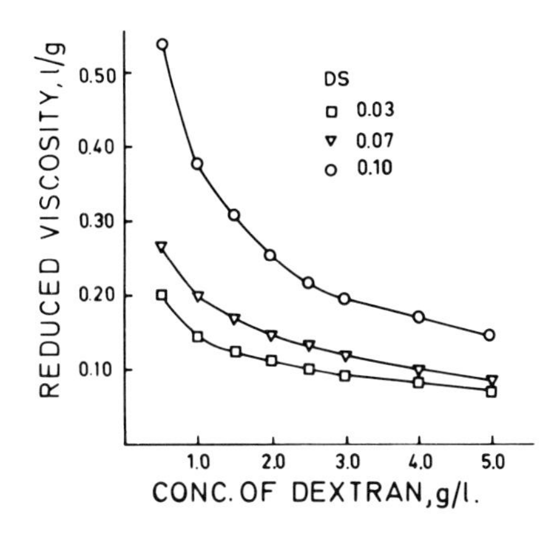

Fig. 6. Reduced viscosity versus concentration of cationic dextrans of different degrees of substitution.

3. pH

The influence of the pH-value on the efficiency of the cationic dextran in recharging the sol is illustreated in Fig. 7. As seen in the figure, a decrease in pH decreases the amount required for charge neutralization. This behaviour may be interpreted in the following way.

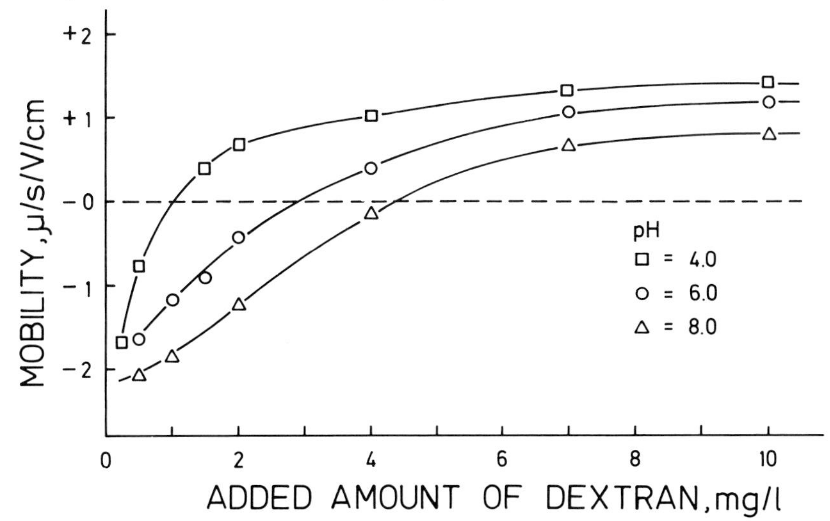

Fig. 7. Mobility of microcrystalline cellulose versus added amount of cationic dextran ($M_w = 1.5 \cdot 10^5$, D.S. = 0.1) at different pH values.

With increasing pH from 4 to 8 the degree of substitution decreases due to a deprotonization of the tertiary amino groups (8) on the dextran molecule. This also means that the polymer tends to coil. The influence of pH on the reduced viscosity is illustrated in Fig. 8. These two factors will imply a higher addition of dextran to reach charge neutrality of the particles when the pH is increased. Furthermore, at a pH of 4, the cellulose is more or less protonized, and thus the surface charge decreases (21). This also contributes to mean that less polyelectrolyte is required for charge neutralization.

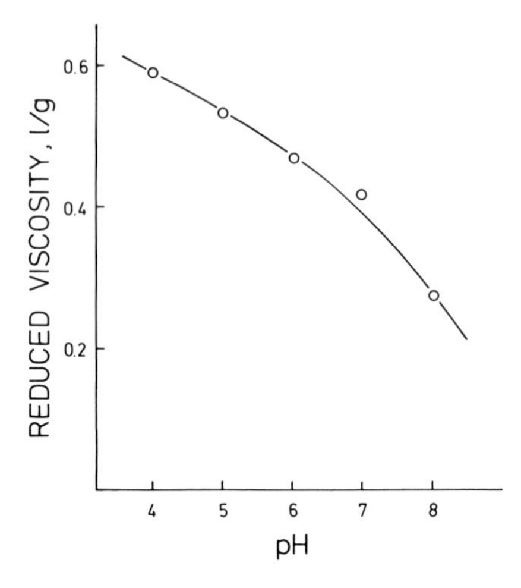

Fig. 8. Reduced viscosity versus concentration of cationic dextran (M_w = = 1.5 • 10^5, D.S. = 0.1) at different pH values.

4. *Salt concentration*

To illustrate the effect of ionic strength on cationic demand, sodium chloride has been used. As seen in Fig. 9, there is an increase in the amount of polyelectrolyte for charge neutralization with increasing amount of salt. When salt is added to the polyelectrolyte the cationic charges on the dextran are shielded and the polymer tends to "coil" (Fig.10), i.e. the cationic demand increases, since the polymer might be imagined to be adsorbed in a more compressed conformation.

5. *Sol concentration*

Finally, the influence of the concentration of the cellulose dispersion on the amount required for charge neutrality has been studied. The results are summarized in Table 1.

TABLE 1
Added amount of cationic dextran ($M_W = 1.5 \cdot 10^5$; D.S. = 0.1) required for charge neutrality at different concentrations of cellulose dispersion.

Cellulose Concentration g/l	0.1	1	5	10	20	50	100
Added amount %	0.018	0.018	0.020	0.022	0.032	0.038	0.042

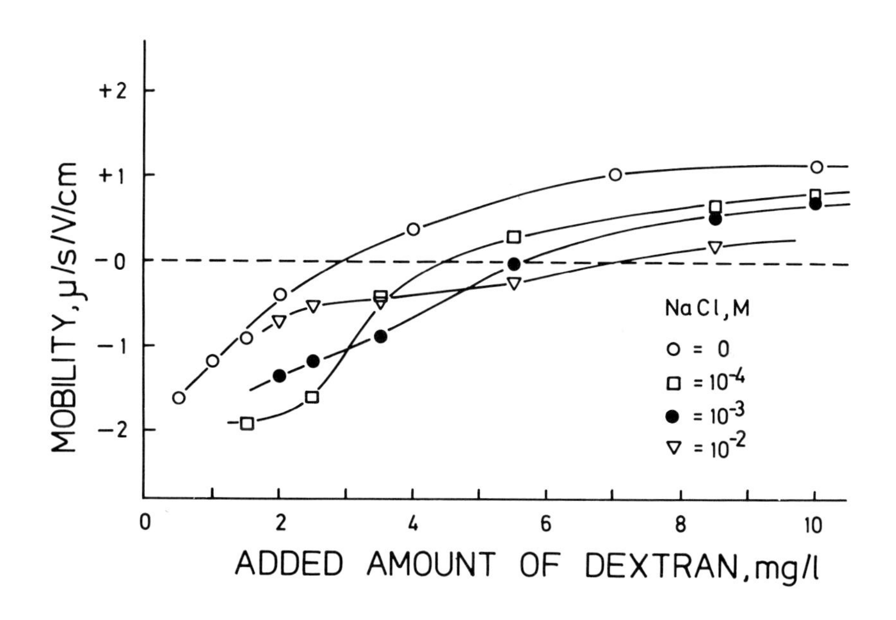

Fig. 9. Mobility of microcrystalline cellulose versus added amount of cationic dextran ($M_w = 1.5 \cdot 10^5$, D.S. = 0.1) at different additions of NaCl.

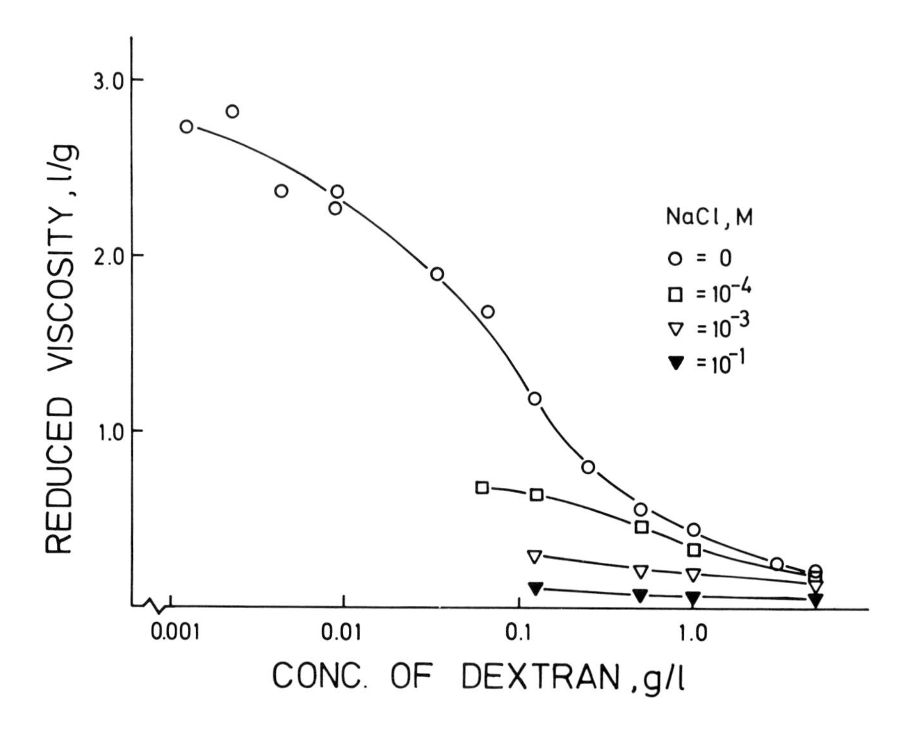

Fig. 10. Reduced viscosity versus concentration of cationic dextran ($M_w = 1.5 \cdot 10^5$, D.S. = 0.1) at different additions of NaCl.

As seen in the table, more polymer (calculated as weight-%) has to be added with increasing concentration of the cellulose dispersion in order to obtain charge neutrality.

A qualitative explanation for this can be found in Fig. 10, where the reduced viscosity of the polymer at different concentrations is shown. The lower the concentration of the dispersion, the less polymer is required because of the more expanded conformation of the polymer in solution.

B. Destabilization

In a separate series of experiments the influence of molecular weight and degree of substitution of the cationic dextrans on the destabilization and restabilization of microcrystalline cellulose was investigated. The stable sol had a final concentration of 100 mg/l and was prepared as described above.

1. *Molecular Weight*

The stability domains of the cellulose sol with respect to molecular weight are shown in Fig. 11.

The hatched area is the flocculation region. The open area at low dextran dosage is the region in which flocculation does not occur while the upper open area is the restabilization region. The broken line through the hatched area shows the condition for zero point of charge.

Compared with the non-disintegrated system (c.f. Table 1) the amount of polyelectrolyte required for charge neutralization, calculated on the basis of the weight of the solid, is increased to almost two orders of magnitude owing to the greater specific surface area of the stable sol.

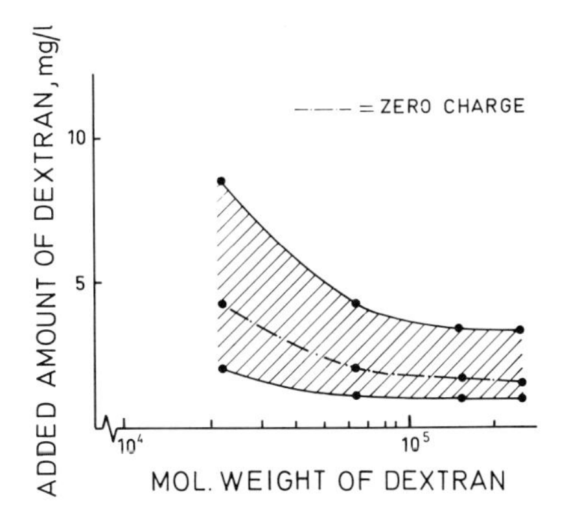

Fig. 11. Flocculation domain of microcrystalline cellulose (100 mg/l) versus molecular weight of cationic dextrans (D.S. = 0.1).

As seen in Table 2, the sol is destabilized symmetrically around the point of zero charge.

TABLE 2
Mobilities at the boundaries of destabilization and restabilization of sol of microcrystalline cellulose (100 mg/l) at different molecular weights and degrees of substitution of the polyelectrolytes.

Polyelectrolyte		Mobility, μ/s/V/cm	
M_W	D.S.	Destabilization region	Restabilization region
$2,2 \cdot 10^4$	0.10	- 1.0	+ 1.0
$6,3 \cdot 10^4$	0.10	- 1.2	+ 1.0
$1,5 \cdot 10^5$	0.10	- 1.1	+ 1.0
$2,4 \cdot 10^5$	0.10	- 1.1	+ 1.0
$1,5 \cdot 10^5$	0.03	- 1.1	+ 1.0
$1,5 \cdot 10^5$	0.07	- 1.0	+ 0.9

As regards the added amount of dextran, however, the conditions for charge neutrality occur much closer to the destabilization than to the restabilization boundary (Fig. 11). This asymmetric behaviour must be due to a more restricted adsorption of the dextran above zero charge.

2. *Degree of Substitution*
The stability domains of the sol as a function of degree of substitution are shown in Fig. 12.

Corresponding mobility data at the destabilization and restabilization boundaries are presented in Table 2. As in the case of molecular weight the sol is flocculated between mobilities of about - 1.0 and + 1.0 μ/s/V/cm.

As regards the added amount, the flocculation region around zero charge is somewhat more symmetrical the lower the degree of substitution. This indicates that the adsorption above zero point of charge is less restricted at low D.S. Furthermore, it can be seen that the flocculation region around the point of zero charge is more restricted at high D.S., although the charge at the boundaries of the region is independent of D.S.

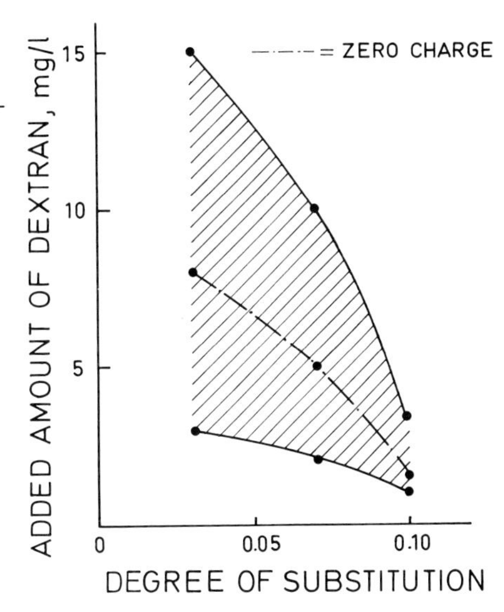

Fig. 12. Flocculation domain of microcrystalline cellulose (100 mg/l) versus degree of substitution of cationic dextrans ($M_w = 1.5 \cdot 10^5$).

V. FINAL COMMENTS

All data presented have been interpreted according to the conformation of the polymer in aqueous solutions, i.e. the more expanded the polymer is the lower is the cationic demand. That this interpretation may be adequate is supported by the fact that the saturation adsorption of the polymer decreases with increasing expansion of the polymer in solution.

The results from this study indicate that the flocculation mechanism is due to a charge neutralization process. This is primarily based on the fact that the destabilization occurs symmetrically around the zero point of charge, irrespective of molecular weight and degree of substitution.

VI. CONCLUSIONS

The amount of cationic dextrans with tertiary amino-groups required for charge neutralization of microcrystalline cellulose decreases when

- molecular weight is increased
- degree of substitution is increased
- pH is decreased
- salt concentration is decreased
- sol concentration is decreased

Furthermore the flocculation of a stable sol of microcrystalline cellulose seems to be due to an overall charge neutralization of the particle, throughout the ranges of molecular weights and degrees of substitution investigated.

VII. ACKNOWLEDGEMENTS

Thanks are due to Mrs Gunborg Glad-Nordmark for her skillful and careful assistance with the experimental work throughout this study.

VIII. REFERENCES

1. Dixon, J.K., La Mer, V.K., Li, C., Messinger, S., and Linford, H.B., J. Colloid Interface Sci. 23, 465 (1967).
2. Slater, R.W., Clark, J.P., and Kitchener, J.A., Proc. Brit. Ceram. Soc. 13, 1 (1969).
3. Gregory, J., Trans. Faraday Soc. 65, 2260 (1969).
4. Shyluk, W.P., and Smith, R.W., J. Polym. Sci. A-2, 7, 27 (1969).
5. Kragh, A.M., and Langston, W.B., J. Colloid Sci. 17, 101 (1962).
6. Black, A.P., Birkner, F.B., and Morgan, J.J., J. Colloid Interface Sci. 21, 626 (1966).
7. Ries, H.E., Jr., and Meyers, B.L., J. Appl. Polym. Sci. 15, 2023 (1971).
8. Lindquist, G.M., and Stratton, R.A., J. Colloid Interface Sci. 55, 45 (1976).

9. Chang, M.Y., and Robertson, A.A., Pulp Paper Mag. Can. 68, T-438 (1967).
10. Lindström, T., and Söremark, Ch., J. Colloid Interface Sci. 55, 305 (1976).
11. Lindström, T., Söremark, Ch., Heinegård, C., and Martin-Löf, S., Tappi 57, 94 (1974).
12. Sandell L.S., and Luner, P., J. Appl. Polym. Sci. 18, 2075 (1974).
13. Painter, T.J., J. Chem. Soc. Sect. C., 922 (1967).
14. Wilson, K., Svensk Papperstidn., 69, 386 (1966).
15. Greene, B.W., J. Colloid Interface, Sci. 37, 144 (1971)
16. Chan, F., and Robertson, A., J. Colloid Interface Sci. 53, 586 (1970).
17. Lipatov, Y.U.S., and Sergeeva, L.M., "Adsorption of Polymers", Wiley, New York, 1974.
18. Peyser, P., and Ullman, R., J. Polym. Sci. A3, 3165 (1965).
19. Schmidt, W., and Erich, F.R., J. Phys. Chem. 66, 1907 (1962).
20. Lindström, T., and Söremark, Ch., Das Papier 29, 519 (1975).
21. Edelson, M.R., and Hermans, Jr., I., J. Polym. Sci. C 2, 145 (1963).

Comparison of the Adsorption on Hydrous Oxide Surfaces of Aquo Metal ions with that of Inert Complex Cations.

Felix Dalang and Werner Stumm.
EAWAG Swiss Federal Institute of Technology
Zürich, Switzerland.

I. SYNOPSIS

The properties of the phase bundary between an oxide surface and an electrolyte solution are dependent (i) on the forces operating on ions and H_2O molecules by the electrified surface; and, (ii) on the forces operating on the solid surface by the electrolyte. In order to gain better insight into the energies of electrostatic and chemical interaction we wish to compare the adsorption behavior at hydrous oxide surfaces of aquo metal ions (transition metals cations) with that of small robust cation complexes such as $Co(NH_3)_6^{3+}$, $Co(en)_3^{3+}$, $Co(NH_3)_5Cl^{2+}$ (these robust complexes are kinetically inert cations which do not dissociate or hydrolyze and which cannot form covalent bonds). Heavy metal cations are specifically adsorbed at hydrous oxide surfaces, i.e., they form chemical bonds with the $\equiv MeO^-$ surface groups; the pH dependence of the interaction can be explained by the pH dependence of the activity of the $\equiv MeO^-$ groups and the affinity of these groups to coordinate with the metal ions. (1-4). The Co(III) complexes, on the other hand, are adsorbed primarily electrostatically to form "outer sphere" type of adducts; adsorption is almost exclusively a function of the surface charge density $\{\equiv MeO^-\}$. These complexes cannot cause charge reversal; they can be used conveniently to determine the maximum exchange capacity of $\equiv MeOH$ groups.

II. AMPHOTERIC PROPERTIES OF THE $\equiv Me-OH$ GROUP

The surfaces of metal or metalloid hydrous oxides are generally covered with OH groups. The acid-base or amphoteric properties of the hydrous oxide surface can be characterized with the help of alkalimetric or acidimetric titration curves in analogy with soluble monoprotic or polyprotic acids; the pH-dependent charge of an oxide results from proton transfers at the surface:

$$\equiv MeOH_2^+ \rightleftarrows \equiv MeOH + H^+ \;;\; K_{a_1}^s = \{\equiv MeOH\}[H^+]/\{\equiv MeOH_2^+\} \quad (1)$$

$$\equiv MeOH \rightleftarrows \equiv MeO^- + H^+ \;;\; K_{a_2}^s = \{\equiv MeO^-\}[H^+]/\{\equiv MeOH\} \quad (2)$$

where [ ] and { } indicate concentrations of species in the aqueous phase (moles dm^{-3}) and concentrations of surface species (moles kg^{-3}), respectively. That portion of the charge due to specific interaction with H^+ and OH^- ions, corresponds to the difference of protonated and deprotonated ≡MeOH groups and is available from the titration curve on the basis of the proton condition (or charge balance). The free energy of deprotonation consists of the dissociation as measured by an intrinsic acidity constant, K_a^s (intr.) and the removal of the proton from the site of the dissociation into the bulk of the solution as expressed by the Boltzman factor; thus

$$K_{a_1}^s = K_{a_1}^s \text{ (intr.) } \exp(F\Psi_s/RT) \quad (3)$$

where Ψ_s is the effective potential difference between the surface site and the bulk solution.(2,5,6) K_a^s (intr.) is the acidity constant of an acid group in a hypothetically completely chargeless surrounding. There is no direct way to obtain Ψ_s theoretically or experimentally. It is possible, however, to determine the microscopic constants experimentally and to extrapolate these constants to zero surface charge in order to obtain intrinsic constants.[6]

Fig. 1 gives the intrinsic values for the acidity constants for γ-Al_2O_3 (ionic strength I=0,1). The zero point of charge[7] is given by

$$pH_{ZPC} = \frac{1}{2} [pK_{a_1}^s \text{ (intr.) } + pK_{a_2}^s \text{ (intr.)}] \quad (4)$$

Table 1 lists intrinsic constants for some hydrous oxides. The pH_{ZPC} obtained this way should agree with the common intersection point of the titration curves obtained with different concentrations of

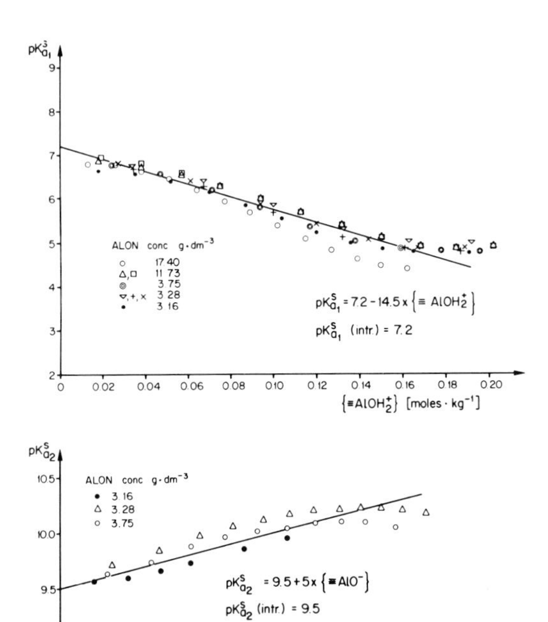

Fig.1. Acidity constants of the surface groups $\equiv AlOH_2^+ \rightleftarrows AlOH + H^+$; $K^s_{a_1}$ and $\equiv AlOH \rightleftarrows \equiv AlO^- + H^+$; $K^s_{a_2}$

The microscopic acidity constants are a function of the surface charge, i.e. of $\{\equiv AlOH_2^+\}$ and $\{\equiv AlO^-\}$ in the acid and alkaline region, respectively. Extrapolation to zero charge conditions gives intrinsic acidity constants.

inert salt.

TABLE 1

Acidity Constants of Surface OH-Groups (25°C)

Group	Solid phase	Electro-lyte	$pK^s_{a_1}$ (intr.)	$pK^s_{a_2}$ (intr.)	Ref.
SiOH	amorph. SiO_2	$.1MNaClO_4$	-	6.8	5
		$.5MNaClO_4$	-	7.2	5
		$1MNaClO_4$	-	7.2	5
MnOH	$\delta-MnO_2$	$0.1MNaClO_4$	-	6.8	1
TiOH	Anatase	$3MNaClO_4$	4.98	7.8	6
AlOH	$\gamma-Al_2O_3$	$.1MNaClO_4$	7.2	9.5	3

III. INTERACTION OF AQUO METAL IONS WITH HYDROUS OXIDE SURFACES

As has been pointed out by James and Healy[8] and others, metal ion adsorption on oxides cannot be accounted for in terms of a simple electric double layer model (Gouy-Chapman theory). Because many metal cations can also be adsorbed to hydrous oxides even against electrostatic repulsion,in these cases, the chemical interaction energy offen must predominate over the coulombic interaction energy.

Fig. 2 illustrates the binding (adsorption) of Pb^{2+} on hydrous $\gamma-Al_2O_3$[3]; although positively charged under these pH conditions, the Al_2O_3-surface removes most efficiently Pb^{2+} cations from the solution. In the pH range considered Pb^{2+} does not hydrolyze to any substantial extent.

The complex forming properties of the oxide can be characterized quantitatively by the following surface coordination reactions

$$\equiv MeOH + M^{z+} \rightleftarrows \equiv MeOM^{(z-1)} + H^+$$

$$*K_1^s = \{\equiv MeOM^{(z-1)}\}[H^+]/\{\equiv MeOH\}[M^{z+}] \quad (5)$$

$$2\equiv MeOH + M^{z+} \rightleftarrows (\equiv MeO)_2M^{(z-2)} + 2H^+$$

$$*\beta_2^s = \{(\equiv MeO)_2M^{(z-2)}\}[H^+]^2/\{\equiv MeOH\}^2[M^{z+}] \quad (6)$$

The concentration terms in Eq. (5) and (6) are accessible experimentally, e.g.,by direct measurement of the M^{z+} uptake by the surface; hence constants for monodentate($*K_1^s$) and bidentate ($*\beta_2^s$) complex formation can be determined.

Fig. 3 (from Schindler) gives the pH-dependence of the surface coordination of Fe(III), Pb(II), Cu(II) and Cd(II) on amorphous SiO_2. Table 2 lists some of the stability constants for cation interaction with hydrous oxides.

TABLE 2

Stability Constants of Surface Complexes (modified from Schindler [9])

Oxides	Metal		log $*K_1^s$	log $*\beta_2^s$	I	Ref.
SiO_2 (amorphous)	Mg^{2+}	*)	-8.1	-16.7	1M $NaClO_4$	10
	Ca^{2+}	*)	-7.3	-14.7	"	10
	Fe^{3+}		-1.8	- 4.2	"	4
	Cu^{2+}		-5.5	-11.2	"	4
	Cd^{2+}		-6.1	-14.2	"	4
	Pb^{2+}		-5.1	-10.7	"	4
TiO_2 (Rutile)	Cu^{2+}	*)	-1.5	- 5.0	"	11
	Cd^{2+}	*)	-3.2	-10.5	"	11
	Pb^{2+}	*)	0.2	- 2.0	"	11
δ-MnO_2	Ca^{2+}	*)	-5.5	-	0.1M $NaNO_3$	1
γ-Al_2O_3	Ca^{2+}	*)	-6.1	-	"	2
	Mg^{2+}	*)	-5.4	-	"	2
	Ba^{2+}	*)	-6.6	-	"	2
	Pb^{2+}		-2.2	- 8.1	0.1M $NaClO_4$	3
	Cu^{2+}	*)	-2.1	- 7.0	"	3

*) preliminary results

Hydrolysis need not be invoked to account for the pH-dependence of metal ion adsorption to the hydrous oxide surface. This dependence can be explained with the pH dependence of the activity of the $\equiv MeO^-$ groups and the affinity of this group for the metal ion.[1-5]

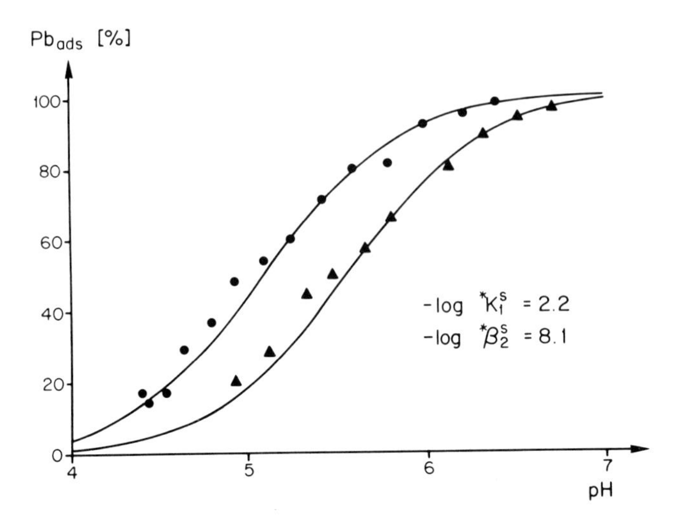

Fig.2. pH-depencence of adsorption of Pb(II) to γ-Al_2O_3 % Adsorption (relative to Pb(II) in solution) measured in batch experiments (48 hrs equilibration time). The lines are theoretical (i.e. calculated with the help of the complex formation constants, $*K_1^S$ and $*\beta_2^S$) determined.
● 11.72 g alon dm^{-3}, $Pb(II)_\tau$ $=2.94 \times 10^{-4}$ M;
▲ 3.18 g alon dm^{-3}, $Pb(II)_\tau$ $=9.8 \times 10^{-5}$ M.

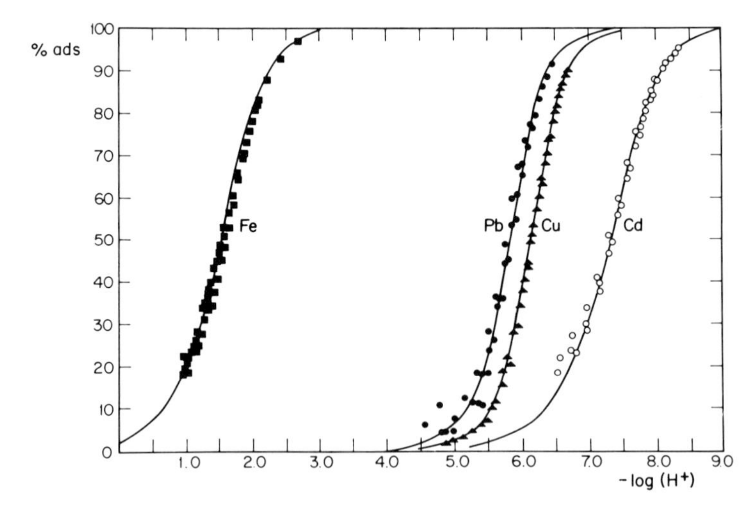

Fig.3. Adsorption of metal ions on amorphous silica as a function of -log [H+] from Schindler et al.[4]

IV. THE ADSORPTION OF INERT Co-COMPLEXES

Figure 4 gives isotherms for the adsorption of $Co(NH_3)_6^{3+}$ on SiO_2. (Minusil). The insert shows the effect of adsorbed $Co(NH_3)_6^{3+}$ on the electrophoretic mobility of the SiO_2 particles. It was not possible to reverse the surface charge of silica particles by $Co(NH_3)_6^{3+}$, $Co(en)_3^{3+}$, or $Co(NH_3)_5Cl^{2+}$, even in the presence of great excess of these complexes. In Figure 5 the surface charge resulting from the proton unbalance at the surface, $\{\equiv SiO^-\}$, is plotted as a function of pH for various concentrations of a trivalent and bivalent complex, respectively.

Calculations have been carried out on the energy of electrostatic interaction of these robust complexes with the charged surface.[12]

In these calculations we combined the protolysis equilibrium of the surface ≡MeOH groups with fixed exchange capacity (vz.Eq.2) with the 3rd Maxwell and the Stephan-Boltzman law.[12] (similar approach as in Gouy-Chapman theory). They show that the pH-dependent adsorption of trivalent and bivalent robust complexes can be accounted for almost solely by electrostatic interaction; i.e. there is no specific adsorption *) (adsorption other than that caused by coulombic interaction). Correspondingly, these complexes cannot cause charge reversal.

The remarkable difference in the adsorption behavior between an aquo metal cation and a robust cation is displayed in Fig. 6; acidimetric and alkalimetric titration curves of γ-Al_2O_3 in presence of either robust cations or Pb^{2+} are plotted. That the cobalt complexes are not specifically adsorbed follows from the common intersection point for titration curves carried out in the presence of various complex concentrations. Pb^{2+}, although less charged than $[Co(NH_3)_6]^{3+}$, exerts a

*) Robust complexes with large ligands such as phenanthrolin or dipyridyl etc. are specifically adsorbed because the large ligand molecules contribute significantly by their van der Waals cohesive energy to the interaction energy.[13]

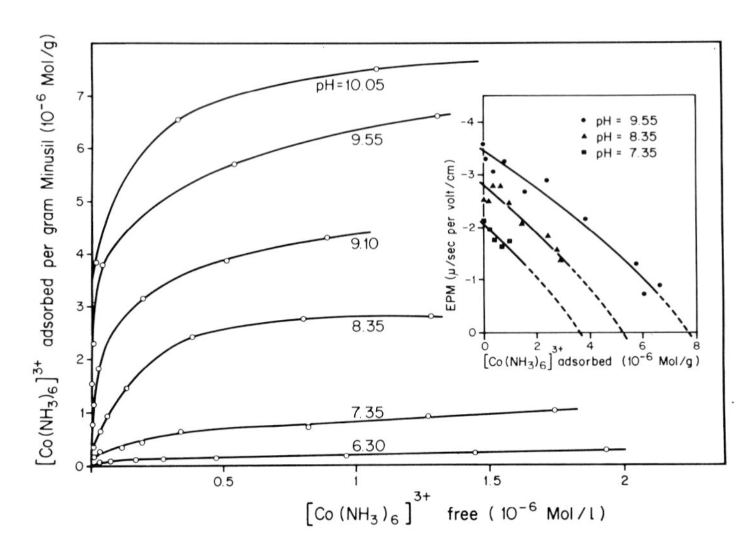

Fig.4. Isotherms for the adsorption of $[Co(NH_3)_6]^{3+}$ on SiO_2 (Minusil). Unlike Me^{z+}·aq ions, robust complexes like $[Co(NH_3)_6]^{3+}$ are not specifically adsorbed on SiO_2.

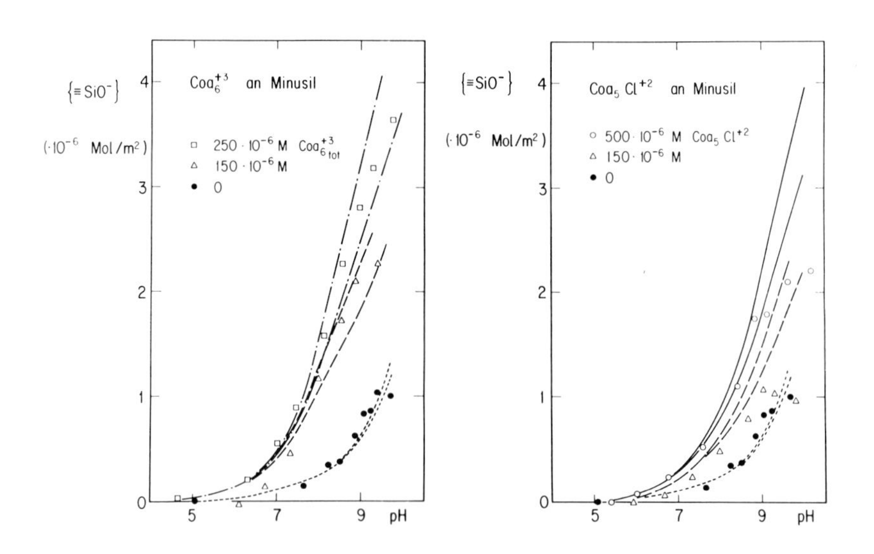

Fig. 5. pH-dependence of surface charge of SiO_2
a) in presence and absence of $Co(NH_3)_6^{3+}$
b) in presence and absence of $Co(NH_3)_5Cl^{2+}$

pronounced displacement of the titration curve. Obviously a nonelectrostatic effect, most likely a partial replacement in the hydration sheath of the cation, dominates the interaction with Pb^{2+} [3]. Apparently Pb^{2+} ions can penetrate efficiently the structured water layers adjacent to the Al_2O_3 surface to form a "chemical" bond between the $\equiv AlO^-$ groups and the metal ion **). The Co(III) central ion in $[Co(NH_3)_6]^{3+}$ and $Co(NH_3)_5Cl^{2+}$, on the other hand, remains shielded by the NH_3 ligands; thus, these complexes more likely form outer sphere type adducts with the anionic groups of the hydrous oxides.

Co(III)-complexes for determining the total number of hydrous Oxide surface OH Groups

In Fig. 7, the charge resulting from the deprotonation of the $\equiv SiOH$ surface is plotted vs the quantity of $Co(NH_3)_6^{3+}$ adsorbed. The slope of 3 indicates, that three protons are released per $Co(NH_3)_6^{3+}$ adsorbed; the same plot for $Co(NH_3)_5Cl^{2+}$ gives a slope of 2; i.e., the charge resulting from deprotonation is compensated by adsorption of the Co-complex. Hence, these complex model cations can be used conveniently to determine the total number of hydrous oxide surface OH groups $\{\equiv MeO_T\}$. The extent of maximum adsorption of the Co-complex is determined at a few pH values in the more alkaline pH-range (e.g. most conveniently by measuring spectrophotometrically the residual complex concentration) Table 3 gives a few representative results:

**) It is tempting to speak of an "inner sphere" complex but we have not established the existence of a direct covalent bond between $\equiv AlO^-$ and Pb^{2+}. Olson and O'Melia provide evidence that the association between Fe^{3+} and $\equiv SiO^-$ is a Fe-O-Si inner sphere, high spin surface complex.

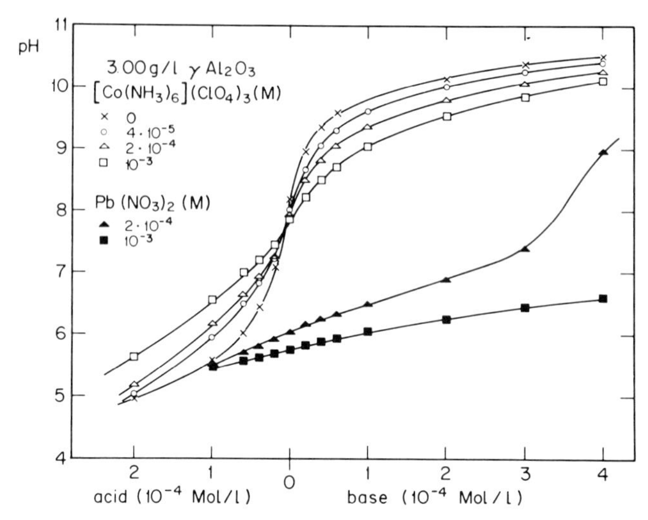

Fig.6. Comparison of the interaction of $[Co(NH_3)_6]^{3+}$ and Pb^{2+} with γAl_2O_3. While $[Co(NH_3)_6]^{3+}$ is not specifically adsorbed (common intersection of alkalimetric titration curves for various concentrations of the complex), Pb^{2+} ions displace markedly the titration curves.

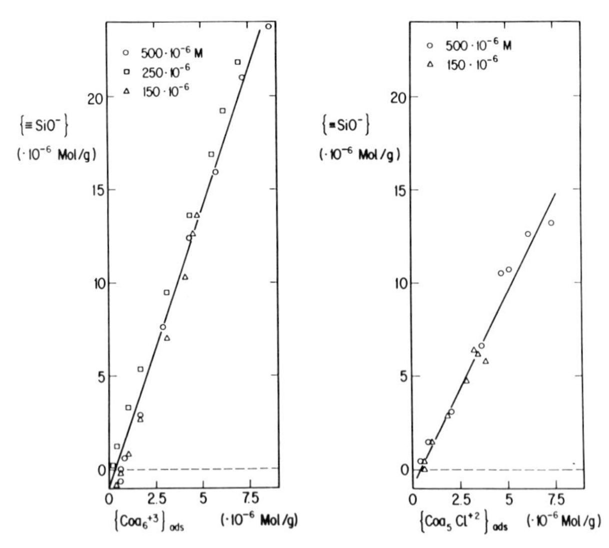

Fig.7. Surface charge of SiO_2 resulting from the deprotonation of ≡SiOH groups as a function of $Co(NH_3)_6^{3+}$ and $Co(NH_3)_5Cl^{2+}$, respectively, adsorbed.

TABLE 3

Determination of exchange capacity $\{\equiv MeO_T\}$, of hydrous oxide surfaces.

Oxide		Adsorption of $Co(NH_3)_6^{3+}$ Mol $g^{-1} \cdot 10^6$	Titration[(1)] Mol $g^{-1} \cdot 10^6$	Ref.
SiO_2	Minusil	30	41	12
	Aerosil	900	1400	14
Al_2O_3	Alon (γ)	125	250	3
			100	2
TiO_2	Rutile	88	150	2

(1) Determined by adding excess base, backtitration of supernatant and correcting for dissolved oxide.

V. CONCLUSIONS

At hydrous oxide surfaces the adsorptive behavior of inert xomplex cations such as $Co(NH_3)_6^{3+}$, $Co(en)_3^{3+}$ and $Co(NH_3)_5Cl^{2+}$ differs significantly from that of transition metal cations. The adsorption of the inert cation complexes is governed almost solely by electrostatic interaction while chemical coordinative interaction dominates the adsorption of transition metal cations. Simple chemical models can be used to describe quantitatively extent and pH-dependence of aquo metal ion adsorption on hydrous oxide surfaces. $Co(NH_3)_6^{3+}$ can be used conveniently to determine the total number of hydrous oxide surface OH-groups $\{\equiv MeO_T\}$.

VI. REFERENCES

Stumm,W., Huang C.P.,and Jenkins,S.R.Croatica Chem. Acta 42, 223 (1970).

Huang,C.P.,and Stumm,W. J.Colloid Interface Sci. 43, 409 (1973).

Hohl,H. and Stumm,W.,J.Colloid Interface Sci., 55,281 (1976).

Schindler,P.W.,Fuerst,B.,Dick,R.,and Wolf,P. J.Colloid Interface Sci.55,469 (1976).

Schindler,P.W.and Kamber,H.R.,Helv.Chim.Acta
53,1781 (1968).
Schindler,P.W.,and Gamsjäger,H.,Kolloid Z.,
u.Z.f.Polymere 250, 759 (1972).
Parks,G.A., in "Equilibrium Concepts in Natural
Waters", Adv. Chem. Ser. 67, 121 (1967).
James,R.O., and Healy,T.W.,J.Colloid Inter-
face Sci. 40, 42, 53, 65 (1972).
Schindler,P.W., The Regulation of Trace Metal
Concentrations in Natural Water Systems.
In press.
Sigg,L., as quoted in Ref.9
Schindler,P.W. and Fuerst,B.,as quoted in
Ref. 9
Dalang,F.,Ph.D. dissertation (unpublished),
ETH Zürich,1976
Matijević,E., Kolloid Z.u.Z.f. Polymere 250,
646 (1972)
Fuerst,B., Ph.D.thesis, Univ. of Berne.

ELECTRON MICROSCOPE OBSERVATIONS ON THE GROWTH OF $Mg(OH)_2$ CRYSTALS

V. A. Phillips, J. L. Kolbe and H. Opperhauser
Martin Marietta Laboratories

I. ABSTRACT

Magnesium hydroxide has been precipitated at various pH levels at 140°F from magnesium chloride and either ammonia, sodium hydroxide, or calcium hydroxide. Particle morphology, perfection and size were studied by transmission electron microscopy. Constant pH runs with ammonia showed that increase in pH favored growth parallel to the basal plane relative to growth along the c-axis. This resulted in a transition from a plate habit over the pH range 11.0 to 9.0, to dense equiaxed particles at pH 8.75. Anisotropy of growth was also observed in the basal plane at pH 9.0 and 9.5, giving plates of hexagonal shape, but near pH 10.0 this tended to disappear yielding larger nearly circular plates with indented outline. Substitution of 30% wt/wt calcium chloride for water in the reactor tended to give pyramidal particles. Precipitation in an aqueous solution of calcium hydroxide yielded imperfect plates of poorly developed hexagonal shape. Precipitation with calcium hydroxide in 30% calcium chloride solution yielded very large imperfect plates. It is concluded that under conditions where the solubility limit of calcium hydroxide in water or calcium chloride solution is approached, Ca^{++} ions retard growth of magnesium hydroxide along the c-axis, probably by adsorption on the basal surface, followed by cationic substitution in the crystal, leading to crystal imperfection.

II. INTRODUCTION

The habit of a crystal, which may be defined[1] as the type of external shape which results from different rates of growth of various faces, can often be modified by changing the conditions of crystallization. The habit may be affected by factors including: the choice of solvent, the pH of the solution, degree of supersaturation, crystallization temperature, presence of impurities, and degree of stirring[1-4].

Habit modification is of great commercial importance because it affects a multitude of processing operations. Although there is a large literature, much of it is concerned with crystallization out of a supersaturated solvent as the temperature decreases, very little relates to hydroxides or to crystals formed by continuous chemical reaction under conditions of sparing solubility. The kinetics of growth of magnesium hydroxide from a seeded supersaturated mixture of magnesium chloride and sodium hydroxide have been studied[5], but no morphological observations were made. After an initial surge probably due to additional nucleation, the growth rate was found proportional to the first power of the relative supersaturation. The crystallization rate constant was reduced by small amounts of the additives nitrilo-tri (methylenephosphonic acid) (NTMP), and N, N, N'N' ethylenediaminetetra (methylenephosphonic acid) (ENTMP), but to a lesser degree than by calcium sulfate (unpublished work)[5] or by calcium carbonate[6].

Crystal growth of $Mg(OH)_2$ from magnesium chloride and sodium hydroxide mixtures containing excess OH^- anions showed a larger crystallization rate constant than the stoichiometric mixture, whereas mixtures containing excess magnesium ions behaved much like the stoichiometric mixture[5]. Precipitated magnesium hydroxide particles will carry a charge below the isoelectric point which occurs at about pH 12(7). Experiments show that, as expected, the sign is positive, indicating that in supersaturated solutions, magnesium ions are more strongly adsorbed by the surface of the crystal than hydroxyl ions. Liu and Nancollas, who worked below pH 12, argued[5] that excess hydroxyl ions would increase the rate constant for crystal growth, whereas excess magnesium ions had little effect.

The present study was concerned with setting up a model system which could be used to evaluate singly the effects of variables, such as pH and the presence of excess reagents or foreign ion species, on magnesium hydroxide particle morphology. Magnesium chloride was used as the source of the magnesium, and in most of the experiments described here, ammonium hydroxide was used as the source of the anions. A somewhat similar study was reported recently by Copperthwaite and Brett[8], in which potassium hydroxide was added to a stirred solution of magnesium chloride in amounts corresponding to deficient, equivalent and excess alkali, monitoring the final pH. This approach has the disadvantage that initially there is a large excess of magnesium chloride which dwindles away so that the concentration of Mg^{++} ion is constantly changing and the pH is not constant. Average magnesium hydroxide plate thickness and diameter were determined by x-ray diffraction. Whereas the average plate diameter was relatively insensitive to final pH, average plate thickness showed an initial increase from about 135 to 155 Å with pH increase from 9.65 to 10.2, then levelled off up to pH 12.15.

III. MATERIALS AND METHODS

Reactions were carried out in a one litre flask with three necks, fitted with pH electrodes, and equipped with a magnetic stirrer. In most of the experiments, the pH was recorded. The pH meter was provided with a preset manual control for bath temperature compensation. Bath temperature was controlled at $140^{\circ} \pm 2^{\circ}$F using a water bath.

In the constant pH experiments with ammonia, the two reagents, 0.75 M magnesium chloride hexahydrate and 1.50 M ammonium hydroxide, both of Analar purity and at ambient temperature, were separately added from burettes into 800 mℓ of distilled water at 140°F in the reaction vessel. Two metering pumps (Milton Roy Co.) connected by stainless steel tubing were used to control the addition rates at about 1.5 mℓ/min. The amounts added were read off from the burettes. The reagent concentrations were chosen so that equal addition rates gave the correct stoichiometric ratio for $Mg(OH)_2$ formation. Alternatively, the rate of one could be varied to hold the pH constant. The run time was about one hour. Using excess of one reagent, it was found possible to regulate the reactor pH during the run at a desired value between 8.75 and 11.0, generally within closer limits than $\pm$ 0.1. Once precipitation started, the magnesium hydroxide crystals were growing in a seeded environment, since stirring kept the solids in suspension. Since the reagents were dripped into a much larger volume, growth occurred under very low effective supersaturation. Except where otherwise stated, crystals were aged in the bath liquor for a further hour at 140°F.

In the experiment with sodium hydroxide, the procedure was the same as above, but a 1.50 M solution of caustic soda was substituted for the ammonia. In two experiments, solutions of either 0.040 M at 3.46 M calcium chloride were substituted for water in the reaction vessel. In one experiment, a 0.009 M solution of calcium hydroxide was used instead of water. In the final experiment, calcium hydroxide (3.0 g/ℓ) dissolved in 30% calcium chloride solution was used as the base in place of ammonia, with a bath of the same composition. In this case, stoichiometric addition rates gave very little variation in pH during the test.

The point at which crystallization started, indicated by turbidity, was noted on the pH record. The crystals were filtered with suction, and washed repeatedly with hot distilled water. The wet filter cake was redispersed in distilled water and a microdrop allowed to dry on a carbon support film for transmission electron microscopic examination. The latter was carried out in a JEOLCO JEM 7 instrument at 100 kV and usually 25,000 x direct magnification. Particle diameters were measured on prints at 75,000 x magnification. Particle diameters were measured on prints at 75,000 x magnification, averaging the results on 50 particles in a given area. The diameter/thickness measurements were made on a few (4 to 10) platelets which fortuitously lay parallel to the electron beam.

IV. RESULTS

A. Constant pH Experiments with Ammonia

Transmission electron microscopy (Fig. 1-5, Table I) showed that both the habit and perfection of the magnesium hydroxide particles formed from magnesium chloride and ammonia were affected by the pH during reaction, although other conditions -- temperature (140°F), stirring rate, reagent addition rate, and concentration were nearly constant. The pH was adjusted initially by adding NH_4OH or $MgCl_2$ to the water in the reactor vessel, maintaining it during the run by adding excess ammonia as necessary to counteract changes due to the gradual accumulation of ammonium chloride in the bath.

At pH 11.0, thin plates were obtained with poorly developed hexagonal shape, but relatively good crystal perfection as indicated by the absence of internal image detail (Fig. 1). At pH 10.0, some well-formed plates of hexagonal habit were present, but many plates had an indented outline (Fig. 2). Some crystals were very imperfect. Some crystals appeared to have grown together to form layered crystals. Crystals with indented outline were absent at lower pH of 9.5 (Fig. 3) and 9.0 (Fig. 4). The crystals formed at pH 9.5 had well developed hexagonal shape and good internal crystal perfection. At pH 9.0 the shape was somewhat less

TABLE I.

Effect of pH on Morphology, Perfection and Size of $Mg(OH)_2$ Particles Formed From $MgCl_2/NH_4OH$ During 1 hr. Run at 140°F

Run Details		Avg. Particle Dia., μm		RATIO Diameter/Thickness		Comments on Morphology and Perfection of Crystals
pH		Unaged	Aged 1 hr. at 140°F	Range	Average	
11.0	xs NH_4OH	--	0.090	8-14	10.9	Thin plates, hexagonal morphology poorly developed, good perfection.
10.0	xs NH_4OH	0.183	0.256	2-6	4.3	Thin plates, many of hexagonal shape, but some with indented outline, some very imperfect, some layered crystals.
9.5	xs NH_4OH	0.147	0.164	4-6	5.1	Thin plates, hexagonal morphology well developed, variable perfection.
9.0	xs NH_4OH	0.112	0.112	2-6	4.5	Thin plates, hexagonal morphology well developed, good internal perfection.
8.75	xs $MgCl_2$	0.303	--	1-3	2.3	Dense, fairly equiaxed. Each particle appeared to consist of a stack of plates with staggered edges.

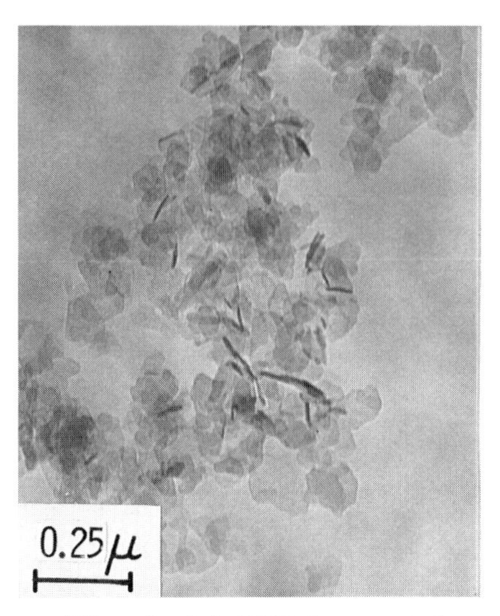

Fig. 1. Mg(OH_2) crystals from $MgCl_2$/NH_4OH xs, 1 hr. run at 140°F, constant pH ~ 11.0. Aged 1 hr. at 140°F.

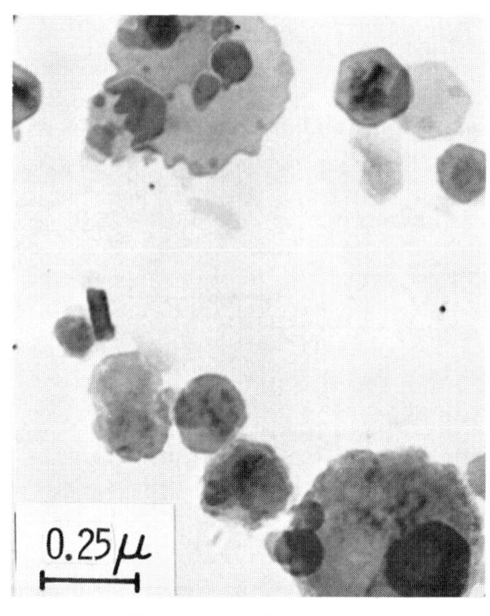

Fig. 2. $Mg(OH)_2$ from $MgCl_2$/NH_4OH xs, 140°F, constant pH ~ 10.0. Unaged.

Fig. 3. $Mg(OH)_2$ from $MgCl_2$/NH_4OH, 140°F, constant pH ~ 9.5, xs $MgCl_2$ used initially to give 9.5 pH, then xs NH_4OH. Aged.

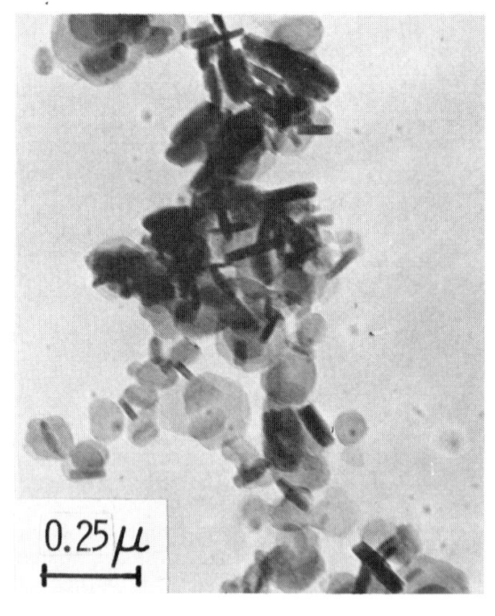

Fig. 4. $Mg(OH)_2$ from $MgCl_2$/NH_4OH, 140°F, constant pH ~ 9.0. Unaged.

Fig. 5. $Mg(OH)_2$ from $MgCl_2$ xs / NH_4OH, 140° F, constant pH ~ 8.75. Aged.

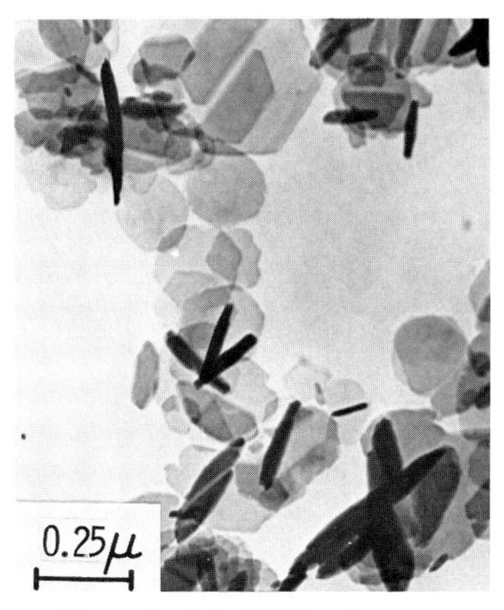

Fig. 6. $Mg(OH)_2$ from stoichiometric $MgCl_2$/NaOH, 140° F, pH 9.9 to 9.35. Unaged.

developed, while at pH 8.75 there was a dramatic change to dense fairly equiaxed particles, which had the appearance of a stack of plates with staggered edges (Fig. 5). Selected area electron diffraction showed that each stack was a single crystal with the "laminations" corresponding to the basal planes.

The average particle diameter was largest at pH 8.75, decreased abruptly at pH 9.0, increased progressively with increase in pH to 9.5 to 10.0, then decreased abruptly at pH 11.0 (Table I). Measurements on particles, fortuitously viewed edgewise in the microscope, indicated that the diameter/thickness ratio averaged over several particles increased progressively from 2.3 to 10.9, with increase in pH from 8.75 to 11.0 (Table I) showing that edgewise plate growth was favored as the hydroxyl ion concentration increased.

The average particle diameter tended to increase on aging in the stirred liquor for 1 hr. at 140° F, particularly at higher pH (Table I). The morphology and internal crystal perfection were not much affected by this short aging treatment. Rough values of the average plate thickness in the aged condition may be computed from the average diameter and ratio data (Table I). These range from 80 Å at pH 11.0, to 600 Å, 320 Å and 250 Å, at pH 10.0, 9.5 and 9.0, respectively.

B. Other Experiments

Particles produced by stoichiometric addition of magnesium chloride and sodium hydroxide solutions to an 800 mℓ bath of water at 140°F gave plates of fairly well developed hexagonal shape and good internal perfection (Fig. 6). The pH was 9.9 initially when precipitation started and gradually decreased during the run to 9.35 as sodium chloride accumulated in the bath. Since the shape and perfection closely resembled those produced from magnesium chloride and ammonia in a similar pH range, the presence of Na^+ ions apparently had no effect on the morphology.

The use of a weak (0.040 M) calcium chloride bath with stoichiometric addition of the magnesium chloride and ammonia gave particles of good hexagonal shape (Fig. 7). The pH was not recorded but was about 8.9 at the end of the run. This result indicated that Ca^{++} ions had little effect at low concentrations. However, use of a strong (3.46 M) bath of calcium chloride tended to give particles of trigonal outline, apparently tending to a layered pyramidal shape (Fig. 8).

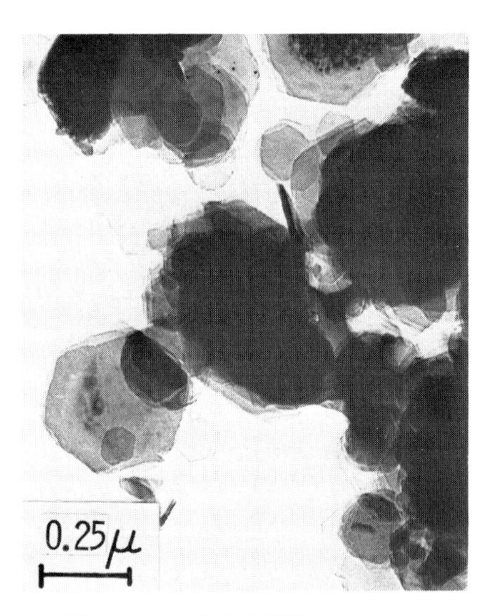

Fig. 7. $Mg(OH)_2$ from stoichiometric $MgCl_2/NH_4OH$, reacted in 0.040 M bath of $CaCl_2$, 140°F. Final pH 8.9. Aged.

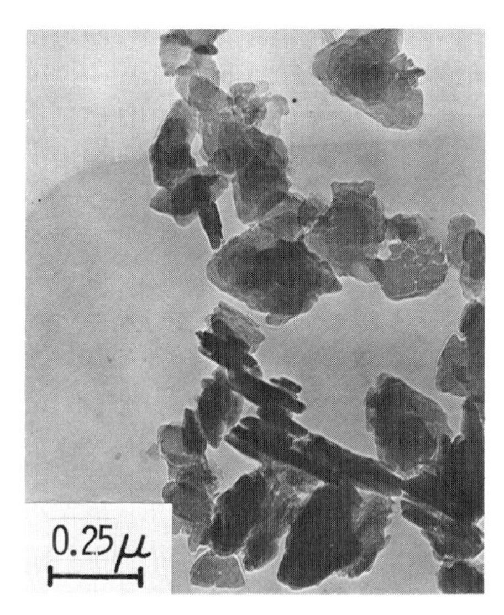

Fig. 8. $Mg(OH)_2$ from stoichiometric $MgCl_2/NH_4OH$ reacted in 3.46 M bath of $CaCl_2$, 140°F, pH 7.85 to 8.55 to 8.35. Aged.

Some irregular plates were also present. On commencing the addition of reagents, the initial bath pH of 7.85 (at 140°F) rapidly increased to 8.55 at which precipitation probably started and then gradually decreased to 8.35 as precipitation continued. We suggest that the pyramidal morphology is associated with precipitation near pH 8.5, rather than being a calcium or chloride ion effect, since a similar dramatic change in morphology in the magnesium chloride/ammonia system occurred at pH 8.75 (Fig. 5) and the tendency to a layered structure was common to both.

The introduction of calcium into the bath as hydroxide rather than chloride was also explored. The stoichiometric addition of magnesium chloride and ammonia into a 0.009 M bath of calcium hydroxide at 140°F yielded plates of rather imperfect hexagonal outline, tending to have low internal perfection (Fig. 9). The initial 11.35 pH of the bath fell rapidly to 9.25 when reagent addition commenced, then more slowly to 9.0 and remained constant. Precipitation started as soon as reagent addition commenced, and thus occurred over a considerable range of pH which may account for the poor shape and internal imperfection of the particles. The size range is similar to that observed in the absence of

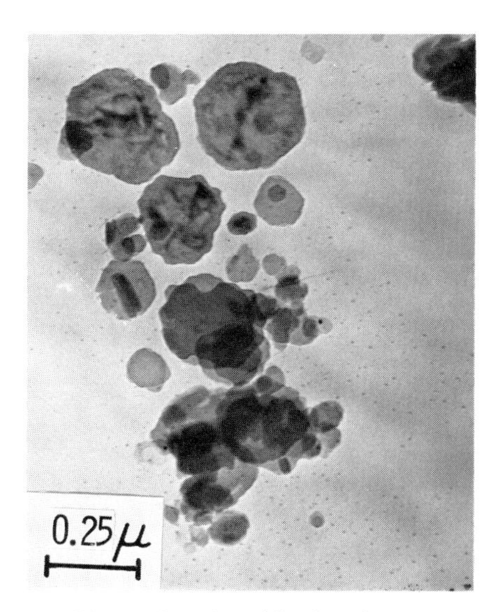

Fig. 9. $Mg(OH)_2$ from stoichiometric $MgCl_2/NH_4OH$ reacted in 0.009 M bath of $Ca(OH)_2$, 140°F, pH 11.35 to 9.25 to 9.0. Aged.

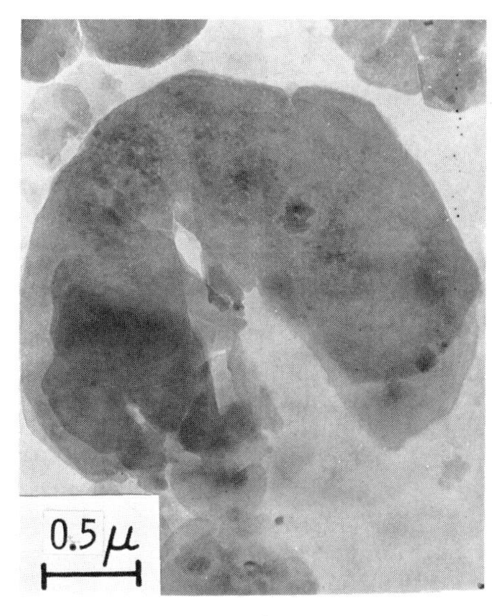

Fig. 10. $Mg(OH)_2$ from stoichiometric $MgCl_2$/ $Ca(OH)_2$-$CaCl_2$ reacted in $Ca(OH)_2$-$CaCl_2$, 140°F, pH 9.25/9.20. Aged.

lime at pH 9.5 - 9.0 (Fig. 3-4). Although the amount of Ca^{++} ion added was only a quarter of the amount present in the dilute calcium chloride bath (Fig. 7), the effect on external shape was larger perhaps associated with the large initial pH increase produced by adding calcium as hydroxide.

Due to the low solubility of calcium hydroxide in water, it was not possible to explore the effect of additions much larger than that used above. Since the solubility in 30% calcium chloride solution is greater, an experiment was carried out in which stoichiometric additions of 0.043 M magnesium chloride solution and a 0.043 M solution of calcium hydroxide in 3.46 M calcium chloride were made to a bath of the same initial $Ca(OH)_2/CaCl_2$ composition at 140°F. The pH remained almost constant at 9.25/9.20. Very large plates of somewhat imperfect hexagonal outline and low internal perfection were obtained (Fig. 10) indicating a large increase in edgewise growth rate or a reduction in nucleation rate. The external shape and size of the crystals was very different from that obtained by addition of magnesium chloride and ammonia to a 3.46 M bath of calcium chloride (Fig. 8). Comparison with Fig. 3-4 also indicates that the results cannot be reconciled on a pH basis alone. The evidence points to a substantial Ca^{++} ion effect on the nucleation and growth of magnesium hydroxide particles under conditions where the calcium ion concentration is approaching that necessary for precipitation of calcium hydroxide. Other experiments, in hand, in which the calcium chloride concentration is maintained constant and the calcium hydroxide concentration varied, indicate that chloride ions do not affect the morphology.

V. DISCUSSION

The constant pH experiments on the magnesium chloride-ammonia system showed that the magnesium hydroxide particle morphology, diameter to thickness ratio, and average diameter varied with the pH, that is, with the OH^- concentration during precipitation. Apart from the slow accumulation of ammonium chloride in solution during a run, the environmental conditions were essentially constant. Although based on relatively few measurements, the diameter/thickness ratio data indicated that edgewise growth in directions lying in the basal plane was favored over growth parallel to the c-axis as the pH increased. Conversely on decreasing the pH to 8.75, on the excess magnesium chloride side, the plate morphology tended to disappear and the particles became nearly equiaxed.

Our measurements on the effect of pH on diameter/thickness ratio show a trend opposite to that indicated by the x-ray data of Copperthwaite and Brett[9] (see Section I) on the magnesium chloride-potassium hydroxide system in experiments where only the final pH was controlled and the lowest pH was 9.65. Size data determined by x-ray line broadening represent a weighted average, which could be affected by size distribution differences between runs. Furthermore, line broadening can also be caused by crystal imperfection and this also varies with pH according to our observations (Table I).

Crystal habit is determined by the crystallographic identity of the slowest growing plane, and regardless of the growth mechanism, can frequently be modified by specific adsorbents, that is, substances which adsorb on a specific crystal surface. Crystals of hexagonal structure are anisotropic in nature with the slowest growth direction along the c-axis, perpendicular to the close-packed basal planes. This leads to the commonly observed plate crystal form, with hexagonal outline due to the formation of prism faces perpendicular to the close-packed directions in the basal plane. Our observations indicate that excess OH^- anions or Mg^+ cations affect the growth rate in different directions to different degrees resulting in the observed morphological changes. This effect is presumably connected with the detailed atomic structure of the crystal and the way in which ions are added on, which involves considerations of bonding and charge. We do not at present propose a detailed model.

Although edge and face ion-adsorption effects may be involved, our explanation of the effect of pH on magnesium hydroxide crystal growth is basically different from that proposed by Liu and Nancollas[5] who argued that Mg^{++} ion adsorption to the crystal face increased the supply of magnesium, accounting for their observed increase in crystallization rate constant with increase in the supply of OH^- ions. This mechanism may well be important in a diffusion limited growth situation, but is not likely to be dominant in the well stirred environment used in our experiments. It should also be pointed out that Liu and Nancollas[5] did not study the morphology of their crystals and that changes in the morphology with pH, including the diameter/thickness ratio, if they occurred in their system, would be expected to change the crystallization rate constant.

The growth of an ionic crystal from solution involves the supply of both anions and cations to the growing crystal surface, and in the case of magnesium hydroxide requires the supply of two OH^- ions for every Mg^{++} ion. There is electrical conductivity evidence[5] for the existence of the $MgOH^+$ ion pair in solution, which raises the question of the possible existence of a solid MgOH phase. Klein et al.[10] concluded from a study of the rate of homogeneous nucleation of magnesium hydroxide that a concentration corresponding to the solubility product for $Mg(OH)_2$ was required to initiate homogeneous nucleation and hence that $Mg(OH)_2$ was the stable nucleus phase. There appears to be no crystallographic data indicating the existence of a solid MgOH intermediate phase either during formation or decomposition of $Mg(OH)_2$. The existence of $MgOH^+$ ion pairs in solution would, however, tend to ease the diffusion supply problem to the growing crystal.

Diameter/thickness and average size measurements are not available for the remaining experiments, so that comments will be limited to crystal shape, perfection and size range. The reaction of magnesium chloride with sodium hydroxide gave particles (Fig. 6) very similar in shape to those obtained in a similar pH range by reaction with ammonia (Fig. 2 and 3). This suggests that there are no specific Na^+ ion effects on the morphology of magnesium hydroxide.

Stoichiometric reaction of magnesium chloride and ammonia in dilute (0.040 M) calcium chloride, did not reveal any specific effect of Ca^{++} ions on crystal shape. The latter was similar to that in the pure system at the same pH, but the platelike crystals apparently had a larger diameter. The use of concentrated (3.46 M) calcium chloride gave a lower pH than the lowest obtained in the pure system, but similar dense fairly equiaxed particles, with a tendency now to pyramidal shape. Evidence of any dramatic change due to calcium was again lacking.

The initial addition of calcium to the reactor water as calcium hydroxide had a much more potent effect than the larger additions of Ca^{++} as calcium chloride in the 0.040 M and 3.46 M baths. Unfortunately this mode of carrying out an experiment resulted in a large drop in pH from 11.35 to 9.25 when reagent addition commenced. However, the poor development of hexagonal shape and high degree of internal imperfection are more characteristic of pH $\sim$ 10.0 in the pure system than of pH 9.25 - 9.0 which existed during most of this calcium hydroxide run, when the bulk of the particles would have formed. This indicates that there was a Ca^{++} ion effect under conditions where the solubility limit for calcium hydroxide was more closely approached than in the calcium chloride experiments. Such a situation would encourage absorption of calcium ions on the crystal facilitating cationic substitution and incorporation of calcium in the growing crystal. Either of these effects or both could lead to internal crystal imperfection.

In the final experiment, a concentrated solution of calcium hydroxide in 30% calcium chloride solution was substituted for ammonia, and the pH maintained almost constant (9.25 - 9.20) by using a reactor bath of the same initial composition. With a stoichiometric addition, the reaction was thus carried out under conditions of nearly constant excess alkalinity (~ 3 g/ℓ $Ca(OH)_2$) and high Ca^{++} ion concentration, which were only slightly reduced (< 10% reduction) during the reaction due to water (100 mℓ) introduced with the magnesium chloride. Under these conditions, a dramatic effect of Ca^{++} ions was observed. The particles (Fig. 10) differed greatly from those produced in the pure system in the same pH range (Fig. 3, 4). The plates ranged up to 3 μm diameter, which is a factor of five larger than the largest size observed at similar pH in the pure system, and tended to be very imperfect internally, particularly at large size. A hexagonal shape was not always well developed (Fig. 10). Since the large plates tend to lay flat on the substrate it is rare to find a plate perpendicular to the substrate on which a diameter/thickness ratio can be measured. A single observation gave a value ≥ 19, the plate being of somewhat irregular thickness. This confirms the conclusion from the calcium hydroxide experiment that under conditions where the solubility limit for calcium hydroxide is approached, Ca^{++} ions favor edgewise growth of magnesium hydroxide plates parallel to the basal plane and tend to reduce the growth rate parallel to the c-axis.

The mechanism of the calcium effect is unknown. We speculate that Ca^{++} ions adsorb on the basal surface of the crystal forming a partial layer, and obstruct growth due to their being larger than the corresponding magnesium ions. As the crystal thickens, these Ca^{++} ions become cationically substituted and incorporated in the growing crystal, causing crystal defects and associated strain fields which account for the imperfection revealed by diffraction constrast in the electron microscope.

VI. REFERENCES

1. Mullins, J. W., "Crystallization", p. 207, C.R.C. Press, Cleveland, 1972.

2. Khamskii, E. V., "Crystallization from Solutions", p. 84, Consultants Bureau, New York, 1969.

3. Shternberg, A. A., in "Growth of Crystals", N. N. Sheftal and E. I. Givargizov, Ed., Vol. 9, p. 34, Consultants Bureau, New York, 1975.

4. Garrett, D. E., Br. Chem. Engng., 4, 673 (1959).

5. Liu, S-T., and Nancollas, G. H., Desalination, 12, 75 (1973).

6. Reddy, M. M. and Nancollas, G. H., Desalination, 12, 61 (1973).

7. Parkes, G. A., Chemical Reviews, 65 (2), 177 (1965).

8. Larson, T. E., and Buswell, A. M., Ind. Eng. Chem., 32, 132 (1940).

9. Copperthwaite, M. and Brett, N. H., Science of Ceramics, 8, p. 85-99, British Ceramic Society, 1976.

10. Klein, D. H., Smith, M. D., and Driy, J. A., Talanta, 14, 937-40 (1967).

MICROSTRUCTURE OF SYNTHETIC LATEX PARTICLES

E. B. Bradford & L. E. Morford
Dow Chemical USA

The only direct method by which the microstructure of latex particles and films can be visualized is by electron microscopy. This method permits the observation of the compositional variation in a dried dispersion of latex particles comprised of soft copolymer and hard seed. A soft styrene-butadiene copolymer can be seen deformed around the hard polystyrene. However, the location of the soft copolymer in the original undried particle is unknown.

In the development of the core-shell theory of emulsion polymerization, Williams and coworkers observed cross sections of individual latex particles which had been coagulated and then redispersed in epoxy resin. Unfortunately, redispersion was very difficult and single particles were uncommon. To circumvent this problem we avoided the coagulation step by mixing the latex with water miscible embedding materials which would suspend the individual particles in a matrix upon drying. Several systems were investigated. The most successful method consisted of mixing the seeded latex with a poly(vinyl acetate) latex and drying into a film; this was then embedded in epoxy resin to hold the film for sectioning. Examples are shown and discussed. A surprising result showed that the styrene-butadiene copolymer was not located symmetrically around the seed particles.

I. INTRODUCTION

The concept of seeding in emulsion polymerization technology was established many years ago.(1) In this technique monomer is added to a previously prepared latex and, if there is no (or very little) free emulsifier in the system, the seed latex grows to a larger size without the initiation of new particles. Many polymerization variables that occur during the particle initiation stage can be avoided by using a seeding technique. In addition, latexes with specific properties may be prepared; e.g., layered latex particles have been formed in which the core is a crystalline polymer, the intermediate layer is partially crystalline and the outer layer is amorphous, or crystallizes only to a small extent.(2) A film from this latex is able to withstand creasing without losing its moisture vapor transmission resistance in the crease and will not adhere to adjacent sheets during stacking.(2) By using specific copolymers in layered latex systems, the film properties may be tailored to impart properties required for particular applications.

The compositional variation in a seeded system where styrene/butadiene(S/B) copolymer is polymerized onto polystyrene (PS) seed latex particles can be observed directly in the electron microscope (EM). In dried dispersions the soft, film forming S/B copolymer was seen deformed about the hard polystyrene seed particles.(3) Another example showed the morphology of seeded latex particles when insufficient time was allowed for the monomer to reach equilibrium swelling of the seed particles; irregularly shaped particles were observed. However, when equilibrium swelling was achieved, spherical particles resulted.(3) In these illustrations, dispersion of the particles were observed after they had been dried onto a substrate film. The actual microstructure of the particles in the aqueous dispersion was not seen.

In the development of the core-shell theory of emulsion polymerization of seeded PS latexes, D. J. Williams and coworkers (4-7) observed ultrathin sections of the particles by electron microscopy. They showed that the second stage copolymer encapsulated the seed particles rather than mixing uniformly throughout the particles. Their ultrathin cross sections were obtained by destabilizing the latex, filtering, washing with methanol and water, then drying. The powder was redispersed in epoxy resin and ultrathin sectioned in the normal manner.(5) Unfortunately, redispersion was very difficult to achieve and single particles were uncommon. Confirming particle microstructure was "akin to looking for the proverbial needle in the haystack".(5)

This paper concerns the method we developed to provide ultrathin sections of individually dispersed latex particles by circumventing the redispersion problem, and describes some preliminary results which show unexpected particle morphology in seeded latex systems.

II. EXPERIMENTAL

A. Water Soluble Materials

Since it is very difficult to redisperse aggregated latex particles in epoxy resin, we attempted to eliminate the drying step. The rationale of this method is to disperse the subject latex in an aqueous solution or dispersion which, upon drying, will encapsulate the individual latex particles in an embedding medium which can be sectioned for observation in the EM.

Several water soluble polymers were investigated. Poly-(vinyl alcohol), PVA, has been used as an intermediary in a two-step replication technique for many years. A dispersion of the latex was prepared in a 15% aqueous PVA solution. Upon drying in a polyethylene capsule designed to provide an embedment to fit in the chuck of a microtome, the embedment shrunk and dried around the edge of the capsule leaving a hole in the center; also, the drying time was exceedingly long (several days). This drying problem was reduced by casting a film of the dispersed latex in PVA which subsequently could be embedded in epoxy resin for sectioning. Other water soluble materials investigated were polyacrylic acid, methyl cellulose, gelatin, a water soluble modified glycol methacrylate resin, modified melamine resin, and water soluble epoxy resins. In some of these systems crosslinking agents were used to impart insolubility to the dried films. Unfortunately, all of these systems were unsatisfactory. Some coagulated the latex, others formed films that were either too water sensitive or not sufficiently hard to permit microtoming good ultrathin sections.

B. Other Embedding Mediums

Mixing a few percent latex in various mixtures of DER[1]-732 and DER-332 epoxy resins resulted in embedments which crumbled during the ultrathin sectioning process; this caused gross distortion of the latex particles. Blends of butyl and methyl methacrylate swelled the latex particles and upon polymerization produced an embedment which was soft, mainly due to the entrapped water.

Water soluble sulfonium zwitterion compounds (8) did not coagulate the latex upon mixing; but elevated temperatures were required to cure the resin, e.g., 80°C/four hours. Ultrathin sections were obtained from these composites, but individual latex particle morphology was not uniform. Considerable distortion of the particles was observed and some of them did not even appear spheroidal in the EM. The zwitterion compound without the addition of latex sectioned well. Unfortunately, none of the water soluble compounds produced films which would provide ultrathin sections that displayed consistent microstructure of individual latex particles.

1. Trademark of The Dow Chemical Company

C. Poly(Vinyl Acetate) Latex

If a small amount of an anionic latex is mixed with a second anionic latex, one might expect little or no agglomeration of the first latex. If the second latex forms a continuous film at room temperature, the film could contain a dispersion of individual spheres of the first latex.

A few percent of a S/B copolymer latex was mixed with a poly(vinyl acetate) latex (PVAc) and dried into a film at room temperature. A portion of this film was embedded in epoxy resin and stained (9) with osmium tetroxide. Ultrathin sections showed that PVAc can be cut thin enough for observation in the EM and that the film contained a dispersion of individual S/B latex particles.

In the normal sectioning procedure the ultrathin sections are floated onto the surface of distilled water (or distilled water containing about 10% acetone) as they are cut with a glass or diamond knife. Unfortunately, thin PVAc films are sensitive to water and many holes may be seen in the sections. An example is shown in Fig. 1-A. The dark particles are ultrathin sections of an OsO_4 stained S/B copolymer latex. The light areas are holes in the section resulting from floating on the water surface for less than two minutes. To avoid this problem, sections were cut with no water in the trough, but these sections folded in a corrugated manner at the knife edge. Attempts to flatten these sections were not successful.

Other liquids were used in the trough. The sections were compressed and folded when ethylene glycol was used (heating did not permit flattening). In addition to compression wrinkles, the sections did not float when n-heptane, chlorotrifluoroethylene, or diethyltetrachlorobenzene were used; 2-3 dichloro-octafluorobutane was so volatile that liquid contact with the knife edge was impossible to maintain during sectioning. Methanol wet the tip of the specimen and apparently softened it. Water containing 50% cesium carbonate allowed sections to float without excessive wrinkling, but the solution dried very slowly from the sections and left them heavily contaminated. Adding 10% sodium chloride to the water did not eliminate the problem, but using 20% NaCl in the water and then submerging the dried sections in distilled water for ~10 seconds to remove salt deposits minimized the number of holes in the ultrathin sections.

Another attempt to reduce the sensitivity of PVAc sections to the water in the trough was made by dialyzing the PVAc latex. The dialysis might remove the impurities which could increase the sensitivity of the dried film to water exposure for a few seconds. Unfortunately, dialysis did not reduce this sensitivity.

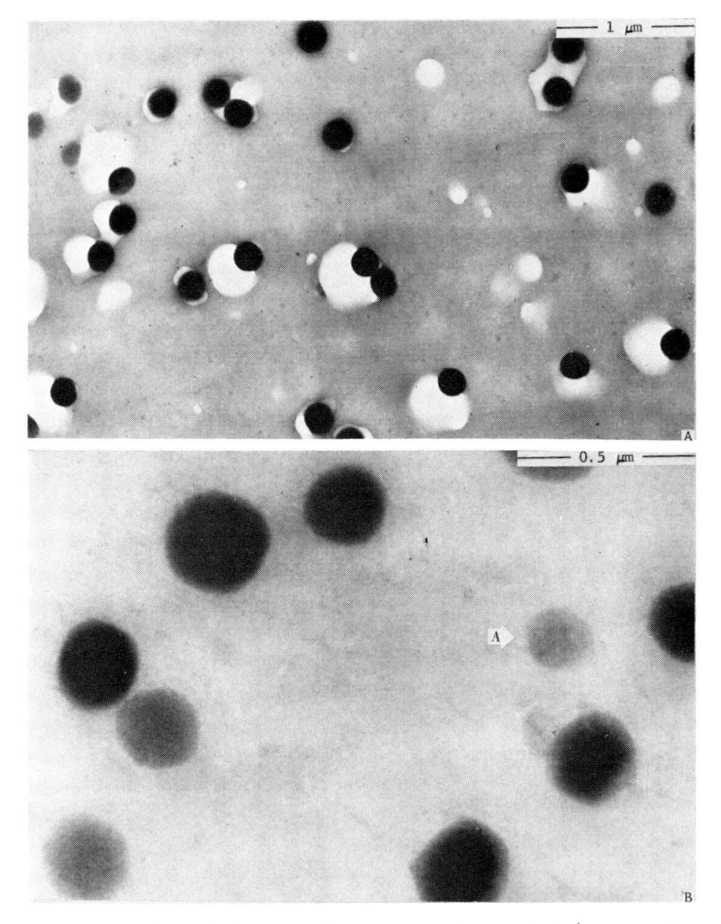

Fig. 1. Ultrathin sections of stained S/B copolymer latexes embedded in PVAc: A. Showing sensitivity of PVAc to water. B. Showing the homogeneity throughout unseeded 60S/40B copolymer latex particles.

III. TECHNIQUE EMPLOYED

Four percent of the subject S/B copolymer latex is added to the PVAc latex. After gently shaking a few minutes, a film is cast on a watch glass and allowed to dry at room temperature. A triangular shaped piece of the film is glued to a small piece of cardboard which is held in a #00 gelatin capsule with a copper wire so that the tip is near the base of the capsule. It is then filled with a mixture consisting of eight parts DER-332 epoxy resin, one part DER-732 epoxy resin and one part diethylenetriamine. After hardening overnight at room temperature, the epoxy resin is trimmed away from the tip of the specimen film. It is then faced off in the microtome and stained with OsO_4, by exposing to the vapors of 1% aqueous OsO_4 solution in a small closed chamber for 24 hours. The PVAc will swell slightly in the vapors but will harden again after removing and storing overnight. The specimen is positioned in the chuck of the Ultrotome[1] III microtome and sectioned at a thickness of about 700 A with a diamond knife. Glass knives can be used but they reduce somewhat the proportion of good ultrathin sections. The sections are floated onto the surface of distilled water containing 20% NaCl. The sections are caught on a carbon substrate film supported by a 200 count copper specimen screen, dried, carefully soaked in distilled water for 5-10 seconds, and then redried (overnight). An example of a cross sectional view of 60S/40B copolymer latex particles prepared without seeding is shown in Fig. 1-B. The particles are dispersed well in the PVAc film. The internal microstructure appears homogeneous, indicating random chain molecular structure throughout the particle. Arrow A points to a section of a smaller, lighter particle. This probably results from sectioning the particle near an edge rather than through its central region.

IV. MORPHOLOGY OF SEEDED LATEX PARTICLES

A simple way to observe the compositional variation is by direct electron microscopic observation of dispersed particles. Fig. 2-A shows particles comprised of three parts of 60 vinyltoluene (VT)/40B copolymer polymerized onto one part PS seed latex. Since the particles were not stained the soft copolymer can be seen deformed about the hard, darker PS seed particles. In some particles the copolymer surrounds the PS uniformly, but in other particles the copolymer is unsymmetrical. This lack of symmetry may be the result of the manner in which the surface tension forces of drying distort the soft copolymer or it may be the result of incompatibility between copolymer and PS. In any event, the dried particles may not show the actual microstructure of the wet particles in the latex.

A similar experiment was carried out by polymerizing three parts of 60S/40B copolymer onto one part PS seed. A dispersion of these particles (Fig. 2-B) does not show the copolymer as distinctly as when VT was used in the copolymer. This indicates that the S/B copolymer may be more compatible with PS than VT/B copolymer. Fig. 2-B also shows particles in which the PS seed is not symmetrically surrounded by the copolymer.

---

1. Trademark of LKB Instruments, Inc.

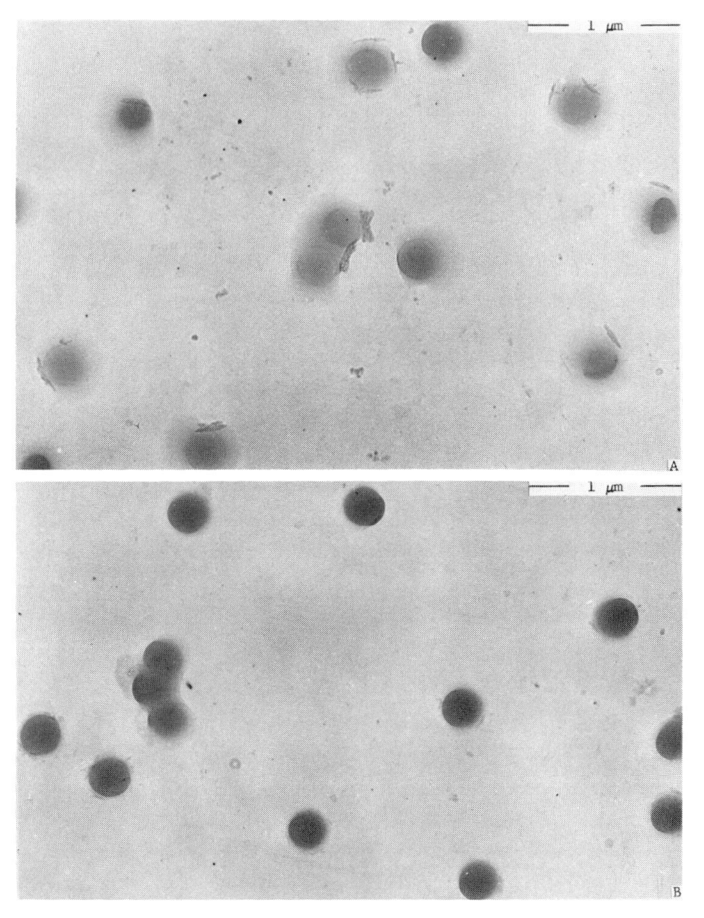

Fig. 2. Dispersions of seeded latex particles: A. Three parts 60VT/40B copolymer polymerized onto one part polystyrene seed latex. B. Three parts 60S/40B copolymer polymerized onto one part polystyrene seed latex.

Soft copolymer seed particles cannot be observed in dispersions when hard polymer is polymerized onto them. Fig. 3-A shows particles comprised of three parts VT polymerized onto one part 50S/50B copolymer seed latex. Compositional gradients are not seen in these particles and they do not appear round. They appear as almost completely coalesced doublet particles. Particles of a similar morphology were observed when the same experiment was carried out after replacing the VT with PS (Fig. 3-B). Here, again, the soft copolymer cannot be distinguished because the hard polymer (PS) apparently encapsulates the seed particles sufficiently to scatter the electrons in the electron beam uniformly throughout the particles.

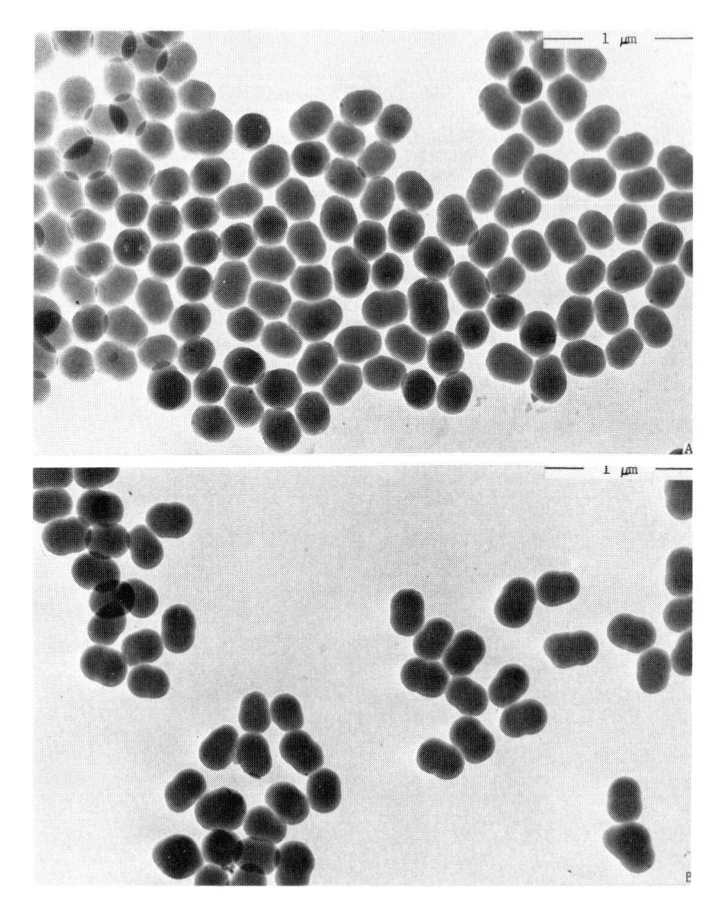

Fig. 3. Dispersions of seeded latex particles: A. Three parts VT polymerized onto one part 50S/50B copolymer seed latex. B. Three parts PS polymerized onto one part 50S/50B copolymer seed latex.

Ultrathin sections should show particle morphology in more detail. However, the conditions of staining with OsO_4 might affect the structure observed. To determine if the S/B copolymer was distorted when dried with the PVAc latex, samples of the seeded particles were stained with OsO_4 before and after casting the films. One part 60S/40B copolymer was polymerized onto one and one-half parts PS seed latex. This latex was mixed with PVAc latex, dried, embedded in epoxy resin, stained and sectioned as described above. The subject latex particles did not deform appreciably during the specimen preparation procedure and the stained copolymer can be readily observed in the EM (Fig. 4-A). It may be noted that the

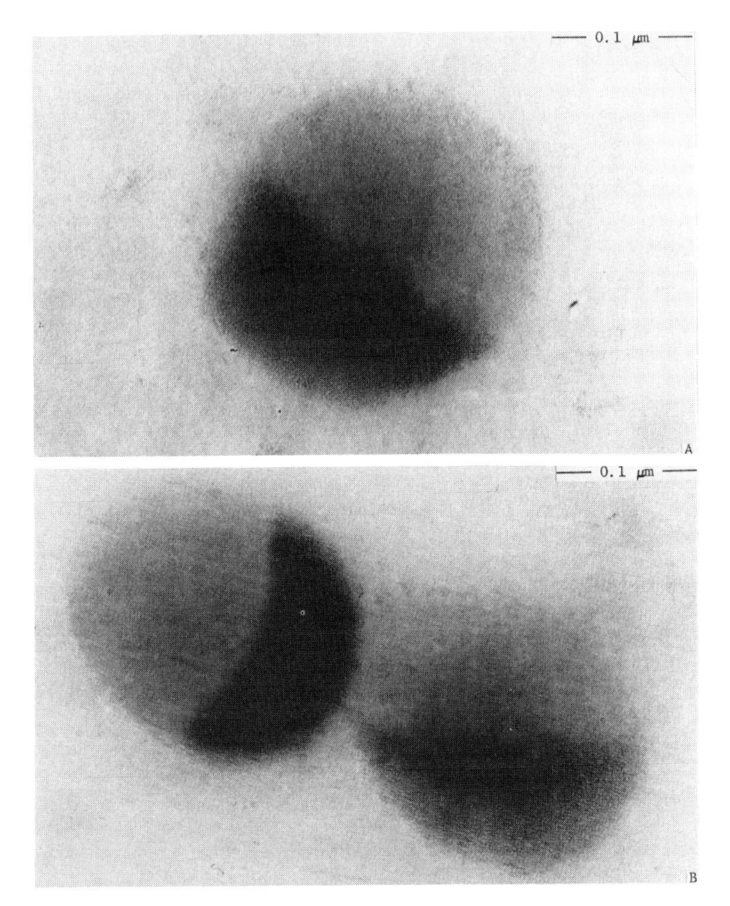

Fig. 4. Effect of OsO_4 staining conditions on the microstructure of sectioned particles composed of one part 60S/40B copolymer polymerized onto one and one-half parts PS seed latex: A. OsO_4 stained after embedment. B. OsO_4 stained before drying.

copolymer is not surrounding the unstained PS seed particle symmetrically. Similar results are seen (Fig. 4-B) when the subject latex was stained by OsO_4 vapors; i.e., staining the subject latex after it was mixed with the PVAc latex but before drying.

If the copolymer polymerized onto the seed particle in a true core-shell fashion, one would expect the particle to ap pear as a doughnut, with the stained copolymer symmetrically surrounding the PS seed. One might argue that the dark region is an artifact caused by the sectioning procedure. If this were true the dark regions should appear on the same side of all particles. This is not the case, e.g., in Fig. 4-B the dark regions of adjacent particles are almost at right angles to each other. It is apparent from Figs. 4-A and B that the sectioned particles which had been stained before and after embedment appear very similar; this indicates that these latex particles do not deform appreciably when the PVAc latex particles coalesce around them to form the embedment film.

Another experiment was carried out to investigate the morphology of seeded latex particles when the seed was a minor portion of the particle. Nine parts of 60S/40B copolymer were polymerized onto one part PS seed latex. A dispersion of these particles shows the darker PS seed particles embedded in the soft copolymer (Fig. 5-A). Again the PS seed is often,

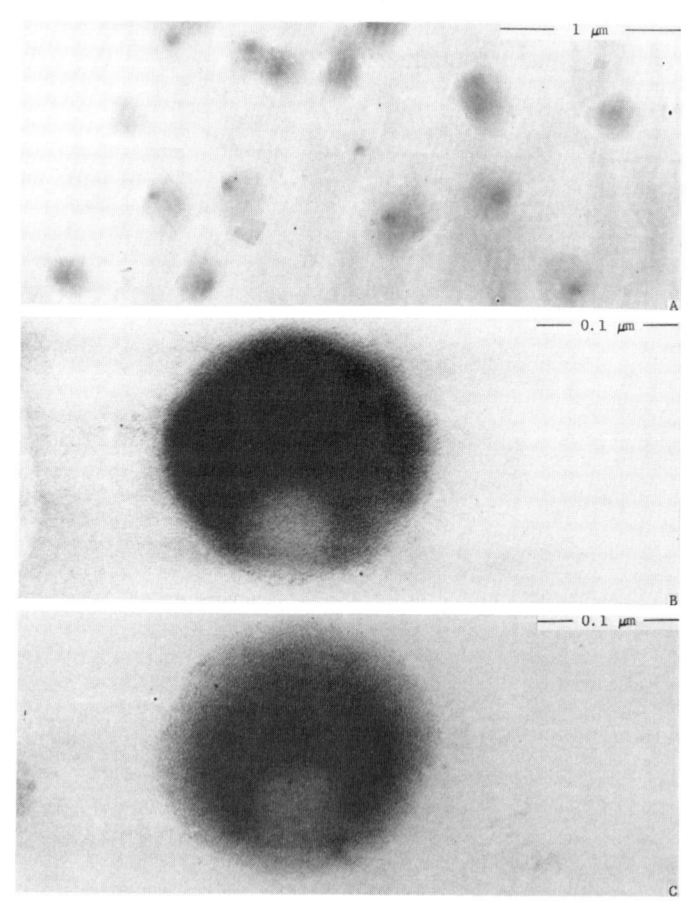

Fig. 5. A. Dispersion of latex particles containing nine parts 60S/40B copolymer polymerized onto one part PS seed latex. B. Cross section of a particle of the same latex stained after embedment. C. Cross section of a particle of the same latex stained before and after drying.

but not always, located near the edge of the particles. It is more difficult to see the seed particles in this system because the seed comprises only about one-tenth of the final particle.

Embedding these particles in PVAc and then staining showed the particle microstructure in more detail (Fig. 5-B). The light seed particle is readily observed near the edge of the stained S/B copolymer. Another sample of this latex was stained before embedment and the ultrathin section was restained. Hardening the particle before embedment and restaining the ultrathin section one hour in OsO_4 vapors did not alter the microstructure of the particles (Fig. 5-C). The seed particle of PS is seen near the edge of the stained copolymer.

A stained film cast at room temperature from seeded latex particles composed of one part 60S/40B copolymer polymerized onto one part PS seed showed a heterogeneous structure (Fig. 6-A). Ultrathin sections do not show PS seed particles stuck together by the stained copolymer as might be expected if all of the PS seed particles were uniformly coated with copolymer. Even in the cast film evidence of crescent shaped copolymer portions of particles are observed (see arrow B in Fig. 6-A). When another portion of the film was heated for one hour at 150°C before embedment, staining and sectioning, the microstructure was altered significantly (Fig. 6-B). This ultrathin section displays larger regions of coalesced PS and copolymer. Some of these regions are much larger than the primary latex particles, although the heat treatment did not cause complete coalescence into two distinct phases. A dispersion (Fig. 6-C) of the original latex displays the PS seed particles associated with the copolymer in a manner similar to the other seeded latex samples shown earlier. Again, most of the copolymer is not symmetrically surrounding the PS seed particles.

V. DISCUSSION

A method is described which permits electron microscopic observation of ultrathin sections of individual latex particles which have not been dried and redispersed in an embedding medium. Since redispersion of dried latex particles is extremely difficult, this new method allows many individual particles to be readily observed. The artifacts which may result from drying and redispersing the particles are eliminated using this technique.

Previous microscopic observations of latex particles composed of various ratios of S/B copolymer polymerized onto a PS seed latex showed doughnut-type morphology.(4-7) This indicates that the S/B copolymer is symmetrically encapsulating the seed particle. The detailed mechanism by which this

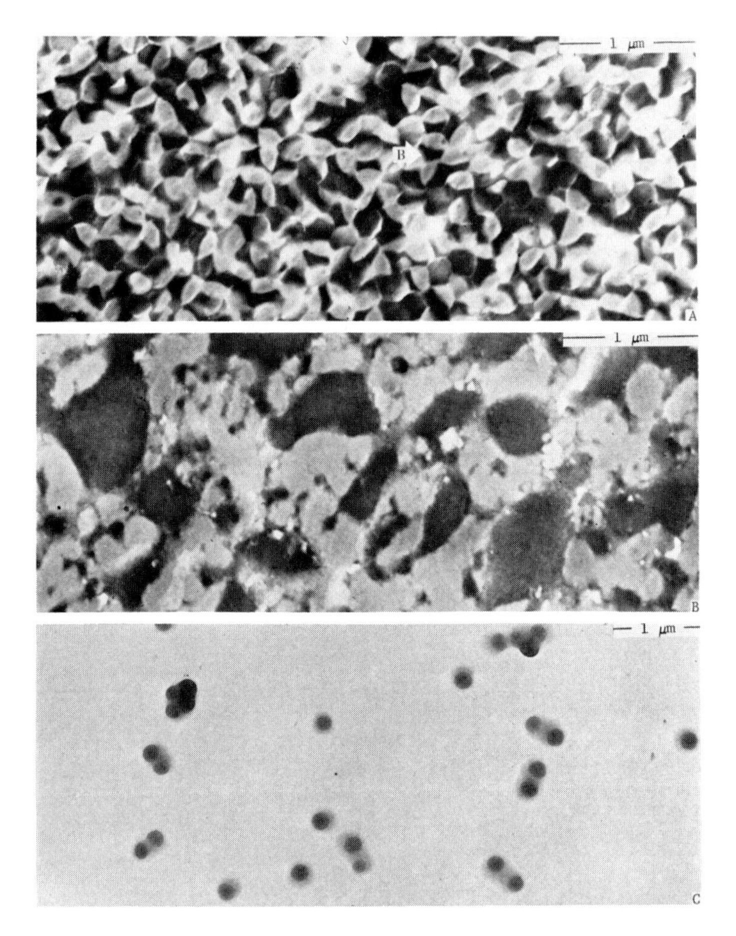

Fig. 6. A. Cross section of a stained film from latex particles composed of one part 60S/40B copolymer polymerized onto one part PS seed latex. B. Effect of heating the film at 150°C/1 hr. C. Dispersion of particles from the same latex.

occurs may not be known with certainty yet but many factors may contribute to it. Since the previously mentioned (4-7) electron micrographs show the second stage polymerization as a symmetrical shell around the seed particle, it is somewhat surprising that this symmetry was not observed in most of the particles observed in this work. One would expect that the incompatibility of the second-step copolymer would not alter the particle symmetry appreciably since a doughnut-type morphology was observed by Keusch, Prince and Williams (6) when the second-step copolymer of S/B contained 50% B.

The observation of ultrathin sections showing copolymer located at the edge of the particles lends support to the validity of particle dispersion observations which often show seed particles that are not encapsulated uniformly with copolymer. An explanation for this lack of particle symmetry will not be attempted at this time. The mechanism of particle morphology will be the subject of a future report. The microstructure of seeded latex particles prepared under various experimental conditions is currently being studied using this technique.

VI. ACKNOWLEDGMENTS

The authors gratefully acknowledge the suggestions and comments of several co-workers during this investigation. Especially appreciated are the suggestions of Dr. J. A. Schmitt concerning the use of poly(vinyl acetate) latex, the latex samples provided by Dr. D. I. Lee and Dr. R. D. Hansen, and the photographic assistance of W. D. Keiser.

VII. REFERENCES

1. Smith, W. V., J. Am. Chem. Soc. 70, 3695 (1948).
2. U.S. Patent 3,291,768 to H. L. Pfluger and C. G. Gebelein (The Borden Company) 12-13-66.
3. Bradford, E. B. and Vanderhoff, J. W., J. Polymer Sci. C-3, 41 (1963).
4. Grancio, M. R. and Williams, D. J., J. Polymer Sci. A-1, 8, 2617 (1970).
5. Keusch, P. and Williams, D. J., J. Polymer Sci. 11, 143 (1973).
6. Keusch, P., Prince, J., and Williams, D. J., J. Macromol. Sci.-Chem. A-7 (3), 623 (1973).
7. Keusch, P., Graff, R. A., and Williams, D. J., Macromolecules 7, 304 (1974).
8. Hatch, M. J., Yoshimine, M., Schmidt, D. L., and Smith, H. B., J. Am. Chem. Soc. 93, 4617 (1971); Schmidt, D. L., Smith, H. B., Yoshimine, M., and Hatch, M. J., J. Polymer Sci.-Chem. 10, 2951 (1972); Schmidt, D. L., Smith, H. B., and Broxterman, W. E., J. Paint Technol. 46, 41 (1974).
9. Kato, K., Polymer Eng. Sci. 7, 38 (1967).

ELECTROKINETIC STUDIES ON COLLOIDAL AND PRECIPITATED NICKEL HYDROXIDE*

V. Pravdić and N. Bonacci
Laboratory of Electrochemistry and Surface Phenomena,
"Rudjer Bošković" Institute
Zagreb, Croatia, Yugoslavia

ABSTRACT

The electrokinetic properties of nickel hydroxide, prepared by slow hydrolysis from $NiCl_2$ solutions, were studied using microelectrophoresis and streaming potential techniques.

The particles of a stable $Ni(OH)_2$ colloidal dispersion were found to be negatively charged throughout the pH range from 12 to 3 where electrophoretic mobility measurements were made. The isoelectric points (IEP) were determined by extrapolation of the mobility -pH curves. They were: 1.50 in HCl (variable ionic strength), 1.79 with NaCl, and 2.77 with $CaCl_2$ as background electrolytes (constant ionic strength). Experiments in the liotropic series of alkali chlorides showed decreasing mobilities at constant pH, and increasing values of the IEPs of 1.64, 1.77, 2.04, and 2.49 at constant, 10^{-3}M/l of LiCl, NaCl, KCl, and CsCl, respectively.

The IEP of precipitated, washed and dried (100^oC in vacuo) $Ni(OH)_2$ was found shifted to alkaline region: 9.7 in NaCl and 9.4 in CaCl solutions.

NiO, produced by thermal decomposition (250^oC in vacuo) of these precipitates was found to have the IEP at pH 11.0 in 10^{-3}M/l NaCl solutions.

The interpretation suggests that the acidic surface OH groups are structural elements of $Ni(OH)_2$. Successive dehydration results both in the aggregation of particles and in the appearance of hydrophobic parts of the surface.

---

*Taken in part from the M.Sc. Thesis of N. Bonacci. Work supported through Grant GF-31530 (P.L.480) by the National Science Foundation, Washington, D. C., and by a grant from the Council for Scientific Research of the Republic of Croatia.

I. INTRODUCTION

Nickel hydroxide and its thermal decomposition product nickel oxide, in their disperse, polycrystalline, high specific surface area form have been studied and utilized as materials for positive plates in some secondary battery systems (1), and as catalysts in various oxidation processes of hydrogen, carbon monoxide, hydrocarbons, and ammonia (2). Further, still unexplored, possibilities of application have prompted research on bulk and surface properties of these compounds in view of their ready availability and moderate cost.

Most of the available data on NiO and $Ni(OH)_2$ in contact with electrolyte solutions have resulted from research on the kinetics of electrochemical reactions (3,4,5,6). Considerably less work has been done and published on the double layer properties of the solid/solution interface. This paper is describing the results obtained studying the electrokinetic phenomena at the nickel hydroxide/solution interface, and a few experiments involving the oxide as the solid phase. The work described here is also a part of a more extensive study of preparation and surface characterization of these materials at the solid/gas and solid/liquid interface (7,8,9,10).

II. EXPERIMENTAL

A. Sample preparation

1. *Colloidal $Ni(OH)_2$.*

Nickel hydroxide was prepared by titrating 0.1M/l $NiCl_2$ solutions with 2M/l NaOH to pH 6.7. The solution was vigorously mixed. After two hours the colloid was centrifuged, washed repeatedly with redistilled water. Washings were continued until no change in the conductivity of the supernatant was measured. Immediately after the washings the hydrolizate was resuspended in the electrolyte solution required for the electrophoretic experiments, making it 10^{-3}M/l in $Ni(OH)_2$. The solutions were left to equilibrate for 24 hours before measurements were made. The final pH, referred to in figures, was measured immediately before filling the electrophoretic cell. The lowest pH attainable was at 3.0, since below this value rapid dissolution of $Ni(OH)_2$ colloid was observed.

2. *Precipitates of $Ni(OH)_2$.*

The hydrolizate, obtained as above, was centrifuged, dried at 100°C in vacuo for 24 hours, then brought to open air, ground in an agate mortar, and stored. When needed, portions of it were resuspended in the electrolyte solution to be measured. These samples were used in streaming potential measurements only. One sample, taken from a batch used in most of the experiments had a specific surface area (argon, BET, 0.168 nm^2/molecule) of 167 m^2/g.

3. *Nickel oxide, NiO.*

The hydrolizate, obtained as above, was transferred into a glass sample tube, evacuated and heated to 100°C for approximately 24 hours. After that the temperature was brought swiftly to 250°C maintaining a dynamic vacuum of 10^{-3} Pa or better. The conversion to a final product of the composition NiO x 0.12 H_2O was completed in 40 hours (9). Prepared from the same batch of $Ni(OH)_2$ of 167 m^2/g the product showed a specific surface area of 217 m^2/g.

4. *Chemicals.*

All the chemicals used were analytical purity grade (Merck, Darmstadt, W. Germany) and were used without further purification. Water was distilled four times, twice from a glass still under vacuum, and then twice in an all-quartz still. Dynamic surface tension measurements were used to ascertain absence of surface active impurities, and conductivity measurements to test for electrolytes.

Other methods of colloid preparation and purification were tried. However, any treatment which required the colloid to remain in an electrolyte solution for more than 72 hours before measurements, resulted in irreproducible electrophoretic mobility - pH relations. This was attributed to incipient recrystallization and change in the habitus of particles, as observed by electron microscopy and X-ray diffraction analysis.

B. Techniques

1. *Microelectrophoresis.*

The electrophoretic mobilities of colloidal nickel hydroxide were measured in a commercially available microelectrophoresis apparatus (Rank Brothers, Cambridge, England). The mobility of particles was observed using a flat, narrow slit, rectangular quartz cell in the light transmission mode. The microscope viewing field was focused to the profile where the electroosmotic contribution was eliminated. Non-gassing Ag/AgCl electrodes and field strength between 3.9 and 9.7 V/cm were used. Each point in the mobility determination has been the mean value of 10 measurements in both directions for at least three different particles.

2. *Streaming potential techniques.*

The electrokinetic potential of coarse particles of $Ni(OH)_2$ precipitates and of its thermal decomposition product NiO x 0.12H_2O was measured using the streaming potential (or current) techniques (11). The result of the measurement is a trace of the potential - pressure relationship on an X-Y recorder, obtained by a pulse of liquid, in which the hydrostatic pressure decays with time. Each point in the figures represents the mean of 4 measurements where linear dependence of streaming potential vs. pressure was obtained. The streaming cell featuring a fritted glass disc as support, was filled with the material studied by suspension settling, minimizing thus possible contamination of the sample. Short liquid pulses prevented electrolysis effects on the material studied, as well as polarization of the platinized platinum electrodes.

3. *Computations.*

Most of the results of electrophoretic studies are expressed as mobilities in dimensions of cm^2/V sec. If needed (cf. Table II), the mobilities were converted into electrokinetic potentials using the tables of Ottewill and Shaw (12). Streaming potential data were converted into the electrokinetic potentials using the simple Helmholtz-Smoluchowsky equation, bulk conductance values, as measured in a separate cell, and bulk values for the viscosity and dielectric constant of water. No corrections for surface conductivity were applied, as all the measurements were done in solutions of considerable electrolyte concentration and conductivity.

III. RESULTS

A. Electrophoretic mobilities of colloidal $Ni(OH)_2$

Results of measurements of electrophoretic mobilities of colloidal $Ni(OH)_2$ are shown in Figs. 1 to 3. Fig. 1 shows the results of measurements of mobilities in dependence of pH in solutions with constant concentrations of NaCl. Fig. 2 shows the same relationship in solutions where either no neutral electrolyte was added, or in solutions with constant concentration of $CaCl_2$. The mobilities indicate a negative surface charge throughout the pH range between 12 and 3. The points were fitted to an empirical relation of the form.

$$\mu = A * (\ln pH - \ln B) \tag{1}$$

where μ is the mobility, A is a constant, and B is, by virtue, the pH of the isoelectric point (IEP). Values of IEP were obtained by numerical extrapolation.

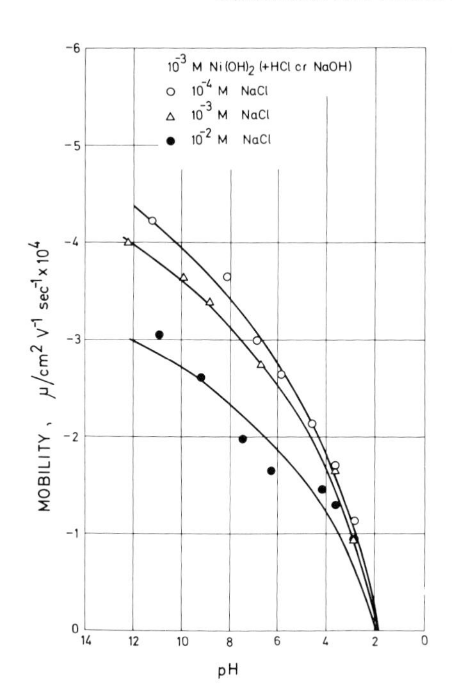

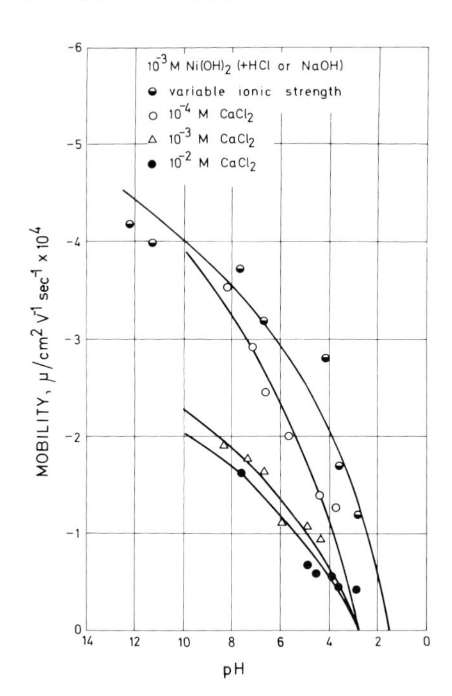

Fig. 1. The pH dependence of the electrophoretic mobility of colloidal $Ni(OH)_2$ in solutions with constant concentrations of NaCl.

Fig. 2. The pH dependence of the electrophoretic mobility of colloidal $Ni(OH)_2$ in solutions with variable ionic strength (HCl added to pH), and in solutions with constant concentrations of $CaCl_2$. Solid lines are least square fittings to Eq. 1 (see text).

Fig. 3 shows the results of mobility measurements in solutions in which the concentration of chlorides of Li, Na, K, and Cs was held constant at 10^{-3}M/l. The points show a considerable amount of scatter. Fitted to the empirical relation, Eq. (1), by least squares curvilinear regression (shown by the solid lines) a "normal" liotropic sequence emerges, i.e. the mobilities at constant pH decrease in the sequence Li, Na, K, Cs.

Table I summarizes the calculated values of A and B (Eq. 1) obtained for the experiments described in Figs. 1 to 3.

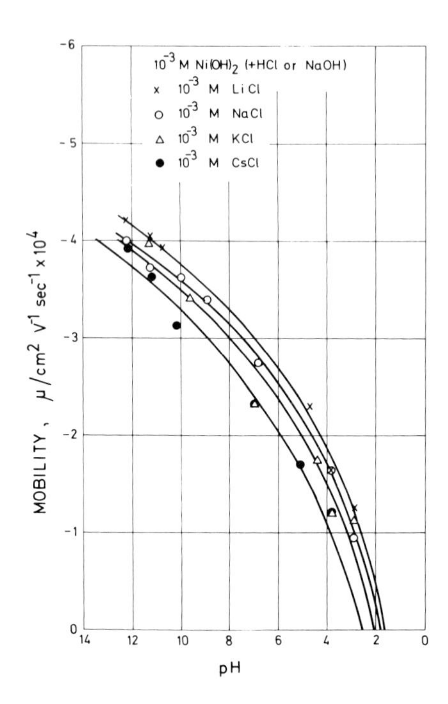

Fig. 3. The pH dependence of the electrophoretic mobilities of colloidal $Ni(OH)_2$ in 10^{-3}M/l solutions of alkali chlorides (liotropic series).

TABLE I
Electrophoretic Mobilities of Colloidal $Ni(OH)_2$ and Extrapolated Values of the IEP. Parameters of the Equation: $\mu = A^*(\ln pH - \ln B)$

Electrolyte	Concentration M/l	A 10^4 cm^2/V sec	B pH
HCl	variable to pH	2.13	1.50
NaCl	10^{-4}	2.33	1.83
	10^{-3}	2.08	1.77
	10^{-2}	1.56	1.78
$CaCl_2$	10^{-4}	3.13	2.76
	10^{-3}	1.79	2.78
	10^{-2}	1.46	2.78
Liotropic sequence:			
LiCl	10^{-3}	2.08	1.64
NaCl	10^{-3}	2.08	1.77
KCl	10^{-3}	2.22	2.04
CsCl	10^{-3}	2.38	2.49

B. Electrokinetic Potentials of Precipitates of $Ni(OH)_2$

Results of measurements of electrokinetic potentials of precipitates of $Ni(OH)_2$ and of NiO are shown in Figs. 4 to 6. Fig. 4 shows the results obtained in solutions with NaCl as neutral electrolete, Fig. 5 for those in $CaCl_2$. Two observations can be made on inspecting the figures. First, that the potential - pH relationship is a linear function for the range of 2 pH units below and above the IEP. Second, that the IEP is found at pH values in the alkaline region, a shift of some 7 pH units from the values observed for colloidal $Ni(OH)_2$.

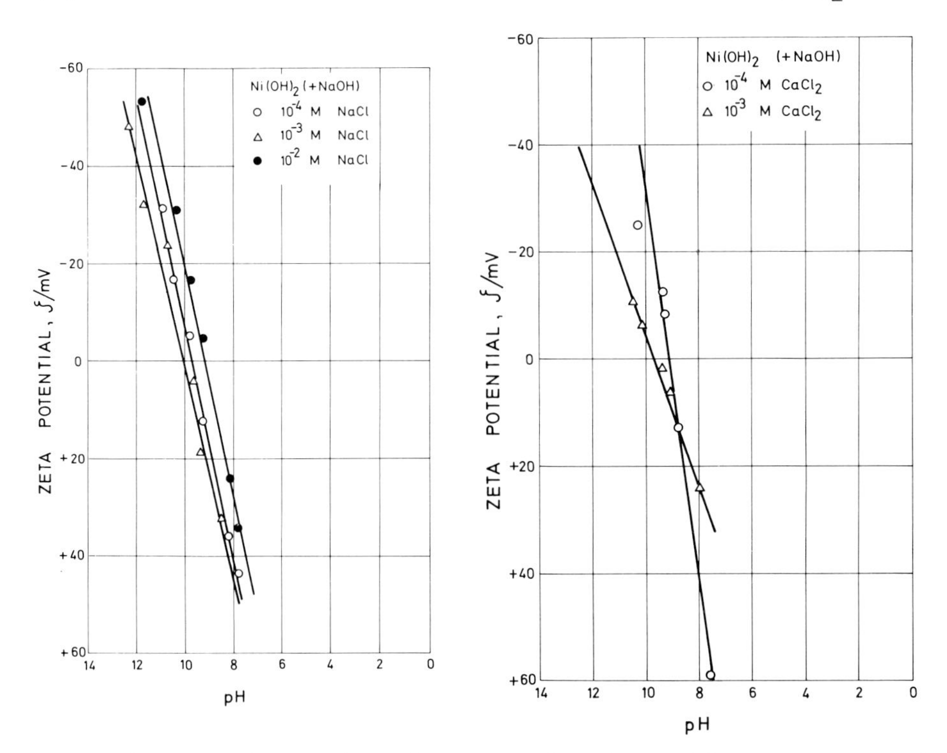

Fig. 4. The pH dependence of the electrokinetic potential of precipitates of $Ni(OH)_2$ in solutions with NaCl. The pH is adjusted by additions of NaOH.

Fig. 5. The pH dependence of the electrokinetic potentials of precipitates of $Ni(OH)_2$ in solutions with $CaCl_2$.

Only one series of experiments has been done with the NiO thermal decomposition product, and the results are shown in Fig. 6, next to the results for $Ni(OH)_2$. The slope of the potential vs. pH is linear but the range of linearity is narrower. There is a change in slope at lower pH. It is also indicative that under identical conditions of measurements NiO has an IEP shifted to even higher pH values than those observed for precipitates of $Ni(OH)_2$.

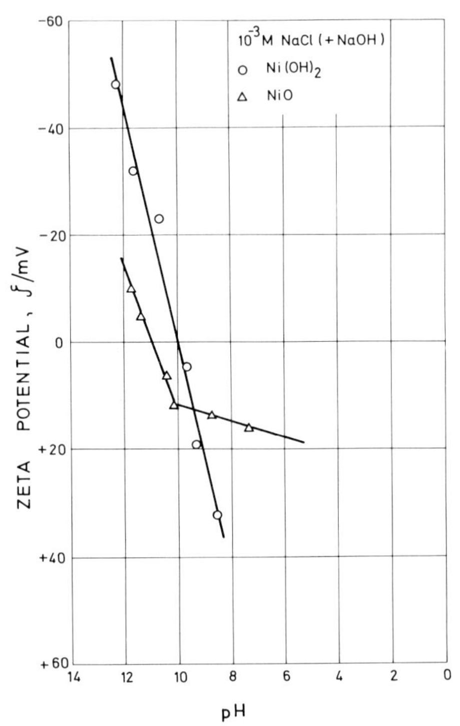

Fig. 6. The pH dependence of electrokinetic potentials for $Ni(OH)_2$ and for its thermal decomposition product, NiO x $0.12H_2O$ in solutions of NaCl.

C. Summary of results

From the results described in the preceeding sections, the most notable features are the shift of the IEP with pretreatment of the hydroxide, and the accompanying, and significant, decrease in the $d\xi/d$ pH values. Table II is an overview of all the results obtained for the three forms of nickel compounds studied in the present work.

IV. DISCUSSION

Few data are available in the literature on the electrokinetic properties of the nickel oxide/hydroxide system. There is reference in Parks' review (13) to some unpublished data and to some older work (14). The only other reference is the paper by Herczynska and Proszynska (15) for NiO formed on Ni metal. An IEP of 7.9 was found. However, the experimental

TABLE II
The derivative dξ/dpH at the Isoelectric Points for Colloidal and Precipitated $Ni(OH)_2$, and for NiO x $0.12H_2O$

Electrolyte	Concentration M/l	IEP pH	dξ/dpH mV/pH
	1. Colloidal $Ni(OH)_2$		
HCl	variable to pH	1.50	-33*
NaCl (+ HCl)	10^{-4}	1.83	-28
	10^{-3}	1.77	-30
	10^{-2}	1.78	-27
$CaCl_2$ (+ HCl)	10^{-4}	2.76	-27
	10^{-3}	2.78	-14
	10^{-2}	2.78	-11
	2. Precipitated $Ni(OH)_2$		
NaCl (+ NaOH)	10^{-4}	9.70	-24
	10^{-3}	10.01	-21
	10^{-2}	9.25	-23
$CaCl_2$ (+ NaOH)	10^{-4}	9.25	-34
	10^{-3}	9.57	-11
	3. Nickel oxide, NiO x $0.12H_2O$		
NaCl (+ NaOH)	10^{-3}	11.0	-13 (at pH >10) - 1.5 (at pH <10)

*Values in this column were taken from the tangent of the mobility vs. pH curve at its maximum slope near the extrapolated IEP. Conversion to electrokinetic potential was done using tables (12).

conditions of sample preparation used in these experiments are either unknown, or have produced films of NiO rather than a bulk phase.

All the results reported in the present work have been obtained using the two types of electrokinetic techniques described. Points of zero charge (PZC) would be of more interest and interpretative value than the IEP values obtained here. However, if the PZCs are anywhere close to the IEPs (and this is a reasonable assumption), then the potentiometric titration experiments would be difficult to perform with the required precision. Indeed, the need to develop techniques of precise measurements of adsorption of H^+ or OH^- near pH 2 or 11, has probably caused a conspicuous absence of reported work in a widely used, and theoretically important material. With these shortcomings of techniques in mind, the electrokinetic measurements did provide some new insight into this solid/liquid interface, and will be discussed briefly here.

A. The origin of charge at the nickel hydroxide/electrolyte solution interface

Results of measurements of both the electrophoretic mobilities and of streaming potentials have been recalculated into the electrokinetic potentials. The slopes, $d\xi/dH$ in vicinity of the IEPs were estimated graphically. These values are used in discussion of the most probable models for the solid/liquid interface.

Using the simple Gouy-Chapman model for the electrochemical double layer at the oxide-electrolyte interface, Anderson (16) and Hunter and Wright (17) have, inter alia, developed models of constant surface charge and that of pH dependent adsorption at constant ionic strength.

The first model predicts a $d\xi/dpH = 59$ mV/pH. This model is appealing on the basis of purely qualitative considerations of a constant concentration of surface OH groups, capable of supporting proton transfer along the surface. Such a surface would function as an ideal hydrogen electrode with a Nernstian response to changes in pH of the solution. There is no support for this model as shown in Table II. Rather values close or less than 30 mV/pH were found. Thus the more probable choice is the second model, that of pH dependent adsorption of potential determining ions. The same type of analysis revealed a similar mechanism for quartz and vitreous silica surfaces (18). In the derivation of expression for this model (16) assumption was made that the adsorption of potential determining ions follows an idealized Langmuir type adsorption isotherm, and, implicitly, that the resulting surface coverage is low. With increasing coverage and with a higher degree of irreversible behavior, the values of $d\xi/dpH$ are expected to decrease. $Ni(OH)_2$ could satisfy this model assuming a low statistical average number of surface sites for adsorption of potential determining ions. In many cases this has been shown as unrealistic. Cronan (10) has advanced a hypothesis that in microcrystalline $Ni(OH)_2$ the hexagonal platelet surfaces (corresponding to a Miller index (0001) plane) are hydrophobic, and the edges hydrophilic. With the spontaneous growth of particles of $Ni(OH)_2$ in the course of preparation of this material from aqueous solutions, there is a spontaneous tendency toward increased hydrophobicity. For NiO it has been shown (10) that this material is almost 100% hydrophilic, and the observed low values for the slope $d\xi/dpH$ are in rough accordance with this model. Another possibility is that some Ni^{2+} hydrolytic species, formed by dissolution of the solid phase, might become potential determining. Thus, a model should be developed incorporating both a dual surface mechanism and two simultaneously operative potential determining ions. Present lack of reliable adsorption data makes further model development useless.

B. Electrostatic surface field model and the IEP

The IEP of nickel hydroxide is shifted from the low values observed for colloidal modification to appreciably higher values for particles which exhibit microcrystallinity. Healy, Herring and Fuerstenau (19) and Murray (20) have reported similar observations in their studies of the MnO_2 system. Amorphous δ-MnO_2 has been shown to have an IEP near pH=1.5, and with increased crystallinity the IEP, as measured by electrophoresis and by coagulation and sedimentation experiments shifts to values between 4.6 and 7.3. Parks (13) used equation (2) to account for the crystal field stabilization energy and the contributions of ionic charge to the IEP:

$$IEP = A - B(Z/R + 0.0029\ C + a) \qquad (2)$$

A and B are arbitrary constants, Z is the charge of the cation, $R = r_+ + 2r_o$, with r_+ being the radius of the cation and r_o the radius of oxygen, C is the crystal field stabilization energy, and a is an arbitrary constant accounting for the coordination number and the degree of hydration. Using values for A and B of 18.6 and 11.5 (derived from the work on oxides like La_2O_3, Fe_2O_3 and WO_3) for Z = 2, r_+ = 0.091 nm, r_o = 0.14nm, and a value of 42 kcal obtained from values reported for the $Ni(H_2O)_6^{2+}$ ion of a coordination number 6, for which a = 0, an IEP of 11.0 was predicted for NiO. Indeed, this prediction is borne out by experiments. In the above deduction, however, there are no data which would indicate what is the difference in the constants A and B, and in the value of the crystal field stabilization energy for NiO and $Ni(OH)_2$, respectively. The latter comes much closer to the symmetry of the $Ni(H_2O)_6^{2+}$ ion, on which the calculations of the CFSE have been based. And the IEP of the hydroxide is, as shown, several pH units away from this prediction. Therefore this treatment is also highly qualitative. A further understanding of the energetics and the structure of the interface would require independent calculations of the mentioned constants.

A further qualitative correlation is satisfied. Healy and Fuerstenau (21), basing their observations on the theory of Zettlemoyer and coworkers (23), have shown that there is a linear correlation between the heats of immersion and the IEP for a number of oxides. The heats of immersion for precipitated $Ni(OH)_2$ and NiO x $0.13H_2O$, have been measured by Cronan (10) as 330 and 560 mJ/m^2, respectively. The shift of the IEP from low energy surface of precipitated $Ni(OH)_2$, to the high energy surface of NiO is in the predicted direction. Quantitative correlation is missing, since this difference in the heats of immersion would account for a shift of approximately 3 units of pH in the PZC. The almost twice as large a shift cannot be possibly accounted for in the difference between the PZC and the IEP.

Healy, Herring and Fuerstenau (19) have shown that the PZC is inversely proportional to $V^{2/3}$ (V is the volume per unit cell). For $Ni(OH)_2$ V = 0.039 nm^3, for NiO 0.018 nm^3 (10). Thus the PZC, and accordingly the IEP, is expected to increase for NiO with respect to $Ni(OH)_2$. This prediction is corroborated by experiments. In addition the normal liotropic sequence observed for the colloidal $Ni(OH)_2$ is, once again, pointing out the similarities between this material and nonporous silica, and is in accordance with the general aspects of the Eisenman theory as interpreted by Stumm (23), and Dumont and Watillon (24).

As indicated above, further advancement of the knowledge on the structure and the energetics of the interfacial layer of nickel oxide and nickel hydroxide in electrolyte solutions would require precise measurements of the adsorption of potential determining ions. The role of the Ni^{2+} hydrolytic species at this interface should be elucidated, as there is indication that it might become potential determining besides the H^+ and OH^- ions. Concerning the electrostatic contribution to the interfacial energy, the hydrophobic platelet-hydrophilic edge model is equivalent to a p-n junction of a thin film semiconductor. Thus the presently used and widely discussed models for the oxide/electrolyte interface (25) will have to be supplemented to account for this horizontal electrostatic field component.

Acknowledgment. The skillful help of Mr. F. Matijevac in surface area determinations is gratefully acknowledged. The authors had many invaluable discussions with Professors H. Leidheiser, Jr. and F. J. Micale on several occasions, both in Zagreb and in Bethlehem. Dr. C. L. Cronan has given us access to much of his unpublished work and supplied information facilitating understanding of the nickel oxide and hydroxide systems. The paper was written during the tenure of a National Science Foundation Senior Scholarship awarded to V. P. at Lehigh University, Bethlehem, PA. USA.

V. REFERENCES

1. Collins, D. H., Batteries 2, Symposium Publ. Div. Pergamon Press, Oxford 1965.
2. Gravelle, P. C. and Teichner, S. J., Advan. Catalysis 20, 167 (1969).
3. Yohe, D., Riga, A., Greef, R., and Yeager, E., Electrochim. Acta 13, 1351 (1968).
4. Vijh, A. K., "Electrochemistry of Metals and Semiconductors", p. 117 ff. Marcel Dekker, Inc. New York, 1973.
5. Conway, B. E., in "Techniques of Electrochemistry" (E. Yeager and A. J. Salkind, Eds.) Vol. 1, p. 427 ff. Wiley-Interscience, New York, 1972.
6. Vertes, G., Horanyi, G., and Nagy, F., Croat. Chem. Acta 44, 21 (1972).
7. Bonacci, N. and Novak, D. M., Croat. Chem. Acta 45, 531 (1973).
8. Topic, M., Micale, F. J., Cronan, C. L., Leidheiser, H. Jr., and Zettlemoyer, A. C., ACS Symposium Ser. 8, 225 (1975).
9. Bonacci, N., M. Sc. Thesis, Univ. of Zagreb (Yugoslavia), 1975.
10. Cronan, C. L., Ph.D. Thesis, Lehigh University, Bethlehem, Pa., 1976.
11. Pravdić, V., Croat. Chem. Acta 35, 233 (1963).
12. Ottewill, R. H., and Shaw, J. N., J. Electroanal. Chem. 37, 133 (1972).
13. Parks, G. A., Chem. Rev. 65, 177 (1965).
14. Mattson, S., and Pugh, A. J., Soil Sci. 38, 229 (1934); cit. ref. 13.
15. Herczynska, E., and Proszynska, K., Naturwiss. 50, 351 (1963).
16. Anderson, P. J., Trans. Faraday Soc. 54, 562 (1958).
17. Hunter, R. J., and Wright, H. J. L., J. Colloid Interf. Sci. 37, 564 (1971).
18. Jednačak, J., Pravdić, V., and Haller, W., J. Colloid Interf. Sci. 49, 16 (1974).
19. Healy, T. W., Herring, A. P., and Fuerstenau, D. W., J. Colloid Interf. Sci. 21, 435 (1966).
20. Murray, J. W., J. Colloid Interf. Sci. 46, 357 (1974).
21. Healy, T. W., and Fuerstenau, D. W., J. Colloid Sci. 20, 376 (1965).
22. Chessick, J. J., and Zettlemoyer, A. C., Advan. Catalysis 11, 263 (1959).
23. Stumm, W., Huang, C. P., and Jenkins, S. R., Croat. Chem. Acta 42, 223 (1970).
24. Dumont, F., and Watillon, A., Disc. Faraday Soc. 52, 352 (1971). cf. also the discussion with Th. F. Tadros and J. Lyklema, Ibid. p. 372 ff.
25. Lyklema, J., Croat. Chem. Acta 43, 249 (1971).

STUDY OF THE DISSOLUTION AND ELECTROKINETIC BEHAVIOR OF TRICALCIUM ALUMINATE

Maher E. Tadros, Wanda Y. Jackson,
and Jan Skalny
Martin Marietta Laboratories

ABSTRACT

Tricalcium aluminate dissolves incongruently in water or aqueous HCl solutions and yields an aluminum enriched surface which adsorbs calcium ions from alkaline solutions and becomes positively charged. The dissolution in HCl solutions is associated with an induction period. The activation energy for the processes which take place during the induction period was found (15°-$46^{o}C$) to be 24.6 ± 1 kcal $mole^{-1}$. Sulfate ions adsorb on the positively charged surface following a Langmuir isotherm. This adsorption is associated with an increase in the duration of the induction period and a concurrent decrease in the surface charge. However, no charge reversal occurred. A mechanism is proposed for the reduced reactivity of tricalcium aluminate in the presence of gypsum, which does not involve the formation of 6-calcium aluminate trisulfate-32-hydrate (ettringite).

I. INTRODUCTION

Tricalcium aluminate, $Ca_3Al_2O_6$, is one of the four major constituents of portland cement. The early setting characteristics of portland cement are predominately determined by the chemical interactions involving $Ca_3Al_2O_6$ and water. The hydrolysis and hydration of $Ca_3Al_2O_6$, as well as the other constituents of cement, are associated with a number of complex surface and colloid chemical phenomena which affect the setting and hardening properties (1). In spite of the obvious need for a comprehensive study of the colloidal properties of these systems, only limited information is reported in the literature. $Ca_3Al_2O_6$ reacts with water, with the evolution of heat, to form a number of hydrates. The most stable hydrate is tricalcium aluminate hexahydrate, $Ca_3Al_2(OH)_{12}$. In the presence of $CaSO_4 \cdot 2H_2O$ (gypsum) the reactivity of $Ca_3Al_2O_6$ is reduced, and it is for this reason that gypsum is added to portland cement to regulate its early setting (2). Two compounds are produced from the $Ca_3Al_2O_3$ - gypsum - water interaction, namely calcium aluminate trisulfate hydrate, $\{Ca_6[Al(OH)_6]_2 \cdot 24H_2O\}[(SO_4)_3 \cdot 1\frac{1}{2}H_2O]$, commonly known as ettringite, and calcium aluminate monosulfate hydrate, $[Ca_2Al(OH)_6][0.5(SO_4) \cdot 3H_2O]$ (2). The reduced reactivity of $Ca_3Al_2O_6$ in the presence of gypsum has been explained in terms of the formation of a layer of ettringite around the $Ca_3Al_2O_6$ particles (3). This layer is thought to be more or less impervious to water.

This paper reports on the dissolution and electrokinetic behavior of $Ca_3Al_2O_6$ in water and in sulfate-containing solutions. Dissolution experiments were carried out both in water and in hydrochloric acid solutions. The acid solution was chosen to allow monitoring of the dissolution process, which is extremely fast in neutral pH water. The effect of sulfate and other ions on the rates of dissolution and the zeta-potential of reacting $Ca_3Al_2O_6$ particles was also determined. No ettringite formation was detected under the conditions employed. Evidence for the adsorption of sulfate ions was found and is proposed to account for the reduction in the reactivity of $Ca_3Al_2O_6$.

II. EXPERIMENTAL

A. Materials

Tricalcium aluminate was obtained from the Portland Cement Association, Skokie, Illinois (Batch Number B-299, Blaine surface area of 3420 cm^2g^{-1}). Analytical reagent grade chemicals and doubly-distilled deionized water were used.

B. Methods

Dissolution experiments were carried out in a double wall thermostatually controlled pyrex glass vessel. Two g of $Ca_3Al_2O_6$ and 75 ml of HCl solutions were employed. N_2 gas was bubbled into the solution throughout the experiments to prevent carbonation, and the solution was stirred by a teflon coated magnetic bar. In some experiments, various amounts of a 1.16×10^{-2} M gypsum solution were added, but the total volume of the final solution was kept to 75 ml. The reaction was initiated by the addition of $Ca_3Al_2O_6$. Subsequent changes in pH as a function of time were recorded using the Orion Model 601 pH Meter.

When needed, the calcium and aluminum ion concentrations were determined by withdrawing samples after suitable time intervals, quickly filtering through a fine pore fritted pyrex glass filter, and analyzing the filtrate. Calcium was determined by titration with EDTA; the aluminum, colorimetrically using the dye eriochrome cyanine R.

Zeta potential measurements were carried out using a Zeta-Meter apparatus made by Zeta-Meter Inc., New York. Suspensions of 0.1 g $Ca_3Al_2O_6$ in 25 ml of solution were used at room temperature.

ESCA and Auger spectroscopic methods were used to examine the surface compositions of partially reacted and unreacted $Ca_3Al_2O_6$. A Physical Electronics Model 548 instrument was used.

Microscopic examinations were conducted on JEOL JSM-U3 scanning electron microscope.

III. RESULTS AND DISCUSSION

A. Dissolution

The liquid phase compositions of mixtures of $Ca_3Al_2O_6$ and water (water:$Ca_3Al_2O_6$ between 25 and 250) were determined after one minute of mixing. The $[Ca^{2+}]:[Al^{3+}]$ in the filtrates was found to be about 1.9 ($[Ca^{2+}] \sim 1.05 \times 10^{-2}$ M). This ionic molar ratio value is higher than the stoichiometric value of 1.5, eq. [1]:

$$Ca_3Al_2O_6 + 6H_2O = 3Ca^{2+} + 2Al(OH)_4^- + 4OH^-. \qquad [1]$$

The surface composition of the solid residue was examined by ESCA and Auger spectroscopic methods. The Ca to Al ratio was found to be about 1:1. Surface milling by argon ion sputtering for 30 minutes resulted in a gradual change of the Ca:Al to the expected value of 1.5 for the bulk material. These results suggest that $Ca_3Al_2O_6$ dissolves incongruently in water. The solid residue was examined by scanning electron microscopy (SEM) and none of the known hydration products of $Ca_3Al_2O_6$ was observed.

The dissolution of tricalcium aluminate in aqueous HCl was studied in order to obtain more information on its very rapid dissolution in water. Typical curves for the variation of pH as a function of time are shown in Fig. 1. The pH

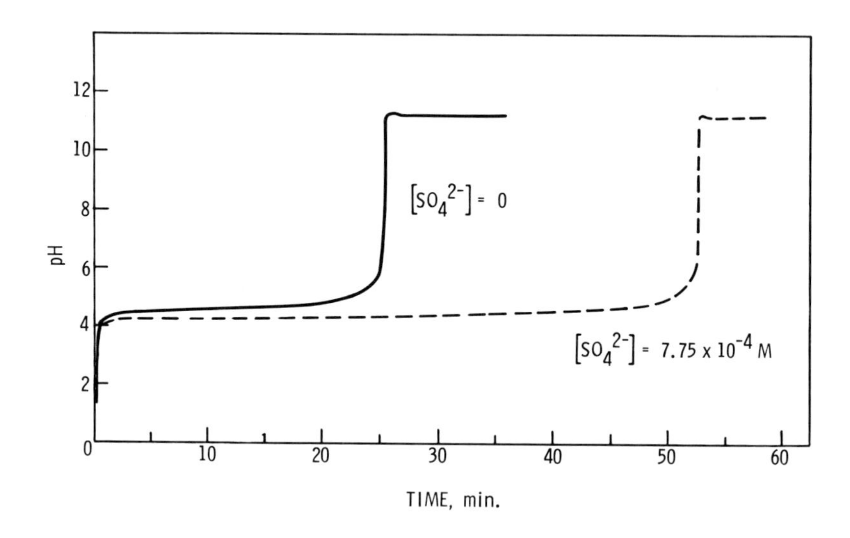

Fig. 1. Plots of rates of dissolution of 2 g $Ca_3Al_2O_6$ in 75 ml of 0.166 M HCl solution. Effect of gypum.

increases abruptly from about 1 to about 4.2, and then increases very slowly. After an apparent induction period, the pH again increases quickly to about 11.2. The duration of the induction period was found to be independent of the stirring rate between 200 and 900 rpm, suggesting that the rate determining process is not diffusion through the bulk of the solution. The corresponding changes in the calcium- and aluminum-ion concentrations are plotted in Fig. 2. These data indicate that most of the original acid was consumed during the first one or two minutes. The initial Ca:Al was 2, indicating an incongruent dissolution. During the induction period the $[Al^{3+}]$ decreases and the $[Ca^{2+}]$ increases. About one to three minutes prior to the end of the induction period a sharp increase in $[Ca^{2+}]$ takes place and a transient decrease in the $[Al^{3+}]$ is observed. These changes appear to be associated with the hydrolysis of aluminum ions to yield polynuclear complexes, or perhaps $Al(OH)_3$, and hydrogen ions, eq. [2]:

$$xAl^{3+} + yH_2O = Al_x(OH)_y^{3x-y} + yH^+. \qquad [2]$$

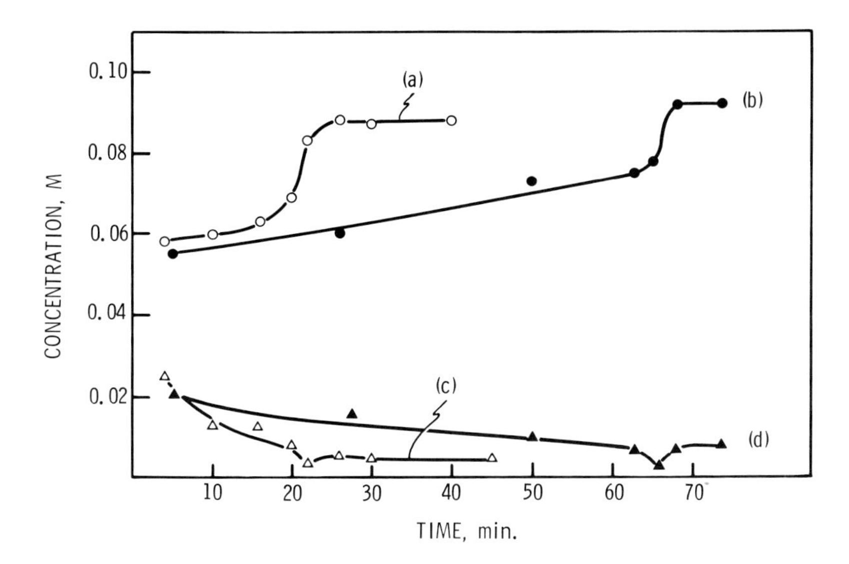

Fig. 2. Dissolution of 2 g $Ca_3Al_2O_6$ in 75 ml of 0.166 M HCl solutions. Plots of $[Ca^{2+}]$ and [aluminum] against time. (a) and (c) for $[Ca^{2+}]$ and [aluminum] in the absence of $CaSO_4$, respectively; (b) and (d) for $[Ca^{2+}]$ and [aluminum] in the presence of 1.55×10^{-3} M $CaSO_4$, respectively.

The acid produced will then dissolve additional calcium. The dependence of the duration of the induction period on the original [HCl] is shown in Fig. 3.

Experiments were performed at temperatures between 15° and 46^oC, using an initial [HCl] of 0.166 M. The reciprocal of the induction period, 1/t (related to the rate constant of the rate determining process), was used to construct the Arrhenius plot shown in Fig. 4. From the plot an activation energy of 24.6 ± 1 kcal $mole^{-1}$ was calculated. This value suggests that the rate controlling process is chemical.

In the presence of small concentrations of calcium sulfate the induction period t was found to increase (Fig. 1, molar $CaSO_4:Ca_3Al_2O_6 = 7.85 \times 10^{-3}$). The variation of the reciprocal of the induction period as a function of $[SO_4^{2-}]$ is

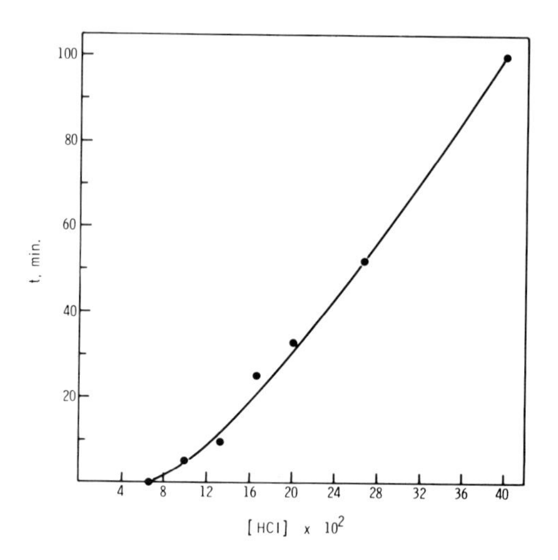

Fig. 3. Dissolution of 2 g $Ca_3Al_2O_6$ in 75 ml solutions of various concentrations. Plot of t against [HCl].

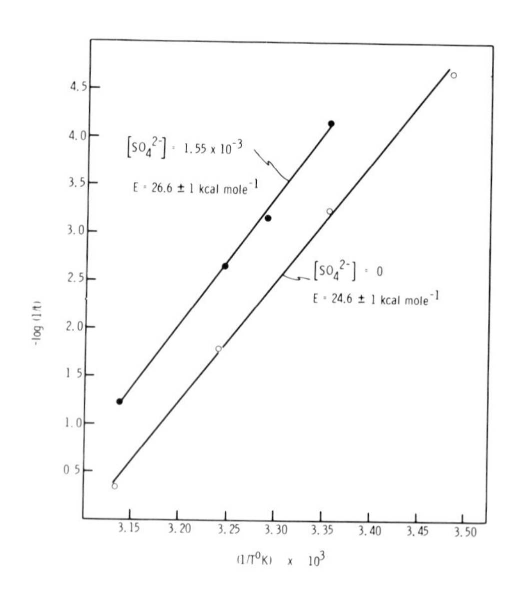

Fig. 4. Arrhenius plots of log 1/t against $1/T^oK$ for dissolution of $Ca_3Al_2O_6$ from 15° to 46°C. Effect of gypsum.

depicted in Fig. 5. If the increase in the length of the

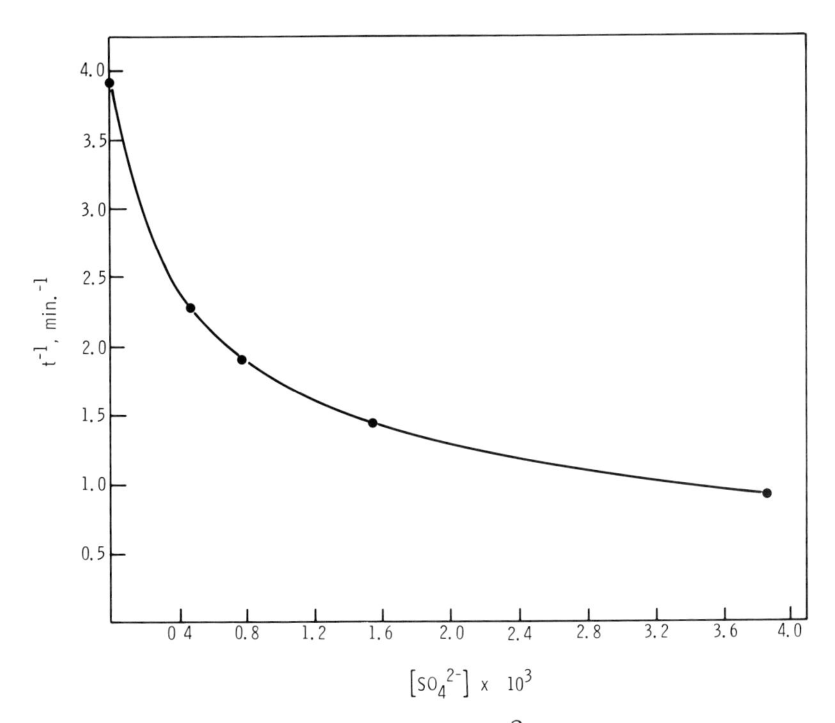

Fig. 5. Plot of 1/t against $[SO_4^{2-}]$ for the dissolution of $Ca_3Al_2O_6$ in HCl solutions in the presence of gypsum.

induction period is a result of simple adsorption of sulfate ions at the active dissolution sites on the surface, then some form of adsorption isotherm should be applicable. Suppose that a fraction α of dissolution sites on the surface is covered by adsorbed ions of molar concentration $[SO_4^{2-}]$. The desorption rate of these ions is $k_2\alpha$ and the adsorption rate may be written as $k_1 [SO_4^{2-}] (1-\alpha)$, where k_2 and k_1 are the rate constants for desorption and adsorption, respectively. At equilibrium these rates are equal, and $\alpha = k_1 [SO_4^{2-}]/(k_2 + k_1 [SO_4^{2-}])$. If the rate of the process occurring during the induction period is proportional to the bare sites, then $1/t = (1/t_o)(1-\alpha)$, where t and t_o are the induction periods in the presence and in the absence of SO_4^{2-}, respectively. Substitution for α yields

$$(1/t_o)(1/t_o - 1/t)^{-1} = 1 + k_2 (k_1 [SO_4^{2-}])^{-1}. \qquad [3]$$

In Fig. 6, $(1/t_o)(1/t_o - 1/t)^{-1}$ is plotted against $(SO_4^{2-})^{-1}$ and

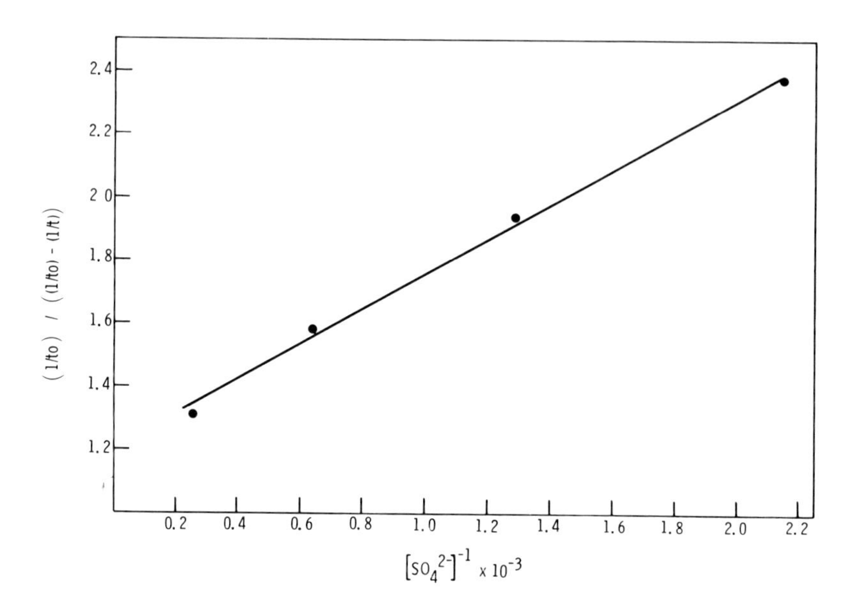

Fig. 6. Langmuir isotherm plot of $(1/t_o)/((1/t_o)-(1/t))$ against $[SO_4^{2-}]$; t_o and t, duration of induction periods in absence and in presence of gypsum, respectively.

it is seen that the Langmuir isotherm is satisfactorily obeyed. This appears to explain the inhibition effects in terms of a monolayer blocking by adsorbed sulfate ions at active dissolution sites.

A series of experiments using initial [HCl] and $[SO_4^{2-}]$ of 0.166 and 1.55×10^{-3} M respectively, were made at temperatures between 25° and 40°C. An Arrhenius plot was constructed by plotting the log of the inverse of the induction period against T^{-1}. The results are shown in Fig. 4 and the calculated activation energy is 26.4 ± 1 kcal $mole^{-1}$. This value is, within experimental error, similar to the value obtained in the absence of SO_4^{2-}. The decrease in reaction rate thus appears to correspond to a decrease in the pre-exponential term, and implies that the sulfate ions do not alter the reaction path.

The composition of the liquid phase during an experiment in the presence of 1.55×10^{-3} M $CaSO_4$ is shown in Fig. 2. The results are similar to those obtained in the absence of gypsum.

The solid products isolated following experiments conducted in the presence of $CaSO_4$ were examined by SEM. No acicular ettringite crystals were observed.

Similar retarding effects were observed when K_2SO_4 was used in place of $CaSO_4$, whereas $CaCl_2$ was found not to influence the induction period. This indicates that the active species are the sulfate ions.

B. Zeta-Potential

The particles in a suspension of 0.1 g of $Ca_3Al_2O_6$ in 25 ml water were found to be positively charged (ζ = 34 mV). This charge did not vary significantly over a period of five hours ($\pm$ 2 mV). Suspensions of 2 g $Ca_3Al_2O_6$ in 25 ml water exhibited a similar charge. The pH values of these solutions were about 11.7. The dependence of the surface charge upon the pH was examined by mixing $Ca_3Al_2O_6$ with acidic (HCl) and basic (NaOH) aqueous solutions. In the cases where $Ca_3Al_2O_6$ was added to HCl solutions, the ζ-potential was found to change with time. The results obtained after two minutes of mixing are shown in Fig. 7. The zero-point of

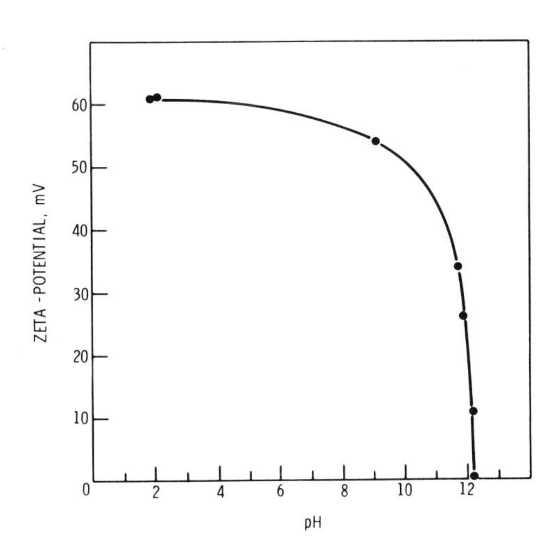

Fig. 7. ζ-Potential of $Ca_3Al_2O_6$ measured after two minutes of mixing with aqueous solutions as a function of pH.

ζ-potential is reached at pH = 12.2. This value is higher than those previously reported for some aluminas and aluminum hydroxide -- between a pH of 5.5 and 9.5 (e.g., Ref. 4, 5). The high observed isoelectric point is attributed to the adsorption of Ca^{2+} ions on the aluminum-rich surface. The specific adsorption of Ca^{2+} on alumina and its charge reversal capability were documented by Huang and Stumm, and was explained in terms of a simple ion exchange model (6).

An experiment was carried out in an HCl solution in which both the ζ-potential and pH were followed as a function of time. The results are shown in Fig. 8 and also in Fig. 9

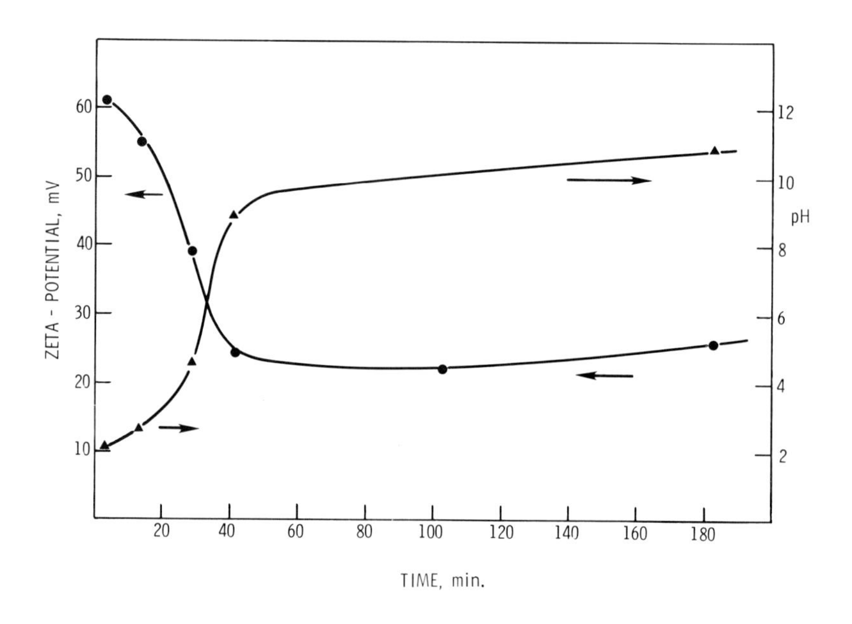

Fig. 8. Plots of changes in ζ-potential and pH as a function of time for a suspension of 0.1 g $Ca_3Al_2O_6$ in 25 ml of 0.02 M HCl.

where the change in zeta-potential is plotted against pH. It is of interest to note the increase in ζ-potential at pH values above 10. This increase is probably due to the additional adsorption of Ca^{2+} whose concentration increases near the end of the induction period.

Small amounts of $CaSO_4$ in solution effectively reduced

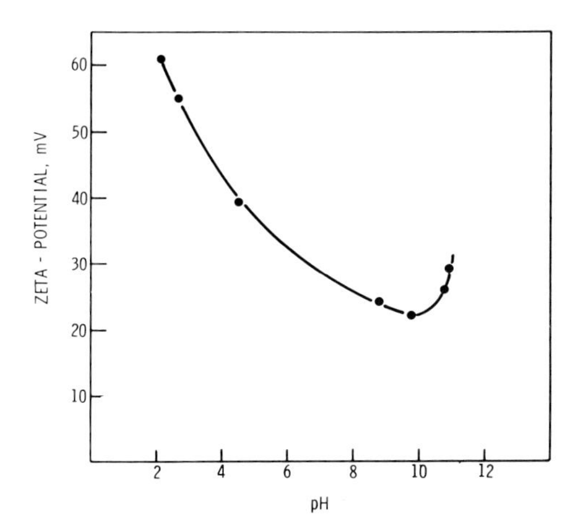

Fig. 9. Plot of ζ-potential against pH. Data from Fig. 8.

the surface charge. A plot of ζ-potential as a function of log $[SO_4^{2-}]$ is shown in Fig. 10. The ζ-potential is reduced to a

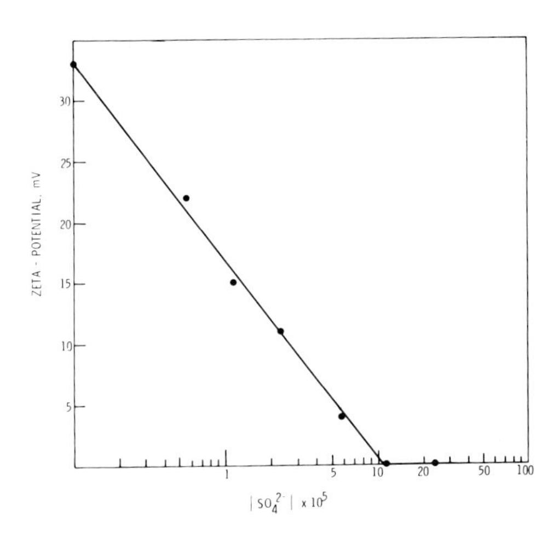

Fig. 10. Plot of ζ-potential against $[SO_4^{2-}]$ for suspensions of 0.1 g $Ca_3Al_2O_6$ in 25 ml of gypsum-containing water.

zero value at $[SO_4^{2-}]$ of approximately 10^{-4}, but a further increase in $[SO_4^{2-}]$ did not result in a charge reversal. Similar results were obtained when K_2SO_4 was employed instead of $CaSO_4$. Equivalent concentrations of $CaCl_2$ or NaCl did not influence the charge by more than $\pm$ 2 mV, suggesting that, in the cases of $CaSO_4$ and K_2SO_4, the observed effects are not merely due to changes in ionic strength. These data indicate that SO_4^{2-} ions are preferentially adsorbed on the positive surface, but are unable to further adsorb on the neutral surface.

IV. CONCLUSIONS AND REMARKS

Evidence is given for the incongruent dissolution of $Ca_3Al_2O_6$ resulting in the formation of an aluminum enriched surface. Analogous results were found for the dissolution of tricalcium silicate (7). Calcium ions adsorb on the aluminum-rich layer, forming positively-charged particles at pH values as high as 11.5. The ability of Ca^{2+} to reverse the charge of negatively-charged hydrous γ-Al_2O_3 at a pH of 11 was reported by Huang and Stumm (6). Similar results with calcium aluminosilicates in 0.01 M NaOH solution were reported by Stein and co-workers (8). The formation of a structure, consisting of an aluminum-rich surface with adsorbed calcium ions, appears to minimize the active dissolution sites and results in reduced rates of dissolution.

In the presence of sulfate ions, adsorption on the positively charged sites further reduces the rates of dissolution. This effect is probably achieved via the blocking effects of sulfate ions on coordination sites which would otherwise be occupied by hydrogen or hydroxyl ions, known to be able to catalyze the dissolution of alumina (9) and other oxides.

Decrease in solubilities due to adsorbed ions has been previously observed for a number of systems. Iler (10) reported that adsorption of small amounts of alumina on amorphous silica reduces the rate of solution as well as the equilibrium solubility of silica in water. Similar solubility effects in mixed oxide dispersions were also observed (11).

The fact that no ettringite was formed under the conditions employed in this study indicates that the previously-proposed (3) mechanism for the reduced reactivity of $Ca_3Al_2O_6$ in the presence of gypsum, which was based on the formation of a protective impervious layer of ettringite around $Ca_3Al_2O_6$ particles, has to be re-evaluated.

This work has been supported by the Master Builders Division of Martin Marietta Corporation. Thanks are due to J. Chen and T. Sun for their assistance with ESCA/Auger experiments.

V. REFERENCES

1. Rebinder, P. A., et al., 6th Int. Congress Chem. Cement, Supplementary Paper, Section II, Moscow 1974, preprint.
2. Steinour, H. H., Portland Cement Association, Bulletin 34, 1951.
3. Schwiete, H. E., Ludwig, U. and Jager, P., Symp. Struct. Portland Cement Paste and Concrete, SR 90, pp. 353-367, HRB, 1966.
4. Yopps, J. A., and Fuerstenau, D. W., J. Colloid Sci. 19, 61 (1964).
5. Robinson, M., Pask, J. A., and Fuerstenau, D. W., J. Amer. Ceramic Soc. 47, 516 (1964).
6. Huang, C. P., and Stumm, W., J. Colloid Interface Sci. 43, 409 (1973).
7. Tadros, M. E., Skalny, J. and Kalyoncu, R., J. Amer. Ceramic Soc. 59, (1976).
8. Siskens, C. A. M., Stein, H. N., and Stevels, J. M., J. Colloid Interface Sci. 52, 244 (1975).
9. Packter, A., and Dhillon, H. S., J. Chem. Soc. A, 2588 (1969); Ibid. 1266 (1970).
10. Iler, R. K., J. Colloid Interface Sci. 43, 399 (1973).
11. Wiese, G. R., and Healy, T. W., J. Colloid Interface Sci. 52, 452 (1975).

GAPS BETWEEN THE THEORETICAL AND EXPERIMENTAL DETERMINATION OF ZETA (ζ) POTENTIAL

P.D.BHATNAGAR*

[Cause and nature of some gaps and uncertainties in determining the ζ potential from:
a) moving boundary method have been discussed and data obtained from a new set up to measure electrophoretic mobility has been reported indicating the existance of a 'moving boundary potential'. Inadequacies in the theoretical treatment of boundary method due to ignoring this potential have been pinpointed.
b) adsorption measurements calculating σ, ψ_0 and ψ_δ have been discussed to be due to the uncertainties about the nature and location of the slipping plane. Conditions and effect of structuring on ψ_δ and ζ relationship have been re-examined. Three ranges of ψ_δ have been suggested where ψ_δ, ζ & θ potential respectively appear to govern the electrokinetic phenomenon observed for Titanium dioxide-water systems]

The Derjaguin-Verwey-Overbeek theory was subject to a number of tests by Reerink and Overbeek (1) who, apart from their own data, extended the test to experimental results obtained by other workers and it was found reasonably successful in describing the coagulation in many cases. Schenkel (2) considered these tests as only the partial checks and not searching tests because of the reasonable values of A, the Hamkar's constant, ψ_0 the surface potential and ψ_δ the potential at the outer edge of the Stern layer for the particles are very uncertain. In the absence of a reliable and unambiguous method to determine ψ_δ he employed experimentally determined value of the zeta potential assuming it to be roughly equal to ψ_δ. A quantitative test of the DLVO theory was made by Holtzman (3) who calculated the parameters of the equation, bypassing the use of ψ_0 and calculating the surface charge density σ from adsorption data.

Hunter (4) considered Holtzman's procedure of obtaining ψ_0 as somewhat dubious, since surface potentials were not actually measured. He used zeta potential as a measure of ψ_δ and calculated $\underline{V_R}$ from Levine and Suddaby (5) approximation.

Another attempt from the direction of zeta potential measurements has been made by Chowdhury (6) who calculated

*Present Address: Professor and Head of Science Department, Regional College of Education, Ajmer - INDIA.

V_R identifying ψ_0 with zeta potentials of titanium dioxide sol measured by Ghosh and Rakshit (7) in the presnece of coagulating concentration of NaCl. This equivalence is only justified in the extreme case of low ψ_0 and electrolyte concentration where Stern's picture of the double layer(8) approaches that of Gouy(9), but certainly not at coagulating concentration of the electrolyte.

It can thus be seen that the attempts to test DLVO Theory of colloidal stability have taken either of the following approaches for determining the net potential energy for approaching colloidal particles:

(a) Calculations of ψ_0 from adsorption data but not supplemented by actual measurements of potential,

(b) Measurement of zeta potential and identifying it with ψ_0 or ψ_δ without determining the surface charge or potential from adsorption data.

However, recently for the first time, Bhatnagar (10) and Webb (11) have undertaken simultaneous studies of adsorption, electrophoretic velocity and stability of the same system (Titanium dioxide-water) under widely varying conditions of pH and electrolyte (NaCl) concentrations using the concepts of zeta potential and the electrical double layer to predict stability of Titanium dioxide-water system.

Approaches from either sides i.e. (a) calculating it from the adsorption data or (b) its determination from electrophoretic measurements, difficulties have been felt in determining zeta potentials and gaps have been observed in the theoretical and experimental values. It is proposed to identify and discuss the nature of these gaps in the following two sections:

Section A: The Moving Boundary potential:

To begin with, confining here to the determination of electrophoretic velocity (vis-a-vis zeta potential) from the moving boundary method - apparently one of the 'direct' and the 'simplest' one, we see that:

Theoretical approaches on the movement ionic and colloid-ionic boundaries by Kohlrausch (12), Weber (13), Henry and Brittain (14), McInnes & Longsworth (15), Haker (16) and Tiselius (17), have invariably been based on considering the discontinuous transition as a limiting case of continuous transition and assuming the validity of ohms law at the boundaries. A condition of electroneutrality ($\Sigma\sigma = 0$) in any plane perpendicular to the direction of current) has therefore always been applied to solve the differential equations of boundary movement.

Mukherjee (18) closely scrutinize the Kohlrausch-Weber theory considering the transfer of ions across the boundary between two electrolytes AR and BR and showed that both the above considerations need revision. Further, wide variation in observed results lead him (19) to devise a set up for measuring the 'actual potential difference under which the boundary moves'. Neither his technique nor the theoretical conclusions that electric current tends to produce the excess of one of the ions and the consequent charge accumulation should be contemplated in the theory of moving boundary however, received adequate attention in later studies with the moving boundary method.

Electrophoretic investigations in general and the technique in boundary method hitherto developed recently by Price and Lewis (20), Tiselius (lec cit), De Mende & Verganolles (21) have been, as summarised by Allexander (22) are "directed more towards obtaining information of specific systems than to provide fundamental knowledge of the process involved."

In view of the above facts and their experimental results Bhatnagar & Bhattacharya (23) pointed out the need of studying boundary electrophoresis under constant current conditions for obtaining a more cogent picture of the process involved and the evidence of the existance and the magnitude of the 'moving boundary potential' as indicated by the results is presented in this section.

The apparatus consisting of separate electrode vessels, constant current supply and potentio-V.T.V.M. system as used by Bhatnagar and Bhattacharya (23) has been employed.

Theoretical aspects of moving boundary method:

I Kohlrausch-Weber treatment of moving boundaries and Mukherjee's observations:

(a) considering a sharp boundary between strong electrolytes AR & BR, Kohlrausch deduces:

$$\frac{\partial a}{\partial t} = \text{if (a)} \frac{\partial a}{\partial n} \qquad (1)$$

where

∂a is the change in concentration in infinitesimal interval,

and $\frac{\partial a}{\partial n}$ is the rate of change of ion species under consideration and $f(a) = dx/da$

Here he assumes that electroneutrality in the interior of the solution is satisfied if Ohms Law has to remain valid.

He further deduced for a dilute solution of mixture of several electrolytes:

$$\frac{\alpha}{a}+\frac{\beta}{b}+\frac{\gamma}{c}+ \;\ldots\ldots\ldots\; +\frac{\rho}{r} = \text{a const.} \qquad (2)$$

where

$\alpha, \beta, \gamma \ldots$ are the concentrations of anions and cations and a,b,c are the mobilities of the ions, and, for a mixture of electrolytes with an ion of same on sign is common:

$$\frac{\alpha}{n_A}+\frac{\beta}{n_B}+\frac{\gamma}{n_C}+\frac{}{}\ldots\ldots\ldots +\frac{\rho}{n_R} = \text{a const.} \qquad (3)$$

where

$n_A, n_B \ldots$ are the transport numbers of ions $A, B \ldots$.

In this step he assumes a continuous change in concentration from point to point and considers for a boundary between two single electrolytes AR and BR at any section normal to the direction of current in the interior of each electrolyte:

$$\frac{\alpha}{n_A} + \frac{\rho}{n_R} = \text{const.} \qquad (4)$$

$$\frac{\beta}{n_B} + \frac{\rho}{n_R} = \text{const.} \qquad (5)$$

It will be seen from that it does not follow necessarily from Kohlrausch treatment that this relationship will remain valid for the discontinuous transition at the boundary. Mukherjee pointed out this fact that Kohlrausch's consideration of the case of a sharp transition as a limiting case of a continuous transition assumes the validity of Ohms Law at the boundary.

Weber, basing his deductions on these assumptions derives that when faster moving ion follows a slower moving ion, the concentrations on both sides of the boundary automatically adjust themselves and the relations

$$\frac{\alpha}{n_A} = \frac{\beta}{n_B} \qquad (6)$$

must be satisfied.

Mukherjee challenged the validity of Ohms law even in Kohlrausch-Weber sharp boundary. Considering the transfer of anion R across the above mentioned boundary, he showed that layer contigious to the boundary containing slower moving ion will be poorer (-ve sign)in R by

$$- i\,\left[(N_R)^{BR} - (N_R)^{AR}\right]\frac{dx}{\rho_{BR} V_R} \qquad (7)$$

where P_{BR} is the potential gradient in layer BR and V_R is the mobility of ion R per unit pot. grad.

He further showed that the electric density of one kind of charge will be considerable and the fundamental assumption of electro-neutrality by Kohlrausch is not justified.

II. Henry & Brittain's theory: (14)

They extended Kohlrausch-Weber theory for ionic displacements to colloids and developed it in a slightly different form, considering the rate of entry of S ions in a volume of 1 cm^2 cross-sectional area, perpendicular to the direction of current, and obtained:

$$\frac{\partial \sigma}{\partial t} = s\frac{\partial (X\sigma)}{\partial x} = is\frac{\partial (\sigma/k)}{\partial x} \qquad (8)$$

where

$\Sigma\sigma$ = o,
 = SF (S-Conductivity in gm Eq/cc),
F = 1 Faraday,
X = pot. grad. in volts per cm.
k = Sp. Conductance i.e. $s\sigma$,
dx = a small distance in the direction of the current in element of unit cross sections,
i = current density and
s = mobility of S ion

Next, they consider a discontinuity in concentration of S ions in the element and for the motion of this discontinuity through a distance obtain:

$$\frac{dx_0}{dt} = is\frac{\sigma''/k'' - \sigma'/k'}{\sigma'' - \sigma'} \qquad (9)$$

such that $(\sigma''-\sigma')dx_0$ represents the net gain of S ions in the volume element.

where

σ'' & σ' and k' & k'' are the values of σ and k immediately to the right and left of the discontinuity.

All further development of their theory is based on the solution of these equations simultaneously, which will be clearly seen as based on fundamentally different conditions. While Eq.(8) holds for a uniform solutions, implies electro-neutrality and consequently the validity of Ohms Law ie $\Sigma\sigma$=o in any plane perpendicular to the direction of current, Eq.(9) is not much different from

Mukherjee's relation dealing with the excess of one kind of ions which the current tends to produce.

It will therefore be plausible to conclude that though developed in slightly a different form, Henry and Brittain's treatment also suffers from the same inconsistency (which Mukherjee pointed out in Kohlrausch's treatment)of treatment the discontinuity as a limiting case of continuity i.e. stretching the assumption of electroneutrality to hold at the boundary and thus assuming the validity of Ohms law even at the discontinuity where an excess of one kind of ions is bound to be produced by the current.

III. Dole (24) and Svenson's (25) Treatment:

The elaborate treatment extended by the above authors independently seems to recognise the above inconsistencies.and On arriving at the equation (for the increase in amount of an arbitrary ion per sec. and cm.^{2} of the tube).

$$\frac{C_{i1} u_{i1} i}{\varkappa_1} - \frac{C_{i2} u_{i2} i}{\varkappa_2} = (C_{i1} - C_{i2}) V \qquad (10)$$

where

i = current density in amp/cm.^{2} (+ve if current flows in direction of X axis).

$\varkappa_1, \varkappa_2$ = Conductivity inside of boundary as shown in Fig.1

u = ionic mobility(cm/volt/sec/unit (pot grad) with sign of charges

V = Velocity of the boundary cms/sec (+ve if it moves in the direction of X axis)

i = as subscript denotes different ion species

u', u'' = mobility of moving boundary as defined in Eq.(11)

j a subscript = the number of boundary in question counted in the direction of X axis.

It is further defined:

$$V = \frac{u' i}{\varkappa_1} = \frac{u'' i}{\varkappa_2} \qquad (11)$$

so that

$$(C_{i1} - C_{i2})\ (u_{i1} - u') = \left(\frac{\varkappa_1}{\varkappa_2} - \frac{u_{i1}}{u_{i2}}\right)\ (C_{i2} - u_{i2}\ ) \qquad (12)$$

and

$$(C_{i2} - C_{i1})\ (u_{i2} - u'') = \left(\frac{\varkappa_1}{\varkappa_2} - \frac{u_{i2}}{u_{i1}}\right)\ (C_{i1} - u_{i1}\ ) \qquad (13)$$

must be satisfied by every ion on both the sides of the boundary which is possible only when:

$$\frac{u_{11}}{u_{12}} = \frac{u_{21}}{u_{22}} = \frac{u_{31}}{u_{32}} = \ldots\ldots = \frac{u_{n1}}{u_{n2}} \qquad (14)$$

is valid.

It is however admitted that this is not the case, but it is "necessary to assume the above as an approximation for small concentration changes and for univalent ions, only in order to carry the theory further and to reach a step nearer to the truth". The results drawn further, as Svenson states are not exact and errors may be appreciable for polyvalent ions and at fair concentrations.

It is thus surprising, although obvious, that none of the treatments referred above could incorporate the idea of accumulation of charge at the boundary - a fact which was deduced by Mukherjee as early as in 1928. It seems that the absence of experimental evidence of this 'moving boundary potential' for want of a suitable technique is responsible for the present situation, because the important role that the 'junction potential' plays in regulating the movement of ions of different mobility in case of concentration cells with transport is well recognised.

OBSERVATIONS AND DISCUSSIONS

While scanning the results of electrophoretic studies under constant current, an interesting fact that in many cases the potential measured across the subsidiary electrodes as recorded by the V.T.V.M. is quite different for different direction of the constant current after about 70-80 minutes of the boundary migration, attracted the attention of the author and few such observations are summarised in table No.1.

It will be seen from the table that the difference in potential observed across s_1 and s_2 can be explained and serves as an experimental evidence of the 'different' changes accumulated at the boundaries which are bound to develop as reasoned by Mukherjee (18) in expression (7). Considering these charges accumulating at either boundaries to give rise to e.m.f. 'e' volts, the observed potential E' & E" across the subsidiary electrodes s_1 and s_2 for opposite directions of the current, thus seems to be due to 'moving boundary potential' which

Table 1

System A_2S_3 sols, different stages of dialysis) Equiconducting Acetic acid as supernatant liquid	Const Curr. I	Potential (volts) across the subsidiary electrodes S_1 & S_2 At the start $\overrightarrow{S_1 S_2}$ E=IR	$\overleftarrow{S_1 S_2}$ E=IR	After 70-80 minutes $\overrightarrow{S_1 S_2}$ E'	$\overleftarrow{S_1 S_2}$ E"	E'-E" (volts)	'e' (volts)
I	41 uA	34.0	34.0	29.5	27.5	2.0	1.0
II	21 uA	49.0	49.0	42.0	38.0	4.0	2.0
III	22 uA	19.0	19.0	15.0	13.5	2.5	1.25
IV	57 uA	23.0	23.0	23.0	17.0	6.0	3.0 (?)

Note: 1. Coagulation and thickening was observed at the boundaries during migration in case of A_2S_3 and Sb_2S_3 sols to a larger extent.

2. Value of 'e' for various Sb_2S_3 sol samples was within the range of 0-.25 volts and those for Fe_2O_3 and other sols was below the limits of reliable measurements (i.e. below .1 volts the experimental set up not having been designed specifically measuring these potentials).

⇄ directions of current

is additive in one direction of the current and opposing in its reversed direction:

$$E = IR \text{ at the beginning of the experiment} \quad (15)$$

$$E' = IR' + e \text{ in one direction of the const. current} \quad (16)$$

and $$E'' = IR' - e \text{ in the other " " " "} \quad (17)$$

R & R' being the total resistances of the column between the subsidiary electrodes in the beginning and after 70 to 80 minutes of migration of boundaries. R & R' might differ if sol coagulates at the boundaries or in bulk or both during migration. From (15) and (16) the value of 'e' can be easily determined as given in the last column in Table 1. The values of e are considerably high for A S and Sb S sol-equiconducting Acetic acid supernatant solution systems as compared to other sols and liquid junction potentials (range about 50 mv). In these cases, the contribution of sol thickening/coagulation at the boundaries in terms of orientation of colloid micelles/aggregates can also be visualised but further work is necessary in this direction to clarify the phenomenon.

Further studies on electrophoresis under constant current with the help of the new technique further develop-ed by Bhatnagar (26) where potential differences at various places in either sides of the electrophoresis tubes are measured, it is hoped that more data will be available soon and it will be possible for the author to present further results about the 'moving boundary potential' and vis-a-vis the calculation of the electrophoretic velocity and zeta potential.

Section B : The slipping plane, slipping layer and thickness of the immobilised layer.

As mentioned in the beginning studies like that by Haydon (27) are scarcly available in literature where alongwith the determinations of zeta potentials have been carried out from electrophoretic mobility measurements and surface adsorption studies have also been made for the same system. Still scarce are such data like those by Bhatnagar and Webb (10) where in addition to the above interaction energy and stability of the same systems are also available. It is thus not surprising that searching tests for the quantitative relationship between ψ_δ and ζ potentials involved in the structure of the double layer and there interaction could rarely be made.

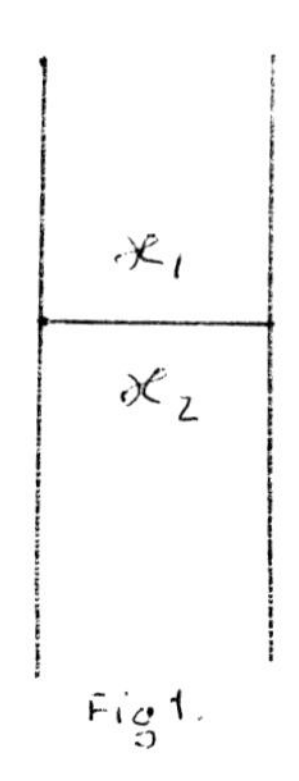

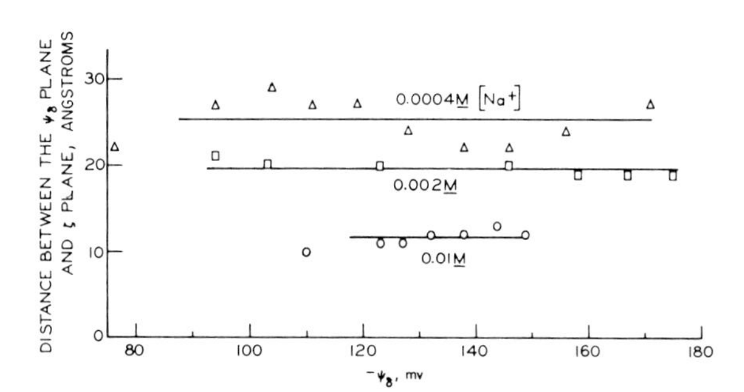

FIG. 2. Distance between the O.H.P. and the plane of shear vs. ψ_δ for various Na^+ concentrations.

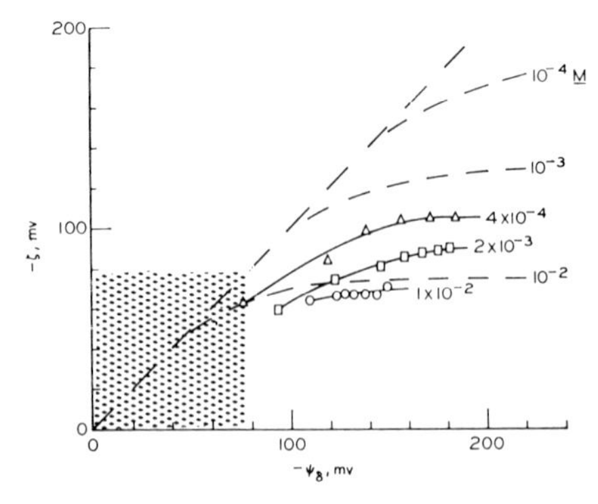

FIG. 3. Relationships between ζ and ψ_δ at various ionic strengths (– – – theoretical predictions of Lyklema and Overbeek (7); ——— experimental results).

In this section, attention is focused on the uncertainties and gaps that exist in calculating the values of potential from adsorption studies and comparing these with those obtained from electrophoretic velocity measurements considering Webb and Bhatnagar's (11) approach to obtain:

a) ζ potential from electrophoretic mobility determined by using Mass Transport analyser Model MIC 1201 manufactured by Numinco and discussed by Oliver (28) and Ross and Long (29). This eliminates the uncertainties discussed in Section A.

b) σ the surface charge density by modified Holtzman's (3) method.

c) ψ_0 the surface potential using Nernst equation (30).

and d) ψ_δ by the numerical solution of the Poisson-Boltzmann equation for spherical particles as published by Loeb (31) and Hunter (32) followed by trial and error calculations of $(\frac{d\psi}{dr})$ at δ plane so that

$$\sigma_1 + \sigma_2 = 0 \tag{18}$$

where, σ_1 = the Stern layer charge density, esu/cm^2

σ_2 = the Gouy Chapman diffuse layer charge density esu/cm^2

ψ = is the electrical potential at some point within the electric double layer

δ = is the hickness of the Stern layer

r = is the distance from the center of a spherical particle

In this way, they avoided the uncertainties involved in the values of ϵ' and δ if the simpler relation(30,p.43) $\sigma = (\epsilon'/4\pi\delta)\,(\psi_0 - \psi_\delta)$ is used. It will be seen that further calculation of zeta potential from $d\psi/dr$ requires the precise knowledge of the location of the classical 'slip plane' or the slipping layer (or both?). This is where, what one could call a gap, exists in extending the calculations further to obtain ζ potential from the values of ψ_0, ψ_δ and $d\psi/dr$ calculated from adsorption measurements.

Apparently, the only way for even knowing the position of the classical discrete plane seems to determine the zeta potential from electrokinetic measurements and locate it on $(\frac{d\psi}{dr})$ plot. This obviously means starting from the other way round as had to be done by Webb (11).

For this purpose, finding the Henry's (33) corrections unsatisfactory, Webb applied W.L.O. correction (34) to the values of obtained from mass transport data and calculated the distances between and zeta planes for different concentrations of NaCl. These calculated distances (12, 20 and 25 A) correspond to 5-10 molecular water layers, a range favourably comparing with that expected by Brube (35) but only about one tenth of that predicted by others. Further it will be observed (Fig.2) that these distances remain constant and independent of the values of for a certain concentration of NaCl, a fact argued (11) to be supporting the concept of water structuring near the surface that could physically control the location of the hydrodynamic shear plane suggested by Lyklema and Overbeek (36). The decrease of structuring with increased concentration of NaCl is pronounced and supports Franks (37) conclusion.

As such, the nature and conditions for the existance of (a) immobilised liquid shear plane or (b) slipping layer or (c) both (a) and (b) require to be examined. Like, Lyklema and Overbeek (36), Webb (11) considers these three cases:

i) The electric field causes the viscosity to vary from zero fluidity at the ψ'_δ plane to bulk fluidity at the ψ'_s plane, therefore, the plane of shear is actually a slipping layer of finite thickness.

ii) The surface influences viscosity by immobilizing liquid at the surface to form a structured layer with the ψ'_δ plane within the structured region and ψ'_s plane at the outer edge.

iii) Viscosity variation affects the description and location of the plane of shear by a combination of cases (i) and (ii) i.e., there is a structured region and a finite slipping layer and the ψ'_δ plane is at the outer edge of the slipping layer.

Taking case (i) to apply on this data, he found that either case (i) was not valid for the visco-electric constant f used by Lyklema and Overbeek in the expression for viscometric effect:

$$\eta = \eta_0 \left[1 + f \left(d\Psi/dx \right)^2 \right] \qquad (19)$$

is not a constant at all. Moreover, the effect of salt concentrations of f is also not known (38).

Alternatively, considering case (ii) to be applicable, the values ζ limit calculated using the values of D, the thickness of the immobilised or structured layer (obtained from complete solution of the Poisson-Boltzmann equation as suggested by Loeb (31) and locating the ζ plane) using the expression

$$\exp \left(e\,\zeta_{lim}/2kT \right) = \frac{1+\exp(-kD)}{1-\exp(-kD)} \qquad (20)$$

and ζ limit found experimentally widely differ from the experimentally determined ones at low electrolyte concentrations.

The observed relationship between ψ_δ and ζ can lead to more precise and plausible picture of structuring or slipping layer if, instead of considering case (i) or case (ii) or combination of the two for all the observations as done by Webb, we consider (Fig.3).

I. Region of low ψ_δ ($\psi_\delta < 50$ mv) shown by the shaded area on the left where case (i) holds i.e. there is no structuring or structuring limited within or upto ψ_δ plane. Here ψ_δ plane is identical with the zeta plane and the values of ψ_δ and ζ tend to converge as shown by the dotted lines. The electrokinetic phenomenon in this region will be governed by ψ_δ.

II. Region of very high ψ_δ ($\psi_\delta > 180$ mv shown by shaded area on the right where case (ii) applies, i.e. structuring extending to several hundreds of molecular diameters as predicted by Webb (39). Here the observed mobility is determined by the thickness of the structured layers, the shear plane being at their edge. In order to distinguish it from the electrokinetic shear plane (ζ plane), it can be identified as structured layer's edge (SE) plane. It is the overlapping or enveloping by the structured layers that will mask any variations in $\frac{d\psi}{dr}$ and consequently resulting into the observed independence of zeta potential from ψ_δ.

In this region, therefore, the observed electrophoretic mobility should <u>not</u> be used to calculate zeta potential. Electrokinetic phenomenon in this region of high will thus be governed by the potential at the edge of the structured layer. In order to distinguish it from 'zeta potential', it is proposed to call it θ potential.

III. Region for 'moderate' values of ψ_δ (50 to 180 mv) where a combination of case (i) and (ii) applies i.e. there is the existance of structured molecular layers, slipping layer and unstructured molecules of dispersion medium. The electrokinetic phenomenon in this region being governed by the zeta potential, the potential at the edge of the slipping layer.

Obviously there can be no sharp lines of demarkation between these regions and we can consider

a) electrokinetic phenomenon to be governed by ψ_0 or ψ_δ in the low ψ_δ range, by zeta potential in the medium ψ_δ range and θ potential in the high ψ_δ ranges.

b) the effect of increased electrolyte conc.(NaCl) contributing to oppose structuring and tending to bring the system where electrophoretic mobility shows dependence of ψ_δ.

The above conclusions, particularly the identification of 'Θ' potential governing the electrophoretic mobility (independence from ψ_δ) can be easily tested by studying the temperature effect which should tend to oppose structuring and the electrophoretic mobility should be reasonably expected to show dependence on ψ_δ at higher temperatures even in the high ψ_δ range.

While it remains to be further worked out as to how far do the above considerations help to obtain the double layer interaction energies and values of Hammaker constant 'A' - consistent with the stability studies of TiO_2-water system, as theoretical approach and some model bridging the gap in calculating the thickness of the structured layer and slip plane on the basis of surface charge density, dielectric and viscoelectric properties of the dispersion medium and the dispersed phase is the fundamental need.

REFERENCES

1. Reerink, H., and Overbeek, J.T.G., Discuss., Faraday Soc. 18, 74 (1954).
2. Schenkel, J.H., and Kitchner, J.A., J. Am. Chem. Soc. 56, 168 (1960).
3. Holtzman, W., J. Colloid Sci., 17, 363(1962).
4. Hunter, Robert J., and Alexander, A.E., J.Colloid Sci., 18, 820 (1963).
5. Lewis, S., Suddaby, A., Proc. Phys. Soc. Lond. A. 64, 287 (1951).
6. Chowdhury, B.K., Nature, 125, 308 (1960).
7. Ghosh & Rakshit, Science and Culture, 18, 498 (1953).
8. Stern O., Z. Elektrochem. 30, 508 (1924).
9. Gouy, G., J. physique (4) 9, 457 (1910).
10. Bhatnagar, P.D., and Williams Dale G., to be published.
11. Webb, Joseph T., Bhatnagar, P.D., and Williams Dale G., J.Colloid Interface Sci., 49, 346 (1974).
12. Kohlrausch, F., Weid, Ann., 62, 207 (1897).
13. Weber H., Stz. Preuss, Akad. Wisc., 936 (1897).
14. Henry, D.C., and Brittain, J., Trans. Faraday Soc., 29, 798 (1935).
15. Mc Innes & Longsworth, Chem. Rev., II, 171 (1932).
16. Haker, W., Koll. Z. 62, 37 (1933) and Kolloid Chem. Beih., 41, 147 (1935).
17. Tisulius A., Trans. Faraday Soc., 33, 529 (1935).
18. Mukherjee J.N., J. Ind. Chem. Society, 5, 593 (1928).
19. Mukherjee J.N., Proc. Royal Soc., A. 102 (1923).
20. Price and Lewis, Trans. Faraday. Soc., A 29,775 (1933).

21. De Mende & Verganolles, Acad. Sci., Paris 227, 1235 (1948).
22. Alexander and Jonson, "Colloid Science" Oxford University Press (1949).
23. Bhatnagar, P.D., and Bhattacharya A.K., Kolloid Z. 170, 29-32 (1960).
24. Dole, V.P., J. Am. Chem. Soc., 67, 119 (1945).
25. Svenson Hary, ARKIV FOR XEMI MINERALOGY O. GEOLOGI, BD. 22-A No.0 10, 1-32.
26. Bhatnagar, P.D., KOLLOEDNI ZHURNAL (USSR) No.4 TOMXXIX, 598 (1967).
27. Haydon, D.A., Proc. Roy. Soc., London Sci., A. 258, 319 (1960).
28. Sonnett, P., and Oliver, J.P., U.S. pat 3, 208-919 (1965); Tappi 52, 153 (1969).
29. Ross, S., and Lang, R.P., Ind. Eng. Chem. 61, 58 (1969)
30. Verway, E.J.M., and Overbeek, J. Th. G., Theory of Stability of Lyophobic colloids" 47 Elsevier, Amskadam (1948).
31. Loeb, A.L., Overbeek, J. Th. G. and Viersema P.H., "The electrical double layer around a colloidal particle" p. 375 M.I.T. Press Cambridge, M.A. 1961.
32. Hunter, R.J., J. Colloid Interface Sci., 22, 231 (1966).
33. Henry, D.C., Proc. Roy Soc. London Ser. A. 133, 106 (1931).
34. Booth, F., Proc. Roy Soc. London Ser. A. 203, 514 (1950).
35. Brube, Y.G., and De Bruyn, P. J. Colloid Interface Sci., 28, 92 (1968).
36. Lyklema, J., and Overbeek, J. Th. J. Colloid Sci., 12, 501 (1961)
37. Frank, F., Chem. Ind. (London) 1968, 560 (1968).
38. Grahme, D.C., Chem. Rev. 41, 441 (1947).
39. Webb, J.T., Doctor's Dissertation Inst. of Paper Chem. Appleton, Wisc, p. 231 (1971).

---

gur/PDB

NEW STATE-OF-THE-ART AUTOMATIC MICROELECTROPHORESIS INSTRUMENTATION

P. J. Goetz and J. G. Penniman
Pen Kem Company

A newly available instrument for automatically measuring mobility is described based on the principle of a "grating microelectrophoresis analyzer." Performance advantages over the laser doppler method are cited. A description is provided of optionally available modules which can provide a variety of sophisticated research, clinical or industrial process control functions.

We will describe an automatic microelectrophoresis instrument, the performance of which is based on the principle of a "grating microelectrophoresis analyzer." Designated the Laser Zee^{tm} System 3000, the instrument is capable of providing mean mobility and mobility histogram data for a wide variety of colloids. We will present certain data, compare the performance of the System 3000 with that of laser doppler instrumentation, and indicate some typical performance parameters.

Recently, after two years of feasibility studies and development work with breadboards, we constructed a preproduction prototype. This working model is currently being used for demonstration purposes, for research, and to offer a "Mobility Measurement Service" to those laboratories in which there is need to increase the quality or quantity of mobility measurements or obtain high resolution histograms.

A block diagram of the instrument configuration used in our development work is shown in Figure 1. The main elements of the System 3000 can be described as follows:

1. A precision chamber of 1 mm bore is immersed in a temperature controlled water bath. Pre-charged palladium electrodes are provided to eliminate electrolysis. The instrument is equally suitable for microelectrophoretic measurement of solvent or aqueous systems.

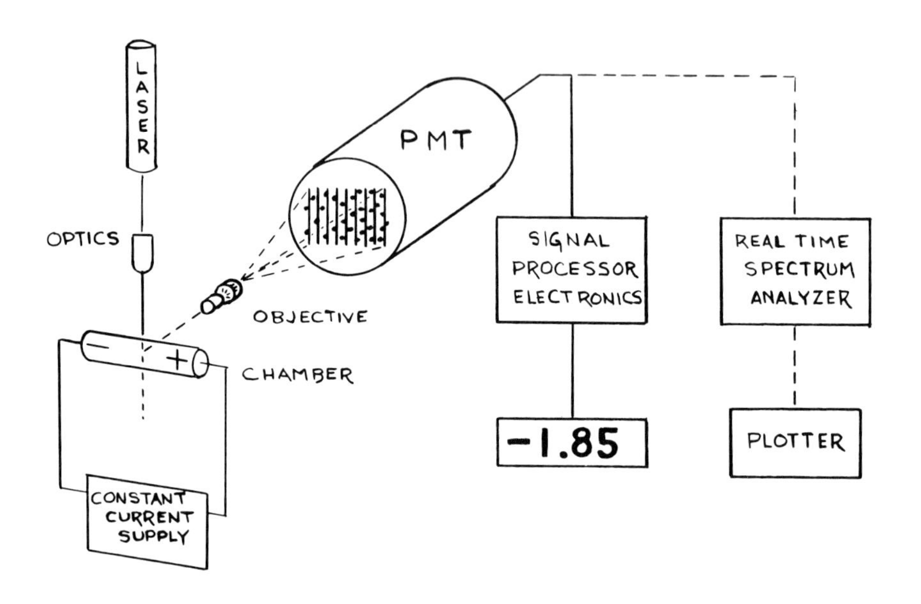

Fig. 1: *Block Diagram of Laser Zee*^{tm} *3000*

2. A 2 mw helium-neon laser and associated optics project a sheet of light at the stationary layer. Means are provided to control the depth of the sheet in order to maximize the number of particles available for measurement and correspondingly speed the generation of data. This comprises a "dark field" system capable of measuring the mobility of particles as small as 100 Å.

3. An objective lens is immersed in the water bath and focused on the colloid at the stationary layer. The lens forms an image of the particles on a grating, which passes the light through to a photodetector. The motion of the particles across the grating produces a modulation of the transmitted light which can be likened to the sound created by running a stick down a picket fence. The mean frequency in either case is proportional to the average speed (of the particle or stick).

4. Electronic circuitry processes the signal from the photodetector, determines the mean frequency and thus the average velocity, divides by the known electric field strength and digitally displays the resultant mean mobility.

5. A real time spectrum analyzer, a Rockland System 512C, can also be used to convert the photodetector output signal to a real time CRT display of power/frequency. For a homogeneous sample, the signal power is proportional to the number of particles and the frequency is proportional to mobility. The analyzer has a built-in averager for accumulating data over extended time periods, thereby reducing the effect of noise and providing a histogram of greater statistical significance.

The real time spectrum analyzer is not part of the basic System 3000, but is available as an option.

A typical output from the grating microelectrophoresis analyzer System 3000 (with the analyzer and plotter option) is shown in Figure 2. The horizontal axis is equal to 5 Hz or 2.5×10^{-4} cm^2 v^{-1} sec^{-1}. The vertical axis is proportional to the power spectral density or, for a homogeneous system, the number of particles.

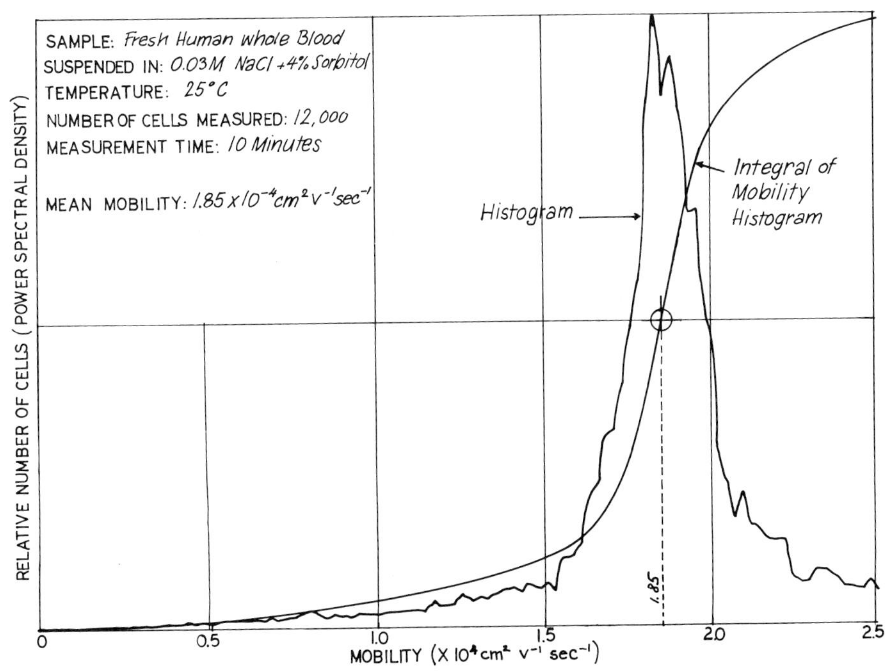

Fig. 2: *Mobility Histogram*

The bell-shaped curve represents the mobility histogram. The upward sweeping curve represents its integral, and the mid-point, indicated by a bullet, indicates the mean mobility (in this case 1.85×10^{-4} cm^2 v^{-1} sec^{-1}. The colloid is whole human blood suspended in 0.03 M NaCl + 4% Sorbitol, at a temperature of 25° C. Twelve thousand cells were measured in 10 minutes. The value obtained for this blood sample is well within the expected range given by Seaman of $1.90 \pm .09 \times 10^{-4}$ $cm^2\ v^{-1}\ sec^{-1}$. (1)

Advantages of the grating microelectrophoresis analyzer Laser Zee™ System 3000 over the laser doppler technique are numerous. The grating microelectrophoresis analyzer:

1. Can use any light source and does not have to be monochromatic, e.g., UV for fluorescence studies.

2. Does not have to work at small scattering angles to achieve resolution, thus minimizing the effect of stray light.

3. Does not require any critical alignment to obtain heterodyne mixing on the detector surface.

4. Permits direct viewing of particles and is thereby compatible with an image analysis system (which can be added as an option).

There are other advantages of a proprietary nature that we are presently not free to discuss. (However, those interested can visit, witness demonstrations, and receive full explanations upon the execution of a secrecy agreement.)

Typical performance parameters for biological cells dispersed in 0.03 M NaCl + 4% Sorbitol are shown in Figure 3. Most of our development work has been performed in this medium. We are presently also working with physiological saline; comparable peformances are obtained, except that data acquisition takes more time (as a function of the increased ionic strength).

Chamber Volume: (Including Electrode Compartments)	100 λ
Sample Size: (Using Micro-injection technique)	$<10\lambda$ and $<10^5$ cells
Precision of Mean Mobility:	$<1\%$
Measurement Time for Mean Mobility: (Based on 25-100 Cells)	<30 seconds
Measurement Time for Histogram: (Based on 1000 Cells)	<5 minutes
Resolution of Histogram:	$.025 \times 10^{-4}$ cm^2 v^{-1} sec^{-1}
Typical Field Strength:	20 v/cm

Fig. 3: *Typical Performance Parameters for Biological Cells in .03 M NaCl + 4% Sorbitol*

Optional modules for the System 3000 include:

01-Sample Transfer: This option provides capability for automatically sequencing samples through the microelectrophoresis chamber. Standard sample size is 100 λ .

02-Micro-Injection Sample Transfer: For automated applications requiring extremely small sample size, this option is recommended in lieu of 01 and provides a sample size as small as 10λ.

03-Sample Tray: This "lazy susan" module provides a means for loading 40 samples for sequential measurement (normally requires option 06 and either 01 or 02).

04-Model 512C Real Time Analyzer: As previously described, this option provides a mobility histogram of each colloid sample.

05-High Speed Plotter: Where necessary, this plotter augments the CRT display of the analyzer and provides a hard copy of the histogram.

06- H.P. 9825 Calculator and Interface: This option provides an extremely versatile means of customizing and extending the basic capabilities of the System 3000 to meet specialized requirements. Required for use with options 01,02,03,07 and 08.

07-Histogram Analysis Software: This option is furnished on a custom basis for the purpose of automatically analyzing the features of the mobility histogram. It could be used, for example, in a clinical environment to determine the probability of existence of a given disease. The calculator is sufficiently simple to program that the end user can easily write his own software when necessary (requires options 04 and 06).

08-Automatic Calibration and Test Software: This package provides means for automatically checking the accuracy of the system and for removing systematic errors. Features include a capability for automatically locating and focusing precisely on the stationary layer as well as evaluating the efficiency of "zero charge" coatings on the chamber walls.

09-TV Image Analyzer System: Still in the development stage is a TV image analysis system for providing a mobility measurement on individual cells. Feature extraction logic will also allow cross correlations to be made between such parameters as mobility, particle size, and shape.

To enable the efficient measurement of process streams in industrial applications such as paper mill furnish waste effluent, the following modules are also available:

10-Influent Conditioner Module: Used in conjunction with option 01, the Influent Conditioner is used instead of the Sample Tray (option 03) and provides an intermittent sampling of a process stream. The unit provides for de-aeration, dilution, agitation, filtering as appropriate to the characteristics of the process stream being sampled.

11-Process Control Computer: For industrial applications, this option replaces the HP calculator (option 06). In conjunction with option 10, it provides a means to optimize the coagulation or dispersion chemistry of process streams through appropriate adjustments in the feed rate of chemicals such as alum, polyelectrolytes, acid or base (depending on the specifics of the process). For the first time, very difficult colloid chemistry processes (some of which are also subject to rapid change) can be maximized with cost efficiency.

REFERENCES:

1. Seaman, G.V.F., Personal Communication.

ELECTROPHORETIC MOBILITY MEASUREMENTS BY CONTINUOUS PARTICLE ELECTROPHORESIS

F. J. Micale, P. H. Krumrine, and J. W. Vanderhoff
Center for Surface and Coatings Research
Lehigh University

ABSTRACT

The Beckman Continuous Particle Electrophoresis (CPE) instrument was designed as a preparative tool to fractionate colloidal sols according to the electrophoretic mobilities of their particles. A thin stream of sample is injected into an electrolyte curtain moving downward through a thin, flat electrophoresis cell. As the curtain passes between the electrodes, the particle stream is displaced laterally according to the electrophoretic mobilities of the particles. Mixtures of as many as ten different-size monodisperse polystyrene latexes have been separated using this instrument. Thus, the CPE gives excellent <u>relative</u> values of the electrophoretic mobility, but in its present form is not well-adapted to give <u>absolute</u> values. The purpose of this paper is to describe the development of a method to give absolute electrophoretic mobilities. The instrument was modified so that the displacement and width of the particle stream can be recorded with a magnification of 18.2X. A theoretical expression was derived for the electrolyte flow profiles in the separation chamber as a function of the experimental parameters of the system. A series of monodisperse polystyrene latexes of known electrophoretic mobility was used to calibrate the instrument. The experimental results with this series were in good agreement with the theoretical expression. It is concluded that the Beckman CPE can be used to measure the absolute electrophoretic mobilities of colloidal particles under certain well-defined conditions. The advantage of this method is that particles smaller than 0.1μ in diameter can be measured, and complex systems comprising particles of different electrophoretic mobilities may be easily characterized.

I. INTRODUCTION

Continuous particle electrophoresis (CPE), sometimes referred to as continuous flow electrophoresis (CFE), has been developed over the past 15 years primarily by two independent investigators, Hannig (1-3) and Strickler (4-6). During this time, the CPE has been used primarily as a preparative method for the separation of biological materials (7). The Beckman CPE has also been investigated as a method for the separation of latex particles according to size (8). This instrument developed by Strickler was available from Beckman Instruments for a period of time. The commercial instrument, which was designed primarily as a preparative tool, has been modified and used as an analytical tool for the absolute measurement of electrophoretic mobilities. Although the dispersions used in this study were model colloids, i.e., monodisperse polystyrene latexes, the objective of this investigation was the evaluation of the CPE method for the electrokinetic characterization of complex dispersions which are difficult to evaluate by standard techniques.

II. EXPERIMENTAL DETAILS

Figure 1 shows a schematic representation of the Beckman CPE instrument. The 0.15 x 4.5 x 50-cm electrophoresis channel is positioned vertically with the 30.5-cm long electrodes on opposite sides. Electrolyte solution is pumped into the top of the channel to form the electrolyte curtain, which flows downward through the channel and empties at the bottom through 48 1-mm O.D. stainless steel tubes set edge-to-edge, so that the electrolyte curtain is separated into 48 fractions. The electrodes are separated from the electrophoresis channel by semi-permeable membranes, and a separate pumping system circulates another supply of the same electrolyte solution through the electrode compartments to remove any products of electrolysis. The sample to be separated is pumped into the center of the electrolyte curtain near the left side of the cell. This sample stream descends vertically in the absence of a voltage gradient, but when the voltage gradient is applied, it is displaced to the right according to the electrophoretic mobility of the particles. When the sample stream passes beyond the electrodes, it again descends vertically, but now displaced to the right. A viewing port with a millimeter scale allows the measurement of the migration distance, i.e., the lateral displacement of the particle stream. The electrolyte curtain and electrode rinse pumping systems use peristaltic pumps or gravity feed. The sample pumping system uses gas generated by the electrolysis

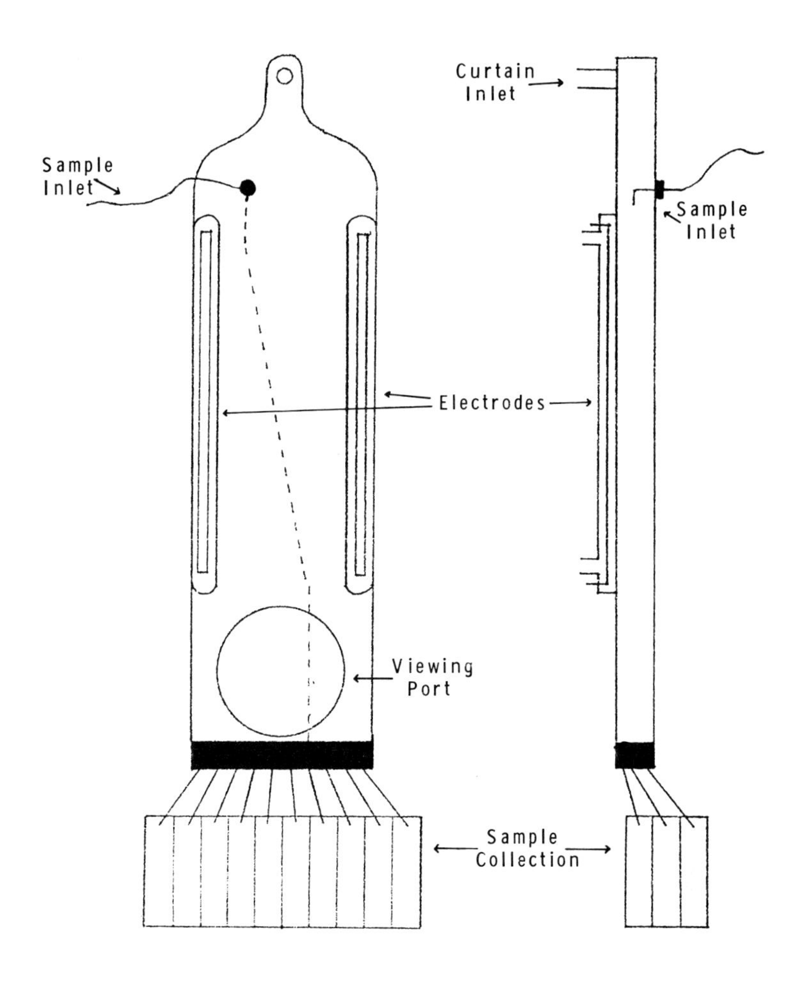

Fig. 1. Schematic representation of the Beckman CPE

of water, which in turn is governed by a rheostat. This instrument was used to separate monodisperse polystyrene latexes according to their electrophoretic mobilities (and hence particle sizes); mixtures of as many as ten different-size latexes were separated successfully (8).

The Beckman CPE Instrument was modified in two ways. The sample delivery system, which was originally controlled by an electrochemically-induced pressure head, was replaced by a micrometer syringe attached to a variable-speed motor. This modification resulted in a more reproducible sample delivery rate, which could be controlled over a wider range of flow rates and which results in a more accurate definition of the sample stream diameter. The second modification concerns the determination of the lateral displacement of the particle stream due to electrophoresis. The Plexiglas viewing port at the bottom of the separation cell, which is used for visual observation of the sample stream by means of reflected light, was replaced by a quartz window which is transparent to ultraviolet light. A small optical bench with an ultraviolet light source and slit for control of beam dimensions was placed in front of the window. The optical bench was attached to a motor, which moves it horizontally across the window to scan the particle stream after the separation, and a resistance box, to establish the precise position of the ultraviolet light beam electronically. A UV-sensitive phototube was located in a fixed position on the other side of the window. A Houston Instruments Model 2000 X-Y recorder was used to record the intensity of ultraviolet radiation on the Y-axis and the horizontal position of the optical bench on the X-axis. The horizontal position on the recorder is magnified by a factor of 18.2X.

The dispersions used in this study were primarily 2% Dow monodisperse polystyrene latexes dispersed in sodium barbital buffer at pH 8.6.

III. THEORETICAL ASPECTS OF CPE

The principles of flow in the CPE system have been described by Strickler (6) and Hannig (3). A schematic representation of the electroosmotic and induced flow in the curtain is presented in Figure 2. The parabolic flow profile due to electroosmosis is in the X-Z plane and is either positive or negative in the X direction, while the induced parabolic flow profile is in the Y-Z plane and is always positive in the Y direction. The sample is injected into the center of the stream in a cylindrical configuration and moves with a constant electrophoretic velocity in the X direction in the presence of an applied potential. The migration of the particles is affected by two factors, both of which are a function of their position in the Z direction: the electroosmosis of the electrolyte medium which affects the net particle velocity; and the induced parabolic flow profile which affects the velocity of the particle in the Y direction and hence the time of exposure of the particle to the electric field. Both factors ultimately affect the migration distance of the particles and the configuration of the sample stream. Both the electroosmotic and induced parabolic flow profiles

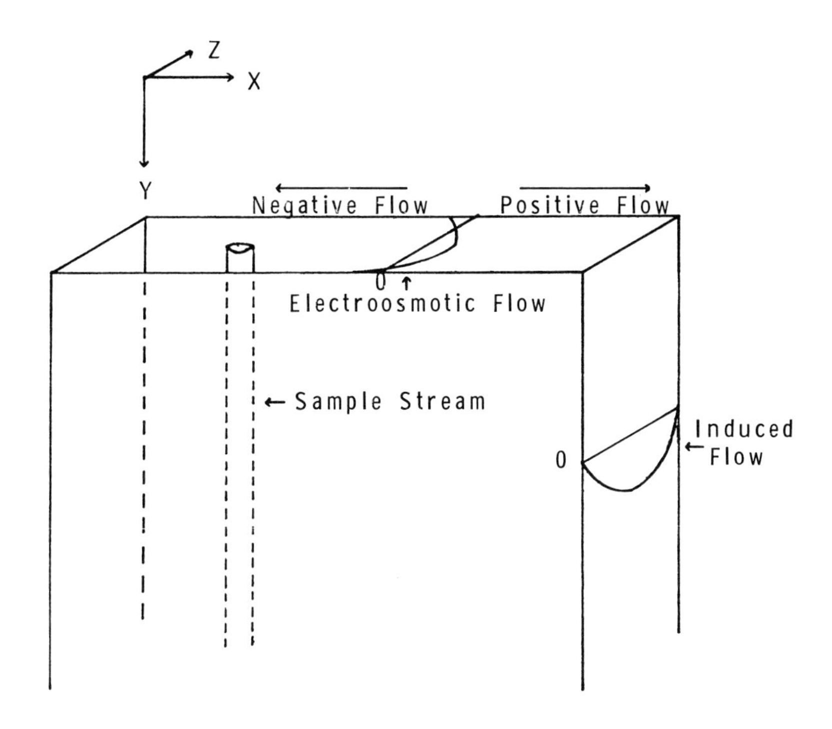

Fig. 2. Schematic Representation of Flow Profiles Present in Electrolyte Curtain of CPE

of the electrolyte medium have compensating effects, i.e., as the particle position moves away from the center of the channel, the net particle velocity decreases in both the $\underline{Y}$ and $\underline{X}$ directions. This means that the slower-moving particles will be subjected to electrophoretic migration for a longer period of time than the faster-moving particles in the center of the curtain flow. In principle, therefore, it is possible to match the electroosmotic flow to the induced flow so that the sample streams will remain undistorted after separation. Under these conditions, the displacement of the sample stream will be controlled by the particles in the center of the curtain where the velocity of buffer in the $\underline{Y}$ direction and the positive velocity of buffer in the $\underline{X}$ direction are both at a maximum.

The migration distance $\underline{X}$ of the particles as a function of position relative to the center of the channel in the $\underline{Z}$-direction and the electroosmotic and induced parabolic flow profiles has been given by Strickler (6) as follows:

$$X = (HE/V_o)\,[\,(U_{os} + U_e)/\{1 - (Z^2/d^2)\} - (3\,U_{os}/2)]\ , \quad (1)$$

where H is the length of the electrodes, E is the potential gradient, V_o is the induced solvent flow velocity in the center of the channel, i.e., at $Z = 0$, and d is one-half the channel thickness. If the sample stream is injected in the center of the channel and the radius of the sample stream is small relative to d (i.e., less than 20%), then for all practical purposes the value of Z in Equation 1 may be set equal to zero, to yield:

$$X = (HE/V_o)\,[U_e - (U_{os}/2)]\ . \quad (2)$$

Equation 2 is valid only if there is no distortion of the sample stream due to the electroosmotic and induced parabolic flow profiles.

The physical dimensions of the electrolyte curtain are such that $V_o = 2.25F/A$, where F is the curtain flow rate and A is the cross-sectional area of the curtain. Substitution of V_o into Equation 2 yields:

$$X = (A\ H\ E/2.25F)\,[U_e - (U_{os}/2)]\ . \quad (3)$$

Substitution of the physical dimensions of the Beckman CPE Instrument (i.e., $A = 0.675\ cm^2$ and $H = 30.5$ cm) into Equation 3 yields:

$$X = (0.0549\ E/F)\ [U_e - (U_{os}/2)]\ . \quad (4)$$

Equation 4 predicts the migration distance of the particles as a function of the instrument parameters E and F, and as a function of the electrophoretic mobility of the particles U_e and the electroosmotic flow at the channel-wall interface U_{os}. Equation 4 may be rearranged to yield:

$$U_e = (XF/0.0549E) + (U_{os}/2)\ , \quad (5)$$

where U_{os} must be evaluated in order to calculate U_e from an experimental measurement of X. In practice, the 1.10μ-diameter monodisperse polystyrene latex was used as a standard to determine U_{os} from microcapillary measurements of U_e.

Another method, which was used to determine the electrophoretic mobility of an unknown dispersion when standard latex particles were mixed with the dispersion, is to measure the difference in migration distance ΔX between the unknown and standard particles. Thus, $\Delta X = X_1 - X_2$ where X_1 and X_2 are the migration distances of the standard and unknown particles, respectively. Equation 4 may be used in this definition of Δ X to yield:

$$U_{e_2} = U_{e_1} - (\Delta X\ F/0.0549\ E) \qquad (6)$$

where $\underline{U}_{e1}$ and $\underline{U}_{e2}$ are the electrophoretic mobilities of the standard and unknown particles, respectively. Thus, Equation 5 may be used to calculate the absolute electrophoretic mobility from the migration distance of the particles $\underline{X}$ and from a knowledge of $\underline{U}_{os}$ determined using the standard particles, and Equation 6 may be used to calculate the absolute electrophoretic mobility from the difference in migration distances $\Delta \underline{X}$ between the unknown and standard particles.

IV. EXPERIMENTAL RESULTS

Figure 3 shows a typical recorder scan of a mixture of three monodisperse latexes (diameters of 0.234, 0.46, and 1.10μ) separated in the Beckman CPE. The electrolyte was 10^{-3} M sodium barbital, and the curtain flow rate was 25 cc/min.

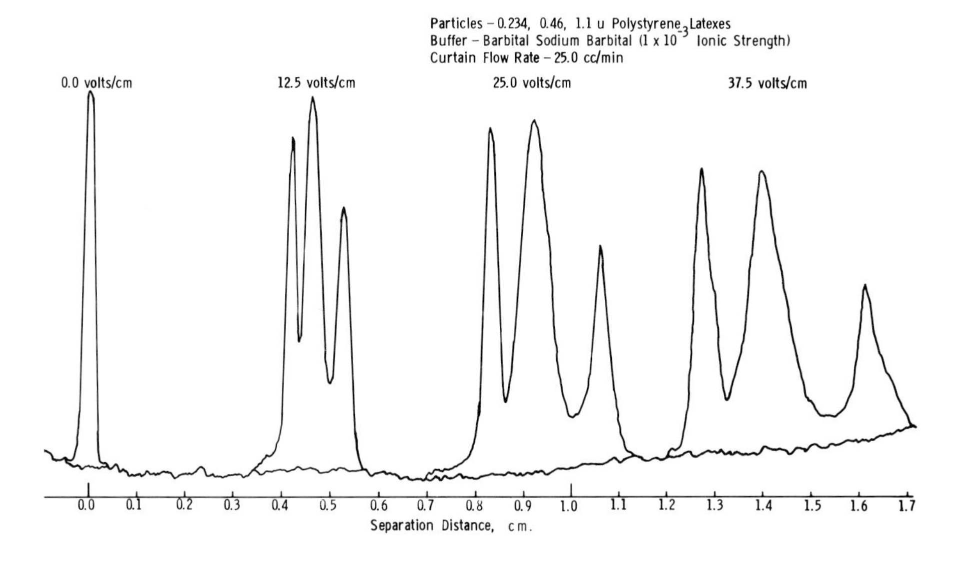

Fig. 3. Recorder Scan of Separation of Monodisperse Polystyrene Latexes in Beckman Continuous Particle Electrophoresis Instrument.

The four different runs presented in Figure 3 were at applied potentials of 0, 12.5, 25.0, and 37.5 volt/cm. The electrophoretic mobilities of these latexes in this medium were 3.74, 4.65, and 5.64μ cm/volt sec for the 0.234, 0.46, and 1.10μ diameter particles, respectively.

A necessary condition to establish that the Beckman CPE is operating properly is that the migration distance must be a linear function of the applied potential which extrapolates back to zero at zero potential. Figure 4 shows the variation of migration distance with $\underline{E}$ for a mixture of three latexes for curtain flow rates of 12.5 and 25.0 cc/min. The plots are linear and are compatible with Equation 4 when the 1.10μ diameter latex is used as a standard to calculate $\underline{U}_{os}$.

These results demonstrate that the Beckman CPE is operating properly according to theory. The feasibility of using this instrument to measure the absolute electrophoretic mobility of particles was determined by the following procedure. The sodium barbital buffer was prepared at five different concentrations in the range of 10^{-4} to 10^{-2} M. The

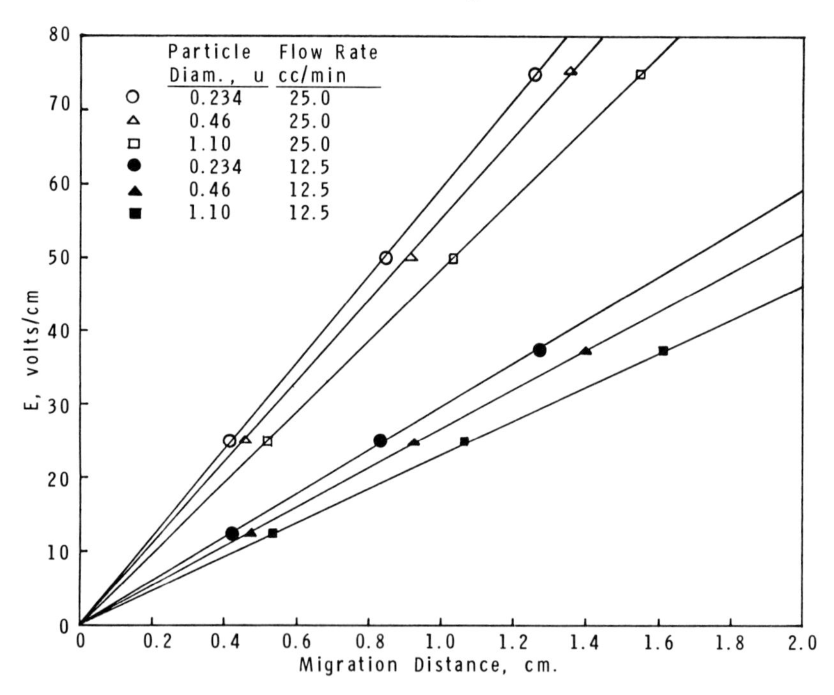

Fig. 4. Variation of Migration Distance of Polystyrene Latex Particles with Voltage Gradient in Continuous Particle Electrophoresis.

electrophoretic mobilities of seven different monodisperse latexes were measured in the sodium barbital buffers by microcapillary electrophoresis. The 1.10μ diameter latex was used as the standard to calculate the electrophoretic mobility of all the other latexes at the different sodium barbital buffer concentrations, according to either Equation 5 or Equation 6. The results are summarized in Table I where column 1 identifies the latex, column 2 the buffer concentration, and columns 3, 4, and 5 the electrophoretic mobility results obtained from Equation 6, Equation 5, and microcapillary electrophoresis, respectively. Column 6 gives the difference between the results obtained from microcapillary electrophoresis and Equation 6, and column 7 gives the electro-osmotic mobility at the channel wall calculated from Equation 4 using the 1.10μ diameter latex as the standard. The agreement between microcapillary electrophoresis and the

TABLE I

Comparison of Electrophoretic Mobilities by Microcapillary and Continuous Particle Electrophoresis

Polystyrene Latex sizes μm	Barbital Na-Barbital Conc.	Electrophoretic Mobility, μm cm/volt sec Continuous Particle Electrophoresis				
		Δ x	x	Microcapillary	MCE-CPE	U_{os}
0.109	0	3.24		3.28	+ 0.04	-39.0
0.357	0	3.32		3.28	- 0.04	
0.109	1×10^{-4}	3.87		4.11	+ 0.24	- 8.76
0.357	1×10^{-4}	4.01		4.03	+ 0.02	
0.045	1×10^{-3}	3.22	3.32			- 7.73
0.088	1×10^{-3}	2.99	2.92			
0.109	1×10^{-3}	2.99		2.64	- 0.35	
0.234	1×10^{-3}	3.76	3.68	3.74	- 0.02	
0.359	1×10^{-3}	3.93		3.90	- 0.03	
0.460	1×10^{-3}	4.47	4.41	4.65	+ 0.18	
0.109	2×10^{-3}	3.06		3.28	+ 0.22	-10.89
0.234	2×10^{-3}	3.98	4.05	4.20	+ 0.22	
0.357	2×10^{-3}	4.17		4.30	+ 0.13	
0.460	2×10^{-3}	5.00	5.03	4.93	- 0.07	
0.80	2×10^{-3}	5.11		5.18	+ 0.07	
2.02	2×10^{-3}	6.94		6.45	- 0.49	
0.234	5×10^{-3}	3.80		4.10	+ 0.30	-11.82
0.357	5×10^{-3}	4.02		4.54	+ 0.52	
0.46	5×10^{-3}	5.27		5.47	- 0.14	
0.109	1×10^{-2}	3.10		3.68	+ 0.58	-12.13
0.357	1×10^{-2}	4.30		4.54	+ 0.24	

CPE (column 6) is generally good and demonstrates that the CPE method is a valid approach for measuring the electrophoretic mobilities of particles.

One important caution is that Equations 5 and 6 assume that the particle stream is not distorted by the electroosmotic and induced parabolic flow profiles of the electrolyte curtain and therefore are not applicable where such distortion is observed. This fact can usually be established by selecting two standard latexes which have electrophoretic mobilities near the upper and lower limits of the range of interest. Also, it is possible to distinguish such distortion by visual observation (5).

There is very little difference between the results obtained from Equations 5 and 6. From practical considerations, it is easier experimentally, and thus possibly more accurate, to use Equation 6, because the zero point does not have to be measured. Also, there is an unexplained phenomenon that affects the use of Equation 5: the large calculated values of $\underline{U}_{os}$ given in column 7 of Table I. The zeta potentials calculated from these values range from 150 to 300mv in the sodium barbital buffer and have a value of about 1000mv in deionized water, i.e., without sodium barbital buffer. Since these values seem unreasonably large, it must be assumed that there is a flow in the same direction of the separation which is not entirely due to electroosmosis along the channel wall. For all practical purposes, however, Equation 5 is not concerned with the source of the flow, but only that this flow be proportional to the applied potential. Since this proportionality has been established experimentally, it may be assumed that the additional flow is electroosmotic in origin and probably occurs in the semi-permeable membranes which separate the electrodes from the channel.

Figure 5 shows the separation of a mixture of seven monodisperse polystyrene latexes measured at applied potentials of 30 and 60 volts/cm. Seven peaks can be identified, and it is possible to estimate the resolution of the CPE method from these results. The differences in the calculated electrophoretic mobilities of latexes 1 and 2, and latexes 3 and 4 are 0.16 and 0.37μ cm/volt sec, respectively. The microcapillary electrophoresis results for latexes 3 and 4 show a difference of 0.10μ cm/volt sec; latex 1 could not be measured by this method because the size of its particles is too small for detection in dark field illumination. Although microcapillary electrophoresis is considered to be more accurate for determining absolute mobilities, the CPE method gives directly the different mobilities of a

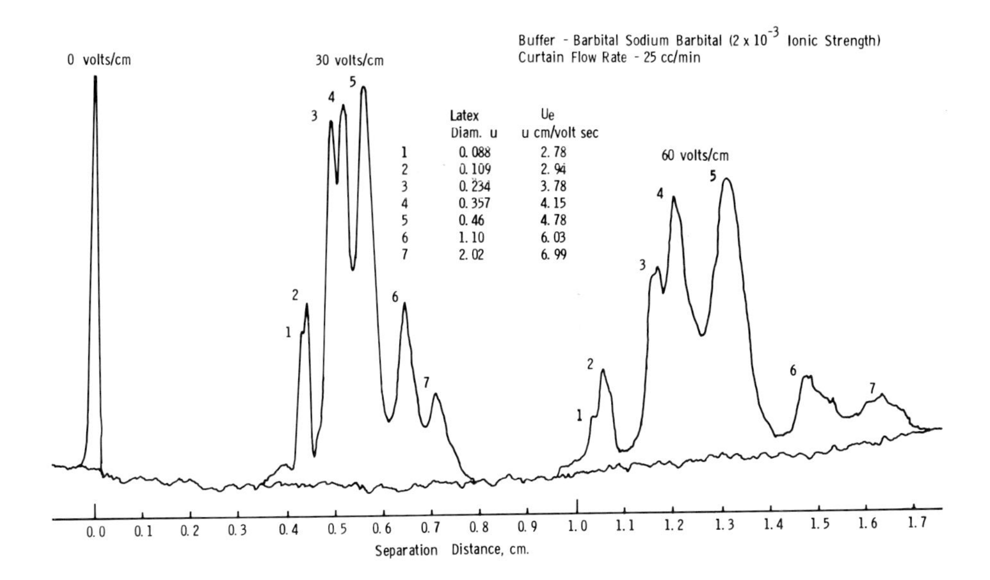

Fig. 5. Separation of Seven Polystyrene Latexes in Beckman Continuous Particle Electrophoresis Instrument

mixture of two samples.

V. CONCLUSIONS

The CPE method compares favorably with microcapillary electrophoresis for measuring the electrophoretic mobility of colloidal particles. The advantages of the CPE method are summarized as follows, in the order of decreasing importance:

1. The CPE method offers the capability of measuring the absolute electrophoretic mobility of particles which are too small to be detected by microscopic techniques;

2. The electrophoretic mobility characteristics of complex dispersions, which contain particles of different mobilities, may be evaluated routinely by the CPE method;

3. The absolute difference in the electrophoretic mobilities of two different dispersions in the same electrolyte can be established more accurately by the CPE method;

4. The absolute electrophoretic mobilities can be measured more rapidly (estimated 5 to 10 determinations/hour) by the CPE method than by microcapillary electrophoresis.

The CPE method has the following disadvantages compared to microcapillary electrophoresis:

1. The CPE is a more complicated instrument and, as a result, takes more time to set up initially;

2. It is difficult to change the electrolyte type and concentration in the CPE;

3. The CPE cannot be operated reliably without using microcapillary electrophoresis as a reference;

4. The Beckman CPE Instrument is no longer in production, and only used instruments or expensive prototypes of a new design are available.

VI. REFERENCES

1. Hannig, K., Z. Anal. Chem. 181, 244 (1961).
2. Hannig, K., Hoppe-Seyler's Z. Physical Chem. 338, 211 (1964).
3. Hannig, K., Wirth, H., Meyer, B. H., and Zeiller, K., Ibid. 356, 1209 (1975).
4. Strickler, A., Kaplan, A., and Vigh, E., Microchem J. 10, 529 (1966).
5. Strickler, A., and Sacks, T., Prep. Biochem. 3, 269 (1973).
6. Strickler, A., and Sacks, T., Ann. N. Y. Acad. Sci. 209, 497 (1973).
7. Hannig, K., and Heidrich, H. G., Methods Enzymol 21, 746 (1974).
8. McCann, G. D., Vanderhoff, J. W., Strickler, A., and Sacks, T., Sep. & Purific. Methods 2, 153 (1973).

NON-EQUILIBRIUM THERMODYNAMIC STUDIES OF ELECTRO-KINETIC EFFECTS-X ETHYLENEGLYCOL + WATER MIXTURES

R.L. Blokhra, M.L. Parmar and V.P. Sharma

ABSTRACT

Experimental results for the measurement of electro-osmosis, electro-osmotic pressure difference, streaming potential for ethyleneglycol-water (EG - H_2O) mixtures (30%, 40%, 50%, 60% and 70% by weight) using pyrex sintered disc (G_3) as a membrane at temperatures 25° to 40°C and at voltages 0-300 volts are reported. The data are analysed in the light of the theory of non-equilibrium thermodynamics. Attempts have been made to explore the domain of validity of linear phenomenological relations in the case of these mixtures. It has been found that the validity of the theory increases with the increase in the viscosity of the mixture and decreases with the increase in the dielectric constant of the mixture. The higher order co-efficients introduced owing to the non-validity of the linear phenomenological relations have been estimated from the data on electro-osmosis and electro-osmotic pressure difference. It is found that higher order straight co-efficients do not contribute effectively to the non-linearity of the relations. Onsager's reciprocity relations have been found to hold for all the mixtures reported here. It has been found that concentration dependence of the co-efficient, L_{22}, representing the volume flow per unit pressure difference and the co-efficient, L_{21} or L_{12}, representing the electro-osmotic velocity do not conform to Spiegler's frictional model. Efficiency of electro-kinetic energy (E) for electro-osmotic flow has been calculated and the results have been found to be in accordance with the non-equilibrium thermodynamic theories. For different compositions of EG - H_2O mixtures at 25°C $(E)_{max}$ was obtained at half the value of electro-osmotic pressure difference.

INTRODUCTION

Blokhra et al.[1-7] have reported the results of electro-kinetic effects for different categories of liquids using a sintered glass disc as a membrane. We found that the phenomenological co-efficients of the phenomenon could be related to the properties of the solvent[2,3] and it was shown that the structural order[5] of the liquids could be derived from these phenomenological co-efficients. In 1966, Blokhra et al.[1] showed that the domain of applicability of non-equilibrium thermodynamics was dependent on the dielectric constant. The higher the value of the dielectric constant, the smaller was the domain of validity of the linear phenomenological relations for the simultaneous transport of matter and electricity, and were not valid where the potential difference, $\Delta\phi$, exceeded a particular limit and the non-linear form[2] of the phenomenological relations could describe the phenomena correctly, and these non-linear forms suggest[7] that the phenomenological co-efficients in the linear relations did not remain constant with the applied potential difference.

In the present communication, we selected a homogeneous mixture in which viscosity can be varied through a large range and it is achieved by selecting a highly viscous liquid, viz., ethyleneglycol (η = 0.173P at 25°C) in conjunction with water (η = 0.8937CP at 25°C). Besides investigating the role of viscosity in the non-validity of the linear phenomenological relations, the other aims of the present study were (i) to evaluate the phenomenological co-efficients from data on electro-osmosis of ethyleneglycol (EG) + water mixtures, (ii) to study the variation of the phenomenological co-efficients with temperature and the composition of the mixtures, (iii) to determine the range of validity of the phenomenological relations, (iv) to evaluate the second-order phenomenological co-efficients, introduced owing to the non-validity of the linear relations, from the measurement of electro-osmotic pressure difference and electro-osmosis, (v) to test the validity of Onsager's symmetry relations from streaming potential measurements, and (vi) to evaluate the efficiencies of electro-kinetic energy conversion (E_e) and $(E_e)_{max.}$ for electro-osmosis and the results thus obtained have been analysed in the light of thermodynamic theories.

EXPERIMENTAL

Materials

Ethyleneglycol (BDH) was purified by vacuum distillation after drying it over calcium oxide, calcium sulphate and sodium metal and stored in sealed bottles. Triple distilled water over alkaline $KMnO_4$ was used for preparing the different compositions of mixtures of ethyleneglycol and water.

Apparatus and Procedure

The design of the apparatus, its experimental set up is shown in Fig. 1.

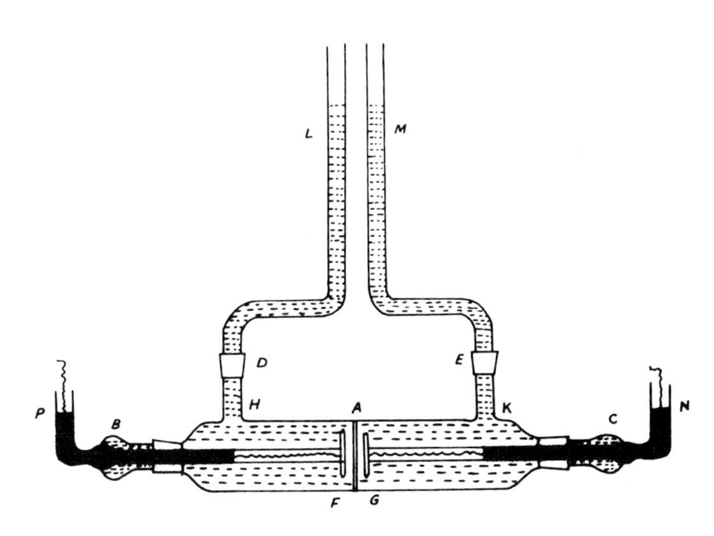

FIG.I – APPARATUS FOR STUDYING ELECTRO-OSMOSIS

The apparatus consists of a pyrex tube of length 28 cms and of diameter 2.5 cms, with a sintered glass disc A of porosity G_3 in the middle. This tube has two female joints B and C at the ends. Perforated platinum discs F and G are fixed to the standard male joints with the help of platinum wires. The whole length of each electrode except the perforated disc is insulated from the liquid medium by sealing it in a glass tube of 0.5 cm in diameter. The length of these glass tubes are adjusted in such a way that when the standard joints are kept in position, the platinum discs approach very close (within 2 mm) to the sintered disc. The main pyrex tube bears two side tubes H and K, to which tubes L and M of about 0.24 cm^2 cross-sectional area are connected, through standard taper joints D and E, in such a way that both the tubes are kept parallel and close together. These tubes are graduated in cm scale.

After cleaning the sintered disc of porosity G_3 with concentrated nitric acid, it was rinsed with water and dried in an oven at 150°C. The filling of the apparatus was done by adding the liquid on one side of the disc and then sucking it to the other side under a pressure gradient by means of a vacuum pump, this ensured the complete filling of the capillaries of the disc. The tubes L and M were then introduced and the liquid under investigation was introduced into them by means of a syringe to bring the levels of the liquid to a desired level. Mercury was introduced into the tubes P and N to make the contacts of the electrodes. The apparatus was suitably mounted inside the thermostat where it was allowed to attain a constant temperature. A constant D.C. voltage was then applied and the levels of the liquid in the tubes L and M were recorded after regular intervals of time by means of a cathetometer.

Streaming Potential Measurements

The experimental assembly designed for carrying out experiments on measurements of streaming potential is essentially similar to that described above for the determination of electro-osmotic flow except for a few necessary modifications. The experimental set up in this case has been illustrated in Fig. 2.

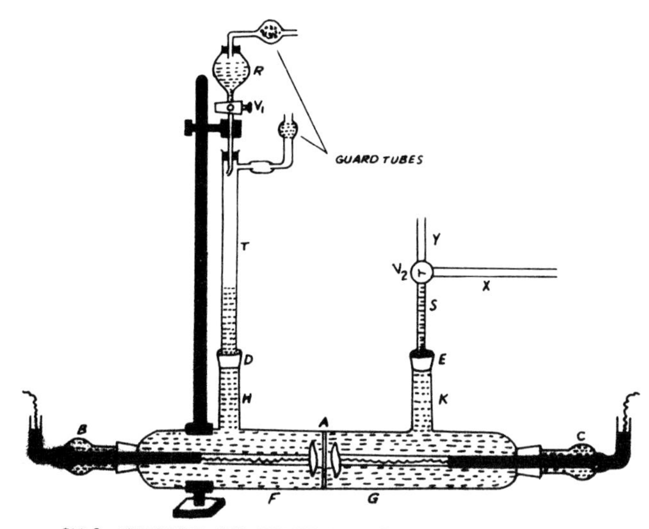

FIG. 2 - APPARATUS FOR MEASURING STREAMING POTENTIAL

For the determination of streaming potential the apparatus shown in Fig. 1 has been modified in respect of the following :

1. The limb L (Fig. 1) was replaced by a graduated tube T of about 80 cm in length and 1.0 cm in diameter. A reservoir R (Fig. 2) was used to introduce the liquid in tube T at any desired rate. There was an arrangement for attaching guard tubes both to the tube T and the reservoir R.
2. The limb M (Fig. 1) was replaced by tube S of narrow bore (5-6 mm), which communicates with a horizontal tube X

and a vertical tube Y through a stop cock V_2. The diameter of the tubes X and Y are about 5 mm. The length of the tube Y was usually about 10 cm and that of X about 15 cm.

The apparatus was assembled and filled in a similar manner as the apparatus for electro-osmotic flow. In this case the liquid was allowed to flow under a constant pressure gradient. Depending on the rate of flow of the liquid from reservoir R (controlled by stop cock V_1) into the limb T, it was possible to adjust a desirable pressure gradient. The streaming potential developed for a given pressure difference, was measured with the help of an electrometer, capable of reading upto 0.01 mV, which was supplied by the Electronics Corporation of India Ltd. The resistance of the system (electro-osmotic cell) with the liquid contained in the apparatus, was measured with the help of a conductivity bridge supplied by Toshniwal Instruments (Bombay). All the experiments on streaming potential measurement were carried out at 25°C.

The data on electro-osmosis and streaming potential are reproducible within ± 2%.

RESULTS AND DISCUSSION

According to the theory of thermodynamics of irreversible processes, the entropy production, σ , for an electro-kinetic phenomenon is given by[8]

$$\sigma = J.\,\Delta P + I.\Delta\phi \qquad (1)$$

where J is the volume flow, I is the current flow, ΔP is the pressure difference, $\Delta\phi$, is the potential difference. We can accordingly write phenomenological equations as the relations :

$$I = L_{11}\cdot\Delta\phi + L_{12}\ \Delta P \qquad (2)$$

$$J = L_{21}\cdot\Delta\phi + L_{22}\ \Delta P \qquad (3)$$

where L's are the phenomenological co-efficients with $L_{12}=L_{21}$ according to Onsager's theorem, L_{11} is related to resistance while L_{22} is related to the permeability of the liquid. The co-efficients L_{12} and L_{21} are the cross-phenomenological co-efficients. Further, according to the theory of irreversible thermodynamics, these co-efficients are considered to be constants. The phenomenological co-efficients L_{21}, L_{22} are calculated as follows :

If x is the rise in the liquid level at time t and A is the area of cross-section of the tube L or M, the volume flow J will be given by

$$J = A\ (dx/dt) \qquad (4)$$

Thus (3) becomes

$$A\ (dx/dt) = L_{21}\cdot\Delta\phi + L_{22}\cdot\Delta P \qquad (5)$$

Since ΔP is the pressure difference across the disc, it is

2x at a particular time. dx/dt is calculated from the plot of x versus t by drawing slopes at different time intervals. Now (5) shows that if we plot J versus ΔP at a fixed temperature and potential difference, we get a straight line. The slope of the straight line will be

$$dJ/d(\Delta P) = L_{22} \qquad (6)$$

and the intercept is

$$L_{21} \cdot \Delta\phi \qquad (7)$$

Thus, by knowing the intercept and slope we can calculate L_{21} and L_{22} simultaneously by using (6) and (7). A sample plot of J versus ΔP for 50 percent EG - H_2O mixture is shown in Fig. 3.

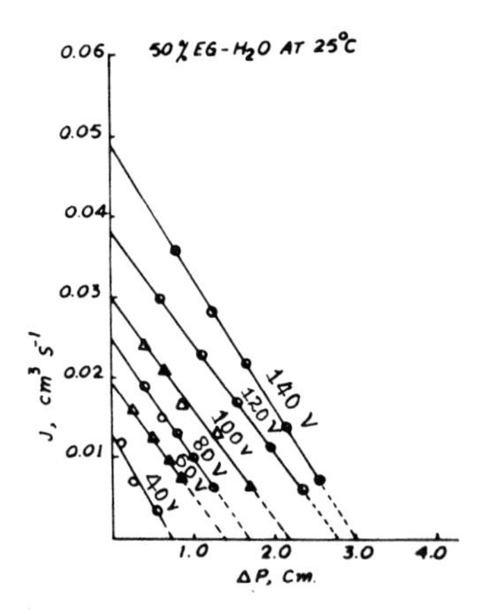

FIG. 3-VARIATION OF J WITH ΔP

The values of L_{21}, $(J)_{\Delta P = 0}$ evaluated from similar figures and the values of, $(\Delta P)_{J = 0}$, electro-osmotic pressure difference at different voltages for different EG - H_2O mixtures are given in Table I.

TABLE I

Electro-osmotic data for different compositions of EG - H_2O mixtures at 25°C

$\Delta\phi$ Volts	$(J)_{\Delta P=0}$ $cm^3\ S^{-1}$	$(\Delta P)_{J=0}$ cm	$L_{21} \times 10^4$ $cm^3\ S^{-1}\ V^{-1}$
	30 percent (W/W)		
40	0.029	1.00	7.24
60	0.044	2.20	7.30
80	0.059	3.00	7.30
100	0.081	4.80	8.11
120	0.102	5.40	8.52
140	0.127	6.40	9.00
160	0.152	7.00	9.51
180	0.184	8.40	10.22
200	0.214	9.20	10.74
220	-	10.40	-
240	0.286	11.60	11.90
260	-	12.80	-
280	0.366	13.60	13.11
300	-	14.60	-
	40 percent (W/W)		
40	0.027	1.00	6.70
60	0.040	2.22	6.66
80	0.054	3.20	6.70
100	0.067	4.00	6.70
120	0.082	4.40	6.80
140	0.098	4.80	7.00
160	0.115	5.60	7.21
180	0.136	6.84	7.52
200	0.172	7.62	8.62
220	-	9.22	-
240	0.216	10.20	9.00
260	-	11.40	-
280	0.270	12.40	9.64
300	0.320	15.60	10.66
	50 percent (W/W)		
40	0.013	0.80	3.12
60	0.019	1.40	3.10
80	0.025	1.80	3.12
100	0.030	2.20	3.00
120	0.038	2.80	3.10

contd.

140	0.049	4.00	3.50
160	0.067	5.40	4.20
180	0.086	0.20	4.71
200	0.106	7.20	5.30
220	0.122	8.00	5.50
240	-	9.00	-
260	0.152	11.00	5.80
280	-	12.40	-
300	0.196	13.40	6.50
	60 percent (W/W)		
40	0.017	1.00	4.25
60	0.025	1.60	4.16
80	0.034	3.00	4.25
100	0.042	3.60	4.20
120	0.051	4.20	4.25
140	0.059	4.80	4.25
160	0.074	5.80	4.60
180	0.092	6.80	5.00
200	-	7.60	-
220	0.124	8.00	5.63
240	-	8.80	-
260	0.160	9.60	6.10
280	-	10.20	-
300	0.200	11.00	6.60
	70 percent (W/W)		
40	0.0034	0.40	0.85
60	0.0052	0.60	0.86
80	0.0068	0.80	0.85
100	0.0085	1.00	0.85
120	0.0102	1.20	0.85
140	0.0120	1.40	0.85
160	0.0134	1.60	0.83
180	0.0153	1.80	0.85
200	0.0174	2.20	0.87
220	0.0202	2.60	0.91
240	0.0232	2.80	0.96
260	-	3.00	-
280	0.0296	3.40	1.05
300	0.0340	3.80	1.13

We get straight lines at all the voltages for every mixture and unparallel lines suggest that L_{22} plays a significant role for explaining the domain of validity of the relation (3). It has been suggested earlier[1-4,7,9,10] that the validity of relations (2) and (3) can be increased by writing these relations in the non-linear form given by (8) and (9) :

$$I = L_{11} \cdot \Delta\phi + L_{12}\,\Delta P + L_{111}(\Delta\phi)^2 + L_{112} \cdot \Delta\phi \cdot \Delta P + L_{122}(\Delta P)^2 + \ldots \quad (8)$$

$$J = L_{21} \cdot \Delta\phi + L_{22}\,\Delta P + L_{211}(\Delta\phi)^2 + L_{212} \cdot \Delta\phi \cdot \Delta P + L_{222}(\Delta P)^2 + \ldots \quad (9)$$

where L_{ikj} are the higher order co-efficients. Before discussing the second-order phenomenological co-efficients, we shall discuss the first-order co-efficients of expression (3).

Electro-Osmosis

At $\Delta\phi = 0$, equation (3) is reduced to

$$(J)_{\Delta\phi=0} = L_{22} \cdot \Delta P \quad (10)$$

Now, the compatibility of (10) with Poissuille's law through a diaphragm of thickness l and n number of capillaries of porosity r shall give

$$L_{22} = \pi \sum_{i=1}^{n} r_i^4 / 8\eta l \quad (11)$$

Since the observed flow is the sum-total of the flows through all the capillaries. Now, viscosity decreases with increasing temperature and hence L_{22} should increase with the rise of temperature, and this is found to be so. For 50 percent EG - H_2O mixture, L_{22} at 25° and 30°C came out as 1.37×10^5 and 2.25×10^5 cm^5 dyn^{-1} S^{-1} respectively. Further, viscosity of the EG - H_2O mixtures increases with the increase in EG content and accordingly L_{22} should decrease with the increase in EG content, and this is clear from Table II.

TABLE II

Values of L_{22} for different compositions of EG - H_2O mixture at 40 volts at 25°C

EG - H_2O (W/W)	$L_{22} \times 10^5$ (cm^5 dyn^{-1} s^{-1})
30	2.36
40	2.27
50	1.57
60	1.42
70	1.19

Fig. 3 is in accordance with equation (10) but the slope changes for different values of $\Delta\phi$, indicating that non-linearity of the phenomenological relations (2) and (3) might arise from the non constancy of the hydraulic co-efficient L_{22}, and it is in accordance with the findings of Srivastava and Awasthi[11].

At $\Delta P = 0$, relation (3) can be written as

$$(J)_{\Delta P=0} = L_{21} \cdot \Delta\phi \tag{12}$$

Therefore, if we plot $(J)_{\Delta P=0}$ versus potential difference we should get a straight line. For the case of 50 percent EG - H_2O mixture, these quantities have been plotted in Fig. 4.

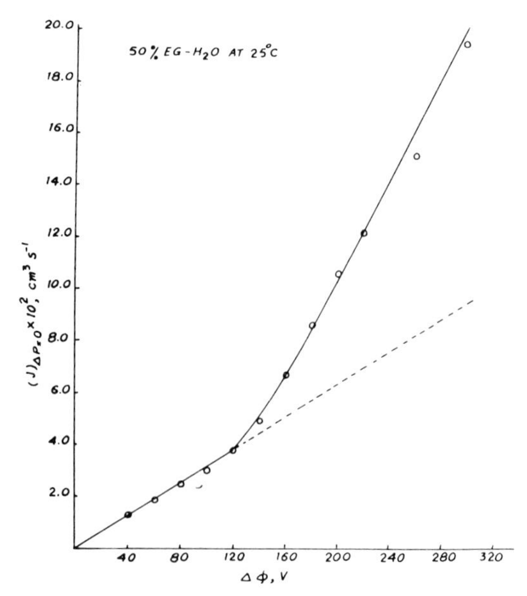

FIG. 4 - VARIATION OF $(J)_{\Delta P=0}$ WITH $\Delta\phi$

The Fig. 4 shows that (12) is valid only upto $\Delta\phi$= 120V, after L_{21} becomes a non-linear function of potential difference. This indicates that the phenomenological relation (2) or (3) is valid upto $\Delta\phi$= 120V for 50 percent EG - H_2O mixture. The values of $\Delta\phi$ at which non-linearity starts for different mixtures alongwith the viscosities and dielectric constant of the mixtures is given in Table III.

TABLE III

Magnitude of $\Delta\phi$ at which non-linearity starts for different EG - H_2O mixtures at 25°C

Mixture (% by weight)	$\Delta\phi$ (V)	Viscosity[12] (P)	Dielectric constant[13] (D)
30	80	0.018	69.80
40	100	0.023	66.60
50	120	0.030	63.20
60	140	0.041	59.40
70	160	0.057	54.70

Values of $\Delta\phi$ corresponding to the point of break of the linear relation are plotted versus viscosity and dielectric constants of the mixtures in Fig. 5.

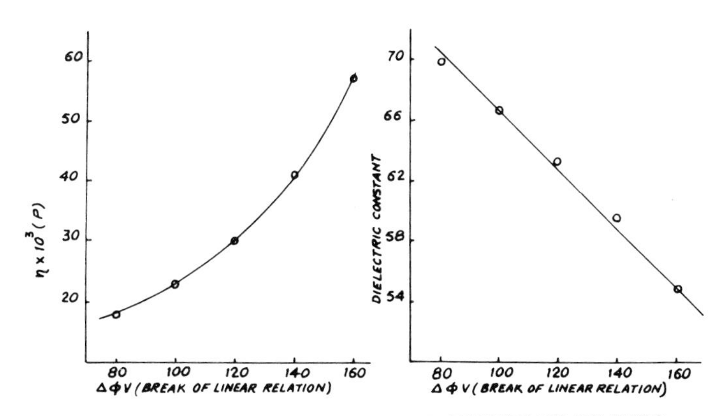

FIG.5 - DEPENDENCE OF BREAK OF LINEARITY ON VISCOSITY AND DIELECTRIC CONSTANT FOR EG - H_2O MITURES AT 25°C

Fig. 5 suggests that the validity of the theory of non-equilibrium thermodynamics as applied to electro-kinetic effects increases non-linearly with the increase in the viscosity of the mixture and decreases linearly with increase in the dielectric constant of the mixture. The effect of dielectric constant on the validity of the phenomenological relations (2) and (3) is in accordance with our earlier findings[1-4].

Over-beek[14] pointed out that the co-efficient L_{21} is directly proportional to the ζ-potential, viz.,

$$L_{21} = K\zeta \tag{13}$$

where $K = \epsilon\zeta / 4\pi\eta l$ = constant for a liquid and symbols have the usual significance. As pointed out earlier in the discussion that non-validity of relations (2) and (3) after a certain magnitude of $\Delta\phi$ suggested that L_{21} becomes a non-linear function of $\Delta\phi$, it can be inferred according to (13) that ζ-potential becomes a non-linear function of potential difference. Rutger and de Smet[15] carried out experiments on electro-osmosis at higher voltages with solutions of different dielectric constants and calculated zeta-potential upto 900V. They found that for lower values of the dielectric constant, the zeta-potential values varies according to

$$\zeta = -0.03 \cdot \Delta\phi - 25.6 \qquad (\epsilon = 2.57) \tag{14}$$

For higher dielectric constants, a power series fits the data

$$\zeta = 77\text{x}10^{-6}\,(\Delta\phi)^2 - 0.15 \cdot \Delta\phi - 74 \quad (\epsilon = 9.97) \tag{15}$$

This shows that in liquids of higher dielectric constants, the linear phenomenological relations (2) and (3) have got lesser domain of validity, and we found so in case of different EG - H_2O mixtures.

Electro-Osmotic Pressure Difference

Values of electro-osmotic pressure difference $(\Delta P)_{J=0}$ can be estimated from Fig. 3 by extrapolation though we chose to carry out the estimation separately, because near the stationary state volume flow could not be noted correctly. In this case the experiment was repeated at 25°C upto the attainment of the stationary state, J = 0. The pressure difference, $(\Delta P)_{J=0}$, on the two sides of the disc corresponding to different potential difference for different mixtures is recorded in Table I.

Now at J = 0 expression (3) gives

$$\Delta P/\Delta\phi \quad = \quad -L_{21}/L_{22} \tag{16}$$

Thus according to (16), the plots of ΔP versus $\Delta\phi$ should be a straight line throughout the range of $\Delta\phi$ reported here unless the co-efficients L_{21}, L_{22} become non-linear function of $\Delta\phi$. In all the cases, we found that non-linearity in the plots of $(\Delta P)_{J=0}$ versus $\Delta\phi$ is observed at values of $\Delta\phi$ close to those reported in Table III. The relative contribution of the non-constancy of L_{21} and L_{22} can be assessed from the estimations of the higher order co-efficients L_{ikj} of L_{21} and L_{22}. The estimation of these co-efficients is described in the followings :

Non-Linear Phenomenological Relations/Higher Order Co-Efficients

The non-validity of the phenomenological relation (3) suggests that non-linear relation of the type (9) may be valid beyond the range of validity for (3). At $\Delta P=0$, equation (9) becomes

$$(J/\Delta\phi)_{\Delta P=0} = L_{21} + L_{211}.\Delta\phi \tag{17}$$

$(J/\Delta\phi)_{\Delta P=0}$ has been ploted against $\Delta\phi$ as a sample plot for 30 percent EG - H_2O mixture at 25°C in Fig. 6.

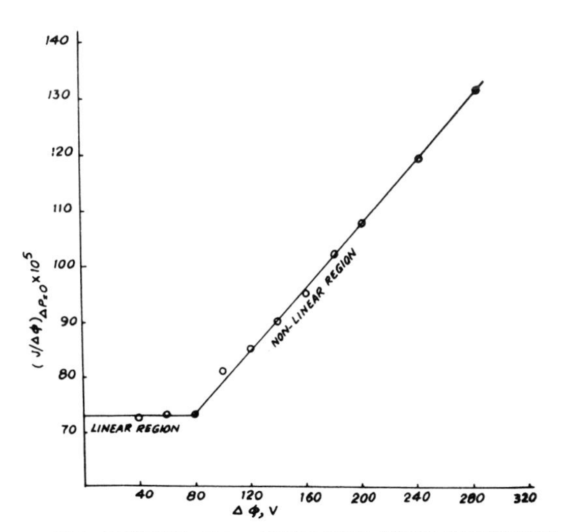

FIG. 6 - ESTIMATION OF L_{211} FROM ELECTRO-OSMOTIC MEASUREMENTS FOR 30 PERCENT EG-H_2O AT 25°C

The plots for different mixtures are in accordance with equation (17). The portion of the plot which is parallel to the $\Delta\phi$-axis represents the region upto which the expression (12) holds. The values of the co-efficients L_{21} and L_{211} could be estimated respectively from the intercept of this straight line (parallel to $\Delta\phi$-axis) on the $(J/\Delta\phi)$ axis and from the slope of the straight line which represents the region where the relation (12) does not hold. Further at $J = 0$, expression (9) becomes

$$L_{21}\cdot\Delta\phi + L_{22}\cdot\Delta P + L_{211}(\Delta\phi)^2 + L_{212}(\Delta\phi\cdot\Delta P) + L_{222}(\Delta P)^2 + \ldots = 0 \quad (18)$$

On rearrangement and using equation (16) we get

$$L_{222}(\Delta P/\Delta\phi)^2 + L_{212}(\Delta P/\Delta\phi) + L_{211} = 0 \quad (19)$$

Equation (19) is true for the non-linear region only. Substituting the value of L_{211} estimated from (17) and the value of $(\Delta P/\Delta\phi)_{J=0}$ corresponding to a particular value of $\Delta\phi$ in the non-linear region we get simultaneous equations and on solving values of L_{212} and L_{222} are obtained. The values of the L_{ikj} co-efficients obtained from (17) and (19) at 25°C are given in Table IV.

TABLE IV

Values of the second-order (L_{ikj}) phenomenological co-efficients for different compositions of EG-H_2O at 25°C

Percent (W/W)	$L_{211} \times 10^6$ $cm^3\,S^{-1}\,V^{-2}$	$L_{212} \times 10^6$ $cm^5 dyn^{-1} S^{-1} V^{-1}$	$L_{222} \times 10^6$ $cm^7 dyn^{-2} S^{-1}$
30	2.73	- 0.108	0.0012
40	2.50	- 0.116	0.0013
50	2.19	- 0.124	0.0018
60	1.43	- 0.105	0.0017
70	0.23	- 0.034	0.0012

Table IV suggests that the magnitude of L_{222} can be neglected in comparision to L_{211} and L_{212} and therefore the non-linear phenomenological equation can be written as

$$J = L_{21}\cdot\Delta\phi + L_{22}\cdot\Delta P + L_{211}(\Delta\phi)^2 + L_{212}(\Delta\phi\cdot\Delta P) \quad (20)$$

Further the values of L_{211} and L_{212} go on decreasing with the decrease in dielectric constant. At very low values it may attain values which can be neglected. As dielectric constant increases, values of these co-efficients become significant and cannot be neglected. This also supports the conclusion that the range of the validity of the relations (2) and (3) increases with the lowering of the dielectric constants.

Streaming Potential Studies

The analysis of the streaming potential data also throws light on the nature of the cross-phenomenological co-efficients. The streaming potential, $\Delta\phi_s$, corresponding to the pressure difference ΔP has been plotted for 30 per cent EG - H_2O mixture in Fig. 7.

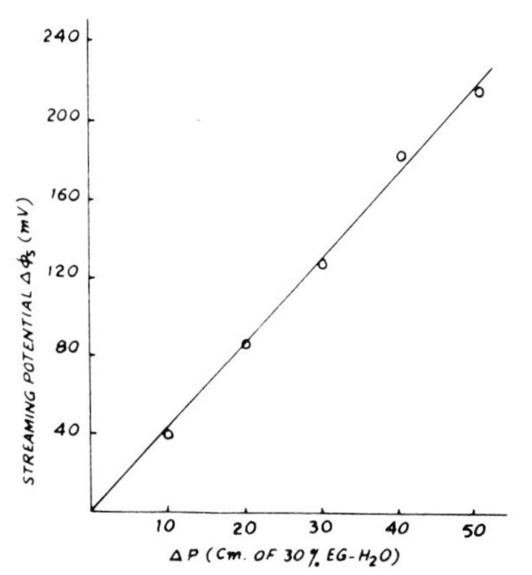

FIG. 7 - ESTIMATION OF L_{12} FROM STREAMING POTENTIAL MEASUREMENTS

A straight line plot indicates that the linear phenomenological relation for flow of current is valid upto $\Delta P = 50$ cm. Similar plots are obtained for all the mixtures. The data on $\Delta\phi_s$ for different values of ΔP for various EG - H_2O mixtures is given in Table V.

TABLE V

Data of the streaming potential across pyrex sintered disc at 25°C

System	Resistance R (ohm)	Pressure difference ΔP (cm)	Streaming potential $\Delta\phi_s$ (mV)
30% EG-H_2O	0.515×10^5	10	39.82
		20	82.44
		30	121.60
		40	165.88
		50	204.26
40% EG-H_2O	0.550×10^5	10	35.65
		20	73.24
		30	104.32
		40	146.16
		50	187.86
50% EG-H_2O	0.605×10^5	10	19.62
		20	43.80
		30	55.22
		40	82.22
		50	107.92
60% EG-H_2O	0.125×10^6	10	53.96
		20	104.08
		30	167.86
		40	215.78
		50	264.22
70% EG-H_2O	0.770×10^5	10	7.94
		20	16.26
		30	25.80
		40	28.42
		50	43.78

At I = 0, equation (2) gives

$$\Delta\phi_s / \Delta P = -L_{12}/L_{11} \qquad (21)$$

where L_{12} has been determined from $\Delta\phi_s$ values for different ΔP as described earlier[6]. It is found that in all the cases L_{12} comes out very close to L_{21} estimated from electro-osmotic data in the linear region. These values are given in Table VI.

TABLE VI

Verification of Onsager's Reciprocity Relations for different mixtures at 25°C

System	$L_{12} \times 10^4$ $cm^3\ AJ^{-1}$	$L_{21} \times 10^4$ $cm^3\ AJ^{-1}$
30% EG-H_2O	7.60	7.30
40% EG-H_2O	6.43	6.70
50% EG-H_2O	3.36	3.12
60% EG-H_2O	4.12	4.25
70% EG-H_2O	0.97	0.85

This Table VI shows that Onsager's reciprocity relation[8] holds good in all the mixtures reported here.

Concentration Dependence of Phenomenological Co-Efficients

The fundamental assumption underlying the general treatment of Spiegler[16,17] is that the frictional forces that counter-balance the thermodynamic driving forces are additive. At $\Delta\phi = 0$, the entropy production, σ for a binary mixture of EG and water would become[16].

$$\sigma = X_G v_G dm_G/dt \cdot (\Delta P) + X_W v_W dm_W/dt \cdot (\Delta P) \qquad (22)$$

where v represents the specific volume, m represents the mass and the subscripts G and W represent the quantities for glycol and water respectively. Taking v·dm/dt as the volume flux of the species denoted by the subscripts, the two conjugate thermodynamic forces are $X_G \cdot \Delta P$ and $X_W \cdot \Delta P$. If X_{GW}, X_{GS} and X_{WS} represent the frictional forces between the species denoted by the subscripts then the hydrodynamic description of these forces is

$$X_{GW} = f_{GW} (V_G - V_W)$$

$$X_{GS} = f_{GS} (V_G - V_S)$$

$$X_{WS} = f_{WS} (V_W - V_S) \qquad (23)$$

In (23) the subscript S stands for the sintered glass membrane, f stands for the frictional co-efficients between the species denoted by the subscripts and V represents the velocities of the various species indicated by the subscripts. Regarding the membrane velocity V_S as the reference and $V_S=0$, the Spiegler's assumption of additivity of frictional forces will lead to the relationship

$$X_G \cdot \Delta P = X_{GW} + X_{GS} = V_G (f_{GW} + f_{GS}) - V_W f_{GW} \qquad (24)$$

and

$$X_W \cdot \Delta P = X_{WG} + X_{WS} = V_W (f_{WG} + f_{WS}) - V_G f_{WG} \qquad (25)$$

Solving (24) and (25) for V_G and V_W, and identifying the total volume flux $(J)_{\Delta\phi=0}$ with $(V_G + V_W)$, the co-efficient L_{22} can be identified with the quantity

$$X_G \cdot \frac{(f_{WG} + f_{WS}) + f_{WG}}{(f_{WG} + f_{WS})(f_{GW} + f_{GS}) - f_{GW} f_{WG}}$$

$$+ X_W \cdot \frac{f_{GW} + (f_{GW} + f_{GS})}{(f_{WG} + f_{WS})(f_{GW} + f_{GS}) - f_{GW} f_{WG}} \qquad (26)$$

Substituting the bracketed terms as $(L_{22})_G$ and $(L_{22})_W$ respectively, (26) can be written as

$$L_{22} = X_G (L_{22})_G + X_W (L_{22})_W$$

or

$$L_{22} = (L_{22})_G + \left[(L_{22})_W - (L_{22})_G\right] \cdot X_W \qquad (27)$$

Equation (27) suggests that L_{22} varies linearly with the mass fraction X_W. The present investigations show that L_{22} does not vary linearly with X_W showing thereby that the frictional model of Spiegler does not fit in the EG-H_2O mixtures. This suggests that the frictional forces are not additive and it may be attributed to the formation of associated complex of EG-H_2O in the liquid state. The plot of specific conductance of different mixtures versus molefraction gives sharp break in the beginning of the plot ($\approx$30% EG - H_2O), suggesting the formation of the associated complex. This indicates that Spiegler's model of frictional co-efficients is valid only in the dilute regions.

The above mentioned considerations in case of L_{21} also do not fit in the frictional model.

The Efficiency of Energy Conversion

Osterle and coworkers[18-21] and Kedem and Caplan[22] have discussed the efficiency of electro-kinetic energy conversion on the basis of non-equilibrium thermodynamics. In case of the phenomena of electro-osmotic flow and streaming potential, energy conversion takes place from one form to another. In the former case, electrical energy is converted into mechanical work while in the latter case, reverse conversion of mechanical work into electrical energy takes place.

The expression for the efficiency of energy conversion (E) in terms of thermodynamic fluxes J and forces X as deduced by Osterle can be written as

$$E = - J_o X_o / J_i X_i \qquad (28)$$

where the subscripts o and i represents the output and input quantities respectively. The negative sign in (28) signifies that output fluxes and output forces are in the direction opposite to that of their input components.

In the phenomenon of electro-osmosis, the applied potential difference, $\Delta\phi$, is the input force and the consequent pressure difference ΔP is the output force. Therefore efficiency energy conversion 'E_e' for electro-osmosis would read as

$$E_e = - J . \Delta P / I . \Delta\phi = - J . \Delta P / (\Delta\phi)^2 / R \qquad (29)$$

where subscript e represent the phenomenon of electro-osmosis and R is the electrical resistance of the system. In an electro-osmosis experiment, the applied potential difference is used to derive the liquid up hill. This liquid, if allowed to accumulate, exerts a pressure difference across the membrane causing a back flow of the liquid. When ΔP equals electro-osmotic pressure difference, the net volume flux J becomes zero. Thus it is clear from (29) that E_e would be zero when either $\Delta P = 0$ or ΔP equals electro-osmotic pressure difference and therefore plot of E_e versus ΔP for a fixed value of input force $\Delta\phi$ would pass through a maximum as ΔP varies from zero to the electro-osmotic pressure. In the present investigations, we find that the value of E_e attained a maximum value when ΔP equals half the value of electro-osmotic pressure difference i.e.

$$\Delta P = \tfrac{1}{2} (\Delta P)_{J=0} \qquad (30)$$

for all the compositions of the mixtures. This is shown in Fig. 8 for 30 percent and 50 percent EG - H_2O mixture.

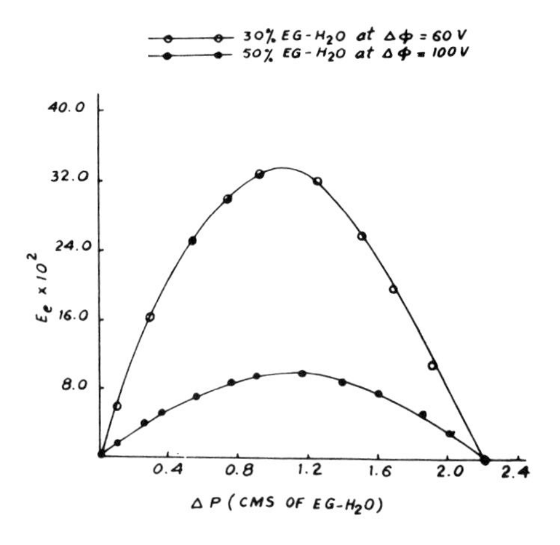

FIG. 8 - DEPENDENCE OF E_e ON OUTPUT FORCE ΔP AT 25°C

This observation can be easily justified on theoretical consideration on substituting the value of J from (3) in equation (29) we get

$$E_e = (L_{21} \cdot \Delta\phi + L_{22} \cdot \Delta P) \cdot \Delta P / I \cdot \Delta\phi \qquad (31)$$

Since the input force is kept constant, the quantity I in equation (31) is constant. Now applying the condition

$$\partial E_e / \partial (\Delta P) = 0 \qquad (32)$$

for a maximum we get

$$2 L_{22} (\Delta P / \Delta\phi) + L_{21} = 0 \qquad (33)$$

But we know from equation (3) that at the steady state where $J = 0$

$$(\Delta P/\Delta\phi)_{J=0} = -L_{21}/L_{22} \tag{34}$$

From equations (33) and (34), we have

$$\Delta P = \tfrac{1}{2}(\Delta P)_{J=0} \tag{35}$$

where $(\Delta P)_{J=0}$ represents the value of electro-osmotic pressure difference at $J=0$.

Acknowledgement

Authors are thankful to Professor M.L. Lakhanpal for encouragement and discussion during the investigations. One of the authors (MLP) is grateful to CSIR (India) for the award of Post-doctoral Fellowship.

REFERENCES

1. Blokhra, R.L., Kaul, C.L., Soni, B.R. and Jalota, S.K., Electrochim. Acta, 12, 773 (1967).
2. Blokhra, R.L., and Singhal, T.C., J. Electroanal. Chem., 48, 353 (1973).
3. Blokhra, R.L., and Singhal, T.C., J. Electroanal. Chem., 57, 19 (1974).
4. Blokhra, R.L., and Parmar, M.L., J. Colloid Interface Sci., 51, 214 (1975).
5. Blokhra, R.L., and Singhal, T.C., Ind. J. Chem., 13, 913 (1975).
6. Blokhra, R.L., and Singhal, T.C., J. Phys. Chem., 78, 2303 (1974).
7. Blokhra, R.L., and Parmar, M.L., J. Electroanal. Chem., 62, 373 (1975).
8. Prigogine, I., "Introduction to Thermodynamics of Irreversible Processes", John Wiley and Sons, New York, (1967).
9. Rastogi, R.P., and Jha, K.M., J. Phys. Chem., 70, 1017 (1966).
10. Rastogi, R.P., and Jha, K.M., Trans. Faraday Soc., 62, 585 (1966).
11. Srivastava, R.C., and Awasthi, P.K., Kolloid Z. Polymer, 250, 253 (1972).

12. Sestra, B., and Berardelli, M.L., Electrochim. Acta, 17, 915 (1972).
13. Harned, H.S., and Owen, B.B., "The Physical Chemistry of Electrolytic Solutions", Reinhold Amsterdam (1957), p. 161.
14. Overbeek, J.Th.G., J. Colloid Sci., 8, 420 (1953).
15. Rutgers, A.I., and de Smet, M., Trans. Faraday Soc., 48, 635 (1952).
16. Katachalsky, A., and Curran, P.F., "Non-equilibrium Thermodynamics in Biophysics", Harvard University Press, Massachusetts (1965), p. 244.
17. Spiegler, K.S., Trans. Faraday Soc., 54, 1408 (1958). (See also ref. 16 Chapter 10).
18. Osterle, J.F., J. Appl. Mecha., 31, 161 (1964).
19. Osterle, J.F., Appl. Sci. Res., 12, 425 (1964).
20. Morrison, F.A., and Osterle, J.F., J. Chem. Phys., 43, 2111 (1965).
21. Gross, R.J., and Osterle, J.F., J. Chem. Phys., 49, 228 (1968).
22. Kedem, O., and Caplan, S.R., Trans. Faraday Soc., 61, 1897 (1965).

****

THE ADSORPTION OF THE POTENTIAL DETERMINING ARSENATE ANION ON OXIDE SURFACES

D.T. Malotky and M.A. Anderson
University of Wisconsin

The arsenate-aluminum hydroxide adsorption system is modeled using a modified stern equation. The stern potential is experimentally derived as a function of the potential determining ions for the system. Results indicate that adsorption of arsenate can be modeled using a single specific adsorption energy (Ø). Adsorption data is quantitatively described in terms of pH and equilibrium arsenate concentration. Constant pH Langmuir isotherms and other adsorption models are discussed in relation to the data presented.

I. INTRODUCTION

Concentrations of inorganic ions in aqueous environments may be controlled by a variety of mechanisms. For trace cations and anions, adsorption reactions may be the principle controlling mechanism in aquatic systems. These adsorption reactions commonly occur on clay, oxide and hydroxide surfaces and are usually pH dependent (1,2).

Recently, several pH-adsorption models based on the application of double layer theory (3,4,5,6) have been proposed. Utilization of these models has been somewhat restricted by the inability of available experimental techniques to measure theoretical parameters.

In a previous paper, Anderson et al. (7) described empirical correlations between adsorption of arsenate on amorphous aluminum hydroxide and the isoelectric pH (pH_{IEP}).

The primary data transformations presented by Anderson et al. (7) are illustrated in Figures 1-3. These figures suggest that: isoelectric pH (pH_{IEP}) is a linear function of adsorption, Figure 1, adsorption at the pH_{IEP} can be described by the Langmuir adsorption isotherm, Figure 2, adsorption consistently decreases as the adsorbent becomes more negatively charged ($(pH_{IEP}-pH) < 0$), Figure 3.

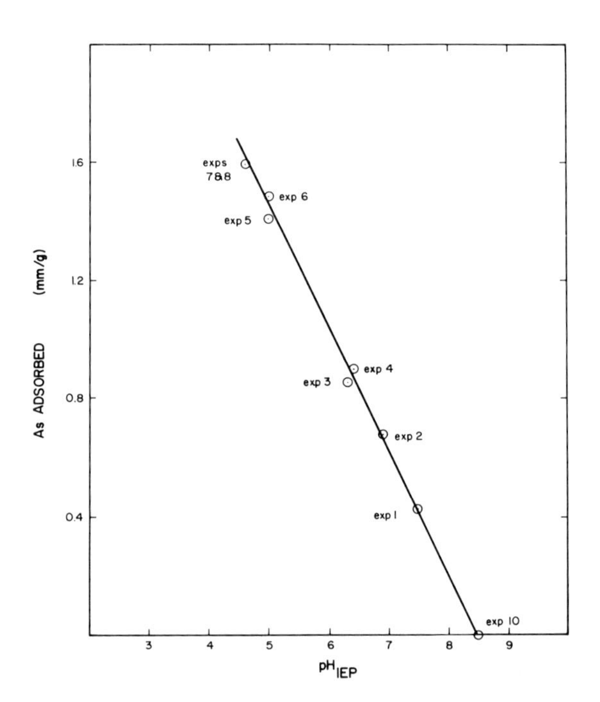

Fig. 1. Arsenate adsorption as a function of isoelectric pH. Initial concentrations of arsenate, 0 to 1600 μmole/ℓ. The concentrations of aluminum hydroxide, 0.13 to 0.16 g/ℓ. $NaClO_4$ = 0.01 moles/ℓ.

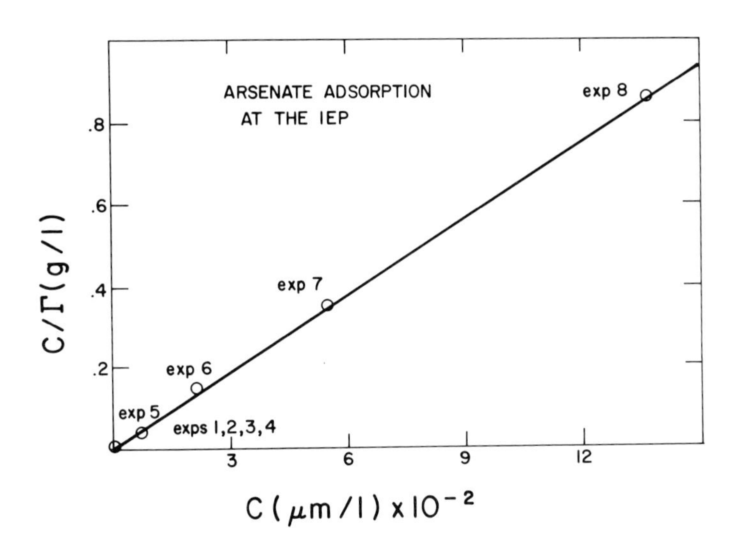

Fig. 2. Equilibrium arsenate adsorption at the isoelectric pH plotted in single reciprocal Langmuirian form. Initial concentrations of arsenate, 0 to 1600 μmole/ℓ; the concentration of aluminum hydroxide, 0.13 to 0.16 g/ℓ.

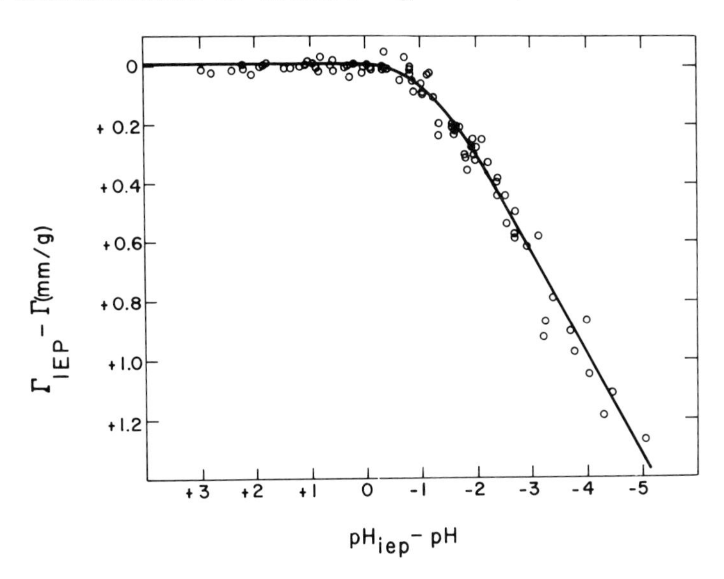

Fig. 3. Reduced arsenate adsorption ($\Gamma - \Gamma_{iep}$) as a function of the reduced variable ($pH_{iep} - pH$).

The unique nature of the data transformations presented in Figures 1-3 has lead to the development of a workable model for anion adsorption consistent with the adsorption data and basic double layer theory.

II. ADSORPTION MODEL

A model adsorption equation consonant with the results illustrated in Figures 1,2, and 3 was sought. The results depicted in Figure 2 imply that the model should be of the Langmuir form. The model should also describe the functional dependence between adsorption and pH_{IEP} as shown in Figure 1. Furthermore, Figure 3 illustrates that (pH_{IEP}-pH) should be included as an independent adsorption variable.

The Stern (8) equation (Equation 1) was found to satisfy the requirements illustrated by Figures 1-3 and was therefore selected as a model equation.

$$\Gamma = \frac{\Gamma_{max}\ C\ exp(-(\psi Ze + \emptyset)/KT)}{1 + C\ exp(-(\psi Ze + \emptyset)/KT)} \qquad \textit{(Equation 1)}$$

ψ = *Stern potential* $\emptyset$ = *chemical bonding energy*

Γ = *Moles adsorbate adsorbed per mass adsorbent*

Γ_{max} = *Maximum adsorption*

C = *Concentration of adsorbate*

Z = *Ionic charge*

K = *Boltzman constant*

T = *Temperature*

e = *Electronic charge*

Although this equation does have conceptual limitations (9), it works well in this application.

At the pH_{IEP}, ψ is equal to zero and the Stern equation reduces to the Langmuir equation with the Langmuir constant (K_L) equal to exp-($\emptyset$/KT).

Traditionally, a linearized data plot is used to determine the Langmuir constants K_L and Γ_{max}. There are three basic lineralizations of the Langmuir equation. Dowd and Riggs (10) have shown that the Langmuir linearizations illustrated in Figures 2 and 4 are superior to the double reciprocal form.

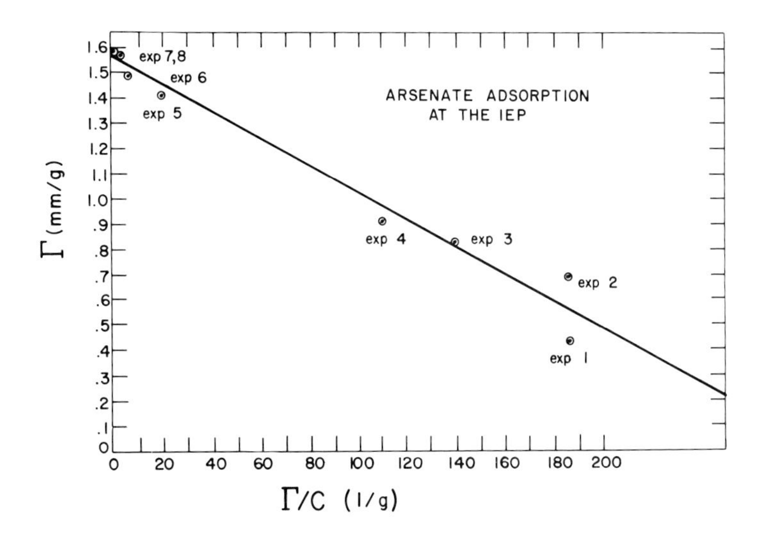

Fig. 4. Equilibrium arsenate adsorption at the isoelectric pH plotted in a linearized Langmuir form. Initial concentrations of arsenate, 0 - 1600 μmole/ℓ; the concentration of aluminum hydroxide 0.13 - 0.16 g/ℓ.

They also suggest that for the analytical uncertainties associated with the data of Anderson et al. (7), the linearization presented in Figure 4 is favored for the estimation of both Langmuir constants. The constants obtained from Figure 4 are:

$$K_L = .186\ \ell/\ \mu\text{mole} = \exp-(\emptyset/KT) = 1.03 \times 10^7\ \frac{\text{moles}}{\text{mole}}$$

Γ_{max} = 1522 μmole/g

$\emptyset$ =-9.56 Kcal/mole

Using $\emptyset$ and Γ_{max} values obtained, experimental adsorption data was used to solve the Stern equation for the Stern potential (ψ). The adsorption data was also correlated with pH_{IEP} (Figure 1) so that ψ could be plotted as a function of pH_{IEP}-pH (Figure 5).

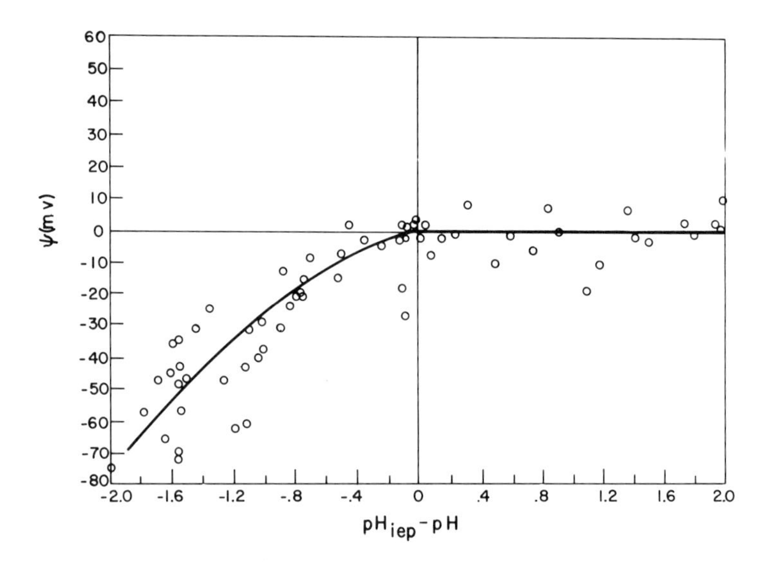

Fig. 5. The effective Stern potential as a function of pH_{IEP}-pH.

The nonlinearity of the data plotted in Figure 5 agrees with the prediction of Wright and Hunter (11) and Levine and Smith (12), that the aluminum hydroxide solid does not obey the Nernst equation. Consequently, the data shown in Figure 5 was described by an empirical equation. By using this empirical expression for ψ in the Stern equation, and the pH_{IEP}, Γ relationship (Figure 1), an iterative computer program was developed to predict adsorption as a function of pH and equilibrium arsenate concentration. This adsorption model will subsequently be referred to as the constant Ø model. The response surface generated by the model is shown for two concentration ranges in Figure 6 and 7. Numeric results are listed in Table 1.

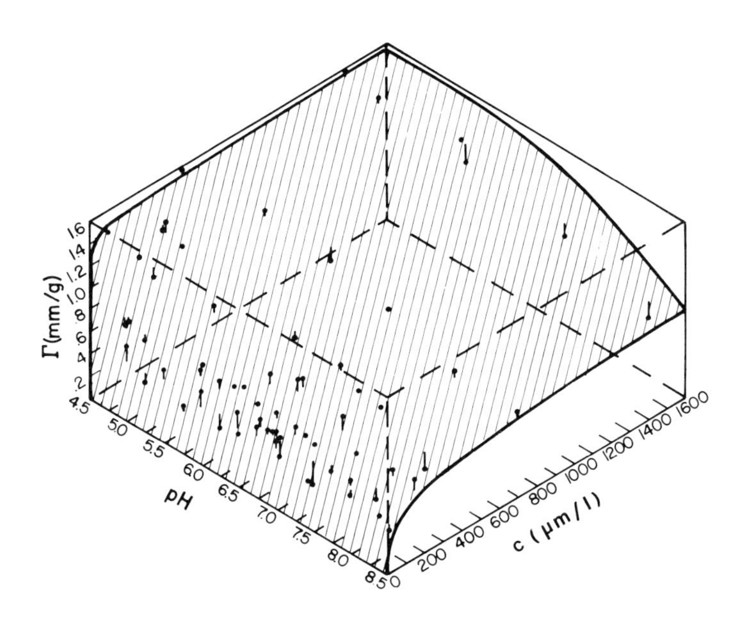

Fig. 6. Equilibrium arsenate adsorption as a function of pH and equilibrium arsenate concentration. Concentration range 0 - 1600 μmole/ℓ. Lines drawn from data points to the single Ø surface illustrate deviations in Γ at a given pH and equilibrium arsenate concentration.

It should be noted here that the model was restricted to the pH range 4.5 - 8.5. In this pH region, 95% or more of the aluminum is present in the solid phase.

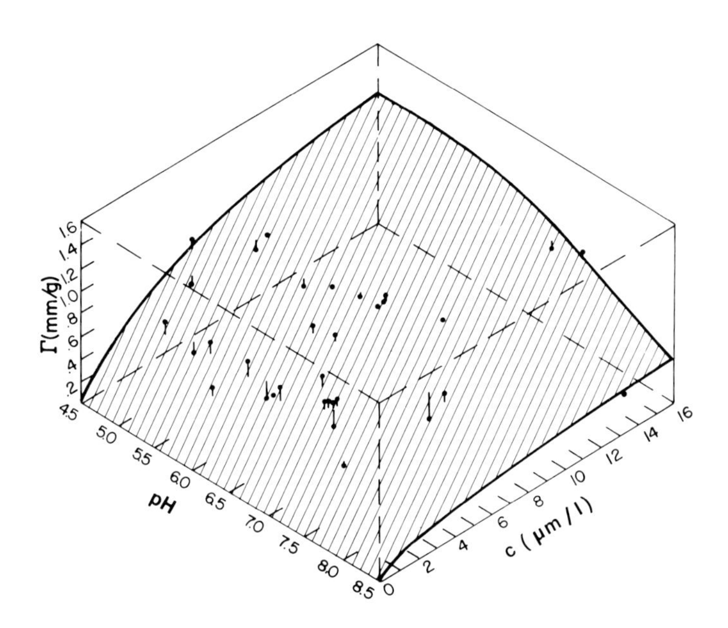

Fig. 7. Equilibrium arsenate adsorption as a function of pH and equilibrium arsenate concentration. Concentration range 0 - 16 μmole/ℓ. Lines drawn from data points to the single Ø surface illustrate deviations in Γ at a given pH and equilibrium arsenate concentration.

Table 1

pH	C μm	Γ_{model} μm/g	Γ_{exp} μm/g	pH_{iep}-pH
7.19	248.00	1107	1169	-1.40
7.51	120.00	917	1049	-1.25
7.49	266.00	1000	1030	-1.44
7.60	124.00	887	1019	-1.27
7.77	279.00	902	931	-1.48
8.27	162.00	680	749	-1.44
8.35	329.00	720	545	-1.60
8.49	169.00	612	696	-1.49
5.14	1370.00	1551	1589	- .44
5.92	1380.00	1521	1514	-1.11
5.52	548.00	1527	1574	- .76
6.12	1410.00	1509	1309	-1.28
6.32	584.00	1447	1349	-1.33
7.36	1440.00	1208	1099	-1.79
7.13	609.00	1218	1192	-1.59
8.20	1490.00	904	759	-1.89
7.83	640.00	955	999	-1.67
8.45	691.00	751	681	-1.77
7.45	2.29	450	424	- .04
6.90	3.60	627	674	- .06
6.40	8.20	912	894	- .11
5.00	213.00	1503	1474	- .15
4.60	1370.00	1552	1589	- .09
8.49	13.30	393	362	- .94
6.52	5.72	805	851	- .00
6.71	7.93	852	836	- .28
6.98	9.42	813	826	- .47
7.44	21.40	772	746	- .81
7.81	36.50	701	645	-1.01
8.28	52.10	579	537	-1.18
6.90	3.62	629	674	- .06
7.66	6.78	573	650	- .56
6.01	73.40	1337	1378	- .78
7.42	6.85	646	398	- .49
6.79	89.40	1148	1261	-1.10

Table 1 Continued

pH	C µm	Γ µm/g model	Γ µm/g exp	pH_{iep}-pH
4.60	546.00	1543	1589	.11
6.70	3.80	647	739	.21
6.82	3.62	629	674	.13
6.97	3.57	626	674	- .00
8.41	30.50	490	481	-1.10
5.00	70.00	1434	1399	.01
6.05	7.53	903	899	.25
6.44	8.27	908	893	- .14
6.66	8.10	867	893	- .28
6.96	15.20	888	841	- .61
7.20	15.80	814	841	- .69
7.35	22.40	807	792	- .81
7.99	38.60	644	678	-1.06
8.21	57.50	611	542	-1.19
5.79	211.00	1457	1451	- .83
5.28	78.20	1433	1349	- .26
4.89	193.00	1515	1591	- .10
5.24	2.96	549	423	1.92
5.70	1.85	400	431	1.81
6.29	3.32	591	422	.77
6.53	2.03	428	430	.91
6.89	3.69	630	419	.08
4.51	5.89	816	849	1.98
5.00	7.56	904	839	1.30
5.79	7.00	875	842	.58
6.25	5.53	792	852	.30
4.65	4.78	728	665	2.08
4.94	2.77	531	679	2.25
5.51	3.08	569	676	1.59
5.98	2.98	557	677	1.15
6.48	2.77	531	678	.71
4.90	8.04	927	893	1.34
4.50	69.70	1434	1404	.51
6.77	242.00	1260	1214	-1.33
7.15	106.00	1037	1149	-1.19

III. DISCUSSION

The constant Ø model is based on the assumption that adsorption at pH_{IEP} can be described by the Langmuir equations. The linearity of the pH_{IEP} data plotted in Figures 2 and 4 support this assumption. The single K_L generated by using the Langmuir equation to describe the pH_{IEP} adsorption data implies a single specific adsorption energy over the entire concentration range. This means that the constant Ø model does not need to consider variable specific adsorption energies with associated multiple adsorption sites, nor is it necessary to have the specific adsorption energy dependent upon the degree of ionization of the adsorbate.

Nonspecific Adsorption

Figure 3 illustrates that $\Gamma = \Gamma_{IEP}$ when pH_{IEP}-pH is equal to zero or positive. Under these circumstances, the surface is positively charged and the nonspecific adsorption of arsenate anions is enhanced. Anderson *et al*. (7) used a .01 M sodium perchlorate concentration to obtain their mobility data. The perchlorate anions compete with the arsenate in nonspecific adsorption, and therefore tend to minimize the nonspecific adsorption of the arsenate. Suppression of nonspecific arsenate adsorption helps account for the pH independence of arsenate adsorption at positive pH_{IEP}-pH values.

Langmuir Isotherms

Adsorption data has traditionally been described by a Langmuir curve fit to data obtained at a fixed pH (13,14, 15,16). Anderson *et al*. (7) previously described arsenate adsorption on aluminum hydroxide with constant pH Langmuir curves. Closer examination of the data has, however, revealed some fundamental problems associated with this procedure.

Adsorption is described by the pH_{IEP} Langmuir function only if $\Gamma = \Gamma_{IEP}$. If $\Gamma \neq \Gamma_{IEP}$ then Γ cannot be described by the isoelectric pH Langmuir constants determined in Figure 4. Figure 3 illustrates that $\Gamma < \Gamma_{IEP}$ when $(pH_{IEP}-pH)$ is negative. At a fixed pH, the value of $pH_{IEP}-pH$ depends upon Γ (Figure 1). Within the range $4.6 < pH < 8.5$, the value of $(pH_{IEP}-pH)$ will be positive for $\Gamma = 0$ ($pH_{IEP} = 8.5$) and negative for Γ_{max} ($pH_{IEP} = 4.6$). The shift from a positive to a negative $pH_{IEP}-pH$ corresponds to a shift from $\Gamma = \Gamma_{IEP}$ to $\Gamma < \Gamma_{IEP}$. Since Γ_{IEP} can be described by the pH_{IEP} Langmuir function and $\Gamma < \Gamma_{IEP}$ cannot, a single Langmuir constant does not describe adsorption at a fixed pH.

Adsorption Sites

In previous anion adsorption models, other authors (17,18) found it necessary to define more than one type of adsorption site to describe their data. The constant Ø model employs only a single specific adsorption energy (Ø) over a wide pH (4.5 - 8.5) and concentration range. The existence of a single specific adsorption energy suggests that multiple adsorption sites are not necessary to describe arsenate adsorption on aluminum hydroxide.

In their adsorption model, Bowden *et al.* (4) used separate specific adsorption energies for each ionization state of the adsorbate. In the constant Ø model, the specific adsorption energy is independent of the degree of ionization of the arsenate anion. The degree of ionization enters into the coulombic repulsion term as Z in the Stern equation but does not affect the specific adsorption energy (Ø).

The Free Energy of Adsorption

James and Healy (3) have shown an increase in the pH_{IEP} of TiO_2 with increasing Co(II) adsorption. These authors determined specific adsorption energy at the pH_{IEP} of the TiO_2. Their results indicate increasing Co(II) concentration (or alternatively Co(II) adsorption) leads to decreasing specific adsorption energy. At a constant pH, specific adsorption of cations increases the positive component of charge on the TiO_2. The decreasing specific adsorption energy calculated by James and Healy (3) may be due to increasing coulombic repulsion and a constant Ø may be applicable in their system as well.

In applying James and Healy's (3) model to anion adsorption, Huang (6) used a fixed pH_{IEP} in the estimation of ψ. The pH_{IEP} corresponded to the isoelectric point of γ-Al_2O_3 and disregarded the effect of the specific adsorption of phosphate anions on surface charge. As shown by the investigation of Anderson _et al_. (7) anion adsorption appears to be a function of the combined system parameter (pH_{IEP}-pH) (Figure 3) rather than a function of pH only.

The Stern Potential

In the arsenate-aluminum hydroxide adsorption system, hydrogen, hydroxyl, and arsenate are "potential determining" ions. Their interactions with the surface of the aluminum hydroxide must be included as an integral part of the ψ term in either the Stern (8) or Grahame (19) adsorption models. Due to the complex origion of the Stern potential the constant Ø model presently uses an empirically derived fit of the ψ, pH_{IEP}-pH function. Further investigation is needed to develop a mechanistic equation for this functional dependence.

IV. SUMMARY

The constant Ø adsorption model for arsenate adsorption on aluminum hydroxide was developed using the data of Anderson et al. and the Stern equation. This model makes use of a Langmuir curve to describe adsorption data obtained at pH_{IEP}. A single specific adsorption energy (Ø) was found sufficient to model this adsorption system. The use of a single specific adsorption energy provides a simplified approach to modeling adsorption as a function of pH.

V. REFERENCES

1. Fox, R.L., in "Soils of the Humid Tropics" (M. Drosdoff, ed.) National Research Council, 1968.
2. Shukla, S.S., Syers, J.K., Williams, J.D.H., Armstrong, D.E., and Harris, R.F., Soil Sci. Soc. Amer. Proc., 35, 244, 1971.
3. James, R.O., and Healy, T.W., J. Colloid Interface Sci., 40 (1), 42, 1972.
4. Bowden, J.W., Bolland, M.D.A., Posner, A.M., and Quirk, J.P., Nature Phy. Sci., 245, 31, 1973.
5. Murray, D.J., Healy, T.W., Fuerstenav, D.W., Adsorption from Aqueous Solution, Advances in Chemistry Series, 79, American Chemical Society, Washington, D.C., 1968.
6. Huang, C.P., J. Colloid Interface Sci., 53 (2), 178, 1975.
7. Anderson, M.A., Ferguson, J., and Gavis, J. J. Colloid Interface Sci., 54 (3), 391, 1976.
8. Stern, O., Z. Elektrochem., 30, 508, 1924.
9. Overbeek, J. Th. G., in Colloid Science (edited by Kruyt, H.R.) 1, Chapter IV (Elsevier, New York).
10. Dowd, John E., and Riggs, Douglas S., J. Biol. Chem., 240 (2) 863, 1965.
11. Wright, H.J.L., and Hunter, R.J., Austr. J. Chem., 26, 1191, 1973.
12. Levine, S., and Smith, A.L., Discuss. Faraday Soc. 52, 290, 1971.
13. Rennie, D.A., and McKercher, R.B., Can. J. Soil Sci. 39, 64, 1959.
14. Cole, C.V., Olse, S.R. and Scott, C.O., Soil Sci. Soc. Amer. Proc. 17, 352, 1953.
15. Fried, M., and Shapiro, R.E., Soil Sci. Soc. Amer. Proc. 20, 471, 1956.

16. Olsen, S.R., and Watanabe, F.S. Soil Sci. Soc. Amer. Proc. 21, 144, 1957.
17. Muljadi, D., Posner, A.M., and Quirk, J.P., J. Soil Sci. 17 (21), 212, 1966.
18. Robarge, W.P., Characterization of sorption sites for phosphate, Ph.D. Thesis, Soil Science Department, University of Wisconsin, Madison, Wisconsin (1975).
19. Grahame, D.C., Chem. Rev. 41, 441, 1947.

The authors would like to thank Professor D.E. Armstrong for his critical evaluation and review of this manuscript. This work has been supported by a research grant from the Agriculture Research Service (Cooperative Agreement No. 12-14-3001-226).

X-RAY DIFFRACTION ANALYSIS OF DIFFERENTLY PREPARED AgI

R. Despotović, N. Filipović-Vinceković and B. Subotić
"Ruđer Bošković" Institute

A systematic X-ray diffraction analysis of differently prepared AgI is described. AgI was prepared by fourteen different procedures. Quantitative data on the cubic/hexagonal ratio and the size of AgI crystallites in suspensions are given. The influence of the concentration of the constitutive and coagulating ions, surface active agents, acidity, aging, temperature and the rate of precipitation has been observed.

I. INTRODUCTION

Silver iodide is as a perfect model widely applied in colloid and surface science investigations. The scientific interest and research in the structural properties of silver iodide is contributing to a better understanding of the basic processes of nucleation, crystallization and recrystallization. From the practical point of view it is well-known that the phenomena of the latent image formation are closely related to the structural silver halide properties.

Silver iodide exists in four different forms: (i)

low-temperature of face-centred cubic structure (zinc blende type), (ii) high-temperature of body-centred cubic structure, (iii) hexagonal structure (wurtzite type) and (iv) high pressure form, of face-centred cubic structure (sodium chloride type). In aqueous media, at room temperature and at normal atmospheric pressure, silver iodide is known to exists in two modifications: the cubic of zinc blende type, and the hexagonal of wurtzite type (1 - 13). According to Cohen and Dobbebburgh (1) differently prepared AgI always produce mixtures of cubic and hexagonal AgI . It is also known that during the long aging periods heterogeneous exchange continues after precipitation (14) . The exchange of Ag^{+} and I^{-} ions with the AgI crystal is caused by Ostwald ripening process and/or by selfdiffusion of Ag^{+} and I^{-} ions respectively (13, 14) . Quantitative data of observations, high differences in the diffusion coeficients (3) and the various the cubic/hexagonal ratio indicate that there is a complex interdependence between mentioned processes. Only a systematic X-ray diffraction study could shed more light onto the role of various parameters in the crystal silver iodide formation. Following these reasoning a systematic X-ray diffraction analysis of differently prepared AgI were applied in an attempt to collect data relevant to the silver iodide crystal formation. Quantitative data regarding the cubic structure and the size of AgI crystallites in dependence on various parameters defining the silver iodide formation are given.

II. EXPERIMENTAL

A. Materials

All inorganic chemicals used were of Analar quality. Sodium n-dodecylsulfate was of a special purity grade supplied by BDH . Sodium n-dodecanate and n-dodecylammonium nitrate were prepared as previously described (10) . Organic solvents used were of Analar grade or high purity supplied by Fluka. Distilled water was twice redistilled from an all Duran 50 apparatus.

B. Preparation of Silver Iodide Sols

All the systems were prepared to contain about 100 mg of AgI , the quantity necessary for the X-ray analysis. The solution to be added was stirred with a magnetic stirrer. The sols were prepared *in statu nascendi* i.e. by direct mixing of equal volume of precipitation components. After mixing of precipitation components, an excess of NaI (expressed as pI) or $AgNO_3$ (expressed as pAg) remained in the colloid system. The aged silver iodide sols were thermostated at 293 K in a Haake ultrathermostat. The stable silver iodide sols were centrifuged using a Sorvall RC2b type centrifuge with a SS-1 rotor. The following designations have been used for silver iodide prepared in this paper:

-Sol I , prepared by mixing of 0.002 M $AgNO_3$ and 0.004 M NaI or 0.02 M $AgNO_3$ and 0.04 M NaI solutions. The details of preparative methods have been given elsewhere (11) . After 100 and 150000 minutes of aging the suspensions were used for centrifugation and X-ray analysis:

-Sol II , prepared in the same way as sol I and coagulated by adding 0.5 ml of saturated $Mg(NO_3)_2$ solution just before centrifugation:

-Sol III , $AgNO_3$ solution was added to a mixture of KI + KNO_3 or by adding KI solution to a mixture of $AgNO_3$ + KNO_3 solution:

-Sol IV , prepared by mixing 1000 ml 0.0020 M KI and 500 ml 0.8 M KNO_3 + 500 ml 0.8 M $AgNO_3$ or 1000 ml 0.0020 M $AgNO_3$ and 500 ml 0.8 M KNO_3 + 500 ml 0.008 M KI . Addition were t_p = 10 , 100 , 1000 minutes and time of aging t_A = 10 , 100 , 1000 minutes:

-Sol V , KI solution was added to the mixture of $AgNO_3 + HNO_3$, KNO_3 , K_2SO_4 or K_3PO_4 solution:

-Sol VI , prepared by adding 50 ml 0.02 M $AgNO_3$ solution to 50 ml 0.04 M NaI solution. Silver nitrate and sodium iodide were dissolved in organic solvents so that systems contained 50% or 80% acetone, dioxane, methanol, or ethanol:

-Sol VII , prepared by mixing 250 ml 0.002 M $AgNO_3$ and 250 ml 0.004 M NaI solution with var. concns. of n-dodecyl ammonium nitrate added before the precipitation or 0.5 and 1500 minutes after mixing of precipitating components. After aging for 100 minutes the suspensions were centrifuged and used for X-ray analysis:

-Sol VIII , prepared by adding 50 ml 0.02 M $AgNO_3$ to 50 ml solution containing 40 ml 0.05 M NaI + 10 ml 0.01 M sodium n-dodecylsulfate. After aging for 10 and 10000 min silver iodide sol was concentrated by centrifuging in a RC2b Sorvall centrifuge:

-Sol IX , 5.87 g of dried AgI were placed into a 25 ml measuring flask: a fresh solution of 4.00 M KI was added and when the AgI completely dissolved up to 25 ml KI solution was added into the flask. Sols were prepared by adding 20 µl of complex solution into 8 ml of 0.00001 M or 0.01 M sodium n-dodecylsulfate. Sols were used for X-ray analysis and for electron microscopy preformed on a Trüb-Tauber Co. KM-4 electron microscope:

-Sol X , prepared by adding 500 ml 0.002 M $AgNO_3$ to 500 ml solution containing 400 ml 0.005 M NaI + 100 ml 0.00001 M , 0.001 M or 0.1 M sodium n-dodecylsulfate. After aging for 1000 minutes silver iodide sols were concentrated using a RC2b Sorvall centrifuge:

-Sol XI , prepared by adding 500 ml 0.002 M $AgNO_3$ to 500 ml NaI solution containing 0.000002 M , 0.0002 M or

0.02 M sodium n-dodecanate. After aging for 100 minutes silver iodide sols were concentrated using a RC2b Sorvall centrifuge:

-Sol XII , prepared in the same way as sol XI but using as a surfactant n-dodecylammonium nitrate:

-Sol XIII , 0.05825 g AgI in 0.25 ml 2.00 M KI was heated to 433 K in a sealed Pyrex ampulla and then cooled continuously (0.2 K / min) to 293 K :

-Sol XIV, prepared by mixing 500 ml 0.002 M KI and 500 ml 0.004 M $AgNO_3$ solutions. The sol was aged for 4000000 minutes.

C. X-Ray Diffraction Analysis

The percentage %C of cubic silver iodide and the crystallyte sizes d (nm) were determined on silver iodide sols in contact with the original supernatant liquid. The %C and d values were determined using the Philips diffractometer with a scintillation counter and a single-channel pulse analyser. Filtered copper Kα radiation was used in all cases. The ratio of the proportions of cubic and hexagonal modification is

$$K / H = \{ | I_{(111)+(002)} - (I_{(100)} / 1.72) | / 4 \} / (I_{(100)} / 1.72)$$

where $I_{(111)+(002)}$ and $I_{(100)}$ are observed intensities of lines at $\Theta = 11.9^{o}$ and $\Theta = 11.2^{o}$ respectively. A series of independent experiments were run with each sample and it was found that the preferred orientation of the crystallites could be neglected, because the fluctuations of the results were less than 2%. This conclusion was confirmed by using the overlapping diffraction lines (220) and (110) at $\Theta = 19.6^{o}$ of cubic and hexagonal structure respectively.

The crystallite size was evaluated according to Scherrer´s formulae $R = \lambda (\beta_i \cos \Theta)^{-1}$ and

$R = 0.9 \lambda (\beta_{1/2} \cos \Theta)^{-1}$ using the integral breadths, β_i, and half-maximum breadths, $\beta_{1/2}$, respectively (λ is the wavelength). The pure diffraction breadths were deduced by the use of Alexander´s correction curves. The accuracy of the determination of the absolute crystallite size is reduced with a decrease of pure diffraction broadening. If the crystallites are larger than about 100 nm the relative rather than the absolute changes of the crystallite size from sample to sample are to be considered.

III. RESULTS AND DISCUSSION

The crystallographic phase equilibrium of silver iodide has been determined by a number of investigators. There are many, very interesting papers which discuss polymorphism of AgI (1, 2, 5 - 12, 15, 16). Meanwhile, it is not easy to compare numerous literature data without mistake because of various different experimental sources and more often because of the absence of the precise description of experiments.

In this paper a systematic X-ray diffraction analysis of differently prepared AgI in aqueous media is described. Fourteen preparation procedures were applied and the obtained results show complex dependence of the crystallographic characteristics on the conditions of preparation of AgI sols. At 8 cited conditions 14 various systems were analysed: 1 - at various concentrations of the constitutive I^- and Ag^+ ions in the systems with stable, coagulated and also in sols coagulated just before analysis (sols I, II, III): 2 - systems with various total amount of the solid phase (sol III): 3 - sols precipitated for different times of precipitation (sol IV): 4 - sols precipitated by various coagulating electrolyte, of various valency and at various concentrations, and at different acidity (sol V): 5 - sols

TABLE 1

Influence of Concentration of Constituent Ag^+ *and* I^- *Ions (pAg and pI) on the Percentage %C of Cubic* AgI *and on Crystallite Size d (nm) of the Stable (S) , the Coagulated Stable (KS) and the Coagulated (K) in statu nascendi* AgI *Sols Aged for* t_A *Minutes*

Sol	t_A	*pI*	*pAg*	*%C*	*d*
I-S	100	2		57	23
I-S	100	3		52	28
I-S	150000	2		83	69
I-S	150000	3		81	35
II-SK	100	2		64	17
II-SK	100	3		69	20
II-SK	150000	2		98	47
II-SK	150000	3		81	35
III-K	1500	0		0	162
III-K	43000	0		5	270
III-K	1500		0	55	350
III-K	43000		0	58	350
III-K	1500	4		63	88
III-K	43000	4		66	165
III-K	1500		4	74	39
III-K	43000		4	68	50
I-S	1500	1		65	
I-S	43000	1		70	
I-S	1500	2		63	
I-S	43000	2		72	
I-S	1500	3		60	
I-S	43000	3		71	
I-S	1500	4		75	
I-S	43000	4		81	

prepared in the organic solvent mixtures (sol VI): 6 - sols prepared in the presence of various surface active agents using different experimental procedures (sols VII - XII): 7 - sols prepared by a temperature changes, i.e. by cooling of samples (sol XIII) and 8 - sols prepared by aging for a long period of time (sol XIV).

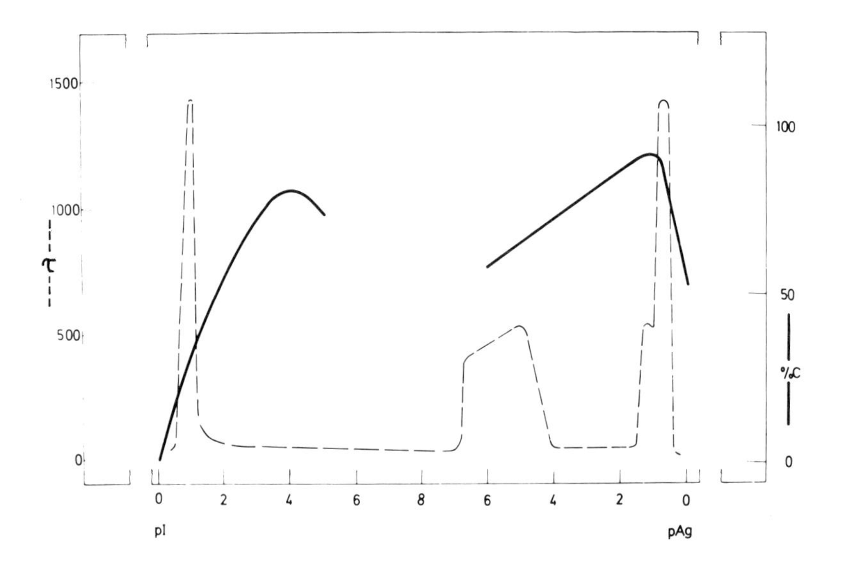

Fig. 1. Tyndallometric value τ *and percent* %C *in cubic form, for silver iodide sols in aqueous media.*

The dependence of the tyndallometric value τ on the concentration of the I^- and Ag^+ ions indicates different crystal growth in silver iodide suspensions. In order to obtain a detailed picture of the crystallite growth and the part of cubic structure at different concentrations of the potential determining ions, the sols of I, II and III types were made. The results (Table 1.) show a strong dependence of the percentage *%C* of cubic AgI and crystallite size d on pI or pAg . In systems containing an excess of

TABLE 2

Influence of the Amount (α)[a] *of* AgI *on the Percentage* %C *of Cubic* AgI *in the Stable (S) and Coagulated (K)* AgI *Sols Aged for* t_A *Minutes at an Excess of Iodide* (pI) *or Silver* (pAg) *Ions*

Sol	t_A	pI	pAg	$\%C_{\alpha=0.1}$	$\%C_{\alpha=1.0}$	$\%C_{\alpha=10}$
I-S	1500	2		58	62	45
I-S	43000	2		77	72	50
I-S	1500	3		56	60	51
I-S	43000	3		70	71	58
III-K	43000	3		55	63	72
III-K	43000		3	76	79	78

a. $\alpha = n^S / n^L$ *where* n^S = *amount of* Ag^+ (I^-) *in the solid phase and* n^L *in the liquid phase.*

TABLE 3

The Percentage %C *of Cubic* AgI *and the Crystallite Size* d (nm) *in the Coagulated Sols Precipitated for Time* t_p *(in minutes) and Aged for the Time* t_A *(in minutes)*

			t_p:	10		100		1000	
Sol	t_A	pI	pAg	%C	d	%C	d	%C	d
IV	10	3		73 ± 5	40	73 ± 5	65	73 ± 5	80
IV	1000	3		73 ± 5	110	73 ± 5	140	73 ± 5	250
IV	10		3	70	67	78	72	77	90
IV	1000		3	62	80	83	89	78	90

Ag^+ or I^- ions the maximum of the " *%C* vs. pI, pAg " curve corresponds to the one of the " τ vs. pI, pAg " curve (Fig. 1.). X-ray data in accordance with tyndallometric values, show that the precipitation system " $AgNO_3$ + NaI " is an asymetric one. The results show an increase of crystallite size via aging in all cases. The hexagonal → cubic transformation by aging is the highest among the stable sols, significant for the sols coagulated just before analysis and nonsignificant or very low for the sols coagulated *in statu nascendi*. After 4000000 minutes of aging in the positive coagulated sol (type XIV) the *%C* was 71 , indicating practicaly the same *%C* value as for 1500 minutes aged systems. The d value is also increased slightly. Very interesting are the results obtained for the coagulated sol, prepared at an excess of iodide ions of pI = 1 . The sol aged for 1 day contains pure hexagonal silver iodide crystallites. Contrary to these conditions, the stable sol prepared at pI = 2, aged for 150000 minutes and coagulated just before analysis contains 2% of the hexagonal structure only.

The amount of AgI in the positive coagulated sols does not influence to *%C* values, while by an increase in the solid phase the percentagewise amount of cubic AgI in negative coagulated sols significantly increases (Table 2.). In stable sols the *%C* value depends differently on the aging process with various amounts of silver iodide present in the system: an increase in the AgI amount in negative stable sols decrease the aging process.

According to the results published so far (17, 18) presented results (Table 3.) show that the crystallite size d and the *%C* are dependent on the mixing rate of the precipitation components. The rate of mixing in systems at pAg = 3 has an influence on the *%C* (1000 minutes aged sol), whereas in those containing an excess of I^- this influence is weak. In all cases the d value increases with the aging time. The acidity of the positive coagulated sol does not play a significant role in the crystallographic properties of the solid phase (Table 4.). The coagulating electrolytes potassium nitrate, sulfate and phosphate influence the *%C* value in accordance with the results obtained by Yamada (6) . The high similarity of the results obtained for KNO_3 and HNO_3 species present separately in the systems indicates that HNO_3 must be observed as an univalent electrolyte.

TABLE 4

Influence of Coagulating Electrolytes KNO_3 , K_2SO_4 *and* K_3PO_4 *and Acidity* pH *(adjusted by* HNO_3 *) on the Percentage %C of Cubic* AgI *in Coagulated Sols Aged for 1500 Minutes at* pAg = 3

Sol	Electrolyte		%C
V	KNO_3		75
V	K_2SO_4		86
V	K_3PO_4		60
V	HNO_3 pH	1	75
		2	79
		3	74
		4	71
		5	78

Crystallographic characteristics of stable silver iodide sols formed in " water - organic solvent " solutions depend on the amount of organic solvent in the mixture (Table 5.). The maximum *%C* , i.e. 100% of cubic AgI is reached in

TABLE 5

Percentage %C of Cubic AgI for Sols Aged for t_A Minutes and Prepared in 50 vol.% (I) and in 80 vol.% (II) Acetone (A), Dioxane (B), Methanol (C), and Ethanol (D) in Aqueous Solution of 0.01 M NaI

Sol	$t_A = 1500$		$t_A = 43000$	
	%C			
	(I)	(II)	(I)	(II)
VI-A	40	24	60	40
VI-B	75	60	95	60
VI-C	75	98	95	100
VI-D	78	98	96	100

the systems with methanol and ethanol, and the lowest value in the systems with acetone. In the systems conatining dioxane the obtained results are slightly lower as compared with those containing alcohols. In all acses, like in the stable sols i aqueous media, the *%C* is increasing with the aging of sol. It is interesting to stress the high rate of hexagonal → cubic transformation and the simple experimental procedure for the preparation of pure cubic silver iodide.

n-Dodecylammonium nitrate of various concentrations does not influences on the *%C* if present in the systems *in statu nascendi* , whereas its addition after mixing of Ag^+ and I^- ions, has an significant influence (Table 6.). The crystallite size d depends in all cases on the surfactant concentrations and on the mode of addition in the system. At critical coagulation concentration of surfactant, the discon-

TABLE 6

Influence of Concentration c of n-Dodecylammonium Nitrate $DDANO_3$ *on the Crystallite Size d (nm) and on the Percentage %C of cubic* AgI . $DDANO_3$ *was Dissolved in* NaI *Solution* ($t_d = 0$ *Minutes*) *or Added in* AgI *sol Aged for 30 seconds or 1500 minutes . Sols VIII.*

	t_d / Minutes					
	0		0.5		1500	
c_{DDNO_3}	%C	d	%C	d	%c	d
0.0000001	48	62	53	70	53	64
0.000001	44	56	52	60	52	77
0.00001	41	62	35	106	26	53
0.00003	49	67	48	58	58	80
0.0001	46	122	46	83	62	77
0.001	43	122	42	82	52	120

tinuities of %C and d versus surfactant concentrations indicates the same phenomena of metaphase transformation as previously described (19, 20) . The high hexagonal to cubic transformation reaches in the sols aged for 10000 minutes at pI = 3 and with surfactant present over critical micellar concentration: 100% of the cubic structure is reached after ripening for one week (Table 7.). Sodium n-dodecyl sulfate neither inhibit crystal growth nor the hexagonal → cubic transformation in the AgI suspensions (sol VIII , Table 7.) but the mode of the sol precipitation plays a very important role. Sols prepared by dillution of complex silver iodide solution in potassium iodide with various

TABLE 7

Influence of Sodium n-Dodecyl sulfate (S_1) , *Sodium n-Dodecanate* (S_2) *and n-Dodecylammonium Nitrate* (S_3) *Solutions on the Crystallite Size d (nm) and Percentage %C of Cubic* AgI *Aged for* t_A *Minutes*

Sol	c_S	pI	t_A	%C	d
VIII-S_1	10^{-3}	2	100	56	60
VIII-S_1	10^{-3}	2	10000	82	280
IX-S_1	10^{-5}	2	10000	78	170
IX-S_1	10^{-2}	2	10000	65	36
X-S_1	10^{-6}	3	1000		44
X-S_1	10^{-4}	3	1000		28
X-S_1	10^{-2}	3	1000		76
XI-S_2	10^{-6}	3	100		57
XI-S_2	10^{-4}	3	100		49
XI-S_2	10^{-2}	3	100		29
XII-S_3	10^{-3}	3	100	35 - 43	
XII-S_3	10^{-2}	3	100	59 - 62	
XII-S_3	10^{-3}	3	10000	60	
XII-S_3	10^{-2}	3	10000	100	

concentrations of sodium n-dodecyl sulfate consist of the crystallites of different sizes (sols IX , Table 7.). *In statu nascendi* prepared sols containing various concentrations of sodium n-dodecyl sulfate show at 0.0001 M surfactant lowest d value (sol X) . At higher surfactant concentration crystallites are of lower dia. Electron micrographs of AgI crystallites aged for 10000 minutes show tetrahedra of sizes from 20 to 800 nm (mostly 400 - 800 nm) at 0.00001 M surfactant and 80 nm to 1300 nm (mostly 700 - 1300 nm) at 0.01 M surfactant. Sodium n-dodecanate shows the same influences as sodium n-dodecyl sulfate (sol XI , Table 7)

Cooling of the hot complex solution of AgI + KI from 433 K to 293 K , clear, transparent AgI crystals (d = 0.=5 to 1 mm) of pure cubic structure were formed probably corresponding to the high temperature cubic modification (3) . Presented results show a complex dependence of crystallographic properties and crystallite sizes on parameters defining the precipitation and aging conditions. Concentration and chemical nature of electrolytes present in the systems are determining factors influencing various hexagonal → cubic equilibria and sols dispersity. This is in good agreement with Težak´s concepts (21) of interrelation between the ions in the liquid phase and the chemical and physical properties of solid phase. Different hexagonal → → cubic changes for coagulated and stable sols show that there are significant differences between the structural properties of stable and coagulated AgI sols, what is important for the elucidation of metaphase → solid phase transformation (19, 20) . The surfactants accumulated at the solid / liquid interphase influence in various ways structural properties of AgI causing also an "irregular" behaviour in comparison with other published data: an acceleration of crystal growth and the promotion of hexagonal → → cubic changes. The role of chemical nature of bulk solution is shown in the systems with mixtures of organic solvents. Each solvent composition has a different solubility equilibrium, with different dielectric constants and surface tension at the crystal / solution interface, causing different properties of the interface layer. At high electrolyte concentrations as at the high temperature of the sols preparation the complex species must play important role in the formation of primary and consequently secondary structure of silver iodide. Finally, we hope that described experiments will throw some light onto the role of the various precipitating conditions in the hexagonal → cubic equilibration as reveald by the nucleation, crystallization and recrystallization phenomena.

IV. REFERENCES

1. Cohen, E. and Van Dobbenburgh, W.J.D., Z. phys. Chem. A 137, 289 (1928).
2. Bloch, H. and Möller, H., Z. phys. Chem. A 152, 245 (1931)
3. Jordan, P. and Pochon, M., Helv. Phys. Acta 30, 33 (1957).
4. Horne, R. W., Matijević, E., Ottewill, R. H. and Weymouth, J.W., Kolloid Z. 161, 50 (1958).
5. Chatterjee, S.N., Intern. Kongr. Elektronenmikroskopie, 4, Berlin, 1958, Verhandel. 1, 453 (1960).
6. Yamada, K., Bull. Soc. Sci. Phot. Japan 11, 1 (1961).
7. Burley, G., Am. Mineral. 48, 1266 (1963).
8. Chateaus, H., Cugnac, A. and Pouradier, J., Compt. Rend. 285, 1548 (1964).
9. Herz, R. H., 10th Photogr. Sci. Symposium, Paris 1965, 68 (1967).
10. Despotović, R., Despotović, Z., Jajetić, M., Popović, S. and Telišman, Ž., Kemija u Industriji (Zagreb) 17, 197 (1968).
11. Despotović, R., Despotović, Z., Jajetić, M., Mirnik, M., Popović, S. and Telišman, Ž., Croat. Chem. Acta 42, 445 (1970).
12. Despotović, R., Despotović, Z., Mirnik, M. and Subotić, B., Croat. Chem. Acta 42, 557 (1970).

13. Despotović, R., Discussion Faraday Soc. 42, 208 (1966).
14. Despotović, R., Horvat, V., Popović, S. and Selir, Z., J. Colloid Interface Sci. 49, 147 (1974).
15. "GMELINS Handbuch der anorganischen Chemie", 8 ed., Silber, Teil B2, Verlag Chemie - GMBH - Weinheim, (1972) 61 (Ag) Hb/B2, 200-221.
16. Byerley, B.L.J. and Hirsch, H., J. Photographic Sci. 18, 53 (1970).
17. Bassett, W.A. and Takahashi, T., Am. Mineral. 50, 1576 (1965).
18. Sieg, L., Naturwissenschaften 4o, 439 (1953).
19. Despotović, R., in "Particle Growth in Suspension" (A. Smith, Ed.), p. 121. Pergamon Press, London, 1973.
20. Despotović, R. and Subotić, B., J. Inorg. Nucl. Chem. 38, 1317 (1976).
21. Težak, B., Discussion Faraday Soc. 42, 175 (1966).

CHARACTERIZATION AND OXIDATION OF COLLOIDAL SILVER

W. J. Miller and A. H. Herz
Research Laboratories, Eastman Kodak Company

Selected particle size and absorption characteristics of aqueous silver dispersions are described; it is shown that changes in their absorption spectra can be used to measure the oxidation rate of silver under various conditions. In agreement with the reaction $2Ag^o + \frac{1}{2}O_2 + 2H^+ = H_2O + 2Ag^+$ *(*ΔF^o *= -20 kcal/mole) aerial oxidation rates varied with partial oxygen pressure and suggested a first-order dependence on metallic silver and on hydrogen and halide ion concentrations. Low pH and excess halide,* X^-*, led to the conversion of colloidal* Ag^o *to AgX dispersions at rates that increased in the expected order* $Cl^- < Br^- < I^-$*. Similarly, oxidation rates of silver were enhanced by ligands that formed stable, soluble* Ag^+*-complex ions and thus shifted the* Ag^o/Ag^+ *electrode to more oxidizing potentials. In contradistinction to such ligands, some additives with little or no effect on the* Ag^o/Ag^+ *electrode, e.g., cyanine dyes and gelatin, diminished oxidation rates by factors of* 10^4*. Slow oxygen diffusion could not be ruled out as a contributor to this behavior, but the major effect of gelatin and other additives in decreasing the oxidation rate of colloidal silver in the presence of halide was associated with a tendency to promote the formation of adhering and passivating AgX films at the* Ag^o*/water interface.*

II. INTRODUCTION

Early investigations on colloidal silver with which the name of M. Carey Lea became intimately connected, were reviewed by Freundlich (1) and by Weiser (2). Apparently the relatively low popularity of this colloidal element was related to difficulties initially encountered in the preparation of reproducible suspensions and their comparative instability.

This chemical instability of colloidal silver is precisely what interested us on the assumption that some problems encountered in photographic silver halide systems can be related to the chemical instability of small silver clusters. The photographically relevant problems include fading of silver latent images (3a,b), the stability of reduction sensitization (4) and, indeed, the stability of the final silver image (5). Previously, oxidation rates of silver were obtained in solutions of cyanide (6), cerric (7) and ferric ions (8) or in hydrogen peroxide (9,10); under some conditions they led to the formation of silver halides (11-14). These measurements often involved deposited silver, silver foils or electrodes but since the cited photographic problems relate to the stability of silver cluster of colloidal dimensions, silver sols with their high surface-to-volume ratios provide a particularly useful model. Dilute Ag^o sols with particle diameters of 5 nm can be obtained in the absence of any organic stabilizers (15); moreover, concentrated and stabilized dispersions with specific surface areas in the range of 30-80 m^2/g Ag are readily prepared and conveniently employed for adsorption determinations (16-23). In addition, such protected sols obey the Beer-Lambert law and their high extinction coefficient (18) facilitates the spectral determination of Ag^o concentrations.

As the point of departure in these experiments, the assumption was made that the instability of aqueous silver in air was caused by an oxidation reaction

$$2Ag^o + \tfrac{1}{2}O_2 + 2H^+ = H_2O + 2Ag^+,$$

for which the equilibrium constant, $K = 4 \times 10^{14}$, and the standard free energy change, $\Delta F^o = -20$ kcal/mole, were calculated for 25°C from appropriate half-cell potentials (3c). The presence of a ligand for Ag^+ further enhances this equilibrium constant for oxidation of silver and, indeed, the computed equilibrium constants in Table 1 show that the formation of either sparingly soluble silver salts or of soluble complex ions facilitates the pH-dependent oxidation of silver by air or other reagents such as benzoquinone. In the absence of any Ag^+-ligands, the thermodynamic stability of the silver-water system was shown to be greatest in the pH 11-12 region (24).

All these equilibrium data lead to the conclusion that the thermodynamic instability of silver in respect to oxidation will be enhanced by increased concentration of oxygen and H^+ and by those compounds that decrease Ag^+-activity. These conclusions are based solely on equilibrium results and it will be one of the purposes of this study to learn to what

TABLE 1

Thermodynamic Equilibrium Constants for Oxidation of Ag^{o} in Water, Standard Conditions, $25^{o}C$. The Oxidant Q is Oxygen or Benzoquinone.

$$2\,Ag^{\circ} + Q + 2\,H^{+} = 2\,Ag^{+} + H_2Q$$

pH	Additive	Product	Equilibrium Constants: Oxygen	Equilibrium Constants: Benzoquinone
0	None	Ag^{+}	10^{14}	0.001
0	Br^{-}	AgBr	10^{39}	10^{21}
7			10^{25}	10^{7}
9			10^{21}	10^{3}
7	I^{-}	AgI	10^{32}	10^{15}
9			10^{28}	10^{11}
7	$S_2O_3^{=}$	$Ag(S_2O_3)_2^{\equiv}$	10^{27}	10^{9}
9			10^{23}	10^{5}

extent oxidation rates of silver are in accord with such thermodynamic considerations.

III. PREPARATION OF COLLOIDAL SILVER

In the course of these studies, properties of various silver dispersions were investigated; however, unless indicated otherwise, all silver dispersions were prepared by a procedure of Carey Lea (22), as adapted by Frens and Overbeek (23), in which a mixture of $FeSO_4$ and sodium citrate is poured into a solution of $AgNO_3$. An Ag^{o} sol forms immediately and just as quickly is flocculated by the high concentrations of dissolved salts. The wet sediment is redispersed in water and then reflocculated again by addition of the concentration sodium citrate solution. This dispersion and flocculation process was repeated twice. Redispersal in water was aided by ultrasonic treatment and the final sol, which was approximately 0.25 M in respect to total silver, was centrifuged at 7000 rpm for 20 minutes; any sediment was discarded. This purification procedure effectively removed Fe^{+2}, Fe^{+3}, SO_4^{-2} and NO_3^{-} from the suspension; only Na^{+} and citrate were reported to be associated with the highly negatively charged Ag^{o} particles that exhibit zeta-potentials of about -80 mV

(23). When freshly prepared at 25°C, the ca. 0.25 M Ag^o sol yielded Ag^+-potentials near pH 7 that ranged between pAg 5.5-6.2. Hence, the sols may contain free ionic Ag^+ in addition to any Ag^+ adsorbed at the silver surface either as ions (25) or as the moderately insoluble silver citrate, pK_{sp} 12.7 (26). This conclusion was supported by the observation that the maximum absorbance of these sols shifted hypsochromically and increased 10-20% upon addition of a reducing agent like $NaBH_4$. A similar increase in absorbance was observed upon boiling such a sol for about two minutes (27); this treatment is known to form β-ketoglutaric acid (28) as an active reducing agent. Indeed, hot $AgNO_3$-citrate solutions were used for the uniform deposition of Ag^o at the surface of preformed Au^o nuclei (18). Hence, citrate can fulfill various functions; its adsorption is reponsible not only for the high negative charge density of these sols but also for their redispersibility after flocculation (23).

IV. SELECTED PROPERTIES OF COLLOIDAL SILVER

A. Size Characterization

Transmission electron micrographs confirmed that the sol particles were neither monodispersed nor strictly spherical (23). Sizing established that in many sols the mean particle diameter was about 8 nm but it sometimes ranged in different preparations from ca. 7 to 12 nm. On the not wholly valid assumption of spherical particle shapes, these values lead to estimates of specific surface areas ranging from about 81 to 48 m^2/g Ag^o. Frens and Overbeek also noted that a specific area of 70 m^2 was generally obtained (23). In addition to electron micrography, *in situ* adsorption determinations with 1,1'-diethyl-2,2'-cyanine or its zwitterionic monosulfobutyl analog were also employed for measuring Ag^o surface areas. As reported elsewhere (18), these dye adsorption determinations on silver required the presence of halide ions and were often carried out in dilute gelatin solutions at pBr 3-4. A yellow silver dispersion prepared by dextrin-reduction of hydrated silver oxide (2) yielded a surface area of about 35 m^2/g Ag^o. This value corresponds roughly to a spherical particle diameter of 17 nm and agreed with the 14-19 nm estimate obtained with independent methods on a similarly prepared Ag^o dispersion (16, 18). A different Ag^o dispersion was obtained in gelatin with $NaBH_4$ as reducing agent.* The surface spectra of the cited 2,2'-cyanine at pBr 3 (18) yielded a spherical particle diameter of 6.9 nm in acceptable agreement with the 7.5 $\pm$ 0.9 nm obtained from electron micrographs. These results further demonstrated that both types of independent area determinations can give concordant results for silver dispersions.

B. Absorption Spectra

Although unprotected silver dispersions prepared by borohydride reduction yield a maximum at 376 nm with a high extinction coefficient (15), the citrate-protected sols utilized in this study exhibit their maximum absorption in water near 393 nm. The position of this maximum is in good agreement with computations (30-32) and can be changed by altering the refractive index of the medium surrounding the metal particles; bromide changes it to 398 nm, gelatin or polyvinyl alcohol shift the maximum to 408 nm and an additional bathochromic change can be caused by thiols (17).

Spectra of these sols occasionally exhibited an additional shoulder near 520 nm which can be assigned to an aggregate of two spherical Ag^o particles having an axial ratio of about 1:2 (15, 23, 33). The ratio of the absorbance at the maximum to that at 500 nm often increased when the sol was first prepared. Apparently, the sol initially contained a substantial number of "dimeric" eliptical particles that may separate with time, on ultrasonic treatment or upon brief boiling; the latter treatment may also lead to further reduction of ionic Ag^+ as previously discussed.

On aging, the citrate-protected sols may flocculate by a light-accelerated process; a sol kept in the dark for a few hours remained unchanged but was destabilized by exposure to light. This light-induced behavior was reversed in the presence of halide and may involve photolytic regeneration of Ag^o from AgBr that had formed in the acidic bromide solution. In order to avoid uncertainties arising from changes in Ag^o sols protected only by citrate, kinetic measurements were not made with sols that were older than a week or that exhibited

---

*We are indebted to our colleague D. Shuman for the preparation and electron micrographic sizing of this dispersion. For size determinations by adsorption of the cited 2,2'-cyanines, limiting areas of 0.58 mm^2/dye molecule were employed (29).

an absorbance ratio, $A_{max}:A_{500\ nm}$, which was less than 12. Generally this ratio varied between 15-20 and yielded a molar absorptivity for the 398 nm maximum of $1.35 \pm 0.15 \times 10^4$ $M^{-1} cm^{-1}$.

C. Colloidal Stability

Although silver particles of colloidal dimensions may slowly recrystallize even in a high vacuum (33), it is unknown if this process was also responsible for the irreversible flocculation encountered upon aging citrate-protected sols (34). It seems more likely that this flocculation which seems to decrease with concentration of the sol (35), involved desorption of the stabilizing citrate or an oxidative formation and adsorption of Ag^+ which was then accompanied by a shift of the zeta-potential to more positive values. At any rate, citrate effectively protects Ag^o against flocculation by 0.001 M acids whereas those Ag^o sols that are stabilized only by adsorbed AgO^-, are easily flocculated at pH 5 or by addition of CO_2 (36).

Gelatin and other polymers strongly stabilize Ag^o sols which, in the absence of these stabilizers, obey the Schulze-Hardy flocculation rule (34). It was therefore surprising that citrate-protected sols that are stable in 2×10^{-3} M 1:1 electrolyte, were rapidly flocculated by 1,1'-diethyl-2,2'-cyanine pts. This flocculation, which did not occur with analogous zwitterionic or anionic cyanines (29), was unexpected because earlier work had led to the conclusion that these and related cyanines were only adsorbed if the surface of the silver had been converted to silver halide (18). Spectral and other measurements again confirmed that the cationic 2,2'-cyanine was not adsorbed to a measurable extent at the Ag^o surface; hence, the reason for its dye-induced instability remains obscure.

V. OXIDATION RATES OF SILVER SOLS

A. Measurement Methods

The technique used in these rate determinations with 10^{-5} to 10^{-3} M Ag^o can be illustrated by Fig. 1A which shows how the absorbance of a gelatin-protected silver sol in moderately acidic KBr decreases with time until no visible absorption remains. It was convenient to plot such spectral

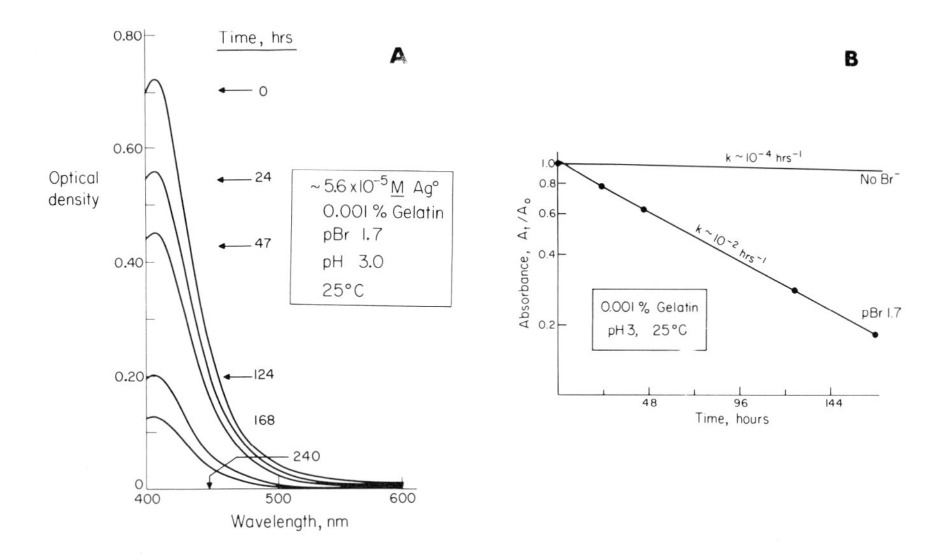

Fig. 1. Spectral changes of Aqueous, Gelatin-Protected Silver Dispersions at pH 3, 25°C. (A) Variation with Time. (B) Influence of Bromide.

changes on a logarithmic scale after normalizing them with the ratio A_t/A_o, where o and t refer to the absorbance at λ_{max} for reaction times zero and t, respectively. The results are shown in Fig. 1B and on the basis of that and similar linear relations, it appears that the reaction can be of first order with respect to the silver substrate. In Fig. 1B the data yield a half-time of 68 hrs and a specific reaction rate (k) of 10^{-2} hrs^{-1}; reproducibility of such kinetic parameters rarely exceeded 25%.

Not all kinetic experiments were of the simple form illustrated in Fig. 1 and in many cases the wavelength of the maximum fluctuated; it moved hypsochromically during the reaction with cyanide but, particularly in the absence of gelatin, it tended towards ca. 10 nm bathochromic shift in acidic citrate and acetate. Instances were encountered, specially in the absence of gelatin, where the time-dependence of the log absorbance exhibited a distinct discontinuity and the data had to be expressed by two separate rate constants. Unless stabilizers were present, reactions that proceeded at a very slow rate were particularly bothersome because the slow appearance of a shoulder near 520 nm, a sign of incipient flocculation, tended to limit the validity of the spectral measurements.

B. Effect of Compositional Variables on Rates

While some aspects of this study are still in progress and will be reported in detail elsewhere, the principal variables are summarized below:

TABLE 2

Specific Reaction Rates for Aerial Oxidation of Ag^o Sols. 5×10^{-5} $\underline{M}$ Ag^o, pBr 2.7, 25^oC)

pH	Electrolyte Millimoles l^{-1}		Spec. Reaction Rate[(a)] hrs^{-1}	Alteration in Conditions
3.0	HNO_3	1	3.7	
	$HClO_4$	1	3.8	
	H-pts.	1	3.5	
	H_2SO_4	0.5	8	
	Na_3 Citrate	6.6	1.8, 2.8	
			9.8×10^{-3} (4×10^{-3})	Without bromide
			2.8×10^{-4}	With 0.5% Gelatin[(b)]
4.0	Na_3 Citrate	6.6	6.8×10^{-2} (0.15)	Restricted air access
		6.6	0.39	
		66.0	1.5	
	Na Acetate	1	0.9	
	Na Acetate	10	0.8	
6.3	Na_3 Citrate	6.6	4.1×10^{-3} (11×10^{-3})	

(a) Initial rates are given in parenthesis.
(b) Bone stock gelatin, isoionic point pH 4.9.

1. *Acidity*

The data of Table 2 show that oxidation rates increased with growing acidity. In the range pH 3-6.3 at constant initial Ag^o and Br^- concentrations, a 20-fold increase in H^+ concentration caused a ca. 10-fold increase in the oxidation rate. Hence, under the given conditions, the reaction was approximately first order in respect to H^+. It is evident that excepting polyvalent sulfate and citrate, the reaction was not sensitive to the anions of the examined acids. However, the influence of ionic strength was not fully evaluated.

2. *Oxidants*

In view of the assumption that oxygen alone can cause oxidation of Ag^o according to the calculated equilibria, an effort was made to demonstrate participation of air in these reactions. This was done by purging an Ag^o sol with nitrogen and then continuing with nitrogen agitation after addition to the sol of acidic bromide. This last step involved temporary admission of air and although total removal of air from the system was therefore not accomplished, a more than 3-fold decrease in rate was observed (Table 2) and illustrated inhibition of the reaction when access of air was limited. Although other oxidants were also examined, e.g. benzoquinone, no experiments were carried out under conditions of rigorous oxygen exclusion.

TABLE 3

Aerial Oxidation of Ag^o Sols in Excess Halide.
(5×10^{-5} M Ag^o, 0.001% Gelatin[(a)] pH 6.8, 25^oC)

Potassium Halide, Millimoles l^{-1}	Specific Reaction Rates, hrs^{-1}
No halide	$< 0.02 \times 10^{-3}$
Cl^-, 0.9	0.25×10^{-3}
Cl^-, 9	3×10^{-3}
Br^-, 9	6×10^{-3}
I^-, 9	15×10^{-3}

(a) Bone stock gelatin, isoionic point pH 4.9.

3. *Halides and Other Ag^+-Ligands*

Compared to the absence of halide at pH 3, the presence of 2×10^{-3} M KBr caused a 200-fold increase in the reaction rate of silver (Table 2). Further, when halide was present in large excess over Ag^o, the reaction rate increased in the sequence $Cl^- < Br^- < I^-$ (Table 3). These rates were directly proportional to Cl^- and Br^- concentration in the range of about 10^{-2} to 10^{-3} M; this first order dependence on halide concentration was also observed at different pH values and in the absence of gelatin. Other experiments established that the reaction with colloidal Ag^o was strongly accelerated by 10^{-3} M ligands that form soluble complex ions, e.g. cyanide, thiosulfate and acid-substituted thiols and thioethers.

Under conditions where, unlike the examples cited so far, the ligand concentration was similar to or smaller than that of Ag^o, the reaction rate often diminished greatly. Examination of this effect is incomplete but it was particularly pronounced with some thiols and with iodide at concentrations near 10^{-5} M.

4. *Reaction Products*

Aerial oxidation of colloidal Ag^o in the presence of halide, X^-, leads to the formation of sparingly soluble AgX and, depending on the concentration of Ag^o and X^-, to the soluble halide complexes. The formation of AgI can be observed from the appearance of its exciton band at 420 nm (37); AgCl and AgBr lack similar spectral characteristics and their formation was established by x-ray methods and by electron micrography. In rapid reactions with acidic bromide and chloride solutions, the initial yellow dispersion took on a blue or pink appearance, respectively, and became more stable towards further oxidation. Formation of these colored products was assigned to small Ag^o nuclei embedded in AgX crystals of high refractive index rather than to F-centers (an electron from Ag^o bound to a halide vacancy). The colored products probably represent the frequently described "photohalides" (1, 38) which were also obtained by co-deposition from the vapor phase of Ag^o with AgBr or with AgCl and yielded absorption bands near 535 and 500 nm, respectively (39-41).

5. *Nonspecific Ag^o Reagents*

In contrast to the previously mentioned cyanine dyes which were not adsorbed at the Ag^o/H_2O interface (18), gelatin and other polymers are often irreversibly bound to that surface (19, 35, 42). These adsorption properties at Ag^o seem to exert a crucial influence on its oxidation behavior. Whereas unadsorbed cyanines had no significant effect on silver oxidation in alkaline KCN, adsorbed gelatin caused an approximately 3-fold decrease in the rate of that reaction. This diminished rate did not appear to involve a slow diffusion of oxygen that was observed in dry gelatin (43) since similar rates were obtained whether the sol was quiescent or strongly agitated. Instead, it is likely that the adsorbed gelatin increased the activation energy for electron transfer at the surface which would be consistent with the observation that adsorbed gelatin lowers the free surface energy of small Ag^o particles by about 500 erg/cm^2 (19).

The situation changed drastically when oxidation of Ag^o was carried out under conditions where the reaction product was not a soluble complex but a sparingly soluble silver halide. In that case oxidation rates were decreased orders of magnitude by the presence of cyanines, gelatin and synthetic compounds like polyacrylates, polyvinyl alcohol and polysaccharides. The magnitude of this retardation varies with the concentration of the additive; the data in Table 2 show that the aerial oxidation of Ag^o at pH 3 and pBr 2.7 is decreased in 0.5% gelatin by a factor of 10^4. Examination of the reaction products by electron micrography revealed AgBr particles larger than 100 nm when the reaction was carried out in the absence of gelatin or cyanines. However, in the presence of such additives, the AgBr crystals had a diameter less than 15 nm and they were thus in a similar size range as the original Ag^o particles.

Because in halide solutions even with gelatin present, the cyanines were particularly effective in reducing both the reaction rate and the size of the resulting silver halide, it was relevant that in agreement with early results, these dyes exhibited perturbed surface spectra in the Ag^o sols only in the presence of halide (18). This change of cyanine spectra was shown to be a consequence of their adsorption (18), hence the effectiveness of cyanines in decreasing oxidation rates can be related to their adsorption at the newly formed Ag^o/AgX substrate. In agreement with this view, it was noted that the anionic cyanines which are more weakly adsorbed at AgBr than their cationic and zwitterionic analogs (29), were also less effective in decreasing Ag^o oxidation rates in bromide.

VI. DISCUSSION AND SUMMARY

It is one of the principal conclusions of this study that the described spectral method has broad utility for the evaluation of corrosion phenomena at silver substrates with large surface areas. Beyond that, the results show that in agreement with equilibrium considerations involving the oxygen electrode, the rates were enhanced by increased partial oxygen pressure and increased concentrations of H^+. Rates were also increased by compounds that shifted the Ag^o/Ag^+ electrode to more oxidizing potentials unless the compounds were present at low concentrations relative to Ag^o and formed sparingly soluble Ag^+-products. In that case, which was exemplified by low levels of thiols or iodide, the formation of a surface film of the Ag^+-salt seemed to inhibit further silver oxidation. However, in accord with equilibrium considerations, excess halide increased these rates in the sequence expected from the solubility products of their Ag^+-salts. Ligands that formed primarily soluble and weakly dissociated Ag^+-complexes, e.g. cyanide and thiosulfate, were particularly effective in enhancing the rate of aerial oxidation of Ag^o dispersions; presumably their soluble reaction products did not form dense protective layers but diffused from the Ag^o surface into the bulk of the solution.

In contrast to the just cited ligands, the behavior of polymers, gelatin and cyanine dyes could not be predicted on thermodynamic grounds; compounds like the cited cyanines neither affect Ag^o/Ag^+ potentials nor are they expected to influence the oxygen electrode. Although gelatin and other polymers may inhibit oxygen diffusion (43) and are known to interfere with the corrosion of some metals (44), no evidence was found that the 10^4-fold decrease caused by gelatin in the oxidation rates of aqueous Ag^o/Br^- systems involved an oxygen-barrier mechanism. Instead, the data lead to the suggestion that polymers, gelatin and dyes inhibit oxidation of Ag^o/halide systems because they primarily facilitate the formation of continuous, adhering and dense passivating silver halide films at the Ag^o surface.

Nonporous and continuous Ag^+-salt layers are important in passivating processes. Even a 10 nm-thick but discontinuous AgCl film on Ag^o failed to fully protect the metal (13, 14). Hence, passivating AgBr layers would not be expected to exist under conditions where soluble complex ions, such as $AgBr_2^-$, may promote recrystallization, Ostwald ripening (45) and the formation of large, nonadhering AgBr crystals. While gelatin retarded both AgBr growth and oxidation in the Ag^o/Br^- system, the behavior of cyanine dyes offered further support for the view that formation of adhering silver halide films at Ag^o surfaces is a necessary condition for inhibition of Ag^o oxidation in the halide solution. These dyes were adsorbed at Ag^o surfaces only in the presence of halide, but in that case, they strongly restrained oxidation and growth of the resulting silver halide. The effectiveness of these dyes in retarding AgBr growth was shown by electron micrographic examination of the Ag^o/Br^- oxidation product. The resulting observations were consistent with more quantitative determinations which demonstrated that cyanines can be at least as efficient as thiols and other restrainers in inhibiting recrystallization of AgBr in gelatin even in the presence of silver halide solvents (45).

VII. REFERENCES

1. Freundlich, H., "Kapillarchemie," Akad. Verlagsgesel., Leipzig, 1922.
2. Weiser, H., "The Colloidal Elements," Vol. 1, p. 119, J. Wiley, N.Y., 1933.
3. (a) Mees, C.E.K., and James, T. E., "The Theory of the Photographic Process," 3rd Ed., The Macmillan Co., N.Y., 1966; (b) Loening, E., in "Fundamental Mechanisms of Photographic Sensitivity," p. 149, Butterworths, London, 1951; (c) Moeller, T., "Inorganic Chemistry, 3rd Ed., J. Wiley, N.Y., 1954.
4. Lowe, W. G., Jones, J. E., and Roberts, H. E., in Ref. 3b, p. 112.
5. Weyde, E., Photogr. Sci. Eng., 16, 283 (1972).
6. Lund, V., Acta Chem. Scand., 5, 555 (1955); Deitz, G., and Halpern, J., J. Metals, 1109 (1953).
7. Salzberg, H., Knoetgen, H., and Malless, A., J. Electrochem. Soc., 98, 31 (1951).
8. Salzberg, H., and King, C., ibid., 97, 290 (1950); King, C., and Lang, F., ibid., 99, 295 (1952).
9. Gossner, K., and Heidrich, H., and Korner, D., Z. Physik. Chem., N. F., 67, 220 (1969); 68, 293 (1969).
10. Vantelon, J-P., and Bernard, M-J., J. Chim. Phys., 73, 57 (1976).
11. Hauffe, K., in "Corrosion of Metals in Gases and Aqueous Solutions," Proceed. 6th Intl. Symp. on Reactivity of Solids, p. 311, Wiley, N.Y., 1969.
12. Gerischer, H., Z. Electrochem., 61, 1159 (1957); 62, 256 (1958); Colomer, J., C. R. Acad. Sci., 246, 1847 (1958).
13. Fleischmann, M., and Thirsk, H., Electrochim. Acta, 1, 146 (1959); 2, 22 (1960).
14. Vermilyea, D., in "Advances in Electrochemistry and Electrochemical Engineering," Vol. 3, p. 211, Interscience, N.Y., 1963.
15. Berry C., and Skillman, D., J. Appl. Phys., 42, 2818 (1971).
16. Stevens, G.W.W., and Block, P., J. Photogr. Sci., 7, 111 (1959); 9, 330 (1961).

17. Evva, F., Z. Wiss. Photogr. Photophys. Photochem., 58, 165 (1965); 60, 178 (1967); J. Aufzeisch. Material., 1, 269 (1973); 4, 43 (1976).
18. Herz, A., Danner, R., and Janusonis, G., in "Adsorption from Aqueous Solution," Advances in Chemistry Series No. 79, p. 173, Am. Chem. Soc., Washington, D.C., 1968.
19. Konstantinov, I., and Malinowski, J., J. Photogr. Sci., 23, 1 (1975).
20. Shuman, D., and James, T. H., Photogr. Sci. Eng., 15, 119 (1971).
21. Wrathall, D., and Gardner, W., in "Temperature, Its Measurement and Control in Science and Industry," Vol. 4, p. 2223, Instmt. Soc. America, Pittsburgh, 1972.
22. Carea Lea, M., Amer. J. Sci., 37, 476 (1889).
23. Frens, G., and Overbeek, J.Th.G., Koll. Z. Z. Polym., 233, 922 (1969).
24. Pourbaix, M., "Atlas of Electrochemical Equilibria in Aqueous Solution," Pergamon Press, London, 1966.
25. King, C., and Schochet, R., J. Phys. Chem., 57, 895 (1953).
26. Kolthoff, I., Rec. trav. chim., 45, 607 (1926).
27. Wrathall, D., private communication.
28. Turkevich, J., Stevenson, P., and Hillier, J., Disc. Faraday Soc., 11, 55 (1951).
29. Gardner, W., and Herz, A., "49th National Colloid Symposium," Am. Chem. Soc., Potsdam, N.Y., June, 1975. To be submitted to Photogr. Sci. Eng.
30. Doremus, R., J. Chem. Phys., 414 (1965); J. Appl. Phys., 37, 2775 (1966); J. Colloid Interface Sci., 27, 412 (1968).
31. Morriss, R. and Collins, L., J. Chem. Phys., 41, 3357 (1964).
32. Klein, E., and Metz, H., Photogr. Sci. Eng., 5, 5 (1961).
33. Skillman, D., and Berry C., J. Chem. Phys., 48, 3297 (1968); J. Opt. Soc. Amer., 63, 707 (1973).
34. Jaeger, H., Mercer, P., and Sherwood R., Surf. Sci., 13, 349 (1969); Hanitzsch, E., and Kahlweit, M., Z. Physik. Chem., N. F., 57, 145 (1968); 65, 290 (1969).
35. Freundlich, H., and Loening, E., Koll. Beihefte, 16, 1 (1922).
36. Kruyt, H., and vanNouhuys, L., Koll. Z., 92, 325 (1940).
37. Berry C., Phys. Rev., 161, 848 (1967).
38. Socher, H., Z. wiss. Photogr. Photophys. Photochem., 37, 51 (1938); Narath, A., ibid., 51, 244 (1956).
39. Rohloff, E., Z. Physik, 132, 643 (1952).
40. Kaiser, W., ibid., 132, 497 (1952).
41. Kleemann, W., ibid., 215, 113 (1968).
42. Groszek, A., and Wood, H., J. Photogr. Sci., 13, 133 (1965).

43. Buettner, A., J. Am. Chem. Soc., 68, 3252 (1964).
44. Evans, U. "An Introduction to Metallic Corrosion," 2nd Ed., p. 155, E. Arnold Ltsl, London, 1963.
45. Oppenheimer, L., James, T. H., and Herz, A., in "Particle Growth in Suspensions," A. Smith, Ed., p. 159, Academic Press, London, 1973.

DYNAMIC INTERACTION IN PARTICLE - BUBBLE ATTACHMENT IN FLOTATION

JANUSZ LEKKI and JANUSZ LASKOWSKI

Wrocław Technical University, Wrocław, Poland

ABSTRACT

The capture of a bubble is discussed with particular emphasis placed upon the stability of the disjoining film between a bubble and a particle. The thermodynamic criterion of flotation, $\theta > o$, must be fulfilled for flotation to be possible, but at the same time the kinetic criterion must be met also. Frothers do not usually affect the contact angle but they speed up all processes in the system and influence the rate of flotation.

The natural floatability of chalcocite was shown to be maximal at pH's close to the IEP's for CuS_2O_3 and $Cu(OH)_2$, that is, at the IEP's of oxidation products on the surface of chalcocite. The effect of frothers was very noticeably far from these values of pH.

It is known that to describe the stabilization of hydrophobic dispersions by nonionic surfactants, that is the change from the very hydrophobic state to the very hydrophilic one, an introduction of the entropic term, ΔG_s, into the general equation was necessary. In the flotation case covering the full range of changes from very hydrophilic to very hydrophobic, the same parameter is necessary to describe the properties of the system. Calculations have shown that the particle-bubble interaction under conditions of constant surface charge leads to an increase in surface potential. This change is high enough to cause desorption of physically adsorbed frother from a solid at the moment of collision with a bubble. Diffusion of large organic molecules across the hydration layer at the very moment of attachment disorientates this layer lowering the value of ΔG_s.

I. INTRODUCTION

Flotation involves the attachment of hydrophobic particles to air bubbles and the subsequent transfer of the particle -laden bubbles to the froth. For this process to be possible some conditions must be fulfilled:

i. mineral particle must collide with a bubble;

ii. during the short time in which the particle is in apparent contact with a bubble, the liquid disjoining film must elongate, then rupture and recede;

iii. the full contact between a bubble and a particle must be established leading to the formation of a stable particle-bubble aggregate capable of withstanding disruptive turbulence in a flotation cell.

Thus, the probability of flotation can be given in the general form

$$P = P_c P_a P_s \qquad /1/$$

where P_c represents the probability of collision between particle and gas bubble, P_a means the probability of adhesion of particle to a bubble in the time of contact and P_s denotes the probability of formation of a stable bubble-particle aggregate.

The parameter P_s can be taken as a function of the angle of contact, the radius of the particle and the bubble, and the density of the particle. The greater the contact angle, the stronger is the particle-bubble aggregate.

The probability of adhesion, P_a, is controlled by kinetic effects and as Sheludko concluded (1) the angle of contact could directly characterize flotation if there were no kinetic resistances to the attachment or if the kinetic resistances depended on the same parameters as the angle of contact. The parameter, P_a, was functionally connected with induction time by Sutherland (2). The induction time is defined as the time required for the disjoining film to thin to such a thickness that rupture can take place. Then P_a cannot be related directly to a contact angle and the contact angle cannot characterize the thinning process of the liquid layer as it is formed after all three stages of approach of the particle to the bubble have been completed.

According to equation /1/ flotation is possible of $\Theta > 0$, because for $\Theta = 0$, $P_s = 0$ and then $P = 0$ also. Thus condition $\Theta > 0$ must be fulfilled for flotation to be possible, but this condition is not enough to describe the flotation process. This can be called the thermodynamic criterion of flotation, but at the same time the kinetic criterion which is not controlled by the same variables must be fulfilled also.

To check these basic principles experiments were carried out with polished vitreous silica plates, with crushed vitreous silica and with quartz which were rendered hydrophobic by methylation with trimethylchlorosilane (3,4). The reaction that take place on the silica surface (5,6)

$$\geq SiOH + (CH_3)_3SiCl \longrightarrow \geq SiOSi(CH_3)_3 + HCl \qquad /2/$$

makes it possible to obtain the hydrophobic model system with a desired degree of hydrophobicity.

The contact angle of the methylated silica plates was found to depend on the pH of solution reaching zero at pH between 10 and 12. However, this was not due to loss of methyl groups for the hydrophobicity could be restored simply by changing the pH. Surprisingly the methylated very hydrophobic and pure hydrophilic silica had practically the same zeta-potential (3). This was later proved by J. Iskra (7) who used quartz and by R. D. Harding (8).

Working with the same system it was found (4) that flotation of the methylated quartz particles in 0.5 M solution of KCl depends on the pH in the same way as the contact angle. On the other hand the contact angle measured at slight acidic pH at various concentrations of KCl was constant while the flotation increased with the rate of KCl concentration. The observations quoted can be divided into two essential conclusions:

i. zeta-potential measurements show that the presence of electrical double layers does not exclude hydrophobicity;

ii. flotation experiments carried out with methylated quartz particles under the conditions of constant Θ revealed that an increase in KCl concentration improves the flotation and this must be correlated with the change of P_a in Eq. /1/ and it proves the validity of this equation.

In the flotation process the mineral particles usually carry molecules of the chemisorbed collector on their surface and the surface of the bubbles is covered with frother. The attachment of those two objects must be proceded by interactions between molecules of these two reagents in the moment of collision. Since the adsorption of frother on the solid under flotation conditions does not usually affect the contact angle, but at the same time frothers influence the rate of flotation, then it must be supposed that this is connected with parameter P_a in Eq. /1/, i.e. these interactions must reduce the energetic barrier between a particle and a bubble.

A maximum on the floatability curves is observed at pHs close to p.z.c. At the pH far from that of p.z.c. the natural floatability of minerals decreases as electrical double layer repulsion hinders the thinning and rupture of the intervening film between a bubble and a particle. Flotation experiments with the use of frothers show that under such conditions far from the p.z.c. a frother can greatly improve flotation. As the zeta-potential of the minerals investigated was not influenced in practice by the frothers then this effect can not be explained simply by a reduction of the electrical double layer repulsion.

II. DYNAMIC INTERACTION

In an earlier work (9) the influence of frother, α-terpineol, on the kinetics of the flotation of chalcocite at various dosages of ethyl xanthate was investigated. It was shown that in the absence of frother the flotation rate at pH 9.7 is very poor but small additions of α-terpineol rapidly increase the rate of flotation. Further experiments indicated quite clearly that the α-terpineol concentration influences the rate of adsorption of ethyl xanthate. The potential-current curves at different levels of α-terpineol concentration and a variable Na_2S concentration revealed that an increase in the concentration of α-terpineol results in an increase in the sulphide diffusion current at the chalcocite electrode. Tuwiner and Korman (10) working with a similar electrode obtained denser diffusion current at the same potential when the conditions of agitation were more intense. It was then feasible to conclude that α-terpineol molecules make it easier for sulphide ions to diffuse toward the surface of the electrode.

These results enabled us to postulate that depending on the surface potential of the mineral which changes continually during flotation, an adsorption or desorption of α-terpineol takes place on the mineral. This process is favoured by the collision of the bubble with the particle. Diffusion of big molecules of frother across the hydration layer of the solid can be compared with mixing and it makes it easier for the collector molecules to diffuse through such a layer. It can also be understood that diffusion of such molecules lowers the viscous resistance of the interventing liquid layer favouring the attachment of a bubble. If so, then the influence of frothers on flotation should be best when the diffusion of frother to or from the solid surface can take place at the moment of collision with the bubble.

III. EXPERIMENTAL

Flotation tests on the -0.2 + 0.075 mm size fractions of quartz previously reacted with trimethylchlorosilane (3,4) and with chalcocite were carried out in a Hallimond tube using 2-g charges. All the experiments were performed under conditions of concstant aeration and agitation. In the first series of experiments the flotation tests were carried out with the use of frothers only at a concentration of 6.10^{-4} M. The pulp was conditioned with the reagents for 1 min, the initial pH in the flotation of chalcocite was fixed at 9, and in the flotation of methylated quartz initial pH was 6.

Further experiments were performed with chalcocite using ethyl xanthate as a collector and α-terpineol. pH was regulated with NaOH and H_2SO_4 and was measured before and after the flotation.

The flotation experiments with chalcocite, ethyl xanthate and propyl acetal were performed in the presence of supporting electrolyte being 10^{-3} M KNO_3. In this case pH was regulated with KOH and HNO_3.

In order to investigate the effect of pH and of acetal on the electrical double layer at the chalcocite solution interface the method of Ahmed was used (11). The oxide layer on the surface of chalcocite has been shown to exist in many papers (12,13) and it enables us to use the same procedure for this system as for oxides. After 80 cm^3 solution of 10^{-3} KNO_3 used as supporting electrolyte attained a steady pH, the sample of chalcocite /1 g of the -0.15+0.075 mm size fraction/ was added and the variation of pH with time was followed. The surface charge and the differential capacity have been calculated from well known relationships (11).

IV. RESULTS

The floatability of chalcocite with various frothers at pH 9 and the floatability of the methylated quartz with various frothers at pH 6 are illustrated by the results reproduced in Fig. 1, where per cent of mineral floated is plotted as a function of the standard molar energy of adsorption of frothers onto mercury, ΔG. The data on ΔG have been extracted from the work by Zembala and Pomianowski (14). As may be seen, there is a very good correlation between the flotation and the standard energies of adsorption. This correlation exists for aliphatic alcohols up to hexanol and become poor for octanol /which in the same conditions yielded a concentrate equal to 40 per cent, the values of ΔG for octanol on mercury being -7.8 kcal/mol (15) and -9.0 kcal/mol (16) at -0.55 V against SCE/.

The effect of ethyl xanthate and α-terpineol and of pH on the flotation of chalcocite is given in Fig. 2. The natural floatability of chalcocite gives curve 1 with two extreme points. Flotation with the use of frother is characterized by curve 2 having three extreme points. As seen the presence of both the collector and the frother eliminates characteristic extreme points on the curves and improves flotation markedly.

Results presented in Fig. 3 were obtained while working with ethyl xanthate and propyl acetal. Similar relations can be seen.

The effect of pH and of acetal on the density of surface charge of chalcocite is shown in Fig. 4.

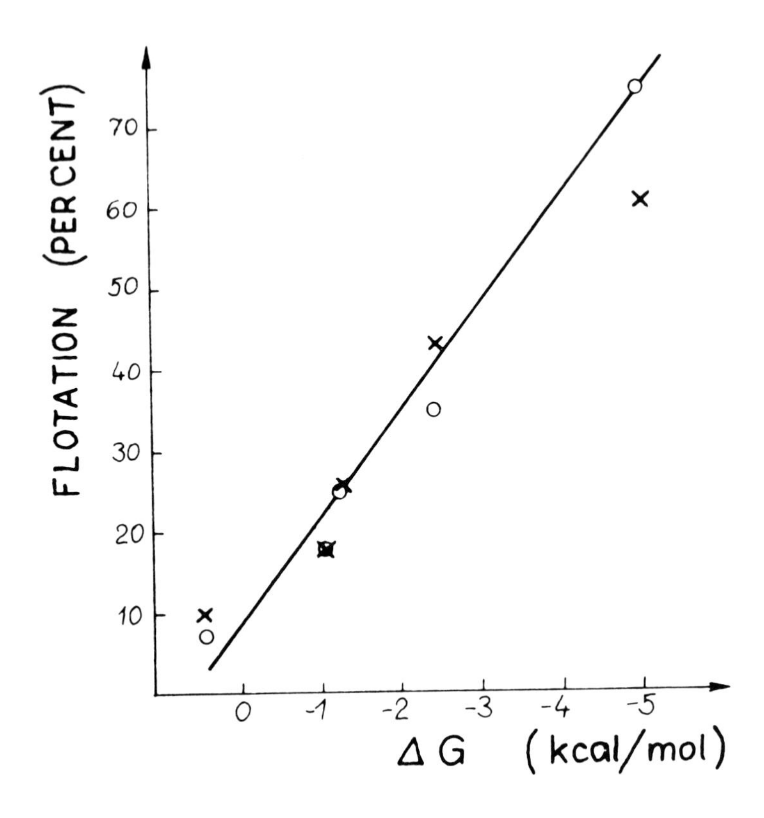

Fig. 1. Correlation between floatability of chalcocite/ o/ and of methylated quartz /x/ and standard Gibbs energy of adsorption of frothers on mercury. Concentration of frothers in flotation was 6.10^{-4} M, pH = 9 /chalcocite/, pH = 6 /methylated quartz/.

Moreover the experiments proved that the time necessary to reache an equilibrium is shorter in presence of acetal. In presence of acetal increases also the speed of adsorption of xanthate. Differential capacity curves shown in Fig. 4b indicate the deep minimum at pH 7.7. At higher values of pH the curves gradually decrease tending to another minimum at pH 9.4.

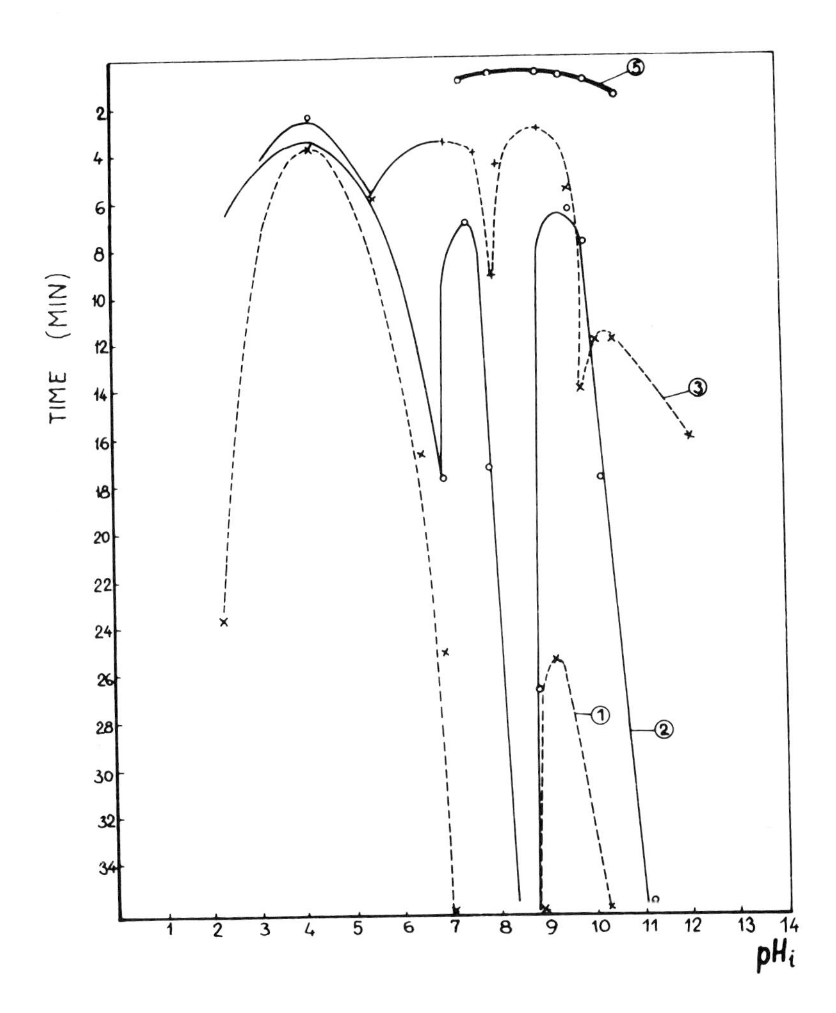

Fig. 2. Relationship between flotation time for 50 per cent recovery of chalcocite and initial pH 1, without reagents; 2, α-terpineol 10^{-5} M, 3, ethyl xanthate 10^{-5} M;5-α-terpineol and ethyl xanthate at concentrations of 10^{-5} M.

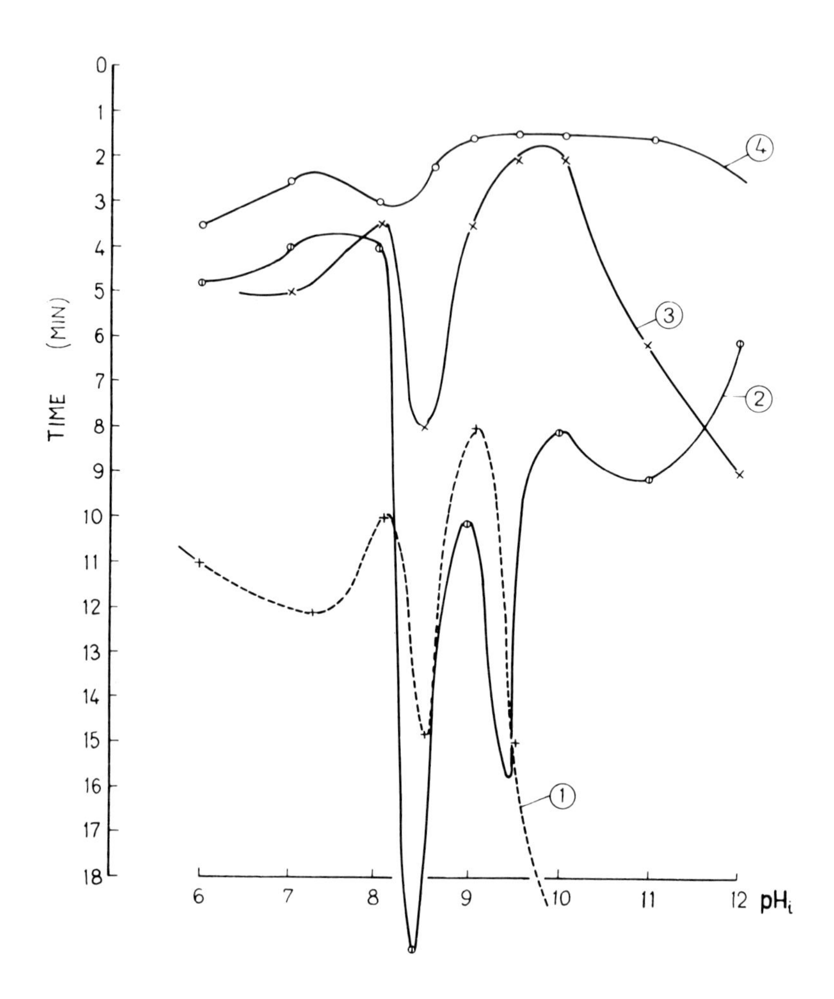

Fig. 3. Relationship between flotation time for 50 per cent recovery of chalcocite and initial pH at 10^{-3} M KNO_3. 1, without reagents; 2, propyl acetal 10^{-5} M; 3, ethyl xanthate 10^{-5} M; 4, propyl acetal and ethyl xanthate at concentrations of 10^{-5} M.

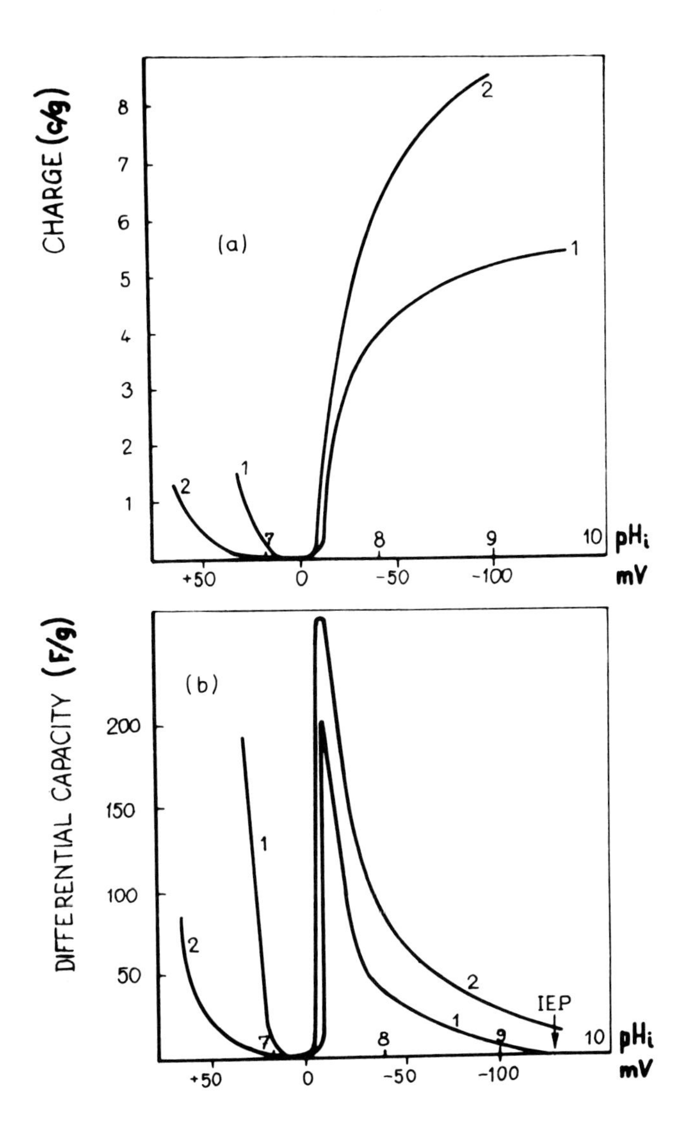

Fig. 4. Effect of pH and of propyl acetal on surface charge density of chalcocite at 10^{-3} M KNO_3. 1, whithout and 2, with propyl acetal at concentration of 10^{-5} M.

V. DISCUSSION

Zeta-potential of the chalcocite at pH 9 (9) and of the methylated quartz at pH 6 (3,7,8) reaches approchimately the same negative values. The experiments performed with different frothers gave very good correlation between the flotation and the standard energies of adsorption of these frothers onto mercury. This relation suggests that the nature of the adsorption forces of frothers on mercury is similar to that on the surfaces of minerals. The fact that this correlation exists for aliphatic alcohols up to hexanol, but does not hold for octanol for which the Gibbs standard molar energy of adsorp - tion is much higher indicates that the flotation effects can not be explained by adsorption of these reagents on solids. As shown in one of the previous works (17) octanol can be rather treated as an additional collector.

Influence of the frother on flotation of chalcocite is especially clear in the experiments at constant ionic strength as shown in Fig. 3.

It is known that the oxidation of chalcocite in aqueous solutions is given by the equation (12)

$$\boxed{yCuS\ xCu_2S} + \frac{1}{2}O_2 + H_2O \rightarrow \boxed{yCuS/x-1/Cu_2S}\ \boxed{CuS}\ \boxed{Cu(OH)_2} \quad /3/$$

with one product, CuS, diffunding into the solid phase and with the another, $Cu(OH)_2$, moving out into the solution. In acid range the solubility of $Cu(OH)_2$ is high and then the surface CuS can be subject to further oxidation.

$$CuS + H_2O \rightarrow Cu(OH)_2 + S \quad /4/$$

with S being oxidized to $S_2O_2^{2-}$, SO_3^{2-} and finally to SO_4^{2-}. Hence on the surface of chalcocite CuS_2O_3 and $Cu(OH)_2$ can be present at the same time. The log concentration - pH diagram for various species in equilibrium with chalcocite is shown in Fig. 5. Two IEPs for CuS_2O_3 at pH 7.7 and for $Cu(OH)_2$ at pH 9.4 both present on the surface of the mineral are seen[x]. At these pHs the natural floatability of chalcocite is at maximum /curve 1, Fig. 3; Fig. 2/. Minima are obtained in Fig. 4b at similar values of pH. It means that local $Cu(OH)_2$ places on the surface of chalcocite are at their i.e.p. at pH round 9, and that the p.z.c. of the mineral lies between IEPs for CuS_2O_3 and $Cu(OH)_2$ calculated from thermodynamic data.

---

x/ ΔG for CuS_2O_3 has been estimated to be 130 kcal/mol. A set of data for the construction of this diagram was extracted from work by Yoon and Salman (18).

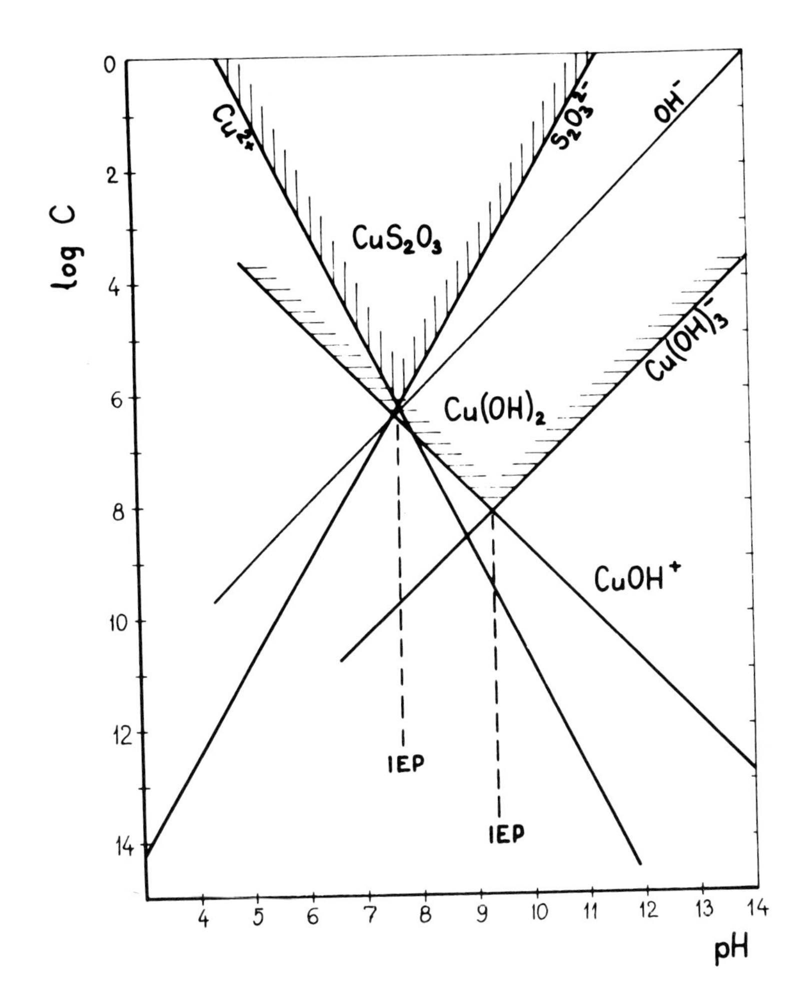

Fig. 5. Solubility diagram for oxidation products of chalcocite: CuS_2O_3 and $Cu(OH)_2$.

Comparison of the curves 1 and 2 in Fig. 3 /and curves in Fig. 2/ reveals that the frother improves greatly flotation at pH > 9.5 and at pH < 8. In the first case on the right from the IEP for $Cu(OH)_2$, the surface is negatively charged and in the second case on the left from the IEP for CuS_2O_3, the positive charge predominates.

The works by Frumkin et al (19) have shown that long-chain alcohols influence the differential capacity of a double layer at the mercury/solution interface. The minimum on the differential capacity curve versus applied potential is appreciable lowered in the presence of, for example, amyl alcohol. At higher potentials the desorption of alcohol from the interface suddenly changes the electrical capacity of a double layer.

It has been considered in the previous sections that the role of frother in the flotation process can be explained on the basis of the dynamic effects. This hypothesis is supported by the results here presented. If the electrical double layer interaction forces between a particle and a bubble is considered with the condition of constant surface charge density then, as seen from Fig. 6, a sharp change of the surface potential occurres at small distances and the curve becomes almost parallel to the surface potential axis. Such a change of potential causes desorption of a frother adsorbed on the solid and its diffusion across the disjoining liquid layer. This process takes place in the moment of collision of the particle to the bubble.

It is now recognized that the adsorption of highly solvated surfactants, particularly those of the ethylene oxide type, converts a lyophobic particle into a lyophilic one (21). To study the stability of such systems the more general relationship is used in which the entropy of interaction is taken into account also.

$$\Delta G_{interaction} = \Delta G_{repulsion} + \Delta G_{attraction} + \Delta G_s \quad /5/$$

This is generally accepted in the case of stabilisation of dispersions by nonionic surfactants (22) but was also proposed in the more general form (23).

Contrary to the case of stabilisation of hydrophobic system with nonionic surfactant, the flotation case covers full range of changes from the hydrophilic to the hydrophobic state. To describe the properties of such a system it is also necessary to use full relationship with the term ΔG_s.

According to Frumkin (24) and Derjaguin (25) the distinction between a wetting and a non-wetting system gives the specific surface energy curve of a planar liquid film on the solid as a function of thickness h. In the Derjaguin's disjoining pressure term, Π, the criterion for a contact angle to develop is that films in a certain range of thicknesses should be subject to a negative disjoining pressure and as it was concluded in one of the earlier works (3) "the instability of water films of a certain thickness on hydrophobic solid is fundamentally due to a deficiency of hydrogen bonding in these films as compared with liquid water".

Read and Kitchener (26) following Derjaguin's works proved that for $h > 20$ nm the most important contribution to the disjoining pressure arises from compression of the diffuse electric double-layer at the silica water interface. As zeta-

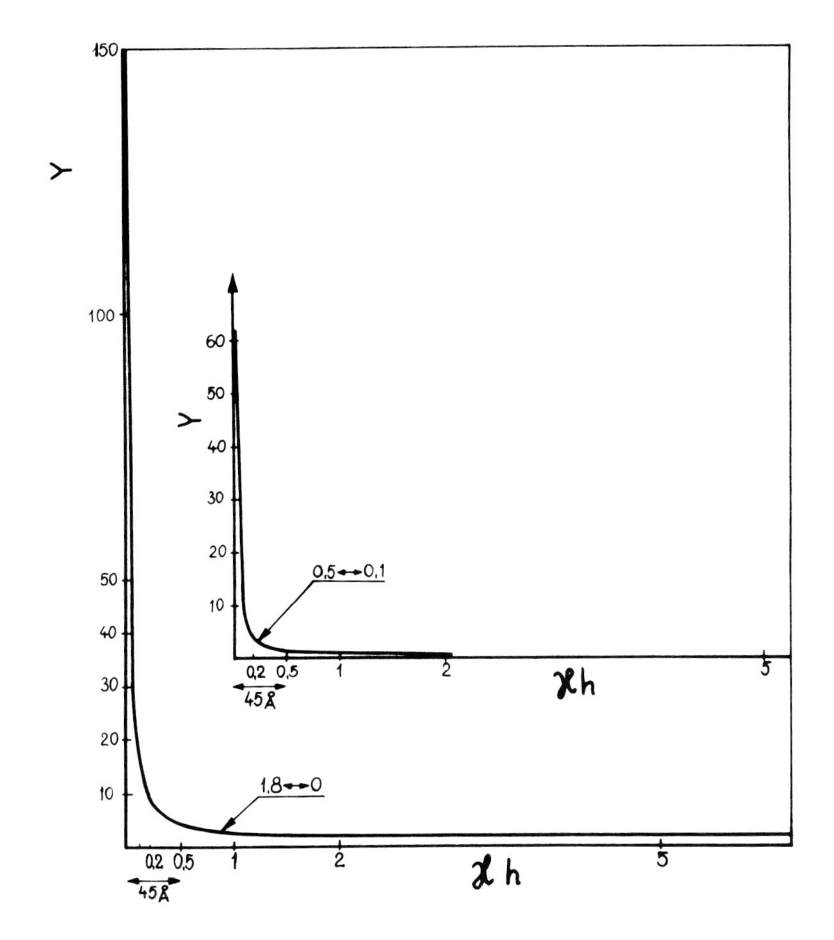

Fig. 6. Effect of distance between two plates $/d_1 = 0.1$ mm, $d_2 = 2$ mm/ on surface potential Y /Y = $ze\Psi/kT$/ at a constant charge. $\varepsilon_1 = 15$, $\varepsilon_2 = 1$, $\varepsilon = 73$, concentration of 1-1 electrolyte 10^{-3} M, values of Y_∞ are given on curves /calculated according to Ohshima (20)/.

potentials of the hydriphobic methylated silica and the hydrophilic pure silica were the same then the double layer repulsion forces should stabilize thick aqueous films on the methylated silica in the same way as they stabilize aqueous films on clean silica. This was recently proved by Blake and

Kitchener (27) who showed that the equilibrium films are also formed on hydrophobic surfaces, collapsing when disturbed and forming a contact angle which may be as large as 90°. This corresponds to $\pi = f(h)$ curve having a maximum and a minimum and it explains the phenomenon of induction time. Blake and Kitchener's experiments proved that spatial extension of hydrophobicity is distinctly smaller than the range of action of electrical double layers.

In view of the results showing the propagation of bounding at the hydrophilic/water interface stronger than that in the liquid water (28-30) as well as a deficiency of hydrogen bonding in water at hydrophobic interfaces (3,31,32), the term ΔG_s appears to be essential for understanding the transition from a fully hydrophilic state to a hydrophobic one.

In our hypothesis the entropic hydration barrier, ΔG_s, is lowered at the moment of attachment of the particle to the bubble if the diffusion of the surfactant molecules across this layer takes place. Hence, the effect of frothers on the flotation should be noticeably at the potentials close to the i.e.p. when an adsorption of frother can take place, but at the same time those potentials should be high enough to reache the value causing desorption at the moment of attachment with bubble.

As shown the correlation between the flotation results and the free energies of adsorption exists up to hexanol and does not hold for octanol. It suggests that when the adsorption energy of frother is too high the frother can not conform to the actual physico-chemical conditions. Then the mechanism of action of frothers can be understood as additional adsorption or desorption in the moment of attachment of the particle to the bubble.

VI. REFERENCES

1. Sheludko, A., Kolloid Z. 191, 52 (1963).
2. Sutherland, K. L., J. Phys. Coll. Chem. 52, 394 (1948).
3. Laskowski, J., and Kitchener, J. A., J. Coll. Interf. Sci. 29, 670 (1969).
4. Laskowski, J., and Iskra, J., Trans, IMM, Sec. C, 79, 6 (1970).
5. Kiselev, A. V., Kovaleva, N. V., Korolev, A. J., and Shcherbakhova, K. A., Dokl. Akad. Nauk SSSR, 124, 617 (1959).
6. Armistead, C. G., and Hockey, J. A., Trans. Farad. Soc. 63, 2549 (1967)
7. Iskra, J., Ph. D. Thesis, Silesian University of Technology, Gliwice, 1969
8. Harding, R. D., J. Coll. Interf. Sci. 35, 172 (1971).

9. Lekki, J., and Laskowski, J., Trans. IMM, Sec. C, 80, 174 (1971).
10. Tuwiner, S. B., and Korman, S., Trans. AIME, 187, 226 (1950).
11. Ahmed, S. M., J. Phys. Chem, 73, 3546 (1969).
12. Lekki, J., and Laskowski, J., "On the Chalcocite/Solution Inteface at Alkaline pH", 22nd Meeting CITCE, Dubrovnik, 1971.
13. Oestreicher, C. A., and Mc Glashan, D. W., "Surface Oxidation of Chalcocite", paper presented at AIME Annual Meeting. Abstract published in Mining Engineering, 23, 75 (Dec. 1971).
14. Zembala, M., and Pomianowski, A., "Adsorption of Flotation Frothers on a Surface of Mercury", 9th Symposium on Physicochemical Problems of Mineral Processing, Silesian University of Technology, Gliwice, 1970.
15. Lorenz, W., Müchel, F., and Muller, W., Z. Phys. Chem. N.F. 25, 145 (1960).
16. Damaskin, B., Petrij, O. A., and Batrakov, V., "Adsorption of Organic Compounds on Electrodes", pp. 126, 131. Izd. Nauka, Moscow, 1968.
17. Lekki, J., and Laskowski, J., "A New Concept of Frothing in Flotation Systems and General Classification of Flotation Prothers", XIth Intern. Mineral Processing Congress, Cagliari, 1975.
18. Yoon, R. H., and Salman, T., Can. Metall, Quart., 10, 171 (1971).
19. Frumkin, A. N., Bagocki, V. S., Ioffa, V. S., and Kabanov, B. N., "Kinetics of the Electrode Processes", University of Moscow, Moscow, 1952.
20. Ohshida, H., Kolloid Z., 252, 158, 257 (1974).
21. Ottewill, R. H., in "Nonionic Surfactants" /M. J. Schick, Ed./, p. 656, Marcel Dekker, New York, 1967.
22. Bagchi, P., J. Coll. Interf. Sci., 47, 86, 100 (1974).
23. Derjaguin, B. V., Gorodetzka, A. V., Titijevska, A. S., and Jachin, V. N., Russ. J. Coll. Chem, 23, 535 (1961)
24. Frumkin, A., Acta Physicochim, URSS, 9, 313 (1938).
25. Derjaguin, B. V., and Obuchov, E., Acta Physicochim. URSS, 5, 1 (1936).
26. Read, A. D., and Kitchener, J. A., J. Coll. Interf. Sci. 30, 391 (1969).
27. Blake, T. D., and Kitchener, J. A., J. Chem, Soc. Farad. Trans. I., 68, 1435 (1972).
28. Derjaguin, B. V., and Kusakov, M. M., Acta Physicochim, URSS 10, 25, 153 (1939).
29. Cafferty, E. Mc., and Zettlemoyer, A. Z., Disc. Farad. Soc. 52, 239, 255 (1971).
30. For a review, cf. F. Franks, Chem. Ind. /London/, 1968,560
31. Pchelin. V. A., Russ. J. Coll. Chem. 34, 783 (1972).
32. Ievleva, V. V., Pchelin, V. A., and Yampolski, B. Y., Russ. J. Coll. Chem. 37, 866 (1975).

FURTHER CONSIDERATIONS ON THE AGING OF POLYDISPERSE SUSPENSIONS (OSTWALD RIPENING)

M. Kahlweit
Max-Planck-Institut für biophysikalische Chemie

ABSTRACT

The driving force of the Ostwald ripening of a polydisperse suspension is the solubility difference between the smaller and the larger particles. This leads to an increase of the mean radius of the particles with time, accompanied by a decrease of their number per unit volume. The first quantitative analysis of this process was published in 1961 being known as LSW theory. It predicts that the aging rate (being a measure for the increase of the mean radius with time) asymptotically approaches a constant positive value, while the size distribution of the particles approaches a time independent shape. As was shown in recent papers (1), this prediction is incorrect. Instead, the aging rate passes through a maximum during the early stages of precipitation to slowly approach zero, while the size distribution continues to change its shape with the right hand shoulder steepening with time. A quantitative prediction of the aging rate cannot be given, since its absolute value will depend on the initial size distribution of the particles, i.e., on the experimental conditions under which the precipitate was formed. In general, however, the aging rate even at its maximum will be much slower than predicted by the LSW theory.

REFERENCES

(1) Kahlweit, M., Adv. Colloid Interface Sci. 5, 1 (1975); Disc. Faraday Society "Precipitation" (1976), in print; Scripta Met. (1976), in print.

SOLUTION OF ORNSTEIN-ZERNIKE EQUATION FOR WALL-PARTICLE DISTRIBUTION FUNCTION

L. Blum, *University of Puerto Rico*
and
G. Stell, *State University of New York at Stony Brook*

ABSTRACT

The Ornstein-Zernike (OZ) equation is considered for the wall-particle distribution function $g_0(x)$ in the case of a flat impenetrable wall at $x = 0$ and a fluid of hard-core particles whose centers are constrained by the wall to occupy the semi-infinite space, $x > \sigma/2$, where σ is the particle diameter. A solution is given in terms of the wall-particle direct correlation function $c_0(x)$ for $x > \sigma/2$ and the particle-particle direct correlation function $c(\underset{\sim}{r}_1, \underset{\sim}{r}_2)$ when the latter is approximated by its bulk value $c_B(t)$, $t = (|\underset{\sim}{r}_1 - \underset{\sim}{r}_2|)$, and $c_B(t)$ satisfies certain asymptotic conditions. Explicit formulae for the contact surface density, total excess surface density, and the Laplace transform of the fluid density near the wall are given. For mean-spherical type approximations, $c_0(x)$ for $x > \sigma/2$ and $c_B(t)$ are both prescribed functions; for this case, a closed-form solution is obtained. A very brief discussion of the case of a charged particle near an electrode is given. The additional equations that enable one to go beyond the approximation $c\ (\underset{\sim}{r}_1, \underset{\sim}{r}_2) = c_B(t)$ are introduced, and a simple iteration scheme to this end is discussed.

ELECTROCHEMICAL PROPERTIES OF HOLLOW SPHERES AND PHOSPHATIDIC ACID VESICLES

Michel Mille, *University of Wisconsin*
and
Garret Vanderkooi, *Northern Illinois University*

ABSTRACT

The nonlinearized Poisson-Boltzmann equation was numerically solved for hollow spherical shells having the dimensions and properties of phosphatidic acid vesicles, and bearing charges on both the inner and outer surfaces of the shell. By assuming electroneutrality inside the shell, we obtained the electrostatic potential inside and outside of the shell, as well as the inner and outer degrees of dissociation of the surface charges. Physically measurable parameters were computed from the potential as a function of vesicle size, salt concentration, etc. The degree of dissociation, under all circumstances, was less on the inside than on the outside surface. At low salt (≤.001M) and low vesicle concentration, the potential is larger inside than outside the vesicle, and co-ions are effectively excluded from the interior. At high particle concentration, the potential becomes larger outside than in; and at still higher particle concentration, the co-ion concentration is greater inside than outside. Comparison of experimental and computed pH titration curves for phosphatidic acid vesicles showed good agreement for salt concentrations ≥0.01M but gave significant unexplained deviations at lower salt concentrations.

THE ROLE OF PARTICLE SIZE AND MOLECULAR WEIGHT ON THE ADSORPTION ON AND FLOCCULATION OF POLYSTYRENE LATEX BY POLY(1,2-DIMETHYL-5-VINYL-PYRIDINIUM BROMIDE)

A. R. Eggert and R. A. Stratton
The Institute of Paper Chemistry

ABSTRACT

The interrelationship between particle size and polyelectrolyte molecular weight on the adsorption and subsequent flocculation of oppositely charged colloidal particles was investigated by the model system of poly(1,2-dimethyl-5-vinylpyridinium bromide)-water-polystyrene latex. Polystyrene latex particles, ranging in diameter from 0.1-1.1 μm, were used in this study and characterized by their surface charge densities and electrophoretic mobilities. Two narrow molecular weight fractions of poly(DMVPB), 3.3×10^4 and 10^6, were prepared and their hydrodynamic molecular diameters were determined by standard diffusion, viscosity and light scattering techniques. Two salt concentrations were employed to vary the solution size of the polyelectrolyte and the double layer thickness of the colloid. At the concentrations of poly(DMVPB) required to neutralize the polystyrene latex particles, complete adsorption of poly(DMVPB) occurred regardless of molecular weight and salt concentration. At still higher poly(DMVPB) concentrations adsorption was a slightly increasing function of salt concentration and molecular weight. The flocculation results indicated that the amount of poly(DMVPB) necessary to initiate flocculation was dependent on molecular weight. However, at optimum flocculation concentrations there was little dependence on molecular weight. The findings from the adsorption-flocculation study were consistent with an electrostatic patch flocculation mechanism.

ADSORPTION OF POLY (VINYL ALCOHOL) ON SILICA AT VARIOUS pH VALUES AND ITS EFFECT ON THE FLOCCULATION OF THE DISPERSION

T. F. Tadros
Jealott's Hill Research

ABSTRACT

The adsorption of poly (vinyl alcohol) (PVA) was investigated at various pH values on three samples of silica (a) precipitated (B.D.H.) (b) fumed (Cabosil and (c) commercial Sol namely "Ludox." The precipitated sample was preheated for ∿ 15 hours at ∿ 350° before being suspended in water for the adsorption experiments. Without any heat treatment this sample showed little or no adsorption of PVA. However, another acid washed sample (B.D.H.) showed relatively small PVA adsorption. With the "Cabosil" and "Ludox" samples significant adsorption occurred without any heat pretreatment. With all the three silica samples, adsorption depends to a large extent on the pH of the equilibrium solution, being higher the lower the pH. Maximum adsorption occurs at or below the point of zero charge of the oxide. Moreover, flocculation by the polymer is more effective the lower the pH. The "optimum" flocculation depends on the pH and the amount of PVA added. At the lower pH values (< pH5), where flocculation is significant the extent of flocculation increases with increase of PVA concentration, reaches a maximum at a certain pH range which depends on the dispersion concentration and type of silica, after which flocculation decreases in magnitude and finally at sufficiently high PVA concentration, the dispersion becomes stable again presumably due to "protection" by the polymer. The results obtained are discussed in terms of the general theories on "bridging flocculation" and particular attention is paid to the effect of pH, the role played by silanol and siloxane bonds for adsorption of the polymer.

THE INFLUENCE OF AN EXTERNAL MAGNETIC FIELD ON THE STABILITY OF A MAGNETIC COLLOID

E.D. Tarapore
University of California (Berkeley)

T.S. Mika
Xerox Corporation

The influence of a uniform external magnetic field on the stability of aqueous dispersions of magnetic colloids is investigated both theoretically and experimentally. In the experiments, dilute dispersions of a narrow size fraction of magnetite (Fe_3O_4), with arithmetic mean particle diameter of approximately 160 nm, are considered. The particles, although irregular, are approximately equidimensional. An optical turbidimetric method, based on a modified Beckman DU-2 Spectrophotometer, is adopted to assess the coagulation behavior of these dispersions.

The experiments on the influence of the external field indicate that:

1) the stability of the dispersion decreases as the magnetic field strength increases with no tendency to aproach an asymptote in the range considered,

2) the stability of the dispersion, for a fixed magnetic field strength, can be significantly altered by control of the aqueous solution pH in a manner which is symmetric about the point-of-zero-charge for the magnetite, and

3) the stability of the dispersion, for a fixed magnetic field strength, can be significantly modified by the presence of surfactants in a manner which is not symmetric about the point-of-zero-charge for the magnetite.

Qualitative interpretation of some of these results is provided.

A theoretical model of the stability of a colloidal dispersion of magnetic particles is developed. It is derived by a modification of the Fuchs-Smoluchowski description of the coagulation of identical spheres which accounts for the orientation-dependent particle-particle magnetic interactions. It is based on the assumption that the spherical particles exhibit single domain behavior, i.e. are uniformly magnetized at the saturation magnetization independent of the magnitude of the external field and the presence of other particles. The orientation of the direction of magnetization of the particles with respect to the external field is assumed to obey the Langevin relation, independent of the presence of other particles.

Some of the results obtained in an extensive computational investigation of this model using parameter values which correspond to properties of the magnetite are described. Semi-quantitative agreement with the experimental results on the influence of pH is observed. The theoretical model predicts that the stability approaches an asymptote as the external magnetic field strength increases, i.e. as the direction of magnetization of the particles becomes completely alligned with the external field. This is not in agreement with the experiments on the influence of the external field on stability. This is interpreted as evidence that the magnetization of the magnetite particles is dependent on the magnitude of the external field, i.e. that those particles deviate from simple single domain behavior.

THE EFFECTS OF CATIONIC SURFACTANTS ON AQUEOUS INORGANIC PIGMENT SUSPENSIONS

Noboru Moriyama and Takeuchi Takashi
Kao Soap Company

ABSTRACT

The effects of cationic surfactants on the stabilities of aqueous inorganic pigment suspensions are examined. Ferric oxide, titanium dioxide, zinc oxide and aluminium hydroxide were used as inorganic pigments. The stabilities were evaluated from sedimentation velocity data. The stabilities are highly dependent upon the concentration and alkyl chain length of the surfactants, and pH of the suspensions, especially, the isoelectric points of the pigments. The stability of the suspensions in the presence of cationic surfactants, e.g., dodecyl ammonium chloride, decreases with increase in pH value, whereas the stability of the suspensions in the presence of anionic surfactants, e.g., Na dodecyl sulfate, increases with increase in pH value. The stability data are discussed in relation to the zeta potential data, the concentration of the surfactants, and the isoelectric point of the pigments.

HETEROCOAGULATION OF AMPHOTERIC LATICES

R. O. James and A. Homola
University of Melbourne

ABSTRACT

The study of heterocoagulation of amphoteric mineral colloids, e.g., the oxides TiO_2 and Al_2O_3 is complicated by specific chemical effects, namely dissolution, adsorption and oxide coating formation.[1-5] *We have recently prepared and characterised amphoteric polystyrene latices with R_3NH^+ and $R\text{-}COO^-$ ionisable surface groups.*[6] *Because of their low solubility, spherical shape and uniform particle size, these latices are ideal models for amphoretic colloid systems. In addition, these latices may be prepared with different iep (or pzc) values according to the relative amounts of amine or carboxyl functional groups. Hence, by using mixtures of latices with different iep values it is possible to study the interaction of dissimilar double layers around colloids of virtually identical bulk composition.*

The colloid stability of these systems has been measured as a function of ionic strength, ionic charge, pH and relative ratio of latex.

1. *T. W. Healy, G. R. Wiese, D. E. Yates and B. V. B. V. Kavanagh. J Colloid Interface Science 42, 647 (1973).*

2-5. *G. R. Wiese and T. W. Healy J. Colloid Interface Science 51, 427 (1975)*
51, 434 (1975)
52, 452 (1975)
52, 458 (1975).

6. *A. Homola and R. O. James. 4th Australian Electrochem Conference, Adelaide, February 1976.*

KINETIC ASPECTS OF ADSORPTION AND FLOCCULATION WITH CATIONIC POLYMERS

John Gregory
University College London

ABSTRACT

In the flocculation and stabilization of suspensions by polymers, three distinct rate processes play a part:

(a) Diffusion of polymer molecules to, and attachment to, a particle surface (Adsorption).
(b) Re-arrangement of adsorbed polymer chains to give an equilibrium configuration (Re-conformation).
(c) Collision of particles, by Brownian motion or velocity gradients, which may result in the formation of aggregates (Flocculation).

In practice, the effect of a polymeric flocculant on the stability of a suspension depends very much on the relative rates of these processes.

Experimental investigations have been carried out, with systems consisting of negative latex particles and cationic polymers, using electrophoretic, turbidimetric and Coulter counting techniques.

With low particle concentrations, the adsorption step (a) may be slow enough to affect the flocculation process, e.g., collisions can occur between particles which are not fully destabilized. Under these conditions, it is possible to follow the adsorption of polycations on a single particle by observing the change of electrophoretic mobility with time.

At much higher particle concentrations the adsorption is more rapid, but in this case the re-conformation step (b) may be slow relative to the collision rate and particles may collide before the equilibrium configuration is achieved, leading to an enhanced possibility of "bridging" by polymer chains.

STABILITY IN MIXTURES OF POLYMER COLLOIDS AND POLYMERS IN A COMMON SOLVENT

A. Vrij

Van 't Hoff Laboratory

Transitorium III, Padualaan 8

Utrecht, The Netherlands

ABSTRACT

Solutions of different polymers in the same solvent are incompatible as a rule and show phase separation when they are mixed. If incompatibility is also to be observed in systems where one of the polymer components is replaced by colloidal particles, sterically stabilized by a cover of polymer chains, will be discussed in this paper. It is predicted that the second virial coefficient of the colloidal particles should decrease and may become negative when the molecular weight and concentration of the polymer are sufficiently large. For high molecular weight polymer this is of the order of a percent or less. Higher concentrations are needed for low molecular weight polymer.

POTENTIAL DISTRIBUTIONS IN PHASE-SEPARATED AQUEOUS POLYMER SOLUTIONS

D. E. Brooks, *University of British Columbia*
G. V. P. Seaman, *University of Oregon Medical School*
and
H. Walter

ABSTRACT

Partition of biological cells in buffered phase-separated aqueous solutions of neutral polymers can be used as a separation procedure which is extremely sensitive to cell surface properties. In dextran-polyethylene glycol (PEG) phase systems, providing certain salts are present, a correlation has been found between cell surface charge density and the quantity of cells partitioning into the PEG-rich phase, suggesting electrostatic interactions can determine partition behaviour. We have therefore attempted to estimate electrostatic potential profiles between the phases under conditions of varying ionic composition and concentration and to correlate these profiles with partition of the salts themselves in these systems. The potential differences between the bulk phases were estimated with salt bridges and reversible electrodes. The potentials between the phase boundary and each of the bulk phases were estimated from the electrophoretic mobilities of droplets of each of the phases suspended in the other, interpreted via Levich's theory. The bulk phase potentials decrease in magnitude with increasing ionic concentration and increasingly disparate salt partition. The zeta potentials, on the other hand, increase in magnitude with increasing ionic strength and are of opposite sign to those expected from the bulk phase potentials. Models for the complex potential variation will be discussed.

ELECTROPHORESIS OF HIGHLY CHARGED CYLINDERS IN SALT SOLUTIONS AND APPLICATION TO DOUBLE STRANDED DNA

Dirk Stigter

U. S. Department of Agriculture

ABSTRACT

The theory of electrophoresis of highly charged cylinders is under study. The method of Wiersema, Loeb and Overbeek (J. Colloid Interf. Sci. 22, 76 (1966)) for the electrophoresis of the colloid sphere is modified for analysis of the motion of the charged cylinder in a transverse electric field. Inclusion of the relaxation effect (deformation of the ionic atmosphere) leads to a system of coupled integral equations whose solution by computer methods is presently attempted. Applications to double stranded DNA are contemplated.

STREAMING POTENTIAL OF FINE PORES

C. P. Bean
General Electric Corporate Research and Development

ABSTRACT

Using etched particle tracks in mica, the streaming potential has been measured--for the first time--in uniform, characterized submicron pores. A new method of data presentation, the resistance-streaming potential plane, gives evidence for the applicability of the Gouy-Chapman theory to the potential distribution within a pore filled with a dilute salt solution. In addition the results imply that the bulk values of dielectric constant and viscosity of water are quite appropriate for pores as small as 70 Å in diameter.

RE-DISTRIBUTION OF CHARGES IN SPACE AND DE-IONIZATION IN PRECIPITATING (POLYELECTROLYTE) COLLOIDS

I. Michaeli
Weizmann Institute of Science

Patterns of phase separation in colloidal systems are discussed in terms of coulombic and non-coulombic forces operative in the system. Emphasis is laid on the distinction between 'particle stability' and 'phase stability', and the role of coulombic interactions is analyzed in terms of the above distinction. It is shown that coulombic forces may favor co-operative re-distributions of the density of charged colloids in solution, and of small ions near a charged surface.

A classification of phase separation phenomena is made, distinguishing between disproportionation in concentration with retention of charge, and (reversible) precipitation due to charge-stripping processes.

Polyelectrolytes are discussed as an interesting class of colloids incorporating a variety of species differing with regard to geometry, flexibility, ionization behavior, and solubility of the non-ionized macromolecule. Thus, the different phase separation patterns discussed above, are to be found in the respective polyelectrolyte systems.

LAMINAR OSCILLATING FLOW STREAMING POTENTIAL MEASUREMENTS (I)

Alan R. Sears and James N. Groves
General Electric Company

ABSTRACT

There exists a large class of materials whose electrokinetic properties cannot be studied by measurement of their electrophoretic mobility. This experimental limitation may be due to inherent properties of the material, e.g., density, size distribution, or from the necessity of using a continuous surface rather than a particulate one. In such cases the measurement of either the streaming potential or current can be used to provide a method of surface characterization. Such measurements have usually been done under conditions of unidirectional Hagen-Poiseuille flow. In aqueous solutions only a small potential difference is produced which requires further amplification. In order to circumvent the experimental difficulties of the unidirectional flow system, e.g., noise, electrode rest potential drift, and D.C. amplification, we have developed an alternating laminar flow system to measure the zeta potential of any material which may be fabricated as a capillary or coated onto a supporting capillary. In this system the sinusoidally oscillating flow streaming potential is measured as a function of both frequency and applied pressure and the zeta potential is determined by using both classical electrokinetic theory and the theory of oscillating laminar flow developed by Uchida. The reliability of this experimental procedure has been confirmed by measuring the streaming potential over a range of frequencies and computing the zeta potential.

SIZES, CONCENTRATIONS, AND ELECTROPHORETIC MOBILITIES OF SUBMICRON PARTICLES BY THE RESISTIVE PULSE TECHNIQUE*

R. W. DeBlois
General Electric Corporate Research and Development

and

R. K. A. Wesley
John L. Smith Memorial for Cancer Research, Pfizer, Inc.

Colloidal particles above about 60 nm in diameter in a liquid may be rapidly sized individually to about ± 1 nm, counted over the range from 10^5 to 10^{11}/ml, and measured for electrophoretic mobilities to about ± 1 x 10^{-6} cm^2/V sec with the Nanopar analyzer, based on the resistive pulse technique of the Coulter Counter and the use of submicron diameter pores made by the etched particle track (Nuclepore) process. Latex sphere size distributions and comparisons are presented, as well as sizes of the T2 bacteriophage (5.10 ± 0.15 x 10^{-16} cm^3), Rauscher murine leukemia virus (122 ± 2 nm), and insect virus Tippula Iridescens (195 ± 2 nm), relative to standard 109 nm latex spheres. In addition, one may monitor the state of colloidal dispersion, follow aggregation and mobility changes, measure absolute particle sizes, and measure pore dimensions, zeta potentials, and surface resistivities.

* *This work has been supported by contracts NO1-CP-3-3231 and NO1-CP-3-3234 within the Virus Cancer Program of the National Cancer Institute.*

BINARY MINERAL FLOTABILITY IN THE PRESENCE OF ARSENIC TRIOXIDE

T. Salman and James McHardy
McGill University

ABSTRACT

The changing flotability of the mineral systems MoS_2-$CuFeS_2$, MoS_2-FeS_2 and MoS_2-$CuFeS_2$:FeS_2 in the presence of As_2O_3 (potassium ethyl xanthate) Z-3, and/or pine oil have been investigated by means of electrophoretic measurements and Hallimond Tube flotation. The various shapes of the zeta curves for these minerals in the different media can be directly correlated with their flotation behaviour.

It has been observed that As_2O_3 in the pH range 9 to 10 will effectively depress each of the sulfide minerals. The addition of frother enables the MoS_2 to float, but the Copper and Iron sulfides remain depressed. As_2O_3 also inhibits the adsorption of xanthates on the sulfide surfaces, but will not remove a xanthate coating if said coating occurs prior to the addition of an As_2O_3 solution.

The common factor among the minerals is the domination of each surface by sulphur atoms. It is postulated that the As_2O_3 solution species interact with these sulphur surfaces making them more hydrophilic. The specific pH range in which maximum depression occurs correlates with the major ion change for dissolved Arsenic trioxide.

DIRECT MEASUREMENT OF LONG-RANGE FORCES BETWEEN TWO MICA SURFACES IN AQUEOUS KNO_3 SOLUTIONS

J.N. Israelachvili and G.E. Adams

The Australian National University

An experiment is described in which the long-range forces between two molecularly smooth mica surfaces, immersed in KNO_3 solutions, are directly measured in the range 10 - 1000 Å. The forces are exponentially repulsive, with a decay length close to the Debye length, though the magnitude of the repulsive forces was found to be different for different mica samples. In 0.1 M and 1 M KNO_3 solutions, the onset of repulsion is preceded by an attractive region for those samples where the repulsion was weak, thereby indicating the existence of a secondary minimum at these concentrations. It is difficult to say whether at very small separations, below about 10 Å, there is also a primary minimum, since at these separations the surfaces have appreciably deformed under the influence of the strong forces, and a more detailed analysis is needed here. The results exhibit all the essential features characteristic of a DLVO-type interaction involving repulsive double-layer forces and attractive van der Waals forces. In particular, we may note that even at 0.1 M KNO_3 the decay length of the exponential repulsive region was equal to the Debye length showing that the Poisson-Boltzmann equation appears to remain valid even at this high concentration (at least for KNO_3). The magnitude of the attractive forces in the region of the secondary minimum are of the order to be expected from the general (DLP) theory of van der Waals forces.

MASS ANALYSIS OF PARTICLES AND MACROMOLECULES BY FIELD-FLOW FRACTIONATION

J. Calvin Giddings, Marcus N. Myers,
Frank J. F. Yang, and LaRell K. Smith
Department of Chemistry
University of Utah
Salt Lake City, Utah 84112

Field-flow fractionation (FFF) is a new separation methodology of considerable scope and versatility. Its range of applicability is enormous, spanning at least a billion-fold mass range from molecules of 500 MW to particles larger than a micron in diameter. Retention in this system is a calculable function of physicochemical parameters, most of which are size and mass dependent. Thus retention measurements can be converted into a mass or size analysis for complex materials. The exact parameter yielded depends on the subtechnique employed. Sedimentation FFF provides a spectrum of effective masses. Flow FFF yields a spectrum of diffusion coefficients, usually convertible into an approximate mass spectrum. Thermal FFF gives thermal diffusion factors, whose molecular weight dependence for linear polymers is now emerging. Electrical FFF can also provide a diffusion coefficient spectrum under some conditions.

In this paper, the basic equations needed to acquire a mass spectrum for each subtechnique are established. The elution spectrum, which is the retention volume versus molecular weight profile needed to get a mass spectrum, is calculated for a number of examples. Experimental results are cited to show that the theoretical curves are valid and the method sound.

I. INTRODUCTION

The concept of field-flow fractionation (FFF) was first published 10 years ago (1). The likelihood of useful applications to macromolecules and colloids was predicted at that time. The interim stages of development have revealed nothing to change that view. Studies of the last few years, particularly, have shown the applicability of FFF to a number of important prototype systems: linear polystyrene polymers, polystyrene latex beads, proteins, viruses and several other particulate materials (2). All present evidence suggests that the scope of the method can be greatly expanded, perhaps to include nearly all colloidal materials where fractionation and mass analysis are important considerations.

At first, FFF was envisioned primarily as a fractionating tool (1). However, *any fractionation device for which the retention spectrum is subject to characterization in terms of some physicochemical parameter can be forced to yield a spectrum of values of that parameter.* If the parameter is mass, then, of course, a mass spectrum can be obtained. The resolvability of the spectrum into its "fine structure" obviously hinges on the intrinsic resolution of the fractionation method.

Exclusion chromatography (EC)--including gel permeation and gel filtration chromatography--is a prominent example of the possibility of extracting size spectra from retention data. It is clear that EC fulfills the stated criterion: retention, by universal accord, is a function of size (3). However, "size" is an elusive concept, and the exact function of size that retention depends on has not been determined. Hydrodynamic volume has been found empirically to work fairly well as a universal size parameter (4); mean external length is a parameter suggested on theoretical grounds (5). However, it can be shown that neither one of these is truly universal, and that pore shape determines the discrepancies (5,6). (Discrepancies appear as unequal retention values for particles having unlike shape but equal values of the proposed universal "size" parameter.) It is likely that no single "size" function exists that will rigorously characterize the retention of diverse material in real EC columns. Only if pore configuration were perfectly controllable and moderately simple could such a goal be hoped for.

For components falling in a narrow class, of course, "size" is adequately definable. Thus molecular weight (MW) serves as a measure of size in a series of linear polystyrene polymers.

FFF is a much more tractable theoretical system than is EC. Separation occurs in an unobstructed channel of well defined dimensions (see Fig. 1). Both retention and peak dispersion (plate height) can be calculated rigorously for the model system (7). Of course, system nonidealities can be expected. However, retention measurements in many systems have been found to agree well with theory. Only peak dispersion parameters have persisted in defying theoretical

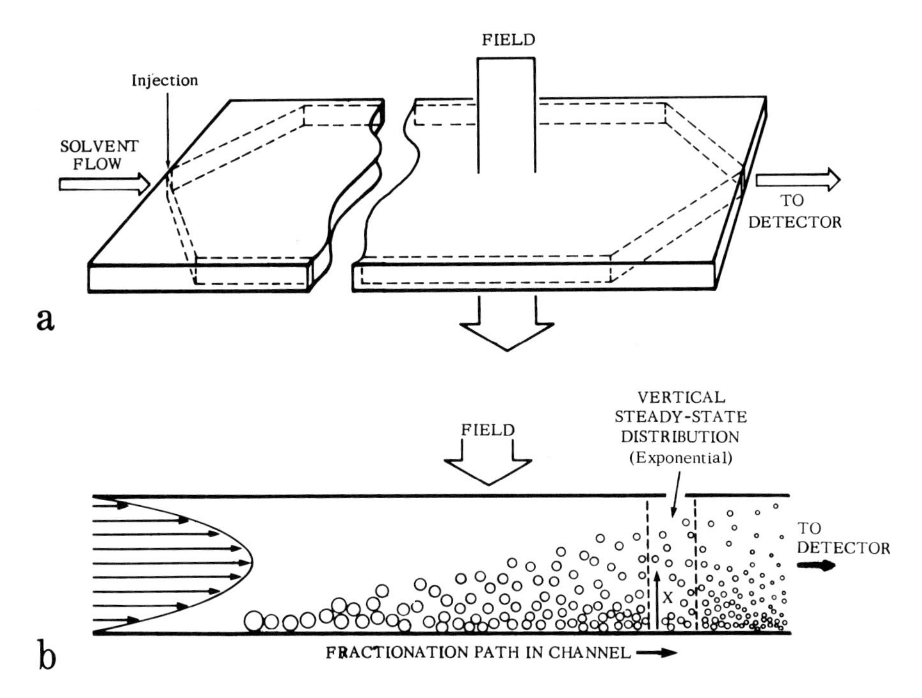

Fig. 1. a) General configuration of FFF channel and the perpendicular field. b) Edge view of channel, showing particle size distribution during fractionation process.

prediction; peak widths are generally much broader than expected. If this discrepancy can be removed, the resolving power of FFF will expand much beyond its present realization.

Mass analysis is extracted from retention data rather than peak dispersion data, and is therefore theoretically amenable. Retention in FFF is a predictable function of some physicochemical parameters (8). The specific parameters involved depend on the subtechnique employed, as will be shown shortly. The spectrum of the applicable parameter group for a heterogeneous material can therefore, in theory, be extracted from the retention (or elution) spectrum without any form of calibration. In practice, some use of calibration or of various internal standards will probably help sharpen the accuracy of the method.

Most of the parameters extractable from FFF are mass or size related, and can therefore yield mass-related spectra. The particular parameters available depend on the subtechnique employed, as will be established in the next section.

We note finally that, aside from its theoretical tractability, FFF has the advantages of being a one-phase system, relatively free of complicating absorptive interaction. The forces of retention are gentle and controllable, and shear degradation of samples is expected to be minimal (2).

II. PRINCIPLES AND THEORY OF FFF

The concept and the present experimental configuration of the FFF channel is illustrated in Fig. 1. Laminar flow is established along the axis of a flat channel. A field or gradient of some kind is applied perpendicular to the flat faces bounding the channel. The "field" forces the solute toward one face, or wall, which we arbitrarily designate as as the "lower wall." After about one relaxation period, a steady state is established between the field-induced flux and back diffusion. At this point, the vertical distribution is exponential

$$(c/c_0) = \exp(-x/\ell) \tag{1}$$

where c is the concentration of solute at distance x above the lower wall, c_0 is the wall concentration at $x = 0$, and ℓ is a characteristic distance sometimes termed the mean layer thickness.

Parameter ℓ is given by the simple expression (8)

$$\ell = \mathcal{D}/U \tag{2}$$

where $\mathcal{D}$ is the solute-solvent diffusion coefficient and U is the mean field-induced velocity. It is best expressed as the dimensionless ratio

$$\lambda = (\ell/w) = \mathcal{D}/Uw \tag{3}$$

where w is the width of the channel--the distance between the flat walls confining the flow.

An alternate form is available by writing $U = F/f$ and $\mathcal{D} = \mathfrak{R}T/f$, where F is the force or effective force induced by the field on a mole of solute, f is the friction coefficient, $\mathfrak{R}$ is the gas constant and T the temperature. The substitution of these two expressions into Equation 3 yields

$$\lambda = \mathfrak{R}T/Fw \tag{4}$$

which is simply the ratio of thermal energy $\mathfrak{R}T$ to the energy change suffered by solute in a full width displacement across the channel.

When we combine Equation 4 with $\ell = \lambda w$ and Equation 1, we get the Boltzmann form of the exponential distribution

$$(c/c_0) = \exp(-Fx/\Re T) \tag{5}$$

An obvious condition for retention--necessary for fractionation or any kind of mass discrimination--is that F be large enough to create a Boltzmann layer that is thin compared to the channel width: $Fw >> \Re T$, $w >> \ell$, or $\lambda << 1$. Not all fields are interactive enough with all solutes to meet this criteria, but one can expect to find an appropriate field for almost any solute system encountered.

The precise criterion for retention can be obtained from the fundamental retention Equation (8)

$$R = 6\lambda\,[\coth(1/2\lambda) - 2\lambda\,] = 6\lambda f(\lambda) \tag{6}$$

where retention ratio R is

$$R = V^0/V_r \tag{7}$$

which is the column void volume (or elution volume of a non-retained solute peak), V^0, divided by the retention volume of the solute fraction under study, V_r. A plot of R versus λ is shown in Fig. 2. Reasonable retention is represented by $R < 0.8$, or preferable $R < 0.5$ for improved resolution (9). This corresponds roughly to $\lambda < 0.25$ and $\lambda < 0.1$, respectively, as can be ascertained from Fig. 2.

Equation 6, then, serves to relate the observable retention ratio, R, to the basic mass-related parameter, λ.

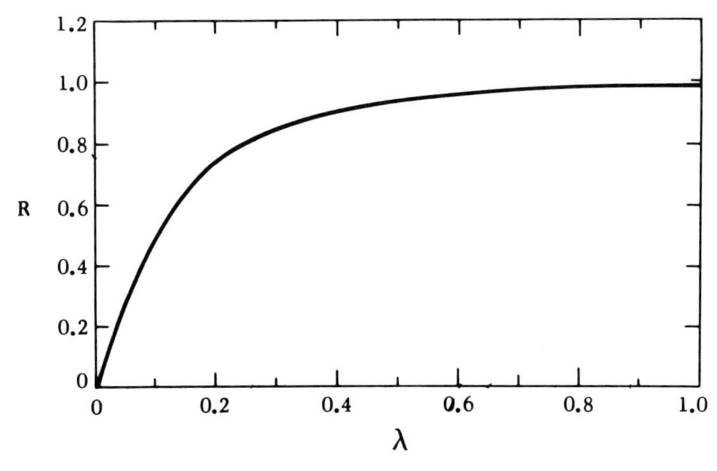

Fig. 2. Retention ratio R versus the basic, dimensionless parameter, λ. Curve is from Equation 6.

It represents the key linkage in extracting mass spectra from empirical retention. In combination with Equation 7, and the explicit relationship of λ to particle or molecular mass

$$\lambda = \lambda(M) \tag{8}$$

it provides a relationship of the form

$$V_r = V_r(M) \tag{9}$$

relating retention volume to the molecular weight, M, of the species eluted at the volume.

If one now acquires a certain empirical elution spectrum--a concentration versus V_r curve--one can write for the amount, mdM, of material in the molecular weight interval between and $M + dM$

$$mdM = c\ dV_r \tag{10}$$

which gives for the molecular weight distribution curve

$$m(M) = c\ dV_r(M)/dM \tag{11}$$

In practice, one should ordinarily allow for the finite dispersion of narrow molecular weight cuts in the column, which causes intermixing. Methods for this have been developed in the literature of gel permeation chromatography (10,11). The goal of this paper, however, is to establish the fundamental relationships of Equations 8 and 9--which is the essential step in ascertaining $dV_r(M)/dM$ in Equation 11. We will not, therefore, pursue the well known details of elution spectrum deconvolution in this paper.

We have noted that FFF has a number of distinctive subclasses, one for each type of field or gradient applied perppendicularly across the channel. Our laboratory has generated considerable data for four "field" types: sedimentation, crossflow, thermal gradients, and electrical fields. The four corresponding subtechniques are termed: sedimentation FFF, flow FFF, thermal FFF, and electrical FFF. Below, we discuss each of these successively, in an attempt to establish the appropriate form of Equation 8.

III. SEDIMENTATION FFF

Sedimentation FFF is perhaps simplest in concept, and it relates most directly to particle mass. However, the experimental implementation is somewhat difficult. It requires a low volume seal to conduct the fluid stream--being spun in a FFF channel coiled against the outside wall of a centrifuge basket--out to a stationary detector. The device constructed in this laboratory functions satisfactorily only up to 4200 rpm--about 1580 g, where g is the earth's gravity. However, this has been sufficient to handle polystyrene latex beads to less than 0.1 μ in diameter (12,13), and the larger viruses (14).

The force, F, acting on Avogadro's number of particles is $F = MG\,(\rho_s - \rho)/\rho_s$ where M is particle molecular weight, G is acceleration, ρ_s is solute density (more technically, the reciprocal partial molar volume), and ρ is solvent density. Substituted in Equation 4, this gives (12)

$$\lambda = \mathcal{R}T/Mgw(\Delta\rho/\rho_s) \qquad (12)$$

where $\Delta\rho$ is the difference between solute and solvent densities. This equation provides the key relationship, $\lambda = \lambda(M)$ of Equation 8.

An alternate form of Equation 12 can be obtained by writing $U = sG$, where s is the sedimentation coefficient. When combined with Equation 3, this gives

$$\lambda = \mathcal{D}/sGw \qquad (13)$$

This equation shows the dependence of λ on the $\mathcal{D}/s$ ratio, a ratio controlling sedimentation equilibrium distributions. Sedimentation FFF can be thought of as a method for differentially displacing equilibrium (or "steady-state") collections of unequal particles along a tube axis according to the individual magnitudes of the $\mathcal{D}/s$ ratio.

Equation 12 is most useful for our purposes; it shows the direct mass (molecular weight) dependence of λ, given the densities of solute and solvent. Its direct dependence, of course, is on effective mass (effective molecular weight), $M\Delta\rho/\rho_s$.

The retention spectrum of Equation 9, $V_r = V_r(M)$, can now be obtained by combining Equation 12 with Equations 6 and 7. We write it as the dimensionless ratio, V_r/V_0 which is the ratio of V_r to void volume V^0

$$(V_r/V^0) = MGw(\Delta\rho/\rho_s)/6\mathcal{R}T\,\mathcal{f}(\lambda) \qquad (14)$$

where $\mathcal{f}(\lambda)$ is the function of λ--often close to unity in practice--expressed in Equation 6. This equation shows that retention volume, V_r, is essentially linear in molecular weight, M, except under the unusual conditions of low reten-

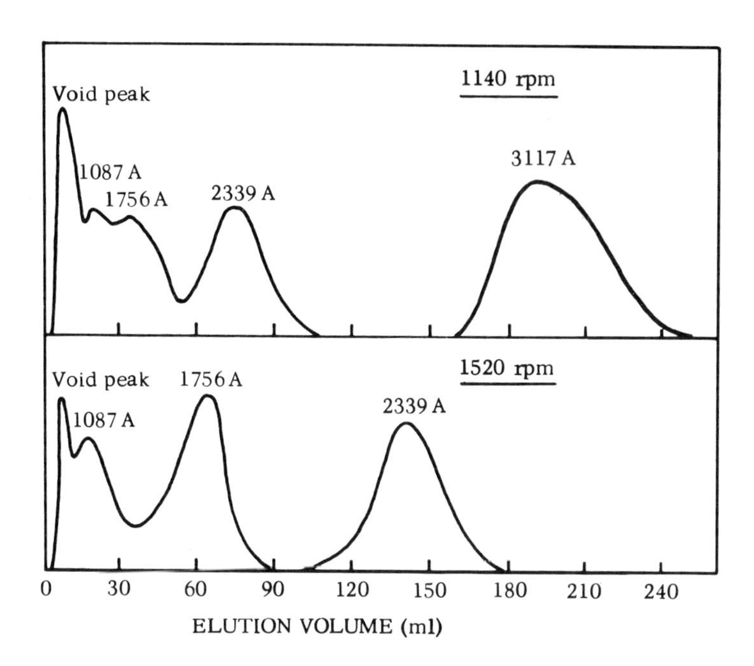

Fig. 3. Separation of polystyrene latex beads of the different diameters indicated. The larger field strength (or rpm) applied in the bottom case is seen to induce more retention and improved resolution for the 1087A and 1756A peaks (12).

tion, when $f(\lambda)$ is important. The linear form is a convenient elution spectrum for mass analysis, and it makes Equation 11 and various deconvolution methods as simple as possible.

Several experimental ties have already been made to verify the mass dependence of Equations 12 and 14. In one study, these equations were used to determine accurately and quickly the molecular weight of a homogeneous species, the T2 virus (14). The result obtained was $M = (236 \pm 7) \times 10^6$ grams/mole.

Sedimentation FFF has also been applied to mixtures of polystyrene latex beads (12,13). Fig. 3 illustrates a simple fractionation of different bead sizes, and shows how the elution spectrum can be controlled and changed through a variation of field strength (12).

The V_r/V^0 ratio of the peaks in Fig. 3 can be plotted as a function of M to yield points on an elution spectrum to test the validity of Equation 14. Figure 4 shows both the theoretical curve (Equation 14) and the experimental points. The M values for the spherical beads has been obtained from bead diameter, d, by the Equation (2)

$$M = (1/6)\ N\pi\rho_s d^3 = 0.31534\ \rho_s d^3 \qquad (15)$$

where, in later expressions, d is in Angstroms and ρ_s g/cm^3. Quantity N is Avodagro's number.

Figure 4 illustrates the excellent predictability, linearity and controllability of retention spectra in sedimentation FFF, and therefore, demonstrates the capabilities of this system to work in mass analysis without special calibration methods or the use of internal standards.

We note before proceeding to other methods that the sedimentation forces of gravity are acting in all the other subtechniques in which the column is arranged horizontally. (In some cases we have intentionally used vertical columns to avoid this.) However, the effect is not generally important for particles of less than 1-2 μm in diameter. When particle size or other conditions are such that two (or more) fields are indeed important, the force, F, may be considered an additive sum and the λ of Equation 4 becomes

$$\lambda = 1/\Sigma\ 1/\lambda_i \qquad (16)$$

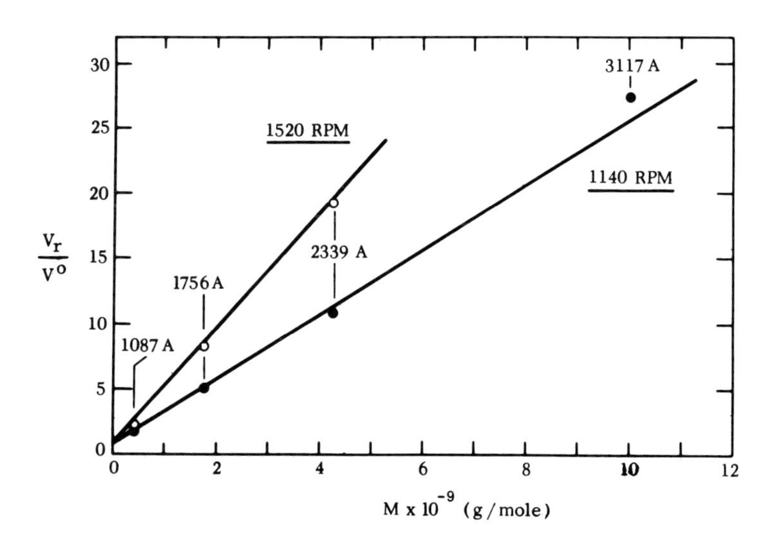

Fig. 4. Theoretical and experimental retention volume-molecular weight curves for the polystyrene latex beads separated in Fig. 3 by sedimentation FFF. The curve can be altered by the variation in field strength, as shown, making possible the choice of conditions to best fit the particulate material needing analysis.

IV. FLOW FFF

Flow FFF utilizes a lateral flow of fluid superimposed on the principal flow along the axis of the column. The cross-flow "field" is implemented by forcing fluid at a constant rate through semipermeable membranes that serve as the 2 principal bounding surfaces of the FFF channel (15,16).

The lateral velocity induced by the cross flow is simply the cross-flow velocity, U. This can be expressed as $\dot{V}_c/aL$, where $\dot{V}_c$ is the volumetric rate of cross flow, a is channel breadth and L channel length. With this, Equation 3, $\lambda = D/Uw$ becomes (16)

$$\lambda = \frac{DaL}{\dot{V}_c w} = \frac{DV^o}{\dot{V}_c w^2} \qquad (17)$$

where the latter expression is derived by writing column void volume, V^o, as the product of the three channel dimensions, awL.

All terms in the above equation are constant for all solutes, except diffusion coefficient, D. Therefore, fractionation is based solely on diffusivity differences. In as much as D relates to friction coefficients f by $D = \mathcal{R}T/f$, λ can be written in terms of

$$\lambda = \mathcal{R}TV^o/f\dot{V}_c w^2 \qquad (18)$$

Only for flow FFF does λ depend explicitly on f. With other subtechniques of FFF, D and U both vary inversely with f, and f therefore drops out of the ratio, D/U of Equation 3. Because U is rigorously constant for all solute species in flow FFF, an explicit f term remains left over from D.

Clearly, there is no universal relationship between f and particle mass, or molecular weight, M. However, for narrow geometrical classes, a precise relationship does often exist, and a mass spectrum can therefore be obtained within the class.

For spherical particles, f can be obtained from Stoke's equation (17)

$$f = 3\pi\eta d \qquad (19)$$

where η is the carrier viscosity and d the particle diameter. With this, Equation 18 becomes

$$\lambda = (\mathcal{R}TV^o/3\pi\eta\dot{V}_c w^2)(1/d) \qquad (20)$$

Finally in that the molecular weight, M, of a spherical

particle is related to d by means of Equation 15, $\lambda = \lambda(M)$ becomes

$$\lambda = \frac{RTV^0}{3\pi\eta V_c \omega^2} \left(\frac{N\pi\rho_s}{6}\right)^{1/3} \left(\frac{1}{M}\right)^{1/3} \quad (21)$$

With this, one can go directly to the retention-MW term $V_r = V_r(M)$ using Equations 6 and 7, as was done for sedimentation FFF (see Equation 14). It is not necessary to show this step again here.

Flow FFF has been recently applied to proteins (15), polystyrene latex beads (15,16), and, in unpublished work, viruses. Structural dissimilarities often exist in the first and the last, so that a realistic mass analysis can only be done with careful attention to structural difference. This point will be addressed later. However, we note here that we are now studying the retention of denatured proteins. These generally exist in a uniform, random-coil configuration that is definitely subject to mass analysis.

For the present we will apply the mass analysis concept only to our results for polystyrene latex beads. Fig. 5 shows the retention spectrum as predicted from Equations 21, 6 and 7, along with experimental points. As with sedimentation FFF, the agreement is good, suggesting that the method

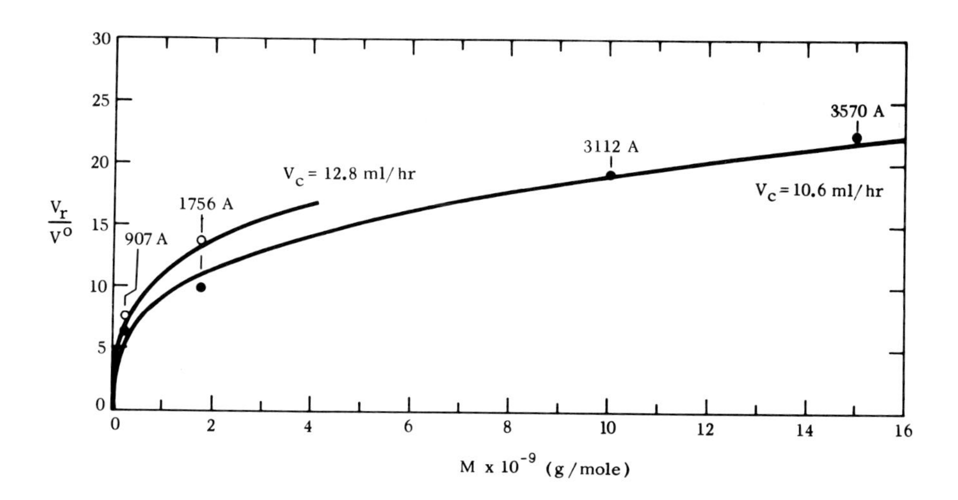

Fig. 5. Theoretical retention-MW curves and experimental points for polystyrene latex beads using flow FFF at two different field strengths, V_r. Experimental conditions are detailed in the paper from which the data have been obtained (16).

is applicable to spherical particles without calibration or standardization.

Nonspherical particles of irregular geometry cannot be treated rigorously, as noted. However, the possibility exists for approximate mass analyses if geometrics do not vary too widely. In many particulate systems, individual particles are generally of similar shape. If we assume that the particles are prolate ellipsoids, the variation in f with axis ratio a/b for particles of equal volume (mass) can be calculated (17). The dependence of f on a/b is weak; from Tanford's plot, we arrive at the following rough approximation for f compared to the value for spheres, fo

$$(f/fo) = 1 + 0.06 \left[\frac{a}{b} - 1 \right] \tag{22}$$

which is valid up to $(a/b) \sim 10$. Thus, for example, if most particles have a/b ratios that lie in the range 1.5 - 2.5, the f/fo ratio would vary only between approximately 1.03 and 1.09, or 1.06 ± 0.03. This 3 percent variation would be negligible compared to the dispersion occurring in most fractionating devices, including those of FFF. Therefore, we conclude that flow FFF could be used quite successfully for the mass analysis of such collections of particles. The expression used would simply be a modification of Equation 21 obtained by employing the mean value, $(\overline{f/fo})$. Thus,

$$\lambda = \frac{\mathcal{R}TV^{0}}{3\pi\eta\dot{V}_{c}w^{2}(\overline{f/fo})}\left(\frac{N\pi\rho_{s}}{6}\right)^{1/3}\left(\frac{1}{M}\right)^{1/3} \tag{23}$$

The remainder of the procedure would follow as before.

V. THERMAL FFF

Thermal FFF is based on the thermal diffusion phenomenon. It is implemented by using the flat side of a heated, metallic bar for one channel surface, and a cooled bar for the other. The details have been described (18). Thermal FFF was the first FFF developed (19); its versatility has increased considerably in recent years (20,21).

The λ value of thermal FFF is obtained by applying the steady-state condition to thermal diffusion and ordinary fluxes (18,22). The equation is

$$\lambda = 1/\left(\frac{\alpha}{T} + \gamma\right)\frac{dT}{dx}w \tag{24}$$

where α is the thermal diffusion factor, γ is the cubical coefficient of thermal expansion of the solution and dT/dx is the temperature gradient. Parameter γ is usually negligible.

Unfortunately, special complications arise in thermal FFF due to the temperature gradient. First, the temperature dependence of viscosity destroys the symmetry of the flow profile and forces a minor (but mathematically complicated) correction to Equation 6 (18). Second, the temperature dependence of heat conductivity leads to a nonconstant dT/dx across channel with w. Thus the effective dT/dx differs from the overall gradient, $\Delta T/w$, by a few percent, and must be corrected for accurate work (23).

A final consideration unique to thermal FFF is that the key parameter, α, has not been very well characterized in the literature regarding its dependence on molecular weight M. In fact, the most extensive studies have been made analyzing the results of thermal FFF itself (23). This has made available a functional M dependence.

While these complications seem oppressive, the final results of thermal FFF studies are quite promising. By combining a series of approximations developed in earlier work, we propose using the following equation for the elution volume-molecular weight characterization curve.

$$\frac{V_r}{V^o} = \frac{1}{6\lambda_a[\coth(1/2\lambda_a) - 2\lambda_a]} \quad (25)$$

which is equivalent to combining Equations 6 and 7, except λ_a is the special, approximate form

$$\lambda_a = \frac{T_c(1 - 0.12\ B\Delta T/T_c^2)\ M^{-s}}{\Delta T[1 + (\Delta T/2k_c)(dk/dT)](q - rT_c)} \quad (26)$$

where B is a viscosity correction term (18) and k is the thermal conductivity of the liquid. Subscript c indicates a value of the parameter at the cold wall temperature. Constants q, r and s are to be determined for each solute-solvent system, until more is known about thermal diffusion in polymer systems.

For polystyrene in ethyl benzene under typical laboratory conditions, the following parameters apply: $(1/2k_c) \cdot (dk/dT) = 9.22 \times 10^{-4}/°C$, $B = 1095°$, $T_c = 295°$, $q = 0.224$, $r = 0.00050$, and exponent $s = 0.6$. With these parameters, we calculate the two retention-MW plots shown in Fig. 6. These agree well with the experimental points. Equation 26 will need to be tested for a wider range of solutes and solvents to establish its general validity.

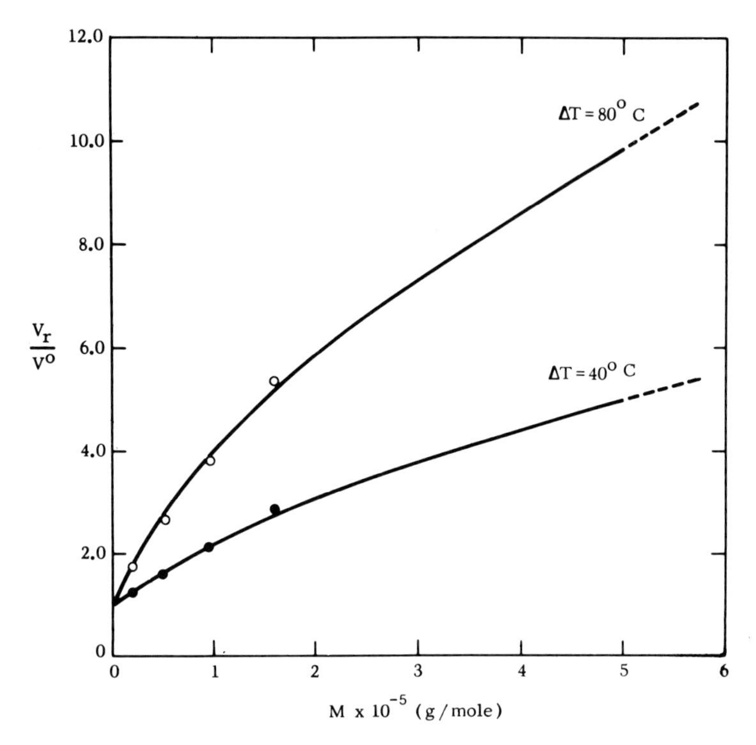

Fig. 6. Retention-MW curves for two ΔT values in thermal FFF according to Equations 25 and 26. The experimental points (18) are in good agreement.

VI. PROGRAMMED THERMAL FFF

The field strength in FFF can be made to vary continuously during a run. If it is gradually decreased, the more massive species that are at first pinned tightly against the wall will be gradually released into a state of active migration. Therefore, the mass range of a single run can be greatly extended by gradually allowing the migration of heavier and heavier species.

Programmed FFF, as outlined above, has been applied both to thermal FFF (21) and sedimentation FFF (13). The mathematics of elution is made more complicated by programming, and it is difficult to obtain explicit equations for retention volume that are accurate. At present, V_r values must be acquired numerically. Nonetheless, full numerical retention-MW plots can be generated under a wide range of conditions, and can, in theory, be applied to mass analysis by the same procedures as we have already presented.

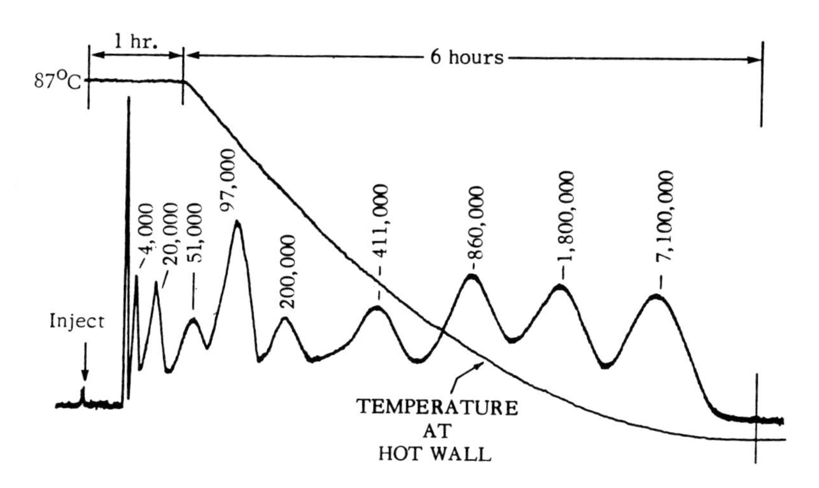

Fig. 7. Programming extends the range of FFF over wide molecular weight limits, as shown by this programmed thermal FFF fractogram of polystyrene polymers.

We have these programmed thermal FFF here to illustrate the general potential of programmed FFF. Fig. 7 illustrates the wide range and good resolving power made possible by programming methods (21). The mass range covered in this single run is 1775 - fold: from M = 4000 to M = 7, 100,000. Obviously, programming promises to be most useful for materials very dispersed in molecular weight.

The retention-MW plot corresponding to Fig. 7 is shown as the upper line in Fig. 8--that is labeled "1 hr time lag". This terminology refers to an interval of time in which ΔT is held at its initial value, ΔT_0, before programming. A curve is also shown for "no time lag". The scale is logarithmic to accommodate the wide range. The experimental points are seen to be in good agreement with the curves. These results are for the so-called parabolic program

$$\Delta T = \Delta T_0[1 - (t/t_p)^2] \tag{27}$$

where t_p is the total program time. A linear program has been found to provide somewhat inferior resolution and also inferior agreement with the theoretical elution spectrum (21).

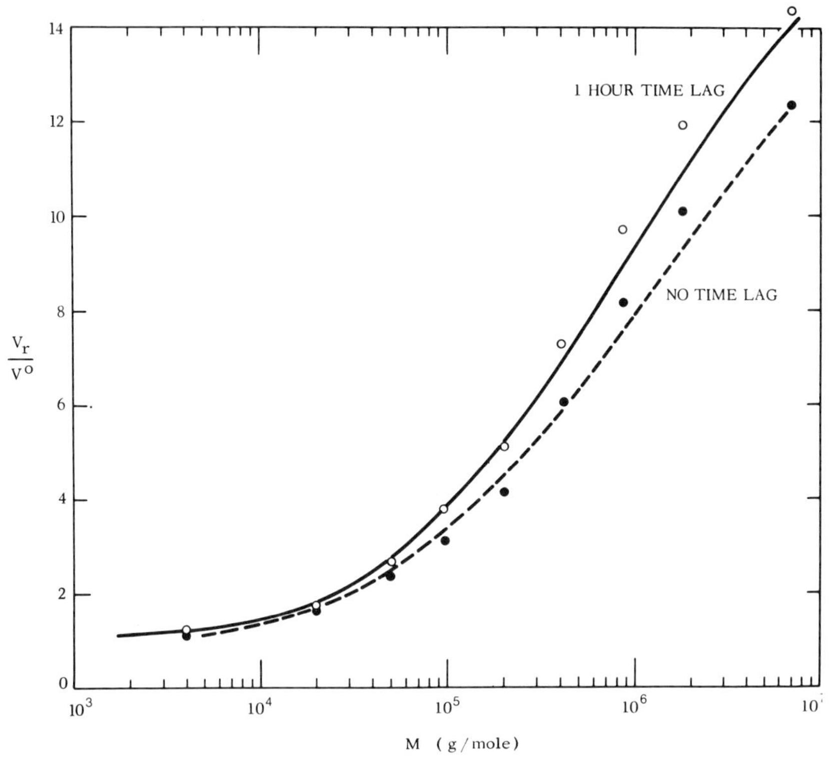

Fig. 8. Calculated retention-MW curves and experimental points for parabolic programmed thermal FFF. Notice logarithmic scale. Time lag is period at beginning before program begins.

VII. ELECTRICAL FFF

Electrical FFF requires species with an electrical charge. Semipermeable membranes are used for the bounding surfaces in order to permit the passage of current, in the form of small ions, between electrodes external to the column.

If velocity, U, is written as μE, equation 3 yields

$$\lambda = \mathcal{D}/\mu E w \tag{28}$$

It has long been known that μ is nearly constant for species of the same chemical composition (surface charge) differing only in particle size (24). This phenomenon has hindered electrophoretic resolution, which requires unequal values of μ. Retention and separation in electrical FFF, by contrast, hinges on the ratio, $\mathcal{D}/\mu$, as shown by the equation. If μ is constant or nearly so, retention will depend almost entirely on $\mathcal{D}$. In this case, retention characteristics and mass analysis will be based on the same factors applicable in flow FFF. As in flow FFF, for example, it should be possible to denature proteins, in which case $\mathcal{D}$ varies $\sim 1/M^{0.5}$. One would, of course, determine the coefficient of M precisely, and perhaps account, too, for the weak dependence of μ on M. With these precautions taken, a good mass analysis should be possible by the procedures detailed earlier in this paper.

Spherical particles, in which D varies as $1/M^{1/3}$, should similarly be tractable to mass analysis.

Electrical FFF has been applied to proteins (25-27), and to polystyrene beads in preliminary studies. However, no data are presently available to check the theoretical elution spectrum as it relates to mass.

VIII. CONCLUSIONS

This paper serves to establish the predictable elution of various types of components as a function of mass (molecular weight). This, then, establishes, under a wide variety of conditions and using a number of subtechniques, the primary criterion stated in the introduction for successful mass analysis, which can be paraphrased: The retention spectrum must be subject to characterization in terms of component mass.

As a result of this successful characterization-combined with the wide applicability and the favorable characteristics of FFF--it is anticipated that many applications of the mass analysis potential of FFF will emerge in future work.

IX. ACKNOWLEDGMENT

This investigation was supported by National Science Foundation CHE74-05260 A04.

X. REFERENCES

1. Giddings, J.C., Sep. Sci., 1, 123 (1966).
2. Giddings, J.C. J. Chromatog., in press.
3. Billmeyer, F.W., Jr., Altgelt, K.J., in "Gel Permeation Chromatography," (K.H. Altgelt and L. Segal, eds.), p. 3. Marcel Dekker, New York, 1971.

4. Benoit, H., Grubisic, Z., Rempp, P., Decker, D., and Zilliop, J.G., J. Chim Phys., 63, 1507 (1966).
5. Giddings, J.C., Kucera, E., Russell, C.P., and Myers, M.N., J. Phys. Chem., 72, 4397 (1968).
6. Cassasa, E.F., J. Polym. Sci., Part B, 5, 773 (1967).
7. Giddings, J.C., Yoon, Y.H., Caldwell, K.D., Myers, M.N., and Hovingh, M.E., Sep. Sci., 10, 447 (1975).
8. Grushka, E., Caldwell, K.D., Myers, M.N., and Giddings, J.C., in "Separation and Purification Methods," (E.S. Perry, C.J. Van Oss, and E. Grushka, Eds.) Vol. 2. Marcel Dekker, New York, 1974.
9. Giddings, J.C., Smith, L.K., and Myers, M.N., Anal. Chem., 47, 2389 (1975).
10. Tung, L.H., loc. cit., ref. 3. p. 73.
11. Duerksen, J.H., ibid., p. 81.
12. Giddings, J.C., Yang, F.J.F., and Myers, M.N., Anal. Chem., 46, 1917 (1974).
13. Yang, F.J.F., Myers, M.N., and Giddings, J.C., Anal. Chem., 46, 1924 (1974).
14. Giddings, J.C., Yang, F.J.F., and Myers, M.N., Sep. Sci., 10, 133 (1975).
15. Giddings, J.C., Yang, F.J., and Myers, M.N., Science, in press.
16. Giddings, J.C., Yang, F.J., and Myers, M.N., Anal. Chem., in press.
17. Tanford, C., "Physical Chemistry of Macromolecules," Ch. 6. John Wiley, New York, 1961.
18. Myers, M.N., Caldwell, K.D., Giddings, J.C., Sep. Sci., 9, 47 (1974).
19. Thompson, G.H., Myers, M.N., Giddings, J.C., Sep. Sci., 2, 797 (1967).
20. Giddings, J.C. Smith, L.K., and Myers, M.N., Anal. Chem., 47, 2389 (1975).
21. Giddings, J.C., Smith, L.K., and Myers, M.N., Anal. Chem., in press.
22. Hovingh, M.E., Thompson, G.H., and Giddings, J.C., Anal. Chem., 42, 195 (1970)
23. Giddings, J.C., Caldwell, K.D., and Myers, M.N., Macromolecules, 9, 106 (1976).
24. Audubert, R., and de Mende, S., "The Principles of Electrophoresis," p. 23. Macmillan, New York, 1960.
25. Caldwell, K.D., Kesner, L.F., Myers, M.N., and Giddings, J.C., Science, 176, 296 (1972).
26. Kesner, L.F., Caldwell, DK.D., Myers, M.N., and Giddings, J.C., Anal. Chem., in press.
27. Giddings, J.C., Lin, G.C., and Myers, M.N., Sep. Sci., in press.

EXPERIMENTAL STUDY OF NONLINEAR SURFACE STRESS-DEFORMATION BEHAVIOR WITH THE DEEP CHANNEL SURFACE VISCOMETER

Lun-yan Wei and John C. Slattery
Northwestern University

I. ABSTRACT

The surface stress-deformation behaviors of three liquid-air interfaces are calculated from measured velocity distributions in these interfaces. The liquids studied are distilled water, a 6% aqueous solution of potassium oleate, and a n-octadecanol monolayer (20.5 Å^2/molecule) over distilled water.

For the interface between air and distilled water, the surface stress is independent of the surface rate of deformation; the surface viscosity is zero.

For the interface between air and a 6% aqueous potassium oleate solution, the surface stress is a linear function of the surface rate of deformation; the surface viscosity is 1.74×10^{-4} dyne sec/cm.

For the interface between air and a n-octadecanol monolayer (20.5 Å^2/molecule) over distilled water, the measured velocity distribution is not described well by assuming that the surface stress is a linear function of the surface rate of deformation. Instead, an analysis suggested by Hegde and Slattery (27) is used to calculate the apparent surface shear viscosity of the phase interface as a function of the surface rate of deformation. The apparent surface shear viscosity decreases as a function of the absolute value of the surface rate of deformation in a manner analogous to that observed with bulk stress-deformation behavior.

II. INTRODUCTION

The majority of the literature concerned with interfacial stress-deformation behavior has focused on the linear Boussinesq surface fluid model (1-3)

$$\tilde{T}^{(\sigma)} = \gamma \tilde{P} + (\kappa - \varepsilon)\left(\mathrm{div}_{(\sigma)}\, \tilde{v}^{(\sigma)}\right) \tilde{P} + 2\,\varepsilon\, \tilde{D}^{(\sigma)} \qquad [1]$$

which describes the surface stress tensor $\tilde{T}^{(\sigma)}$ as a linear function of the surface rate of deformation tensor

$$\tilde{D}^{(\sigma)} \equiv \frac{1}{2}\left[\tilde{P} \cdot \nabla_{(\sigma)}\, \tilde{v}^{(\sigma)} + \left(\nabla_{(\sigma)}\, \tilde{v}^{(\sigma)}\right)^{T} \cdot \tilde{P}\right] \qquad [2]$$

Here γ is the thermodynamic surface tension, $\tilde{P}$ is the projection tensor that transforms any vector on the interface into its tangential component, κ is the surface dilatational viscosity, ε is the surface shear viscosity, $\tilde{v}^{(\sigma)}$ is the surface velocity, $\nabla_{(\sigma)}$ denotes the surface gradient operation, and $\mathrm{div}_{(\sigma)}$ denotes the surface divergence operation. [For a further discussion of the notation used here, see (4, Appendix) and (5, footnote 1).]

Joly (6,7) has summarized approximate values of the surface shear viscosity ε for a wide variety of interfaces. For clean interfaces, the surface shear viscosity is generally small, perhaps zero.

Several experiments have been proposed for measuring the surface shear viscosity at liquid-gas interfaces. The most widely used device at this time is the deep channel surface viscometer shown in Fig. 1 (8-11). The walls are stationary concentric cylinders; the floor of the viscometer moves with a constant angular velocity. The experimentalist is required

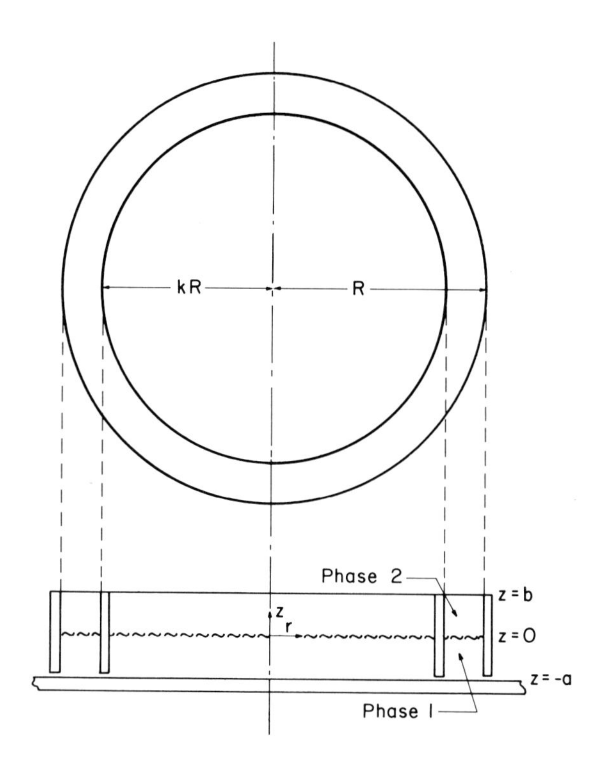

Fig. 1. Deep Channel surface viscometer. The walls of the channel are stationary; the floor rotates with a constant angular velocity Ω .

to measure the velocity of one or more particles in the interface in order to estimate the center-line surface velocity. Other geometries for which suitable analyses have appeared recently are the circular knife-edge surface viscometer (12,13), the blunt knife-edge viscometer (5,14), the disk viscometer (5,15,16), and the rotating wall knife-edge viscometer (17-19).

There have been three suggestions for using the deep channel geometry in measuring the surface shear viscosity at liquid-liquid interfaces (20-22).

Only a few measurements have been reported for a linear

combination of the surface shear viscosity ε and the surface dilatational viscosity κ (4,23).

It is possible to measure a linear combination of the two surface viscosities at a gas-liquid interface by studying the damping characteristics of capillary waves (23). In our opinion it is not an entirely satisfactory technique, since the assumptions necessarily made in describing mass transfer to, from, and within the phase interface obscure the final result.

A linear combination of the surface viscosities at a liquid-liquid interface can be measured by studying the velocity distribution in a drop rotating in a constant shear field (4). The effects of mass transfer drop out in the limit for which the analysis applies.

In the presence of large surfactant molecules, the surface shear viscosity may not be a constant (11, 24-26). These data may be described in terms of a generalized Boussinesq surface fluid model (27) represented by Eq. [1] with the two apparent surface viscosities now understood to be functions

$$\kappa = \kappa\ (\mathrm{div}_{(\sigma)}\ \underset{\sim}{v}^{(\sigma)},\ D^{(\sigma)}) \qquad [3]$$

$$\varepsilon = \varepsilon\ (\mathrm{div}_{(\sigma)}\ \underset{\sim}{v}^{(\sigma)},\ D^{(\sigma)}) \qquad [4]$$

of two scalar invariants of the surface rate of deformation tensor:

$$\mathrm{div}_{(\sigma)}\ \underset{\sim}{v}^{(\sigma)} = \mathrm{tr}\ \underset{\sim}{D}^{(\sigma)} \qquad [5]$$

$$D^{(\sigma)} \equiv \left[\frac{1}{2}\ \mathrm{tr}(\underset{\sim}{D}^{(\sigma)} \cdot \underset{\sim}{D}^{(\sigma)})\right]^{1/2} \qquad [6]$$

Three suggestions have been made for studying nonlinear

behavior at liquid-gas interfaces with the deep channel surface viscometer shown in Fig. 1. Mannheimer and Schechter (20) analyze this flow with the assumption that the interfacial stress-deformation behavior can be described by a Bingham plastic model in which no flow is possible until a critical interfacial stress has been exceeded. Pintar et al. (11) give an approximate analysis in which a Powell-Eyring model is used for the interface. Hegde and Slattery (27) illustrate how the form of Eq. [4] can be determined from measurements of the velocity distribution in the interface, assuming only that the generalized Boussinesq model applies.

In what follows, we demonstrate the practicality of Hegde and Slattery's (27) data analysis. We begin by correcting their solution to account for viscous effects in the gas phase. It is important to account for the viscous stress exerted upon the phase interface by the gas phase when measuring small values of the surface viscosity. The analysis is then applied to three liquid-air systems. The liquids used are distilled water, a 6% aqueous solution of potassium oleate, and a n-octadecanol monolayer (20.5 $\mathring{A}^2$/molecule) over distilled water.

III. SOLUTION

Let us extend the suggestion of Hegde and Slattery (27), taking into account the viscous behavior of the gas phase.

We make the following assumptions.

i) Both the liquid and the gas phases may be described as incompressible, Newtonian fluids.

ii) The surface stress-deformation behavior is described by the generalized Boussinesq surface fluid model.

iii) Inertial effects may be neglected.

iv) The interface is flat.

v) Any surfactant present is uniformly distributed over the phase interface.

vi) There is no mass transfer across the interface.

We have the following boundary conditions imposed by the geometry. The walls and the ceiling of the annular channel are stationary,

$$at\ r = kR : \quad \underset{\sim}{v}^{(1)} = \underset{\sim}{v}^{(2)} = 0 \tag{7}$$

$$at\ r = R : \quad \underset{\sim}{v}^{(1)} = \underset{\sim}{v}^{(2)} = 0 \tag{8}$$

$$at\ z = b : \quad \underset{\sim}{v}^{(2)} = 0 \tag{9}$$

Here $\underset{\sim}{v}^{(1)}$ denotes the velocity distribution in the liquid phase and $\underset{\sim}{v}^{(2)}$ the velocity distribution in the gas phase. The floor of the channel moves with a constant angular velocity Ω ,

$$at\ z = -a : \quad v_{\theta}^{(1)} = r\,\Omega\ , \quad v_{r}^{(1)} = v_{z}^{(2)} = 0 \tag{10}$$

The tangential components of velocity are continuous across the liquid-gas phase interface

$$\begin{aligned} at\ z = 0 : \quad & v_{\theta}^{(1)} = v_{\theta}^{(2)} = v_{\theta}^{(\sigma)} \\ & v_{r}^{(1)} = v_{r}^{(2)} = v_{r}^{(\sigma)} \end{aligned} \tag{11}$$

but, since there is no mass transfer across the phase interface,

$$at\ z = 0 : \quad v_{z}^{(1)} = v_{z}^{(2)} = 0 \tag{12}$$

By $\underset{\sim}{v}^{(\sigma)}$ we mean the velocity distribution in the liquid-gas interface. Because of assumptions v and vi , the jump mass balance at the phase interface (the equation of continuity for the interface; 2, 28) requires

$$at\ z = 0 : \quad \text{div}_{(\sigma)}\ \underset{\sim}{v}^{(\sigma)} = 0 \tag{13}$$

In a standard analysis of this geometry, we would impose the jump balance at the phase interface (28, Eq. 7.9) as a boundary condition to be satisfied and solve for the velocity distribution at the phase interface. Here we assume instead that we have measured the θ component of the velocity distribution in the phase interface as $g(r)$,

$$\text{at } z = 0 : \quad v_\theta^{(\sigma)} = g(r) \tag{14}$$

At the conclusion of our discussion, we will use the jump balance to determine the surface stress distribution in the phase interface.

The boundary conditions [7-13] suggest that we seek a velocity distribution that in each phase takes the form

$$v_\theta = v_\theta(r,z) \qquad v_r = v_z = 0 \tag{15}$$

in the cylindrical coordinate system suggested in Fig. 1. This form of velocity distribution satisfies Eq. [13] and the equation of continuity (29) identically. The equation of motion for an incompressible Newtonian fluid (29) implies

$$\frac{\partial}{\partial r^*}\left[\frac{1}{r^*}\frac{\partial}{\partial r^*}(r^* v_\theta^*)\right] + \frac{\partial^2 v_\theta^*}{\partial z^{*2}} = 0 \tag{16}$$

where we have introduced as dimensionless variables

$$v_\theta^* \equiv \frac{v_\theta}{R\Omega} \tag{17}$$

$$r^* \equiv \frac{r}{R(1 - k)} \tag{18}$$

$$z^* \equiv \frac{r}{R(1 - k)} \tag{19}$$

The solution to Eq. [16] that is consistent with boundary conditions [7-12] and [14] is

$$v_\theta^{(1)*} = \sum_{n=1}^{\infty} \left[A_n \sinh(\lambda_n z^*) + B_n \cosh(\lambda_n z^*) \right] \phi_1(\lambda_n r^*) \qquad [20]$$

$$v_\theta^{(2)*} = \sum_{n=1}^{\infty} B_n \left[-\coth(\lambda_n b^*) \sinh(\lambda_n z^*) + \cosh(\lambda_n z^*) \right] \phi_1(\lambda_n r^*) \qquad [21]$$

where

$$A_n = B_n \coth(\lambda_n a^*) - F_n \operatorname{csch}(\lambda_n a^*) \qquad [22]$$

$$B_n = \frac{2\,(1-k)^2\, g_n}{\left[\phi_0\left(\frac{\lambda_n}{1-k}\right)\right]^2 - k^2 \left[\phi_0\left(\frac{\lambda_n k}{1-k}\right)\right]^2} \qquad [23]$$

$$F_n = \frac{2\,(1-k)\left[k^2\,\phi_0\left(\frac{\lambda_n k}{1-k}\right) - \phi_0\left(\frac{\lambda_n}{1-k}\right)\right]}{\lambda_n\left\{\left[\phi_0\left(\frac{\lambda_n}{1-k}\right)\right]^2 - k^2\left[\phi_0\left(\frac{\lambda_n k}{1-k}\right)\right]^2\right\}} \qquad [24]$$

$$g_n \equiv \int_{\frac{k}{1-k}}^{\frac{1}{1-k}} r^*\, \phi_1(\lambda_n r^*)\, g^*(r^*)\, d\,r^* \qquad [25]$$

$$\phi_p(\lambda_n r^*) \equiv Y_1\left(\frac{\lambda_n k}{1-k}\right) J_p(\lambda_n r^*) - J_1\left(\frac{\lambda_n k}{1-k}\right) Y_p(\lambda_n r^*) \qquad [26]$$

The λ_n $(n = 1,2,\ldots)$ are the roots of (30)

$$\phi_1\left(\frac{\lambda_n}{1-k}\right) = 0 \qquad [27]$$

If we take our surface coordinates as r and θ, it follows that there is only one nonzero component of the dimensionless surface rate of deformation tensor

$$D_{r\theta}^{(\sigma)*} \equiv \frac{(1-k)}{\Omega} D_{r\theta}^{(\sigma)} = \frac{r^*}{2} \frac{d}{dr^*}\left(\frac{g^*}{r^*}\right) \qquad [28]$$

and only one nonzero component of the dimensionless surface shear stress

$$T_{r\theta}^{(\sigma)*} \equiv \frac{T_{r\theta}^{(\sigma)}}{\mu^{(1)} R\Omega} \qquad [29]$$

The jump momentum balance at the phase interface (the equation of motion for the interface; 28, Eq. 7.9) requires

$$\text{at } z^* = 0 : \frac{1}{r^{*2}} \frac{d}{dr^*}\left(r^{*2} T_{r\theta}^{(\sigma)*}\right)$$

$$= \frac{\partial v_\theta^{(1)*}}{\partial z^*} - \frac{\mu^{(2)}}{\mu^{(1)}} \frac{\partial v_\theta^{(2)*}}{\partial z^*} \qquad [30]$$

Let us define c^* to be that value of r^* such that

$$\text{at } r^* = c^* : \frac{d}{dr^*}\left(\frac{g^*}{r^*}\right) = 0 \qquad [31]$$

It follows from Eqs. [1] and [28] that

$$\text{at } r^* = c^* : T_{r\theta}^{(\sigma)*} = 0 \qquad [32]$$

In view of Eqs. [19] and [20], we can integrate Eq. [29] consistent with boundary condition [31] to find

$$T_{r\theta}^{(\sigma)^*} = \sum_{n=1}^{\infty} \left[A_n + B_n \frac{\mu^{(2)}}{\mu^{(1)}} \coth(\lambda_n b^*) \right] \left[\phi_2(\lambda_n r^*) - \left(\frac{c^*}{r^*}\right)^2 \phi_2(\lambda_n c^*) \right] \quad [33]$$

We now are prepared to use the experimental surface velocity distribution $g^*(r^*)$ in Eqs. [28] and [33] in order to determine from Eq. [1] the apparent surface shear viscosity

$$\varepsilon = \mu^{(1)} (1 - k) R \frac{T_{r\theta}^{(\sigma)^*}}{2 D_{r\theta}^{(\sigma)^*}} \quad [34]$$

as a function of the magnitude of the surface rate of deformation tensor

$$D^{(\sigma)} = \frac{\Omega}{1 - k} \frac{r^*}{2} \left| \frac{d}{d r^*} \left(\frac{g^*}{r^*}\right) \right| \quad [35]$$

under conditions such that $\text{div}_{(\sigma)} \underset{\sim}{v}^{(\sigma)} = 0$.

An experiment gives us the surface velocity distribution in the form of discrete measurements rather than a continuous function $g(r)$ (see for example Fig. 5). From Eqs. [11], [20], and [21],

$$v_\theta^{(\sigma)^*} = \sum_{n=1}^{\infty} B_n \phi_1(\lambda_n r^*) \quad [36]$$

which suggests that we assume

$$g^*(r^*) = \sum_{n=1}^{N} \tilde{B}_n \phi_1(\lambda_n r^*) \quad [37]$$

The coefficients $\tilde{B}_n$ can be determined by a least-square-error fit of Eq. [37] to the experimental data. For the optimum value of N, the calculated values of ε as a function of $D^{(\sigma)}$ corresponding to positive values of $D_{r\theta}^{(\sigma)}$ (r less than the center-line radius) fall smoothly on the corresponding curve for negative values of $D_{r\theta}^{(\sigma)}$ (r greater than the center-line radius) as in Fig. 6. One advantage to this form for $\underset{\sim}{g}(r)$ is that we can immediately identify

$B_n = B_n$.

IV. EXPERIMENTAL TECHNIQUE

Our deep channel surface viscometer is very similar to that used by Mannheimer (10,31). The radius of the inner channel wall is 4.990 cm ; the radius of the outer channel wall is 6.450 cm. The values of a and b depend upon the position of the interface in a particular experiment. The details of the design are given elsewhere (26).

In order to determine the velocity distribution $g(r)$, a small number of hollow glass spheres (Eccospheres FTFIS, Emerson and Cuming Inc., Canton, Mass.) roughly ten microns in diameter, were distributed in the phase interface. The radial position of a particular particle was noted on a scale in our 10 X wide field microscope. We were able to measure 1.52 cm. along the width of the channel with 0.0635 mm graduations. The time for this particle to make one complete revolution and to return to the same radial position was then observed.

An interfacial tension gradient in the interface caused the particle to shift its radial position. An interfacial tension gradient might develop in the immediate neighborhood of a particle due to contamination on the surface of the particle. In our case, the particles were spread on a large drop of distilled water before depositing them on the surface. A small amount of water was carried with the particles into the surface, causing them to slowly shift their radial positions. For this reason, we always ran the system overnight before taking any measurements. The maximum shift occuring during a measurement was 0.203 mm , about 1.4% of the channel width. Since the shifts were small, the average radial positions for such points were used.

It is essential that the interface be flat for these measurements. The analysis given in the previous section assumes that the interface is flat. More important, if the interface is not flat, the hollow glass spheres will migrate to the walls or to the center of the channel. For this reason, we machined a 0.254 mm step in each channel wall and we coated each wall with Teflon. In filling the channel, liquid was added until the liquid-gas-Teflon common line reached the sharp corner of the step. At that point, the liquid-gas interface was convex. Liquid was then slowly withdrawn, until the interface appeared to be flat. The common line remained attached to the sharp corner of the step. In this way, the contact angle boundary condition at the common line was avoided and a flat interface was achieved.

We have measured in this manner velocity distributions at three liquid-air interfaces. The liquids used are distilled water, a 6% aqueous solution of potassium oleate, and a n-octadecanol monolayer (20.5 $\mathring{A}^2$/molecule) over distilled water. The results are shown in Figs. 2 through 5. A tabular listing of the experimental data is given elsewhere (26).

V. ANALYSIS OF EXPERIMENTAL DATA

We recommend beginning the analysis of an experimental velocity distribution with the assumption that the surface shear viscosity is a constant. Under these conditions, the surface velocity distribution becomes (31, Eq. III-30)

$$v_\theta^{(\sigma)*} = \sum_{n=1}^{\infty} \frac{F_n \phi_1(\lambda_n r^*)}{\cosh(\lambda_n a^*) + \sinh(\lambda_n a^*)\left\{\varepsilon^* \lambda_n + \frac{\mu^{(2)}}{\mu^{(1)}} \coth(\lambda_n b^*)\right\}} \quad [38]$$

Given the measured center line velocity, we can use Eq. [38] to compute the corresponding dimensionless surface shear viscosity ε^*. With this value of ε^*, Eq. [38] can be employed again to calculate the corresponding surface velocity distribution. If the surface velocity distribution calculated in this manner agrees with the measured velocity distribution, the surface stress-deformation behavior may be described by the linear Boussinesq surface fluid model with a constant surface shear viscosity.

The velocity distributions measured at a distilled water-air interface are shown in Fig. 2 compared with Eq. [38], assuming that $\varepsilon = 0$. There is good agreement, with the exception of the data for $\Omega = 1.99$ rev/min. or for

$$N_{Re} \equiv \frac{2\rho^{(1)} \Omega R^2}{\mu^{(1)}} (1 - k^2) = 384 \quad [39]$$

We believe that these latter data are beginning to show the effects of inertia. Mannheimer (31) observed a similar effect at $N_{Re} = 260$ with a narrower channel.

Figs. 3 and 4 show that an interface between air and a 6% aqueous solutions of potassium oleate is described by the linear Boussinesq surface fluid model with $\varepsilon^* = 0.0117$ and $\varepsilon = 1.74 \times 10^{-4}$ dyne sec/cm. For comparison, the dashed

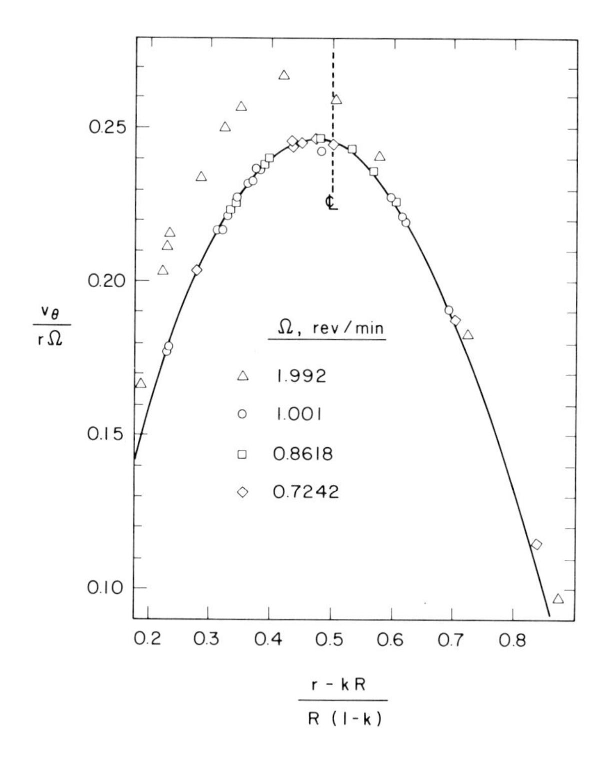

Fig. 2. Velocity distributions at a distilled water-air phase interface corresponding to several values of Ω *. In all experiments* $a = 1.07$ *cm* ($a^* = 0.731$), $b = 3.07$ *cm* ($b^* = 2.10$), *and* $\mu^{(2)}/\mu^{(1)} = 0.0182$ *. The solid curve is predicted from Eq.* [38] *for* $\varepsilon = 0$ *.*

curves in these figures represent the velocity distributions corresponding to $\varepsilon = 0$.

Fig. 5 presents the velocity distribution measured at an interface between air and an n-octadecanol monolayer (20.5 $\mathring{A}^2$/molecule) over distilled water. Given the center-line velocity and assuming that the surface stress-deformation behavior is described by the linear Boussinesq surface fluid model, Eq. [38] requires $\varepsilon^* = 0.414$ and $\varepsilon = 4.70 \times 10^{-3}$ dyne sec/cm. The corresponding surface velocity distribution

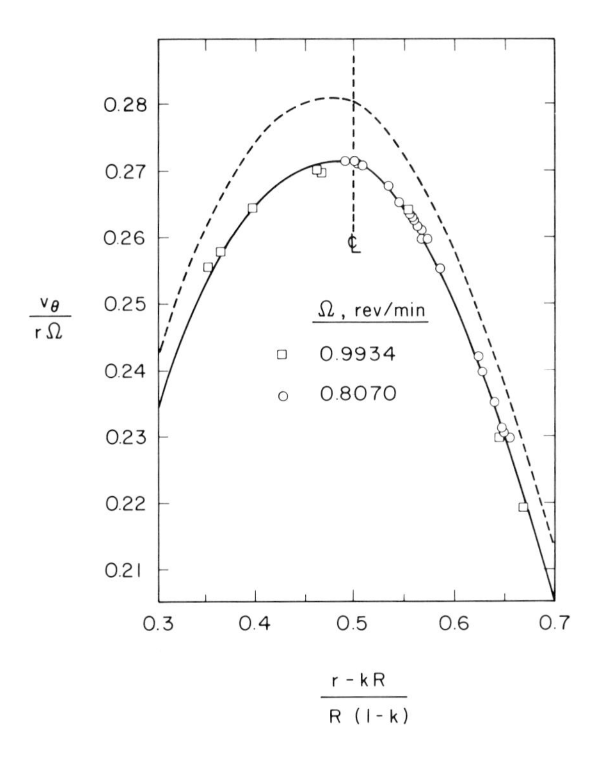

Fig. 3. Velocity distributions at an interface between a 6% aqueous solution of potassium oleate and air corresponding to two values of Ω . *In both experiments* $a = 100$ *cm* $(a^* = 0.687)$, $b = 3.14$ *cm* $(b^* = 2.15)$, *and* $\mu^{(2)}/\mu^{(1)} = 0.0182$. *The solid curve is predicted from Eq. [38] for* $\varepsilon = 1.74 \times 10^{-4}$ *dyne sec/cm ; the dashed curve is predicted for* $\varepsilon = 0$.

predicted by Eq. [38] is shown in Fig. 5 as the dashed curve. This interface is not well described by the linear Boussinesq surface fluid model. Its behavior is better described by the generalized Boussinesq surface fluid model, which allows the apparent surface shear viscosity to be a function of the surface rate of deformation.

When the interface is described by the generalized Boussinesq surface fluid model, the apparent surface shear

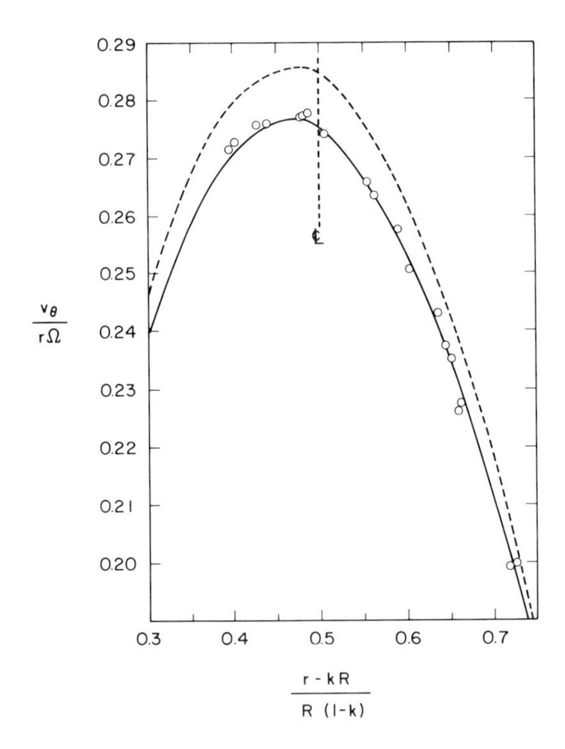

Fig. 4. Velocity distribution at an interface between a 6% aqueous solution of potassium oleate and air. In this experiment, $\Omega = 1.004$ *rev/min ,* $a = 0.996$ *cm* ($a^* = 0.682$), $b = 3.14$ *cm* ($b^* = 2.15$) *, and* $\mu^{(2)}/\mu^{(1)} = 0.0182$ *. The solid curve is predicted from Eq. [38] for* $\varepsilon = 1.74 \times 10^{-4}$ *dyne sec/cm ; the dashed curve is predicted for* $\varepsilon = 0$ *.*

viscosity must be determined from the calculated values of surface shear stress as a function of the surface rate of deformation. Success depends upon the accuracy with which the velocity distribution is determined, since the computation of the surface rate of deformation involves a derivative of the experimental velocity distribution. Some uncertainties were unavoidable in our experimental measurements. This probably explains why a direct least-square-error fit of Eq. [37] to the data shown in Fig. 5 did not give results that were con-

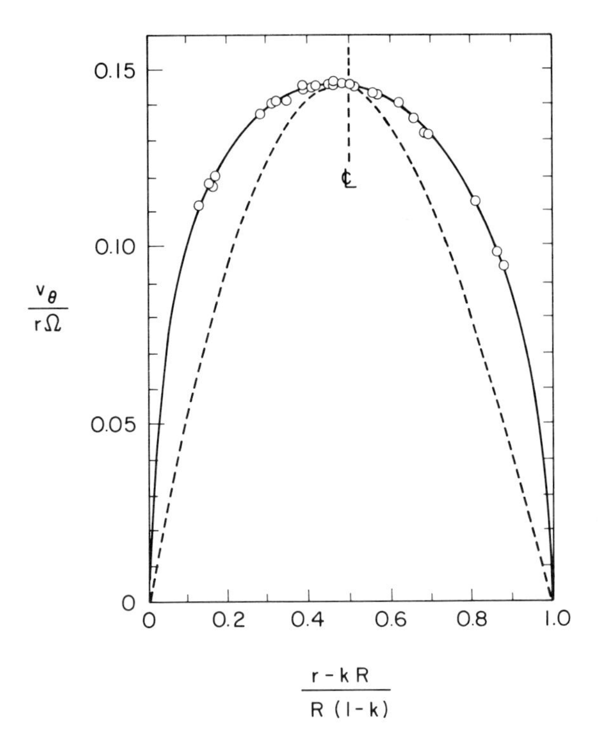

Fig. 5. Velocity distribution at an interface between air and a n-octadecanol monolayer (20.5 Å^2/molecule) over distilled water. In this experiment, $\Omega = 1.004$ *rev/min ,* $a = 0.996$ *cm* ($a^* = 0.682$), $b = 3.14$ *cm* ($b^* = 2.15$) *, and* $\mu^{(2)}/\mu^{(1)} = 0.0182$ *. The solid curve represents a least-square-error fit of Eq. [38] with* $N = 4$ *to the experimental data. The dashed line is predicted from Eq. [38] for* $\varepsilon = 4.70 \times 10^{-3}$ *dyne sec/cm .*

sistent for both positive and negative values of the surface rate of deformation. For this reason, we did a least-square-error fit of Eq. [37] to equally spaced points on a smooth curve drawn through these experimental data. The result for $N = 4$ is shown in Fig. 5. Equations [33-35] were used to compute $|T_{r\theta}^{(\sigma)}|$ and ε as functions of $D^{(\sigma)}$ (under conditions such that $\text{div}_{(\sigma)} \underset{\sim}{v}^{(\sigma)} = 0$). These functions are shown in Fig. 6.

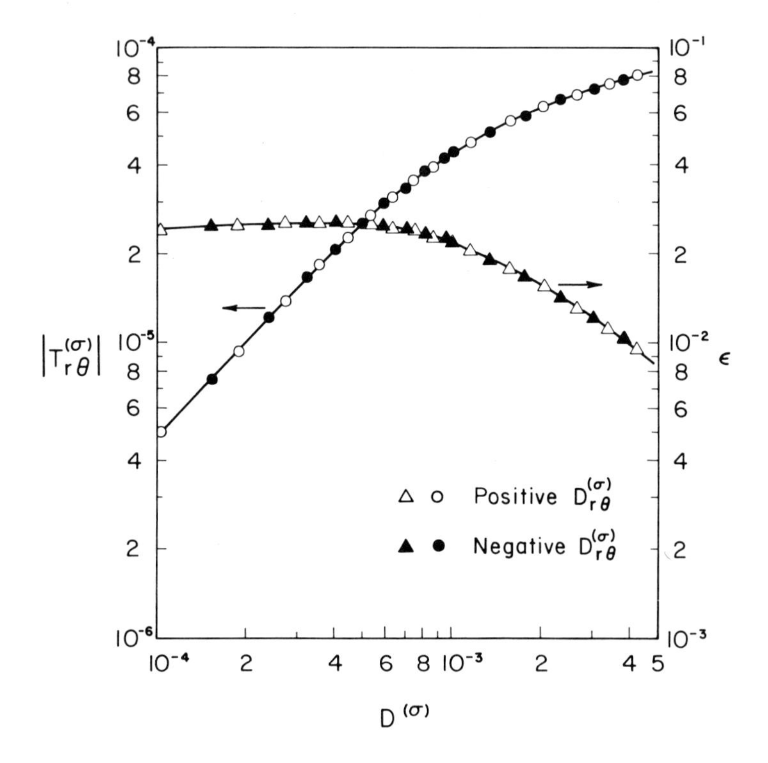

Fig. 6. The absolute value of the surface shear stress $|T^{(\sigma)}_{r\theta}|$ *and the apparent shear viscosity* ε *as functions of* $D^{(\sigma)}$ *for* $\text{div}_{(\sigma)} \underset{\sim}{v}^{(\sigma)} = 0$ *at an interface between air and a n-octadecanol monolayer (20.5* $\overset{\circ}{A}^2$*/molecule) over distilled water.*

Note that in Fig. 6 the apparent surface shear viscosity decreases as the surface rate of deformation increases. There is a direct analogy in bulk stress-deformation behavior: the apparent viscosity is also commonly observed to decrease with increasing rate of deformation.

VI. ACKNOWLEDGEMENT

The authors are grateful for the financial support of the National Science Foundation (GK-450, GK-2950, and ENG75-18783).

VII. REFERENCES

1. Boussinesq, J., C.R. Acad. Sci. 156, 983 (1913)
2. Scriven, L.E., Chem. Eng. Sci. 12, 98 (1960); typographical errors corrected in (3).
3. Slattery, J.C., Chem. Eng. Sci. 19, 379 (1964).
4. Wei, L.-Y., Schmidt, W., and Slattery, J.C., J. Colloid Interface Sci. 48, 1 (1974).
5. Briley, P.B., Deemer, A.R., and Slattery, J.C., accepted for publication J. Colloid Interface Sci.
6. Joly, M., "Recent Progress in Surface Science" (J.F. Danielli, K.G.A. Pankhurst, and A.C. Riddiford, Eds.), Vol. 1, p. 1. Academic Press, New York, 1964.
7. Joly, M., "Surface and Colloid Science" (Egon Matijević, ED.), Vol. 1, pp. 1 and 79. Wiley-Interscience, New York, 1972.
8. Burton, R.A., and Mannheimer, R.J., "Ordered Fluids and Liquid Crystals," Advances in Chemistry Series No. 63, p. 315. American Chemical Society, Washington D.C., 1967.
9. Mannheimer, R.J., and Schechter, R.S., J. Colloid Interface Sci. 27, 324 (1968).
10. Mannheimer, R.J., and Schechter, R.S., J. Colloid Interface Sci. 32, 195 (1970).
11. Pintar, A.J., Israel, A.B., and Wasan, D.T., J. Colloid Interface Sci. 37, 52 (1971).
12. Mannheimer, R.J., and Burton, R.A., J. Colloid Interface Sci. 32, 73 (1970).
13. Lifshutz, N., Hegde, M.G., and Slattery, J.C., J. Colloid Interface Sci. 37, 73 (1971).
14. Goodrich, F.C., and Allen, L.H., J. Colloid Interface Sci. 40, 329 (1972).
15. Goodrich, F.C., Proc. R. Soc. London Ser. A 310, 359 (1969).
16. Goodrich, F.C., and Chatterjee, A.K., J. Colloid Interface Sci. 34, 36 (1970).
17. Goodrich, F.C., Allen, L.H., and Poskanzer, A., J. Colloid Interface Sci. 52, 201 (1975).
18. Poskanzer, A., and Goodrich, F.C., J. Colloid Interface Sci. 52, 213 (1975).

19. Poskanzer, A.M., and Goodrich, F.C., J. Phys. Chem. 79, 2122 (1975).
20. Mannheimer, R.J., and Schechter, R.S., J. Colloid Interface Sci. 32, 212 (1970).
21. Wasan, D.T., Gupta, L., and Vora, M.K., AIChE J. 17, 1287 (1971).
22. Hegde, M.G., Ph.D. dissertation, Northwestern University, Evanston, Illinois, 1971.
23. Hegde, M.G., and Slattery, J.C., J. Colloid Interface Sci. 35, 183 (1971).
24. Brown, A.G., Thuman, W.C., and McBain, J.W., J. Colloid Sci. 8, 491 (1953).
25. Suzuki, A., Kolloid - Z. Z. Polym. 250, 365 (1972).
26. Wei, L.-Y., Ph.D. dissertation, Northwestern University, Evanston, Illinois, 1973.
27. Hegde, M.G., and Slattery, J.C., J. Colloid Interface Sci. 35, 593 (1971).
28. Slattery, J.C., Ind. Eng. Chem. Fundam. 6, 108 (1967); ibid. 7, 672 (1968).
29. Slattery, J.C., "Momentum, Energy, and Mass Transfer in Continua," pp. 59, 61, 630, 635. McGraw-Hill, New York, 1972.
30. Jahnke, Eugene, and Emde, Fritz, "Tables of Functions," 4th ed., p. 205. Dover, New York, 1945.
31. Mannheimer, R.J., Ph.D. dissertation, University of Texas, Austin, Texas, 1969.

VIII. VALUES AND DEFINITIONS OF TERMS

a — depth of phase 1; see Fig. 1.

b — depth of phase 2; see Fig. 1.

A_n, B_n — coefficients in Eqs. [20] and [21], the values of which are defined in Eqs. [22] and [23].

c^* — value of r^* at which surface angular velocity is maximized.

$D^{(\sigma)}$ — scalar invariant of the surface rate of deformation tensor defined by Eq. [6].

$\tilde{D}^{(\sigma)}$ — surface rate of deformation tensor defined by Eq. [2].

$D_{r\theta}^{(\sigma)}$	physical component of $\underset{\sim}{D}^{(\sigma)}$ in cylindrical coordinates.
F_n	coefficients defined by Eq. [24].
$g(r)$	experimentally determined functional dependence of $v_\theta^{(\sigma)}$ upon r .
J_p	p th order ordinary Bessel function of the first kind
k	radius of inner channel wall divided by radius of outer channel wall.
N	number of terms in Eq. [37].
N_{Re}	Reynolds number characteristic of flow in deep channel surface viscometer; defined by Eq. [39].
$\underset{\sim}{P}$	projection tensor that transforms every vector on surface into its tangential component; see (4, Appendix).
r	radial cylindrical coordinate.
R	radius of outer channel wall.
$\underset{\sim}{T}^{(\sigma)}$	surface stress tensor.
$T_{r\theta}^{(\sigma)}$	physical component of $\underset{\sim}{T}^{(\sigma)}$ in cylindrical coordinates.
$\underset{\sim}{v}$	velocity distribution.
$\underset{\sim}{v}^{(1)}$	velocity distribution within phase 1; see Fig. 1.
$v_r^{(1)}, v_\theta^{(1)}, v_z^{(1)}$	cylindrical components of $\underset{\sim}{v}^{(1)}$.
$\underset{\sim}{v}^{(2)}$	velocity distribution within phase 2; see Fig. 1.
$v_r^{(2)}, v_\theta^{(2)}, v_z^{(2)}$	cylindrical components of $\underset{\sim}{v}^{(2)}$.
$\underset{\sim}{v}^{(\sigma)}$	velocity distribution within the liquid-gas phase interface.
$v_r^{(\sigma)}, v_\theta^{(\sigma)}$	cylindrical components of $\underset{\sim}{v}^{(\sigma)}$.

Y_p	p th order ordinary Bessel function of the second kind.
z	axial cylindrical coordinate.
γ	thermodynamic surface tension.
ε	surface shear viscosity.
κ	surface dilatational viscosity.
λ_n	roots of Eq. [27]; see (30).
$\mu^{(1)}$	viscosity of phase 1; see Fig. 1.
$\mu^{(2)}$	viscosity of phase 2; see Fig. 1.
$\rho^{(1)}$	mass density of phase 1; see Fig. 1.
$\phi_p(\lambda_n r^*)$	function defined by Eq. [26].
Ω	angular velocity of the floor of the channel.
$\mathrm{div}_{(\sigma)}$	surface divergence operation; see (4, Appendix).
tr	trace operation; see (29).
$\nabla_{(\sigma)}$	surface gradient operation; see (4, Appendix)
T	superscript denoting the transpose of a tensor; see (29).
*	superscript denoting a dimensionless variable.

EVALUATION OF SURFACE RHEOLOGICAL MODELS

John W. Gardner* and R. S. Schechter
University of Texas at Austin
**Now employed by Shell Development Company*

I. RHEOLOGICAL MODELS FOR AN INTERFACE

It is well established that the interface between two fluid phases can exhibit rheological behavior which is quite different from that of the two adjacent phases. For example, the interface between two Newtonian fluids can be viscoelastic. Both Scriven[1] and Slattery[2] have recognized the importance of the interface and its special rheological properties in governing the flow of multiphase systems. To characterize the special features of the interface, they recommended the inclusion of a surface stress tensor, $\underset{\approx}{\Pi}^{(s)}$, into the boundary conditions which are applied to relate the bulk stress in one phase with that of a second adjacent phase. For example, a simplified form of such a boundary condition has the form

$$\underset{\sim}{N} \cdot \{\hat{\underset{\approx}{\Pi}} - \underset{\approx}{\Pi}\} = \underset{\sim}{\nabla}_s \cdot \underset{\approx}{\Pi}^s \qquad \textit{at the interface} \qquad (1)$$

where $\underset{\sim}{N}$ is an unit normal to the interface, $\hat{\underset{\approx}{\Pi}} - \underset{\approx}{\Pi}$ is the difference between the stress tensors of the two bulk phases evaluated at points immediately adjacent to the interface and $\underset{\sim}{\nabla}_s$ is the surface gradient operator. This boundary condition indicates that there may be a "jump" (discontinuity) in the bulk stresses owing to the mechanical behavior of the interfacial phase. To apply Eqn. *(1)*, a relationship between the surface stress tensor and the deformation history of the interface is required. Gardner and Schechter[3] have developed comprehensive constituitive equations which are not limited to a particular motion of the interface and which are capable of representing complex viscoelastic behavior. These comprehensive equations cannot, however, be applied without experimentally evaluating parameters which appear in them and which depend on the state of the interface - its temperature and composition but not on the deformation nor the deformation rate. The purpose of this paper is to evaluate methods of measurement which use the deep channel surface viscometer[4], [5],[6]. The discussion will, therefore, be limited to two special cases; namely, plane viscometric and plane rheometrical periodic surface flows. These have been given precise mathematical definitions[3], but they will not be presented here.

Plane viscometric surface flows are characterized by fluid motion in a single direction, say x, with speed of movement being dependent on a coordinate position, say y, which is perpendicular to x but in the interface. As is implied, the interface is assumed to remain planar during the course of the motion. If $\upsilon_x(y)$ represents the velocity within the planar interface, the surface stress tensor can, under very general conditions be shown[3] to have the form:

$$\Pi^{(s)}_{xx} = -\sigma + T^{(s)}_{xx} = -\sigma - \psi_1 \left\{\frac{d\upsilon_x}{dy}\right\}^2 \qquad (2)$$

$$\Pi^{(s)}_{xy} = T^{(s)}_{xy} = -\eta \frac{d\upsilon_x}{dy} \qquad (3)$$

and

$$\Pi^{(s)}_{yy} = -\sigma + T^{(s)}_{yy} = -\sigma - \psi_2 \left\{\frac{d\upsilon_x}{dy}\right\}^2 \qquad (4)$$

where σ is the surface tension. Here η is the surface viscosity and ψ_1 and ψ_2 are surface normal stress coefficients. These rheological parameters may all depend on $\gamma = \left|\frac{d\upsilon_x}{dy}\right|$ where γ is often referred to as the shear rate. This theory is the first to suggest the possible existence of surface normal stresses. The magnitudes of these stress coefficients have not, however, been measured.

A planar rheometrical periodic surface flow has been precisely characterized[3], but for a viscoelastic film possessing a fading memory and subjected to small oscillatory motion, the surface stress tensor is given by

$$\Pi^{(s)}_{xy} = -\sigma + Im\{-\eta^* \frac{d\upsilon^*_x}{dy} e^{i\omega t}\} \qquad (5)$$

where the $Im\{\quad\}$ denotes the imaginary part of a complex number. η^* and υ^*_x are complex quantities and ω is the frequency of the small oscillatory motion. The time is designated as t.

The complex viscosity, η^*, can for small oscillations be related to a function, m(t), which can be regarded as a memory function. This relationship is given in Eqn. (6).

$$\eta^* = -\frac{i}{\omega}\int_0^\infty m(s)\ \{1-e^{-i\omega s}\}\ ds \qquad (6)$$

Thus, given the memory function, the complex viscosity can be determined. This quantity will, of course depend on the frequency, ω, and the state of the interfacial phase.

Thus, plane surface viscometric flows are defined by three parameters - the surface viscosity and the two normal stress coefficients. On the other hand, a single function, the memory function, defines the relationship between the stress and the strain rate for a planar small amplitude oscillatory motion. Gardner and Schechter[3] have suggested that the rheological behavior of a wide class of surface films might be represented by a "network rupture model" similar to the model used successfully by Tanner and Simmons[14]. For this model

$$\eta = \int_0^{s_V} s\,m(s)\,ds \tag{7}$$

and

$$\psi_1 = \tfrac{1}{2}\int_0^{s_V} s^2\, m(s)\,ds \tag{8}$$

$$\psi_2 = \tfrac{1}{2}\int_0^{s_V} s^2\, m(s)\,ds \tag{9}$$

where $m(s)$ is the same memory function which is used in Eqn. (6) to yield the complex surface viscosity and s_V is a parameter which defines the strain-rate at which the film ruptures. Thus,

$$s_V = \frac{B}{\gamma} \tag{10}$$

where B is a constant assumed characteristic of the film. If B becomes large, the surface tends to behave as a Newtonian fluid. Using the network rupture model, a constituitive equation has been constructed which is not limited to any particular geometry[3]. The constituitive equation is completely defined by a knowledge of $m(t)$ and the constant B. The problem is, therefore, to characterize a given surface by experimentally determining these two quantities.

The quantities which are readily accessible by experimentation are η^* as a function of the frequency, ω, and η as a function of the shear rate γ. The memory function and rupture constant can in principle be determined; however, the precision of the data must be such to permit the inversion from the measured quantities to the desired rheological parameters. One approach that is often used by rheologists studying constituitive equations to represent bulk phase behavior is to assume a mathematical form for $m(s)$. Gardner and Schechter[3] have used a network rupture model similar to that developed by Bird and Carreau[7] which includes 6 parameters as given by Eqn. *(11)*.

$$m(s) = \frac{\delta(s)}{s}\eta_\infty + \sum_{n=1}^{\infty} \frac{\mu_n}{\lambda_n^2} e^{-s/\lambda_n} \tag{11}$$

where $\delta(s)$ is the delta function, η_∞ is the surface shear viscosity at large shear rates;

$$\mu_n = (\eta_0 - \eta_\infty) \frac{\left(\frac{2}{n+1}\right)^\alpha}{\sum_{p=1}^{\infty}\left(\frac{2}{p+1}\right)^\alpha} \tag{12}$$

and

$$\lambda_n = \lambda\left(\frac{2}{n+1}\right)^\beta \tag{13}$$

η_0 is the surface shear viscosity at small values of the shear rate. The six parameters needed to define the rheological behavior of the surface film are $B, \eta_\infty, \eta_0, \alpha, \beta$ and λ. These are to be determined experimentally. This model is called the Six Parameter Network Rupture Model (SPNR).

II. LINEAR CANAL THEORY

The deep channel surface viscometer has been widely used for measuring surface shear viscosities of Newtonian surface films[4],[5] and of non-Newtonian films as well[6],[7],[8]. The device consists of a moving floor bounded by fixed walls as depicted in *fig. 1*. The actual geometry of the system is an annular one as shown. However, it has been shown that the annular geometry can be assumed to be a linear one if the viscometer is properly designed[10]. It will be assumed in the analysis to follow that the instrument has been designed appropriately.

The following equations are the starting point in analyzing the deep channel surface rheometer:

$$\text{(Equation of Motion)} \quad \frac{\partial v_x}{\partial t} = \nu\left(\frac{\partial^2 v_x}{\partial y^2} + \frac{\partial^2 v_x}{\partial z^2}\right) \tag{14}$$

$$y = 0, \qquad v_x = 0 \tag{15}$$

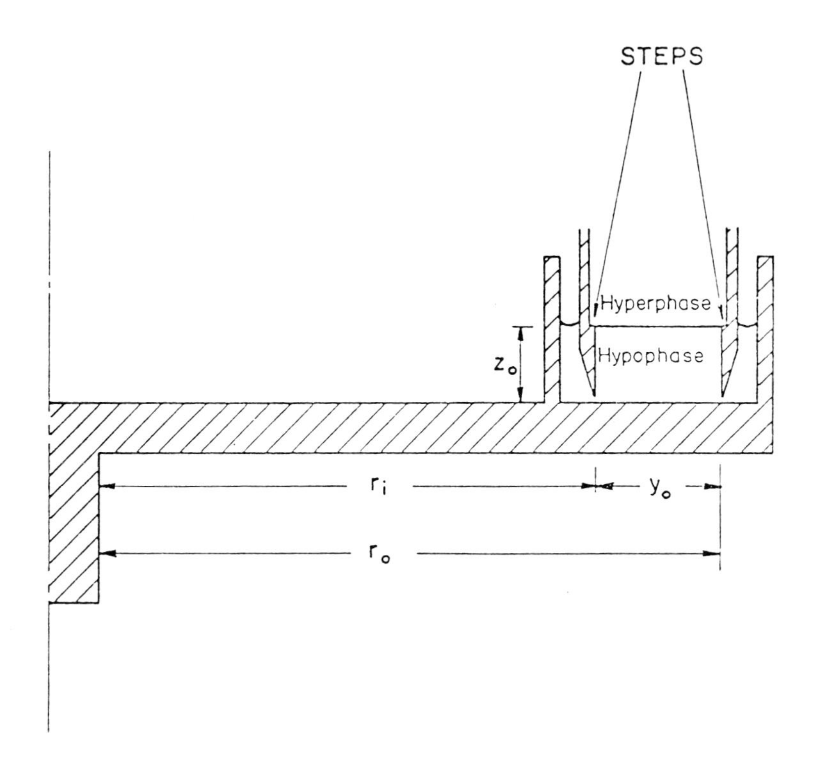

Fig. 1. Schematic of the deep channel surface rheometer.

$$y = y_0, \qquad v_x = 0 \tag{16}$$

$$z = 0, \qquad v_x = v_f(t) \tag{17}$$

$$z = z_0, \qquad \mu \frac{\partial v_x}{\partial z} = - \frac{\partial T^{(s)}_{xy}}{\partial y} \tag{18}$$

where (see *fig. 1*) v_x represents the azimuthal component of the mass average velocity in the hypophase, y is an independent variable whose value is zero at the inner wall and y_0 at the outer wall, z is an independent variable whose value is zero at the floor and z_0 at the interface (assumed to be planar), μ represents the Newtonian shear viscosity of the hypophase, ν represents the kinematic viscosity (μ/ζ) of the hypophase and v_f represents the floor velocity (treated as independent of y).

Eqn. *(18)* is derived by applying the stress jump condition (Eqn. *(1)*). The stresses exerted by the hyperphase have been neglected. This is valid at a gas/liquid interface but will not apply at a liquid/liquid interface.

These equations are written assuming that secondary flow effects can be neglected. There are at least two mechanisms which can produce secondary flows in the deep channel viscometer. It is suggested in this paper that surface normal stresses may exist. These forces would tend to produce a motion in the radial direction or a secondary motion. The normal stresses have not been measured and the magnitude of this motion is not known. The radial motion would produce a surface tension variation in the radial direction which would oppose the normal stresses thereby tending to minimize the secondary motion. The net effect may be expected to be small, perhaps immeasurably so, for most cases. A final judgement must, however, be reserved until the magnitude of the normal stresses has been ascertained. The normal stress coefficients can be calculated assuming a network repture model using Eqns. *(8)* and *(9)*.

A secondary flow may also result because of the fluid inertia, since the flow does not take place in a linear canal but in an annular canal. Consider the T_f/T_p versus Ω data shown in *fig. 2*.

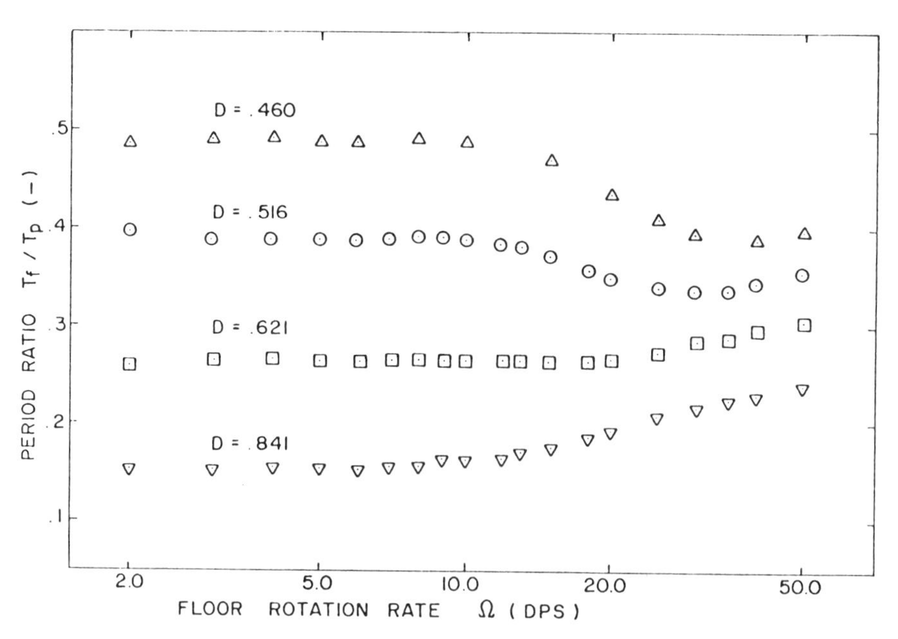

Fig. 2. Experimental data illustrating the effects of secondary flow in the deep channel surface rheometer.

T_f represents the rotational period of the floor, T_p the rotational period of a small particle at the interfacial centerline $(y = Y_0/2,\ z = z_0)$, and Ω the floor rotational rate. The parameter D found in *fig. 2* is defined as

$$D = \frac{z_0}{y_0} \tag{19}$$

The data shown in *fig. 2* are for air-water systems in which no film is present. Note the dependence of T_f/T_p on Ω for values of Ω above 10 Degrees per Second (DPS).

It is interesting to note that for the smallest value of D shown in *fig. 2* (.461) T_f/T_p first decreases and then increases for values of Ω above 10 DPS whereas for the largest value of D (.841) T_f/T_p only increases with Ω for values of Ω greater than 10 DPS.

Other investigators have reported for air-water systems only increases in T_f/T_p with Ω for floor rotation rates above 2-3 RPM[5],[8]. These investigators may have only worked with relatively large values of D (>.600).

It should be noted that the particular deep channel surface rheometer upon which the data shown in *fig. 2* was taken had no steps machined into its canal walls.

The linear canal theory predicts no T_f/T_p dependence on Ω whatsoever in the absence of a film $(T^{(s)}_{xy} = 0)$. The experimentally observed dependence shown in *fig. 2* is therefore attributed to secondary flow effects. Clearly, experiments on air-water systems should not be conducted at floor rotation rates above 10 DPS if they are to be analyzed using the linear canal theory. This critical value may differ if hypophases other than water are used.

III. DETERMINATION OF SURFACE VISCOSITY

For a constant floor motion

$$\upsilon_f = \frac{r_i + r_o}{2}\,\Omega \tag{20}$$

where Ω is the rate of rotation of the turntable (see *fig. 1*). For the resulting plane viscometric surface flow;

$$T^{(s)}_{xy} = -\eta \left.\frac{\partial \upsilon_x}{\partial y}\right|_{Z=Z.} \tag{21}$$

The surface viscosity depends on γ. This dependency will be expressed here as

$$\eta = \eta(\gamma, \alpha_1, \alpha_2, \ldots, \alpha_I)$$

where α_i i = 1, 2, . . ., I are film parameters which depend only on the fluid state. In the case of the six parameter model for example, these film parameters are B, η_0, η_∞, λ, α, and β. Eqns. *(14) - (18)* can be combined with Eqns. *(20)* and *(21)* to yield the following boundary value problem which is expressed in dimensionless form.

$$0 = \frac{\partial^2 v}{\partial y^2} + \frac{\partial^2 v}{\partial z^2} \tag{22}$$

$$y = 0, \quad v = 0 \tag{23}$$

$$y = 1, \quad v = 0 \tag{24}$$

$$z = 0, \quad v = 1 \tag{25}$$

$$z = D, \frac{\partial v}{\partial z} = \frac{\partial}{\partial y} \left(\varepsilon \frac{\partial v}{\partial y}\right) \tag{26}$$

where

$$y = \frac{y}{y_0} \tag{27}$$

$$z = \frac{z}{z_0} \tag{28}$$

$$v = \frac{v_x}{v_f} \tag{29}$$

$$\varepsilon = \frac{\eta}{\mu y_0} \tag{30}$$

The solution of the above boundary value problem can be written in principle as follows:

$$V = \lim_{N\to\infty} \sum_{p=1}^{N} \frac{4}{(2p-1)\pi} \{\cosh\left[(2p-1)\pi z\right] + x_p \sinh\left[(2p-1)\pi z\right]\} \sin\left[(2p-1)\pi y\right] \tag{31}$$

where

$$\begin{Vmatrix} A_{11} & A_{12} & \cdots & A_{1N} \\ A_{21} & A_{22} & \cdots & A_{2N} \\ \vdots & \vdots & & \vdots \\ A_{N1} & A_{N2} & \cdots & A_{NN} \end{Vmatrix} \begin{Vmatrix} X_1 \\ X_2 \\ \vdots \\ X_N \end{Vmatrix} \begin{Vmatrix} C_1 \\ C_2 \\ \vdots \\ C_N \end{Vmatrix} \qquad (32)$$

$$C_p = -\frac{4}{(2p-1)\pi} \tanh\left[(2p-1)\pi D\right] - \sum_{q=1}^{N} \frac{8(2q-1)\cosh\left[(2q-1)\pi D\right]}{(2p-1)\cosh\left[(2p-1)\pi D\right]} I_{2p-1,2q-1} \qquad (33)$$

$$I_{mn} = \frac{1}{2}\int_0^1 \frac{\partial}{\partial\gamma}(\gamma\varepsilon)\cos\left[(m-n)\pi Y\right]dY - \frac{1}{2}\int_0^1 \frac{\partial}{\partial\gamma}(\gamma\varepsilon)\cos\left[(m+n)\pi Y\right]dY \qquad (34)$$

$$A_{pq} = \frac{2(2q-1)^2\pi\sinh\left[(2q-1)\pi D\right]}{(2p-1)\cosh\left[(2p-1)\pi D\right]} I_{2p-1,2q-1} + \delta_{ij} \qquad (35)$$

Note that by Eqn. (31)

$$\frac{T_f}{T_p} = V\Big|_{\substack{Y=\frac{1}{2} \\ Z=D}} = \lim_{N\to\infty}\sum_{p=1}^{N}\left\{\frac{4}{(2p-1)\pi}\cosh\left[(2p-1)\pi D\right] + X_p \sinh\left[(2p-1)\pi D\right]\right\}\sin\left[\frac{(2p-1)\pi}{2}\right] \qquad (36)$$

T_f/T_p is the ratio of the centerline velocity at the interface to the velocity of the floor. This ratio is measurable and can be determined as a function Ω.

It is also possible to measure the interfacial velocity profile. Hedge and Slattery[9] have proposed a method of using this velocity profile to determine the film parameters. This possibility does not seem to have been pursued further.

Methods requiring T_f/T_p versus Ω data have been used successfully to determine the viscometric surface viscosities of Newtonian[11] and Powell-Eyring[8] films.

A generalized method utilizing T_f/T_p versus Ω data in which the values of α_i i=1, 2, . . ., I are determined for which the theoretical T_f/T_p versus Ω curve calculated by the procedure outlined in *fig. 3* best fits the experimental data.

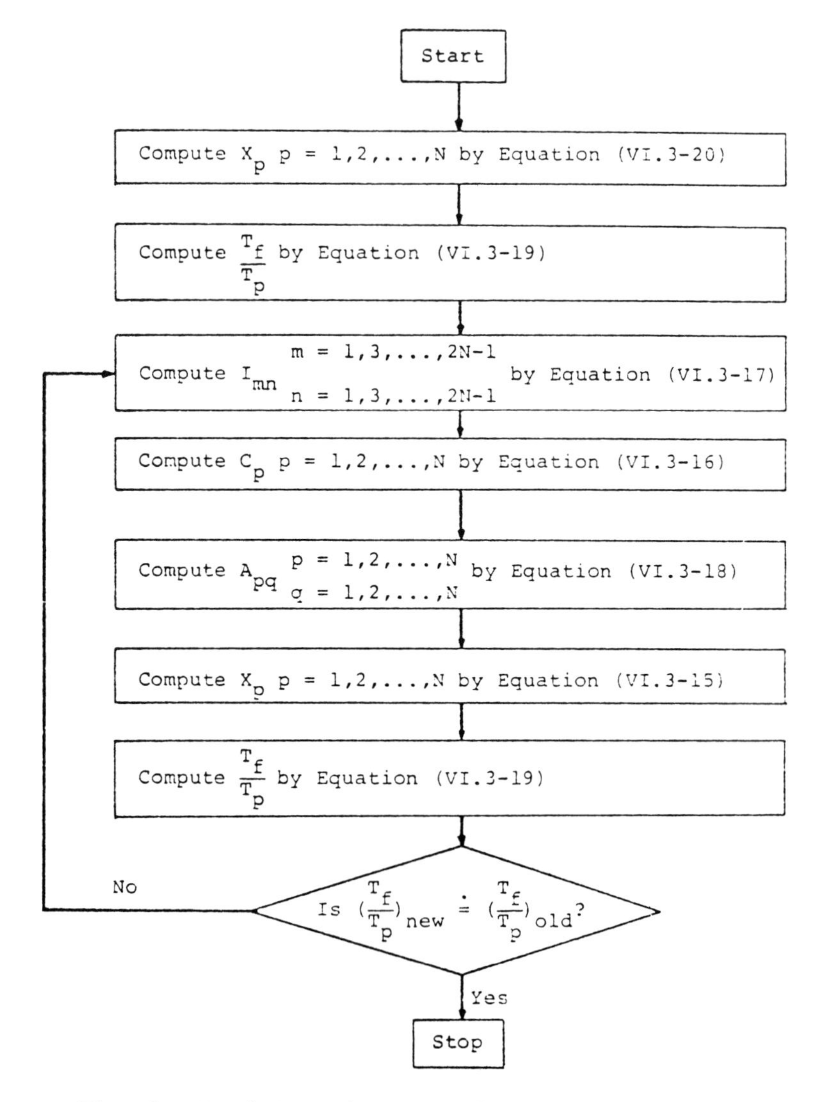

Fig. 3. An interative procedure used in determining the velocity profile in the deep channel surface rheometer.

This method can be used to determine the viscometric surface viscosity but before it can be used routinely, numerical integration techniques more efficient than those tested (see Gardner[12]) will have to be found.

IV. DETERMINATION OF THE COMPLEX VISCOSITY

For a sinusoidal floor motion

$$v_f = C \sin\omega t \tag{37}$$

where C represents a constant and providing that C is small enough a plane periodic surface flow exists such that if a simple fluid film having a fading memory is present at the interface

$$T^{(s)}_{xy} = Im\{-\eta^* \left.\frac{\partial v^*}{\partial y}\right|_{z=z_0} e^{i\omega t}\} \tag{38}$$

where v^* is such that

$$v_x = Im\{v^* e^{i\omega t}\} \tag{39}$$

The complex surface viscosity depends on ω. This dependency will be expressed here as

$$\eta^* = \eta^* (\omega, \beta_1, \beta_2, \ldots, \beta_J) \tag{40}$$

where β_i i=1, 2, . . ., J are film parameters which depend only on Γ. In the case of an SPNR film, for example, these film parameters are η_0, η_∞, λ, α, and β. (Note that for small oscillatory motions, the parameter B does not appear.)

Eqns. (14) - (18) and (37) - (40) can be combined and dimensionless variables introduced so as to obtain the following boundary value problem:

$$V^* = \upsilon^* \left(\frac{\partial^2 V^*}{\partial Y^2} + \frac{\partial^2 V^*}{\partial Z^2}\right) \tag{41}$$

$$Y = 0, \qquad V^* = 0 \tag{42}$$

$$Y = 1, \qquad V^* = 0 \tag{43}$$

$$Z = 0, \qquad V^* = 1 \tag{44}$$

$$Z = D, \qquad \frac{\partial V^*}{\partial Z} = \varepsilon^* \frac{\partial^2 V^*}{\partial Y^2} \tag{45}$$

where

$$V^* = \frac{v^*}{v_f} \tag{46}$$

$$\upsilon^* = \frac{\upsilon}{i\omega y_0^2} \tag{47}$$

$$\varepsilon^* = \frac{\eta^*}{\mu y_0} \tag{48}$$

The solution of the above boundary value problem can be written as follows[13]:

$$V^* = \sum_{p=1}^{\infty} \frac{\phi_p^* \cosh\left[\phi_p^*(D-Z)\right] + \varepsilon^* \sinh\left[\phi_p^*(D-Z)\right]}{\phi_p^* \cosh\left[\phi_p^* D\right] + (2p-1)^2\pi^2\varepsilon^* \sinh\left[\phi_p^* D\right]} \cdot \frac{4\sin\left[(2p-1)\pi Y\right]}{(2p-1)\pi} \tag{49}$$

where

$$\phi_p^* = \left[(2p-1)^2\pi^2 + \frac{1}{\upsilon^*}\right]^{\frac{1}{2}} \tag{50}$$

Note that the interfacial centerline velocity is given by

$$v_c = C\,|V_c^*|\,\sin(\omega t - \theta) \tag{51}$$

where

$$V_c^* = \sum_{p=1}^{\infty} \frac{\dfrac{4}{(2p-1)\pi} \sin\left[\dfrac{(2p-1)\pi}{2}\right]}{\cosh\left[\phi_p^* D\right] + \dfrac{(2p-1)^2\pi^2}{\phi_p^*} \sinh\left[\phi_p^* D\right]\varepsilon^*} \tag{52}$$

and

$$\tan\theta = -\frac{Im\ V_c^*}{Re\ V_c^*} \tag{53}$$

Also note that

$$\frac{A_p}{A_f} = |V^*_c| \tag{54}$$

where A_p represents the amplitude of the motion of a small particle at the interfacial centerline and A_f represents the amplitude of the floor motion.

If no film is present $(\varepsilon^* = 0)$

$$\frac{A_p}{A_f} = \sum_{p=1}^{\infty} \frac{\frac{4}{(2p-1)\pi} \sin\left|\frac{(2p-1)\pi}{2}\right|}{\cosh\left|\phi^*_p D\right|} \tag{55}$$

A_p/A_f versus f curves $(f=2\pi\omega)$ predicted by Eqn. (55) are shown in *fig. 4*.

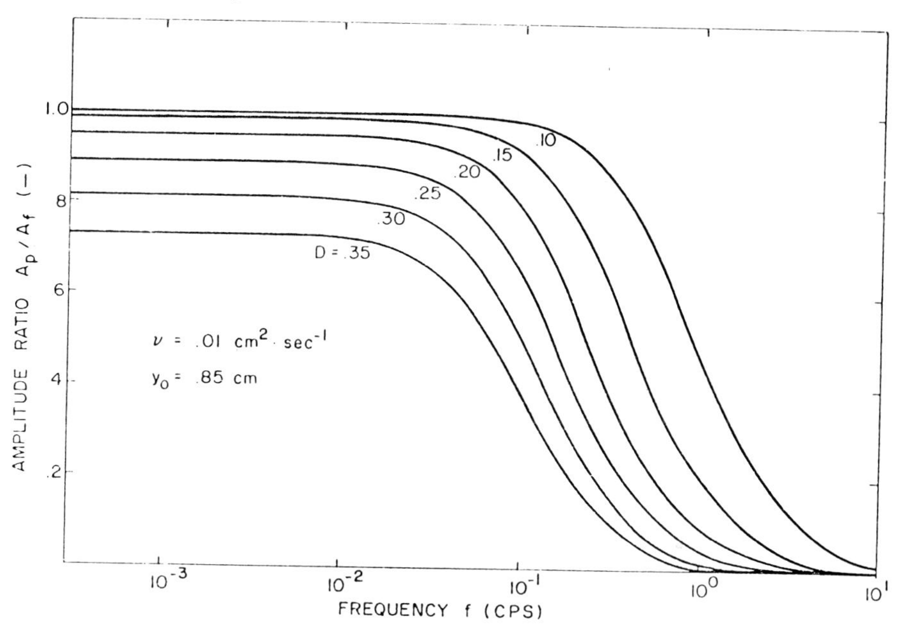

Fig. 4. Frequency dependence of the amplitude ratio predicted by the linear canal theory for air-water systems of different depths.

These particular curves correspond to systems in which the hypophase is water $(\upsilon = .01\ cm^2\ sec^{-1})$. The declines in the curves are due to an increasing inability of the hypophase to transmit stress from the floor to the interface.

Eqn. *(55)* has been experimentally verified. A typical result for an air-water system is shown in *fig. 5*.

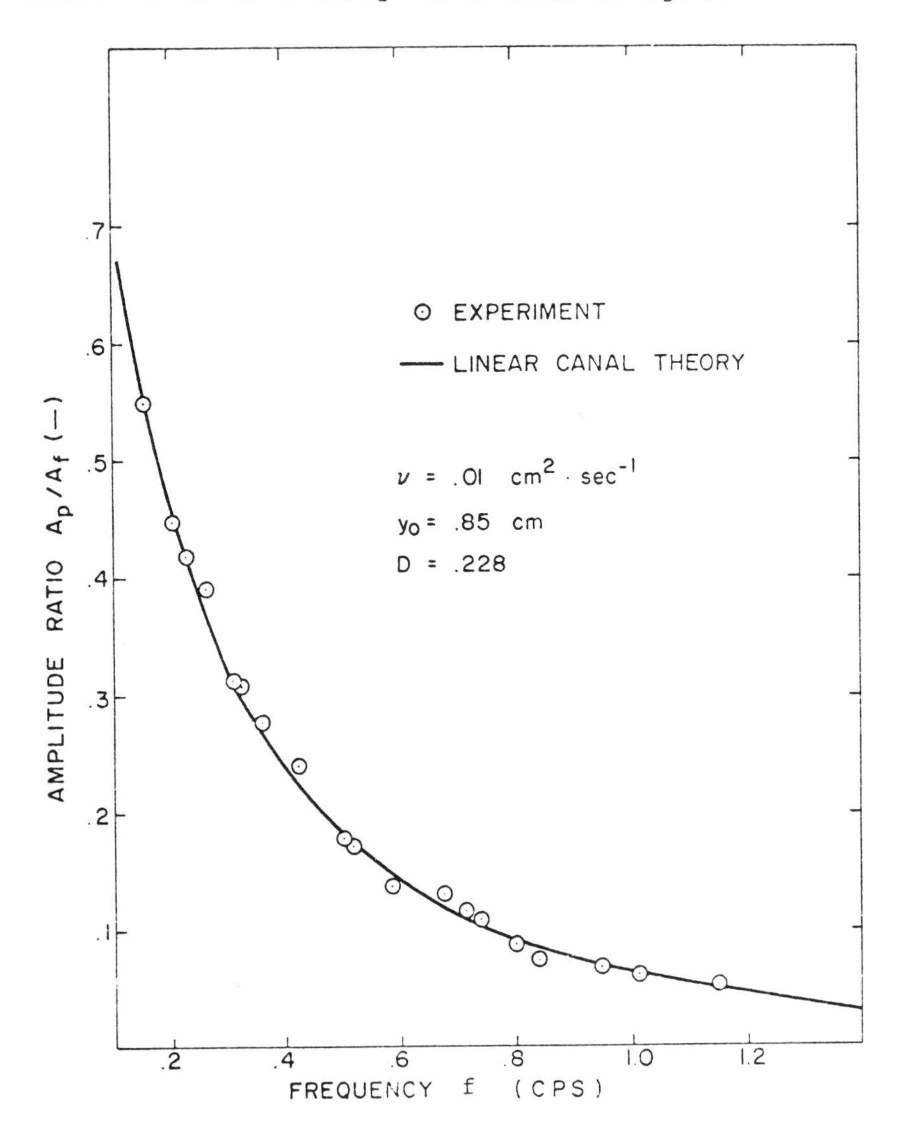

Fig. 5. Experimental and theoretical frequency dependence of the amplitude ratio for an air-water system.

Note that the particular deep channel surface rheometer upon which the data shown in *fig. 5* was taken had no steps machined into its canal walls. When air-cetane systems were examined on this device agreement between theory and experiment was not obtained (see *fig. 6*). The lack of agreement is felt to be due to the presence of a meniscus.

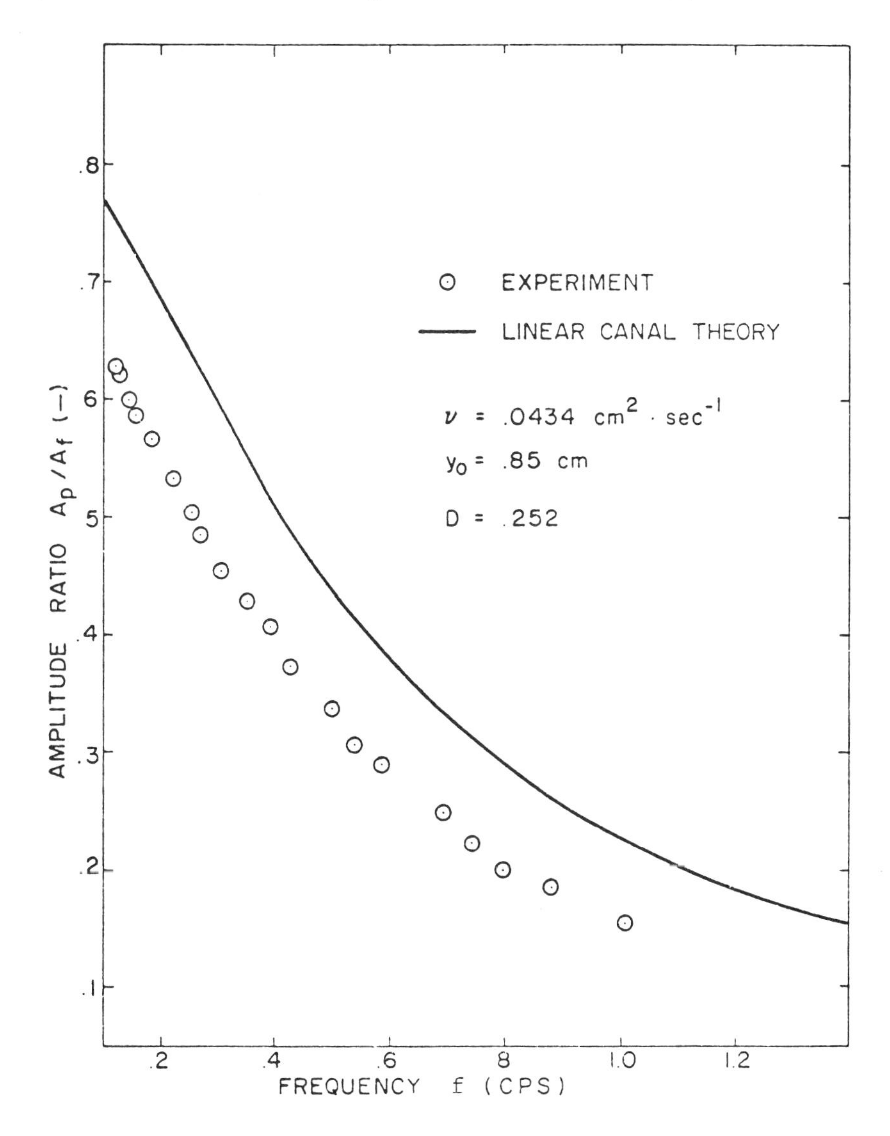

Fig. 6. Experimental and theoretical frequency dependence of the amplitude ratio for an air-cetane system.

The A_p/A_f versus ω curves shown in *fig. 7* were calculated using Eqns. *(52)* and *(54)*. These particular curves are for the SPNR film. Note that the pronounced peak in the case of film D is not due to a maximum in the elastic nature of the film at the peak frequency.

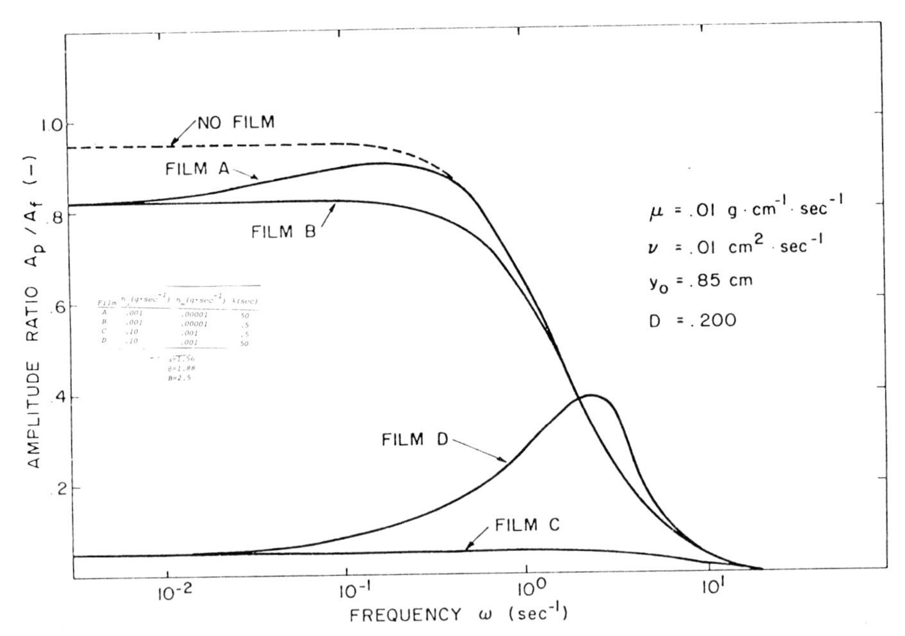

Fig. 7. Frequency dependence of the amplitude ratio predicted by the linear canal theory for air-water systems containing the four six-parameter network rupture films of fig. 1.

Data which can be analyzed to determine the complex surface viscosity are: (1) A_p/A_f versus ω, (2) A_p/A_f and tan θ versus ω, and (3) A_p/A_f versus ω at two or more different values of D.

Methods requiring A_p/A_f versus ω data will be referred to as the generalized amplitude ratio method. In this method the values of β_i i=1, 2, . . ., J are determined for which the theoretical A_p/A_f versus ω most closely approximates the experimental A_p/A_f versus ω curve.

A method requiring A_p/A_f and tan θ versus ω data will be referred to as the direct phase lag method. In this method the values of η' and η" are determined at each experimental frequency for which the theoretical values of A_p/A_f and tan θ calculated agree with the experimental values of A_p/A_f and tan θ.

A generalized method requiring A_p/A_f versus ω data at two different values of D will be referred to as the direct method of different depths. In this method the values of η' and η" are determined at each experimental frequency for which the theoretical values of A_p/A_f calculated agree with the experimental values of A_p/A_f at both values of D.

Extensive sensitivity analyses have been conducted using these techniques. They indicate that if the complex surface viscosity is to be accurately determined the scatter in the A_p/A_f data must be no greater than (1) .00001 for the direct method of different depths, (2) .001 for the direct phase lag method and (3) .01 for the generalized amplitude ratio method. Since the A_p/A_f scatter associated with the deep channel surface rheometer used in this work is about .01 it is felt that the generalized amplitude ratio method can be used successfully to characterize some real films but that the direct method of different depths and direct phase lag method are impractical.

Regarding only the generalized amplitude ratio method sensitivity analyses indicate that for values of D greater than or equal to .200 and for A_p/A_f scatter greater than or equal to .01 the magnitude of the complex surface viscosity must be greater than 10^{-5} g · sec^{-1} if it is to be accurately determined. It also appears that for best results the relaxation time of the film should be considerably greater than the reciprocal of the frequency at which the inefficiency of the hypophase in transmitting stress from the floor to the interface begins to manifest itself. The frequency at which the inefficiency of the hypophase in transmitting stress from the floor to the interface begins to manifest itself is the frequency at which in the absence of any film A_p/A_f begins to decline with increasing ω (see *fig. 4*).

V. CONCLUSIONS

The parameters which appear in comprehensive constituitive equations capable of representing complex viscoelastic behavior of interfaces must be determined experimentally. It is shown here that the deep channel surface viscometer can be used to determine the values of these parameters provided the viscometer is properly designed and the floor of the viscometer is programmed to rotate continuously and in oscillating motion. Continuous rotation is needed to permit the measurement of the surface shear viscosity. The investigator is cautioned to control secondary flow so that reliable measurements are obtained.

The oscillatory experiments have been found to be best analyzed using a method which uses measurements of the amplitude ratio as a function of the frequency.

VI. ACKNOWLEDGEMENT

Acknowledgement is made to the donors of the Petroleum Research Fund, administered by The ACS, for partial support of this project.

VII. REFERENCES

(1) Scriven, L. E., Chem. Eng. Sci. 12, 98, (1960).
(2) Slattery, J. C., Chem. Eng. Sci. 19, 379, (1964).
(3) Gardner, J. W. and R. S. Schechter, "Surface Rheology: Simple Fluid Films", paper submitted for publication (June 1976).
(4) Burton, R. A. and R. J. Mannheimer, "Ordered Fluids and Liquid Crystals", Advances in Chemistry Series, 3, 315.
(5) Mannheimer, R. J. and R. S. Schechter, J. Colloid Interface Sci. 32 (2), 212, (1970a).
(6) Mannheimer, R. J. and R. S. Schechter, J. Colloid Interface Sci. 32 (2), 212, (1970b).
(7) Bird, R. B. and P. J. Carreau, Chem. Eng. Sci. 23, 427, (1968).
(8) Pintar, A. J., A. B. Israel, and D. T. Wasan, J. Colloid Interface Sci. 37 (1), 52, (1971).
(9) Hedge, M. S. and J. C. Slattery, J. Colloid Interface Sci. 35 (2), 183, (1971a).
(10) Mannheimer, R. J. and R. S. Schechter, J. Colloid Interface Sci. 27, 324, (1968).
(11) Gupta, L. and D. T. Wasan, Ind. Eng. Chem. Fundam. 13 (1), 26, (1974).
(12) Gardner, J. W., "On the Mechanics of Non-Newtonian Films", Doctoral Dissertation, The University of Texas at Austin, Austin, Texas, (1975).
(13) Mannheimer, R. J. and R. S. Schechter, J. Colloid Interface Sci. 32 (2), 225, (1970c).
(14) Tanner, R. I. and J. M. Simmons, Chem. Eng. Sci. 22, 1803, (1967).

A NEW METHOD FOR THE MEASUREMENT OF SHEAR VISCOELASTICITY AT LIQUID-LIQUID INTERFACES CONTAINING SURFACTANT AND MACROMOLECULAR FILMS

V. Mohan and D. T. Wasan
Illinois Institute of Technology

A new method for measurement of shear viscoelasticity at liquid-liquid interfaces is presented. The method consists in imparting sinusoidal or on-off motion to the floor of the viscous traction viscometer and observing the centerline displacement either at the liquid-gas or the liquid-liquid interface in a liquid-liquid-gas experiment. It is shown that shear viscoelasticity at a liquid-liquid interface can be determined from displacement data at liquid-gas interface alone, so that this method is particularly useful for opaque systems such as crude oil-aqueous solutions.

A parametric study revealed that the displacement pattern for on-off floor motion may either be increasing with or without a characteristic snap back, or oscillatory. Contrary to normal belief it is concluded that the absence of a snap back does not ensure that the interface is purely viscous. The displacement pattern depends on both the depth of the liquid in the viscometer and the viscoelastic parameters.

The method proposed here was used to obtain viscoelastic parameters of monolayers of crystalline bovine plasma albumin (BPA) spread at a benzene-water interface. The experimental data reveal that, compared to the interfacial pressure, the interfacial shear viscosity and relaxation time are much more sensitive to changes in molecular packing. Specifically, it is shown that the interfacial shear viscosity and relaxation time increase steeply at interfacial concentrations of BPA corresponding to the close-packed structure.

I. INTRODUCTION

Surface active agents, trace impurities and macromolecules at fluid-fluid interfaces exert a significant influence on fluid-fluid phase separation processes by modifying the interfacial mobility and area. The presence of these chemical additives alters the interfacial flow behavior by providing an additional hydrodynamic resistance to flow. For a pure shear in the plane of the interface, the momentum balance equation for simple systems involves only the interfacial shear viscosity. In some instances, the molecules in the plane of the interface exhibit the phenomenon of memory or recoil. As a first approximation, this complex behavior can be represented in terms of viscous and elastic effects.

The importance of rheological behavior of interfaces on the stability of foams and emulsions has been the subject of many papers (1-10). This has led several investigators to measure viscoelastic properties of surfaces and interfaces. The rheology of a surface film has been studied by a number of techniques including the deep channel surface viscometer (11-13). However, the viscoelastic studies on liquid-liquid interfaces is limited and employ the rotational torsion principle or a torsion pendulum (14-20). Quite recently, the theoretical analysis for investigating viscoelastic response in a canal surface viscometer was presented by Mannheimer and Schechter (11). The experimental method (Fig. 1a) involves conducting studies on the interfacial displacement for periodic floor motion. A feature of their technique is that it involves only the liquid-liquid interface under study. However, this technique can not be used when dealing with opaque upper liquids. To treat such situations, it is necessary to have a free surface and observe the effect of interfacial and surface viscoelastic parameters on the surface displacement (Fig. 1b). The interfacial viscoelastic parameters can be determined if the surface viscoelastic parameters are determined from a separate top liquid-gas surface displacement study as described by Mohan, Malviya and Wasan (12). However, in some instances, the surface behavior is purely viscous or exhibits negligible viscosity, so that the surface displacement in the liquid-liquid-gas experiment can be used even with a complicated interfacial viscoelastic model to predict the model parameters. An important feature of our proposed method is that for transparent upper liquids, the interfacial shear viscoelastic parameters can be obtained from either the surface or the interfacial displacement thereby providing a check on internal consistency. For generality we shall assume that both the surface and the interface can be represented by a Voigt-Maxwell model. Such a model has been used previously. Most recently, Gardner and Schechter (21) presented a generalized approach to representing a viscoelastic phase-interface with fading memory.

II. THEORY

A. Model for Surface and Interface Rheology

The rheological behaviors of surfaces (s) and interfaces (i) are assumed to follow the linear viscoelastic model, more specifically the Voigt-Maxwell model in series shown in Figure 2. The quantity E_o represents the modulus of instantaneous surface elasticity of the spring, and ε_o is the

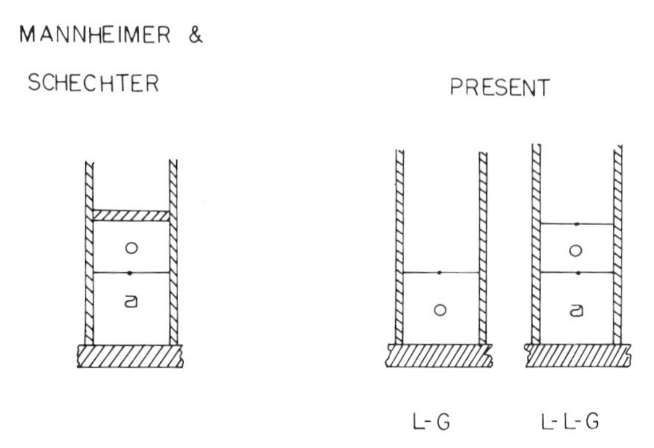

Fig. 1. Comparison between the flow configurations of Mannheimer and Schechter (Fig. 1a) and the present study (Fig. 1b).

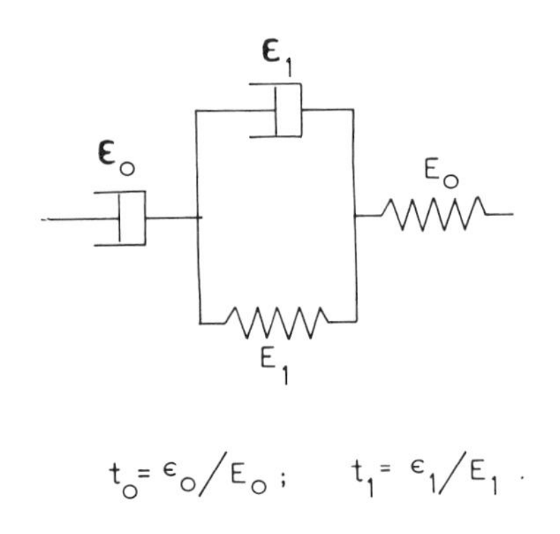

Fig. 2. Model for viscoelastic behavior.

coefficient of viscosity of the dash-pot in the Maxwell model. Similarly, the quantities E_1 and ε_1 signify the retarded surface elasticity and internal viscosity of the Voigt element.

For a periodic forcing function τ_{jk} with period P, the system has a periodic response d_{jk} such that (11)

$$\tau_{jk} = -2\eta \, d_{jk} \tag{1}$$

where,

$$d_{jk} = \tfrac{1}{2} \left(\frac{\partial v_j}{\partial x_k} + \frac{\partial v_k}{\partial x_j}\right) \tag{2}$$

and,

$$\eta = \frac{\varepsilon_0}{(1+j\omega t_0) + \frac{\varepsilon_0}{\varepsilon_1} j\omega t_1/(1+j\omega t_1)} \tag{3}$$

Equation (3) gives the expression for the complex viscosity η in terms of the viscoelastic model parameters $\varepsilon_0, \varepsilon_1$ t_0 and t_1, where t_0 and t_1 are relaxation and retardation times defined as

$$t_0 = \varepsilon_0 / E_0 \tag{4}$$

and,

$$t_1 = \varepsilon_1 / E_1 \tag{5}$$

It should be noted that Equation (3) applies to the surface as well as the interface and the distinction will be indicated by a superscript (s) or (i) to denote the surface or the interface.

B. Solution for the Hydrodynamic Problem

The linear canal geometry is shown in Figure 3b together with the coordinate system employed to analyze the flow. The space between the two fixed walls is filled with a Newtonian aqueous layer 'b', a Newtonian oil phase 'a' and the free space 'c'. In what follows we develop equations for the surface and interfacial displacements for the sinusoidal motion of the floor as well as the on-off motion (Figure 4) in a liquid-liquid-gas experiment. A similar analysis for the liquid-gas experiment (Figure 3a) was presented by Mannheimer and Schechter (11).

1. *Sinusoidal Floor Motion with Frequency* ω

The velocity of the floor v_f is represented by the imaginary part of the complex floor velocity

$$v_f = \bar{v}_f \, e^{j\omega t} \tag{6}$$

where t denotes time and $\bar{v}_f$ represents the time independent part of floor velocity.

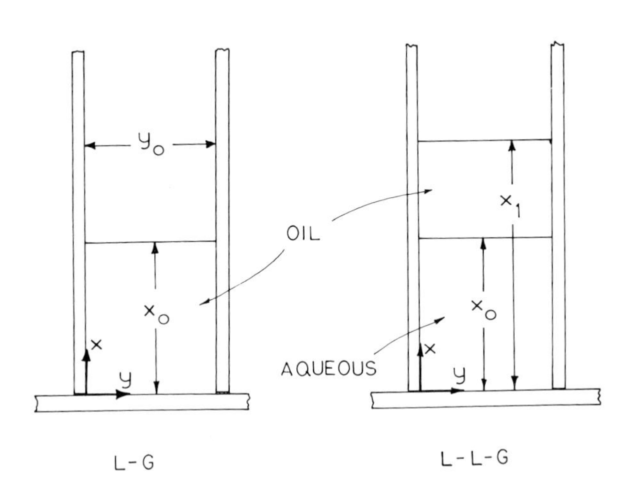

Fig. 3. Cartesian coordinate system in the liquid-gas (Fig. 3a) and the liquid-liquid-gas (Fig. 3b) experiments.

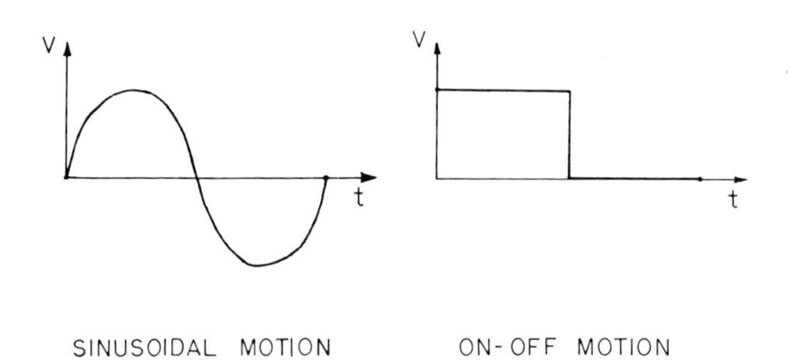

Fig. 4. Floor motion for measurement of shear visco-elasticity.

For the slow flow of the phases, with physical properties

that are uniform and constant in each phase and in the phase boundary, we seek to solve the boundary value problem by first setting down the equation of motion for the phases 'a' and 'b' as

$$\frac{\partial v^i}{\partial t} = \nu^i \left(\frac{\partial^2 v^i}{\partial x^2} + \frac{\partial^2 v^i}{\partial y^2}\right) \qquad (i = a,b) \tag{7}$$

where v represents the velocity in the z direction and ν is the kinematic viscosity. From the sinusoidal nature of the experiment, Equation (7) can be written for the time independent part of the velocity in the form

$$\frac{j\omega}{\nu^i}\overline{v}^i = \left(\frac{\partial^2 \overline{v}^i}{\partial x^2} + \frac{\partial^2 \overline{v}^i}{\partial y^2}\right) \qquad (i = a,b) \tag{8}$$

where

$$v^i = \overline{v}^i e^{j\omega t} \tag{9}$$

Defining the dimensionless groups

$$\overline{V}^i = \overline{v}^i / \overline{v}_f \qquad (i = a,b) \tag{10}$$

$$X = x/y_o \tag{11}$$

and

$$Y = y/y_o, \tag{12}$$

the solution to Equation (8) that satisfies zero velocities at the fixed channels and symmetry about the centerline is

$$\overline{V}^i = \sum_{n=1}^{\infty} \left[\alpha_n^i \sinh(\phi_n^i X) + \beta_n^i \cosh(\phi_n^i X)\right] \sin\left\{(2n-1)\pi Y\right\} \qquad (i = a,b) \tag{13}$$

where

$$\phi_n^i = \sqrt{(2n-1)^2 \pi^2 + j\omega y_o^2/\nu^i} \qquad (i = a,b) \tag{14}$$

In Equations (11) and (12) y_o represents the channel width. The constants α_n^i and β_n^i $(i = a,b)$ are evaluated using the boundary conditions at the floor, the interface and the surface. These boundary conditions can be written down as

$$\overline{V}^b(0,Y) = 1 \tag{15}$$

$$\overline{V}^a(X_o,Y) = \overline{V}^b(X_o,Y) \tag{16}$$

$$-\frac{\overline{v}_f}{y_o}\mu^b\left(\frac{\partial\overline{V}^b}{\partial X}\right) + \frac{\overline{v}_f}{y_o}\mu^a\left(\frac{\partial\overline{V}^a}{\partial X}\right) =$$

$$\frac{d\overline{\tau}^i_{yz}}{dy} = -\frac{\eta^i\overline{v}_f}{y_o^2}\frac{d^2\overline{V}^b}{dY^2} \text{ at } X = X_o = \frac{x_o}{y_o} \tag{17}$$

$$-\frac{\overline{v}_f}{y_o}\mu^a\frac{\partial\overline{V}^a}{\partial X} = \frac{d\overline{\tau}^s_{yz}}{dy} = -\frac{\eta^s\overline{v}_f}{y_o^2}\frac{d^2\overline{V}^a}{dY^2} \text{ at } X = X_1 = \frac{x_1}{y_o} \tag{18}$$

Equation (15) states that the dimensionless velocity of the phase 'b' at the floor equals unity. The continuity of the velocity at the interface is expressed by Equation (16) while Equations (17) and (18) signify the momentum balance at the interface and the surface respectively. The symbols X_o and X_1 represent the dimensionless x coordinate of the surface and the interface respectively. The viscosities η^i and η^s of the interface and the surface are prescribed by Equation (3) with the use of corresponding superscripts on the visco-elastic model parameters.

Substituting Equation (13) into Equations (15) through (18) yields

$$\beta_n^b = \frac{4}{\pi(2n-1)} \tag{19}$$

$$\alpha_n^b = -\beta_n^b G_n \tag{20}$$

$$\beta_n^a = -\left[\beta_n^b \cosh(\phi_n^b X_o) + \alpha_n^b \sinh(\phi_n^b X_o)\right] / \left[F_n \sinh(\phi_n^a X_o) - \cosh(\phi_n^a X_o)\right] \tag{21}$$

and,

$$\alpha_n^a = -\beta_n^a F_n \tag{22}$$

where $F_n = \left[\phi_n^a \tanh(\phi_n^a X_1) + (2n-1)^2 \pi^2 S^{*s}\right] /$

$$\left[\phi_n^a + (2n-1)^2 \pi^2 S^{*s} \tanh(\phi_n^a X_1)\right] \tag{23}$$

$$G_n = \left[\frac{\mu^a}{\mu^b} \phi_n^a \left\{F_n - \tanh(\phi_n^a X_o)\right\} - \left\{\phi_n^b \tanh(\phi_n^b X_o) + (2n-1)^2 \pi^2 S^{*i}\right\} \left\{F_n \tanh(\phi_n^a X_o) - 1\right\}\right] /$$

$$\left[\tanh(\phi_n^b X_o) \frac{\mu^a}{\mu^b} \phi_n^a \left\{F_n - \tanh(\phi_n^a X_o)\right\} - \left\{\phi_n^b + (2n-1)^2 \pi^2 S^{*i} \tanh(\phi_n^b X_o)\right\} \left\{F_n \tanh(\phi_n^a X_o) - 1\right\}\right] \tag{24}$$

$$S^{*s} = \eta^s / \mu^a y_o \tag{25}$$

and

$$S^{*i} = \eta^i / \mu^b y_o \tag{26}$$

The centerline velocity at the surface or the interface can be readily obtained by substituting $Y = ½$, and setting $X = X_o$ and $i = b$ for the interface, or $X = X_1$ and $i = a$ for the surface. This yields

$$\overline{V}^{sc} = \sum_{n=1}^{\infty} (-1)^{(n-1)} \left[\alpha_n^a \sinh(\phi_n^a X_1) + \beta_n^a \cosh(\phi_n^a X_1)\right] \tag{27}$$

and

$$\overline{V}^{ic} = \sum_{n=1}^{\infty} (-1)^{(n-1)} \left[\alpha_n^b \sinh(\phi_n^b X_o) + \beta_n^b \cosh(\phi_n^b X_o)\right] \tag{28}$$

With ω equal to zero, Equations (27) and (28) yield centerline velocities at the surface and the interface for steady motion of the floor. These expressions agree with the results of Wasan, Gupta and Vora (22). In general, however, the centerline velocity given by Equation (27) or (28) is complex and can be written in terms of a phase lag θ and amplitude ratio A

$$\theta^{s} = \tan^{-1}\left[\mathrm{Im}(\overline{V}^{sc})/\mathrm{Re}\ (\overline{V}^{sc})\right] \tag{29}$$

$$\theta^{i} = \tan^{-1}\left[\mathrm{Im}(\overline{V}^{ic})/\mathrm{Re}\ (\overline{V}^{ic})\right] \tag{30}$$

and,

$$A^{s} = \left|\overline{V}^{sc}\right| \tag{31}$$

$$A^{i} = \left|\overline{V}^{ic}\right| \tag{32}$$

Equations (31) and (32) follow from writing the centerline velocity in its dimensionless form as

$$\begin{aligned}\overline{V}^{c} &= \mathrm{Im}\ (\overline{V}^{c}\ e^{j\omega t}) = \mathrm{Im}\left[\left|\overline{V}^{c}\right| e^{j(\omega t+\theta)}\right]\\ &= \left|\overline{V}^{c}\right|\ \sin\ (\omega t + \theta)\end{aligned} \tag{33}$$

and integrating Equation (33) to yield the maximum centerline displacement $\overline{L}^{c}$ given by

$$\overline{L}^{c} = \left|\overline{V}^{c}\right|/\omega \tag{34}$$

The maximum floor displacement can be obtained likewise by recognizing from Equation (6) that

$$\overline{V}_{f} = \mathrm{Im}\ (e^{j\omega t}) = \sin\ (\omega t) \tag{35}$$

so that the maximum floor displacement $\overline{L}_{f}$ is given by

$$\overline{L}_{f} = 1/\omega \tag{36}$$

The ratio of Equation (34) to Equation (36) yields Equations (31) and (32). The analysis indicates that the amplitude ratio and phase lag of the surface and the interface to sinusoidal floor motion are dependent on the viscoelastic parameters of the surface and the interface. Therefore, it is possible, in theory at least, to obtain the viscoelastic parameters from experimental data on the amplitude ratio and phase lag. In practice, however, it is more convenient to impart a periodic on-off floor motion so that, in the following section, we extend the above analysis using the principle of superposition.

2. *On-Off Floor Motion*

In this case, the floor velocity equals $\bar{v}_f$ for half the cycle and equals zero for the rest of the cycle, or in terms of dimensionless velocities,

$$V_f = \begin{cases} 1 & 0 < \tau \leqslant \tfrac{1}{2} \\ 0 & \tfrac{1}{2} < \tau \leqslant 1 \end{cases} \tag{37}$$

where $\tau = t/P$ and P is the period of the cycle. The above periodic floor motion can be expanded in terms of Fourier series to yield

$$V_f = \tfrac{1}{2} + \sum_{m=1}^{\infty} a_m \sin\left[2(2m-1)\pi\tau\right] \tag{38}$$

where

$$a_m = \frac{2}{(2m-1)\pi} \tag{39}$$

Using the principle of superposition, it is possible to construct the solution for the present problem from the solution to the sinusoidal case. Using Equation (38), the solution can be written as

$$V^{sc} = \frac{\left|\bar{V}^{sc}_{\omega=o}\right|}{2} + \frac{2}{\pi}\sum_{m=1}^{\infty} \frac{\left|\bar{V}^{sc}_{\omega=\omega_m}\right|}{(2m-1)} \sin\left\{2(2m-1)\pi\tau+\theta^s_m\right\} \tag{40}$$

and

$$V^{ic} = \frac{\left|\bar{V}^{ic}_{\omega=o}\right|}{2} + \frac{2}{\pi}\sum_{m=1}^{\infty} \frac{\left|\bar{V}^{ic}_{\omega=\omega_m}\right|}{(2m-1)} \sin\left\{2(2m-1)\pi\tau+\theta^i_m\right\} \tag{41}$$

where

$$\omega_m = 2(2m-1)\pi/P \tag{42}$$

and

$$\theta_m = \theta\Big|_{\omega=\omega_m} \tag{43}$$

It is convenient to integrate Equations (40) and (41) to yield expressions for centerline displacements at the surface L^{sc} and the interface L^{ic} given by

$$L^{sc} = \frac{\left|\bar{V}^{sc}_{\omega=o}\right| \tau}{2} + \frac{1}{\pi^2} \sum_{m=1}^{\infty} \frac{\left|\bar{V}^{sc}_{\omega=\omega_m}\right|}{(2m-1)^2} \left[\cos \theta^s_m - \cos\left\{2(2m-1)\pi\tau + \theta^s_m\right\}\right] \tag{44}$$

and

$$L^{ic} = \frac{\left|\bar{V}^{ic}_{\omega=o}\right| \tau}{2} + \frac{1}{\pi^2} \sum_{m=1}^{\infty} \frac{\left|\bar{V}^{ic}_{\omega=\omega_m}\right|}{(2m-1)^2} \left[\cos \theta^i_m - \cos\left\{2(2m-1)\pi\tau + \theta^i_m\right\}\right] \tag{45}$$

where L is the dimensionless displacement defined by

$$L = l/v_f P \tag{46}$$

III. EXPERIMENTAL

A. Apparatus

The deep channel viscous traction viscometer (Figure 5) has been used to measure surface and interfacial viscosities from experiments involving steady floor motion. This instrument was used by Mohan, Malviya and Wasan (12) to obtain surface viscoelastic parameters of polymer solutions by conducting stress relaxation experiments. To the authors' knowledge there exist no measurements of interfacial viscoelastic parameters in a liquid-liquid system using this instrument.

A description of the instrument and motive power is given by Gupta and Wasan (23). It has been shown in literature (24) that the development of the circular channel into a linear canal is a valid procedure for ratios of outer to inner radii close to unity. Such is the case in our instrument.

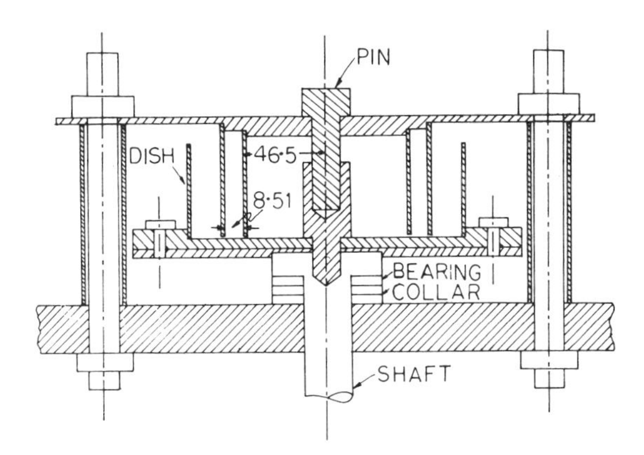

Fig. 5. Schematic of the viscous traction viscometer.

B. Material Studied

Benzene-water interfaces were studied with varying coverage of crystallized Bovine Plasma Albumin (BPA) supplied by Armour Pharmaceutical Company, Kankakee, Illinois. The sample was kept below 5°C for optimal stability.

Monolayers of BPA were spread from a 0.1% solution in 60 percent isopropylalcohol made 0.5 M with sodium acetate. The solution of Bovine Plasma Albumin was prepared by first dissolving the crystals in water and then adding the required amounts of sodium acetate and isopropyl alcohol in that order. The solution was also stored below 5°C.

In addition to monolayer studies, preliminary studies were conducted with adsorbed films of polymer calgon 815 at soltrol 130-water and Gach Saran Crude - water interfaces. These results will be presented in a separate publication.

C. Experimental Procedure

The physical properties of the bulk phases were determined using standard techniques (Table 1). The densities were determined using a specific gravity bottle. A set of

Ostwald Viscometers were used to measure the kinematic viscosities of both water and benzene. The interfacial tension at different BPA concentrations was determined using the Cenco du Nouy tensiometer.

TABLE 1

Physical properties of Benzene and Water

Benzene		*Water*		*Interfacial Tension dyne/cm*
Density gm/cc	*Viscosity poise*	*Density gm/cc*	*Viscosity poise*	
0.8683	0.00555	0.9962	0.00795	32

The aqueous and benzene phases were taken in the viscous traction viscometer and calculated amounts of the BPA solution was spread using a microsyringe. The channel was then placed and fastened. The on-off floor motion was imparted to the dish two minutes after spreading the monolayer. The period of the on-off cycle was fixed at 40 seconds. The surface or interfacial displacement was recorded during consecutive cycles after 10 minutes of operation. All data were obtained at 82 $\pm$ 1°F. The depth of the two phases was measured using a micrometer gauge. The floor speed was determined by timing the dish under steady rotation. In what follows, all quantities are used in cgs units: l, x_o, x_1, y_o, *cm*; t, P, t_o^s, t_o^i, *sec*; v_f, *cm/sec*; ε_o^s, ε_o^i, *s.p.*; ρ^a, ρ^b, *gm/cc*; μ^a, μ^b, poise; π, dyne/cm.

IV. RESULTS AND DISCUSSION

A. Theoretical Results

The two parameter Maxwell model was used to represent

the interface. The results of a parametric study are depicted in Figures 6 to 8. These figures show the surface and interfacial centerline displacements for various values of interfacial viscoelastic parameters. Figure 6 reveals that even at a high value of the interfacial relaxation time (t_o^i = 1000) the surface centerline displacement progressively increases with time at values of interfacial shear viscosity smaller than 0.01; there is no snap back. Similar behavior is also shown on Figures 7 and 8. These show clearly that the absence of a snap back does not necessarily indicate that the interface is purely viscous. At higher values of interfacial shear viscosity, the surface displaces backward at first, and then moves forward. This is followed by a recoil which continues into the next cycle until the effect of the forward floor motion reaches the surface. This explains the initial backward displacement and the delayed forward motion of the surface. Such a behavior was observed in BPA monolayers at 0.5 m^2/mg. The above discussion holds for displacement at a liquid-liquid interface also. What needs to be emphasized is that the delayed forward displacement occurs earlier at the interface than at the surface. This is to be expected since the interface is closer to the floor than the surface, so that the interface receives the impact of the floor motion earlier. This suggests that the displacement characteristics both at the surface and the interface depend on the depth of the aqueous and oil phases in addition to depending on the viscoelastic parameters.

Figure 8 shows the effect of the relaxation time on the displacement patterns of the surface and the interface. A value of 0.1 *sp.* was chosen for the interfacial shear viscosity. An increase in interfacial relaxation time shows a change in displacement pattern from a progressively increasing displacement, to a displacement with characteristic snap back,

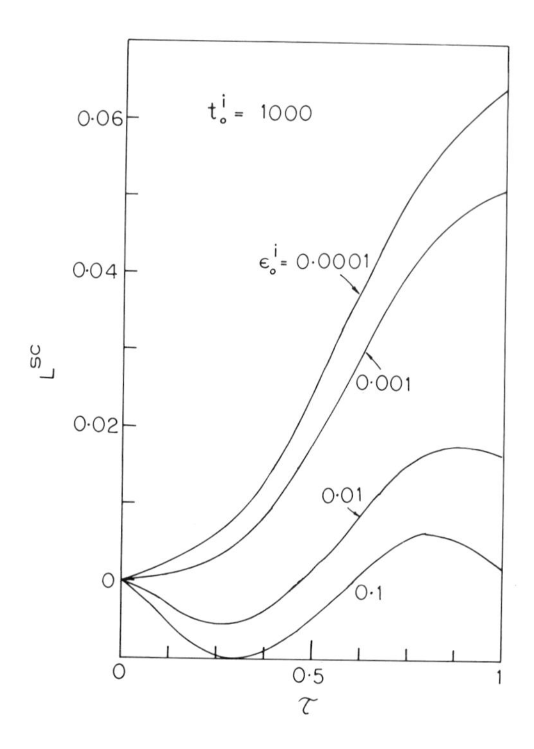

Figure 6. Plot of dimensionless surface centerline displacement versus dimensionless time ($x_o = 0.511$, $x_1 = 0.851$, $y_o = 0.851$, $P = 40$, $\varepsilon_o^s = t_o^s = 0$, $\mu^a = 0.005$, $\mu^b = 0.008$, $\rho^a = 0.8$, $\rho^b = 1$).

to an oscillatory type of displacement.

An important consideration that follows from this parametric study is that for accurate determination of viscoelastic parameters, it is necessary to choose design and operational features (liquid depths) of the viscometer such that the surface or the interfacial displacement exhibits the maximum sensitivity to slight variations in the viscoelastic parameters. It is easy to observe that these optimum features depend on the bulk and interfacial properties of the phases involved. In the light of this complexity, it is suggested that these procedures be adopted for the specific

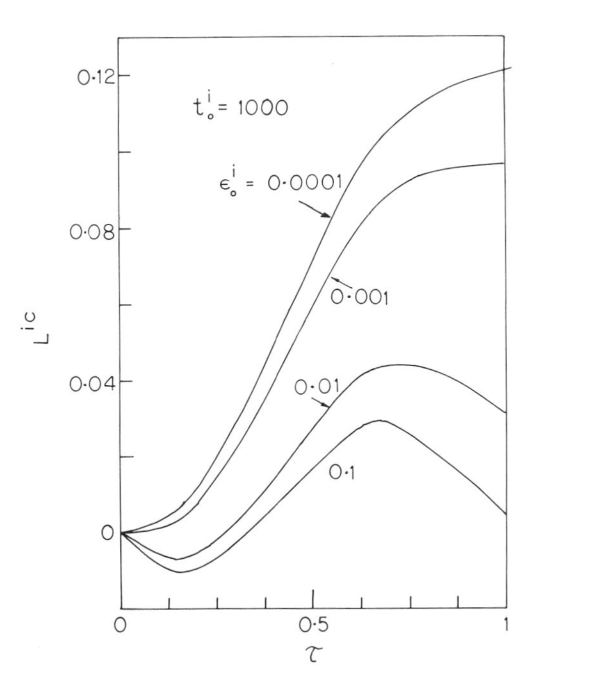

Fig. 7. Plot of dimensionless interfacial centerline displacement versus dimensionless time ($x_o = 0.511$, $x_1 = 0.851$, $y_o = 0.851$, $P = 40$, $\varepsilon_o^s = t_o^s = 0$, $\mu^a = 0.005$, $\mu^b = 0.008$, $\rho^a = 0.8$, $\rho^b = 1$).

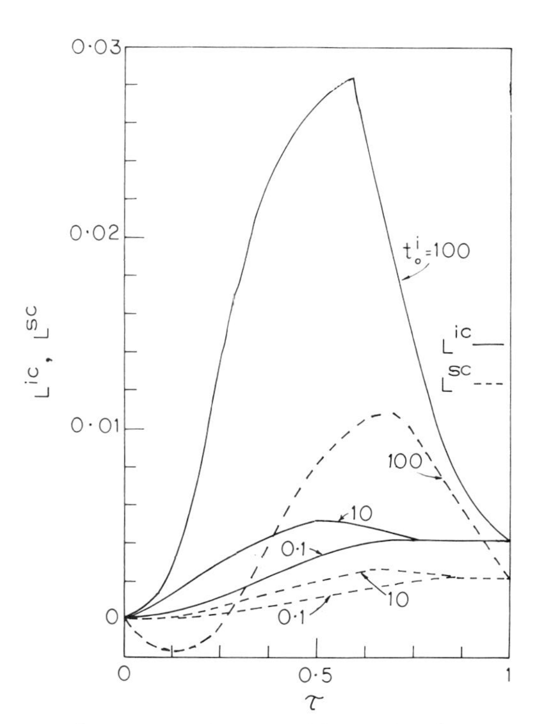

Fig. 8. Plot of dimensionless surface and interfacial centerline displacements versus dimensionless time ($x_o = 0.511$, $x_1 = 0.851$, $y_o = 0.851$, $P = 40$, $\varepsilon_o^s = t_o^s = 0$, $\mu^a = 0.00555$, $\mu^b = 0.00795$, $\rho^a = 0.868$, $\rho^b = 1$).

viscometric design and fluids under consideration.

B. Experimental Results

The surface displacement data for deposited BPA monolayers at benzene-water interface are shown on Figures 9 and 10. It is interesting to observe that at a coverage of $0.7 m^2/mg$ (Figure 9), the surface exhibits the characteristic snap back while at a coverage of $0.8 m^2/mg$ (Figure 10), there is no snap back. The absence of the snap back is also observed in the interfacial displacement as shown in Figure 11.

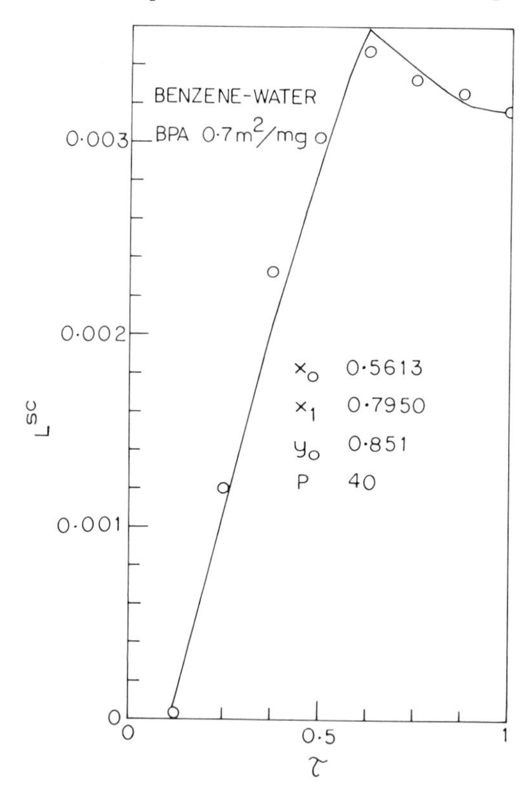

Fig. 9. Experimental data and best fit for BPA coverage of $0.7 m^2/mg$ at benzene-water interface-surface displacement data.

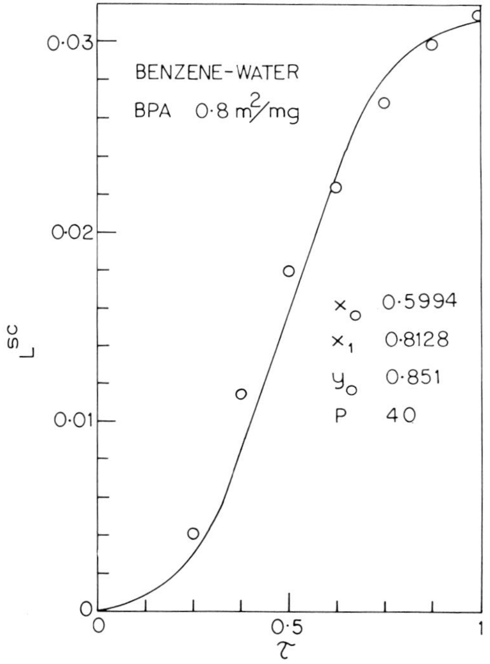

Fig. 10. Experimental data and best fit for BPA coverage of 0.8 m^2/mg at benzene-water interface—surface displacement data.

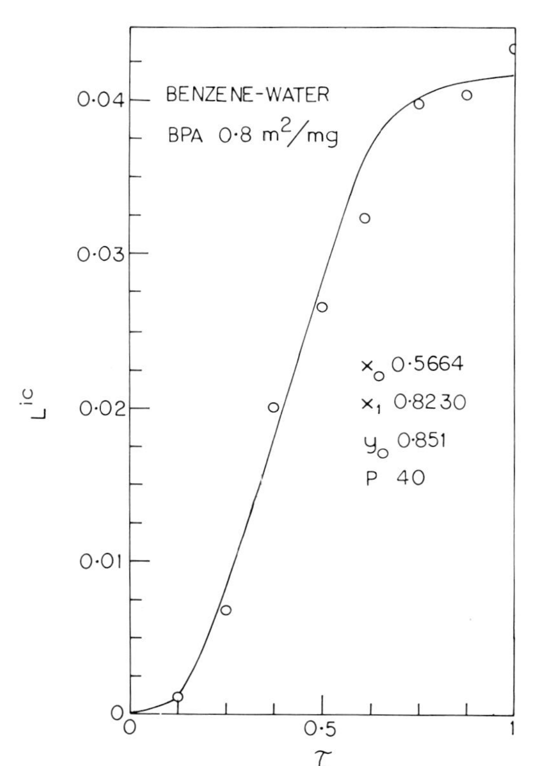

Fig. 11. Experimental data and best fit for BPA coverage of 0.8m^2/mg at benzene-water interface—interface displacement data.

The Maxwell model was employed to obtain the best fit to the experimental data. A two-parameter search was conducted on the interfacial shear viscosity and interfacial relaxation time such that the mean square error between the experimental and theoretical displacements was minimized. In the theoretical calculations, the shear viscosity and relaxation time of the benzene-gas surface were taken to be negligible corresponding to a pure surface. The parameter search was conducted using the method of Rosenbrock (25). An initial guess for the interfacial shear viscosity required in the search procedure was obtained by observing that Equations (45) and (46) yield

$$L^{sc}\Big|_{\tau=1} = \frac{\left|\bar{V}^{sc}_{\omega=o}\right|}{2} \tag{48}$$

and

$$L^{ic}\Big|_{\tau=1} = \frac{\left|\bar{V}^{ic}_{\omega=o}\right|}{2} \tag{49}$$

Therefore, the surface or interfacial displacement at $t=P$ yield the corresponding velocities in a steady experiment. The interfacial shear viscosity may then be readily calculated using the analysis of Wasan, Gupta and Vora (22) with $\varepsilon=0$. The initial guess for the relaxation time was chosen arbitrarily.

The values of the interfacial shear viscosity and relaxation time obtained from the parameter search are plotted in Figure 12 as a function of the interfacial concentration of BPA. The value of interfacial pressure obtained using a Cenco du Nouy tensiometer is also shown in Figure 12 which indicates that the film is of the expanded type. Furthermore, the plots of interfacial shear viscosity and relaxation time show a steep rise at an interfacial BPA concentration of about $1.2 mg/m^2$. An explanation for this stems from the β-keratin structure of proteins at an oil-water interface (14). Using the X-ray data of Astbury (26), the length

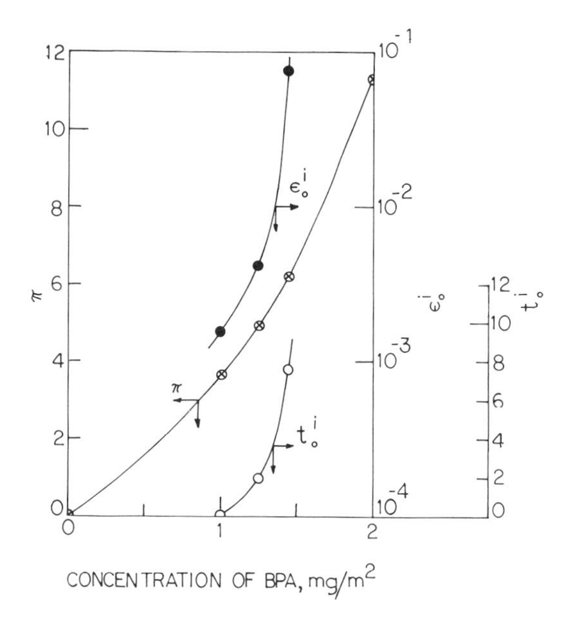

Fig. 12. Plots of interfacial pressure, shear viscosity and relaxation time versus interfacial concentration of BPA at benzene-water interface.

of an amino acid residue along the polypeptide chain is 3.5 Å and the interchain spacing is 4.5 Å. Furthermore, the number of amino acid residues per unit is 588 (27) and the molecular weight of BPA lies in the range 67000-69000. These values indicate that the just-close-packed structure corresponds to an interfacial BPA concentration of 1.2 - 1.4 mg/m^2. It is in this concentration range that the interfacial rheological parameters show a steep rise as shown in Figure 12.

As remarked earlier, the method proposed here permits the prediction of interfacial shear viscoelastic parameters from a liquid-liquid-gas experiment by measuring the displacement either at the liquid-gas interface or at the liquid-liquid interface, thereby providing a check for internal consistency. Such a check was made at a BPA coverage of 0.8 m^2/mg. The viscoelastic parameters obtained using these two

TABLE 2

Comparison of Viscoelastic Parameters obtained from Displacements at Liquid-Liquid and Liquid-Gas Interfaces.

Interfacial Coverage of BPA m^2/mg	*Liquid-Liquid Displacement Study*		*Liquid-Gas Displacement Study*	
	ε_o^i, *s.p.*	t_o^i, *sec*	ε_o^i, *s.p.*	t_o^i, *sec*
0.8	0.0049	1.96	0.0042	2.0

approaches are shown in Table 2. The close agreement between the parameters establishes (i) the validity of the method proposed here and (ii) the possible use of this method for the determination of viscoelastic parameters for systems such as Crude oil-aqueous solutions.

V. CONCLUSIONS

The following conclusions can be drawn from the present study:

(i) a new method for the measurement of shear viscoelastic parameters of liquid-liquid interfaces from displacement studies at *either* the liquid-gas interface *or* the liquid-liquid interface is presented.

(ii) the shear viscoelasticity at a liquid-liquid interface can be determined from displacement data at liquid-gas interfaces *alone*. Therefore, the proposed method is useful for the measurement of interfacial rheological parameters in opaque systems such as crude oil-aqueous solutions.

(iii) the absence of a snap back in the displacement does

not imply that the interface is purely viscous. In fact, the existence of differing displacement patterns is shown both from a parametric study and from an experimental study. The displacement pattern depends on the design and operational features of the instrument in addition to the viscoelastic parameters.

(iv) the need for a sensitivity analysis to obtain accurate values of the viscoelastic parameters is pointed out.

(v) the behavior of BPA monolayers at benzene-water interface indicates that the interfacial shear viscosity and relaxation time are much more sensitive to molecular packing and structural changes than the interfacial pressure. Therefore, the interfacial rheological parameters may be used as diagnostic tools to characterize the physical state of monomolecular films, to study phase transitions in monolayers and to measure the energies of molecular interaction and of deformation in macromolecular films.

Acknowledgement

The authors wish to acknowledge the assistance of Yiu Sum Wong in collecting experimental data and the timely offer of BPA sample by Professor Roth of the Biology Department of the Illinois Institute of Technology. This work was supported by NSF Grant No. GK 43135.

VI. REFERENCES

1. Schulman, J. H. and Cockbain, E. G., Trans. Faraday Soc., 36, 651 (1940).
2. Vold, R. D. and Mittal, K. L., J. Colloid Interface Sci., 38 (2), 451 (1972).

3. Petrov, A. A., Pozdnyzhev, G. N. and Borisor, Khinya i Tekhnologiya Topliv i Masel, 11-4 (Mar. 1969).
4. Levchenko, D. N., Khudyakova, A. D. and Ratich, L. I., ibid., 21-5 (Oct. 1970).
5. Berridge, S. A., Thew, M. T. and Loriston-Clarke, A. G., J. Inst. Petrol., 54 (539), 333-45 (Nov. 1968).
6. Carroll, B. J., "Emulsion Stability and Demulsification: A Literature Review, Project Coordinator, Lucassen, J. PPS73 1226 (BD/See), (July 30, 1973).
7. Clayton, W., "The Theory of Emulsion," J and A Churchill, London (1954).
8. Kung, H. C. and Goddard, E. D., J. Phys. Chem., 67, 1965 (1963);________, ibid., 68, 3465 (1964).
9. Becher, P. and Del Vecchio, A. J., ibid., 68, 3511 (1964).
10. Mysels, K. J., Shinoda, K. and Frankel, S., "Soap Films," Pergamon Press, New York (1959).
11. Mannheimer, R. J. and Schechter, R. S., J. Colloid Interface Sci., 32 (2), 225 (1970).
12. Mohan, V., Malviya, B. K. and Wasan, D. T., Can. J.Chem. Eng., In Press (1976).
13. Schechter, R. S., Kott, A. T., Gardner, J. W. and De Groot, W., J. Colloid Interface Sci., 47, 265 (1974).
14. Cumper, C. W. N. and Alexander, A. E., Trans. Faraday Soc., 46, 235 (1950).
15. Criddle, D. W. and Meader, A. L., J. Appl. Phys., 26 (7), 838 (1955).
16. MacRitchie, F. and Alexander, A. E., J. Colloid Science, 16, 57(1961).
17. Biswas, B. and Haydon, D. A., Proc. Roy. Soc. (London), 271, Ser. A, 317 (1963).
18. Motomura, K., J. Phys. Chem., 68, 2826 (1964).
19. Trapeznikov, A. A. and Zotova, K. V., Kolloydnyi Zhurnal, 27 (4), 614 (1965).
20. Boyd, J. and Sherman, P., J. Colloid Science,34,76(1970).
21. Gardner, J. and Schechter, R. S., Paper presented at the International Conference on Colloids and Surfaces, San Juan, Puerto Rico (June 21-25, 1976).
22. Wasan, D. T., Gupta, L. and Vora, M. K., AIChE J., 17, 1287 (1971).
23. Gupta, L. and Wasan, D. T., Ind. Eng. Chem. Fundam., 13, 26 (Feb. 1974).
24. Pintar, A. J., Israel, A. B. and Wasan, D. T., J. Colloid Interface Sci., 37(1), 52 (1971) .
25. Rosenbrock, H. H. and Storey, C., "Computational Methods for Chemical Engineers," Pergamon Press (1966).
26. Astbury, Trans. Faraday Soc., 29, 133 (1933); "Fundamentals of Fibre Structure," Oxford Press (1933).

27. Waugh, D. F., "Advances in Protein Chemistry," (Eds. Anson, M. L., Bailey, K. and Edsall, J. T.), Vol. IX, Academic Press (1954).

APPLICATION OF EINSTEIN'S THEORY TO THE FLOW OF CONCENTRATED SUSPENSIONS

Thomas Gillespie
Saginaw Valley State College

ABSTRACT

Einstein's equation for the relative viscosity of a suspension can be expressed as

$$\eta_r = \frac{1 + \phi_{eff}/2}{(1 - \phi_{eff})^2}$$

where η_r is the relative viscosity and ϕ_{eff} is the effective value of the fractional volume solids concentration ϕ. Examination of literature data using this equation has indicated that

$$\phi_{eff} = \phi + k\phi^2$$

where k is a constant. This expression can be derived from kinetic considerations utilizing the concept of liquid being occluded between interacting particles. At a critical value of the fractional volume solids concentration ϕ_M, all of the liquid will be occluded. For literature data obtained with uniform spheres ϕ_M ranges from 0.56 to 0.64 and agrees reasonably well with sedimentation volume measurements and packing experiments with macroscopic spheres.

I. INTRODUCTION

In 1911 Einstein published a brief article (1) in which he corrected his prior analysis of the effect of dispersed solids on the flow of a liquid (2). Using both of these publications one can easily derive the following equation:

$$\eta_r = \frac{1 + \phi/2}{(1 - \phi)^2} \qquad [1]$$

where η_r is the relative viscosity and ϕ is the fractional volume solids concentration.

When the concentration is very low, Equation [1] reduces to Einstein's well known equation

$$\eta_r = 1 + 2.5\ \phi \qquad [2]$$

This very useful equation has been the starting point in many attempts to develop equations for suspensions at moderate and high concentrations. Such attempts introduce new constants which one would like to relate to important variables such as particle size distribution and particle aggregation. In addition, it would be of value to determine these constants in independent experiments. For example, the recent work of Lewis and Nielsen (3) suggests the possibility of relating suspension viscosity data and data from sedimentation volume measurements.

In their work, Lewis and Nielsen used the Mooney equation which is one of the equations which have been derived using Equation [2] and can be expressed in the form

$$\eta_r = \exp\left[\frac{k_E\ \phi}{1 - \phi/\phi_M}\right] \qquad [3]$$

where ϕ_M is the maximum allowable fractional solids volume. k_E is the Einstein coefficient and it is equal to 2.5 under the ideal conditions specified in Einstein's analysis (1,2). Although Equation [3] fits considerable viscosity data up to fractional volume solids concentrations of the order of 0.4, the values for the maximum solids concentration, ϕ_M have in many cases been too large to be realistic.

It is possible that the difficulty lies in the reduction of Equation [1] to the linear form (Equation [2]) before attempts to introduce the effects of concentration. When such effects are introduced into Equation [1] directly simple equations result which fit suspension data and give reasonable values for the maximum solids volume fraction.

II. EXTENSION OF EINSTEIN'S ANALYSIS

Equation [1] does not take into account hydrodynamic interactions between particles in a flowing suspension. One possible method of taking such effects into account is to follow the suggestion of Vand (4) that liquid will be occluded when particles interact and that this will make the effective value of the solids volume fraction ϕ_{eff} larger than ϕ. If ν is the coordination number, i.e. the number of particles in contact with a given particle and one neglects Brownian movement, electric double layer effects and particle size distribution (4,5)

$$\nu = 8\ \phi \qquad [4]$$

If v_o is the volume of liquid occluded between two particles in contact, each of which has a volume v, then the effective solids volume fraction would be given by

$$\phi_{eff} = \phi + k\phi^2 \qquad [5]$$

$$k = 4\ v_o/v \qquad [6]$$

Substituting in Equation [1]

$$\eta_r = \frac{1 + \phi_{eff}/2}{(1 - \phi_{eff})^2} \qquad [7]$$

As the effective solids volume fraction approaches unity ϕ should approach ϕ_M. Hence ϕ_M can be determined from

$$k = \frac{1 - \phi_M}{\phi_M^{\ 2}} \qquad [8]$$

III. ANALYSIS OF EXPERIMENTAL DATA

The value of ϕ_{eff} for a particular value of ϕ can be obtained from measurements of the relative viscosity and Equation [1] in the form

$$\phi_{eff} = 1 + \frac{1}{4\ \eta_r}\left[1 - (24\ \eta_r + 1)^{1/2}\right] \qquad [9]$$

According to Equation [5] a plot of ϕ_{eff}/ϕ should be linear as illustrated in Figure 1. The data is for 45-60 micron glass beads in oil (3).

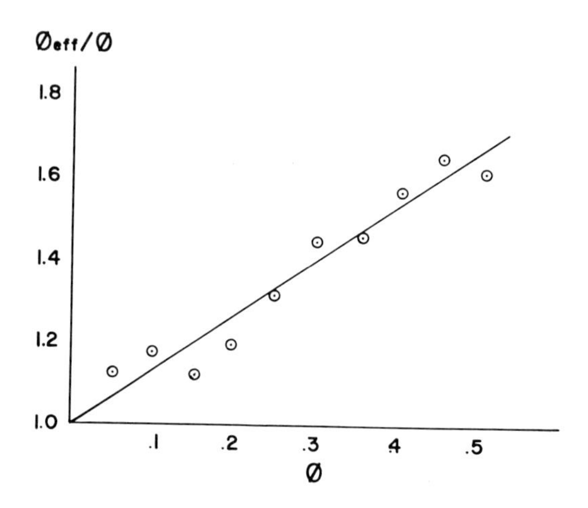

Fig. 1. Illustration of the proportionality between ϕ_{eff}/ϕ and ϕ as suggested by Equation [5].

When some of the data in the literature are treated in this way, the intercept in plots such as Figure 1 is greater than unity. Equation [5] is based on the assumption that the Einstein coefficient, k_E, is equal to 2.5. To take into account variations from this value Equation [5] may be written

$$\phi_{eff} = \alpha\phi + k\phi^2 \qquad [10]$$

$$\alpha = k_E/2.5 \qquad [11]$$

Modifying Equation [8] in a similar manner to take into account variations in k_E,

$$k = \frac{1 - \alpha\phi_M}{\phi_M^2} \qquad [12]$$

The maximum solids volume fraction can be calculated from

$$\phi_M = \frac{-\alpha + (\alpha^2 + 4k)^{1/2}}{2k} \qquad [13]$$

From the slope and intercept of plots such as Figure 1 the constants α, k_E, and k can be determined. In some cases scatter at low concentration made it preferable to determine k_E and k using the usual method of extrapolation to infinite dilution of plots suggested by one of the many viscosity - concentration relations such as Equation [2]. Typical data are given in Table 1.

TABLE 1

Typical Constants Derived from Literature Viscosity Data

Particle Diameter (microns)	k_E	α	k	ϕ_M Mooney Equation	ϕ_M Equation [7]	Reference
0.099	2.54	1.02	1.34	0.74	0.56	6
0.148	2.53	1.01	1.18	0.81	0.58	6
0.249	2.55	1.02	0.96	0.84	0.63	6
0.342	2.51	1.01	0.88	0.87	0.63	6
0.424	2.53	1.01	0.85	0.88	0.64	6
0.871	2.55	1.02	0.80	0.89	0.64	6
3	2.50	1.00	0.94	0.86	0.63	Present work
45-60	2.52	1.01	1.29	0.76	0.57	3
20-30	2.50	1.00	0.98	0.83	0.62	7

The first six systems referred to in Table 1 were monodisperse polystyrene latex. The seventh system was cross-linked polystyrene beads in dioctyl phthalate. The last two systems were glass beads in an oil.

IV. DISCUSSION

All of the values in Table 1 for the maximum solids volume fraction obtained by using the Mooney Equation are too large to be reasonable. Using Equation [7] the values for ϕ_M are in the range 0.56-0.64 which is similar to the values obtained in sedimentation volume experiments and packing experiments with macroscopic spheres (8,9). For the 45-60 micron diameter system in Table 1, Lewis and Nielsen (3) measured a sedimentation volume of 0.59 as compared with the value of 0.57 for ϕ_M. Using the value of 0.59 for ϕ_M the viscosities predicted by the Mooney Equation and Equation [7] were calculated and compared with the experimental data of Lewis and Nielsen (3). Figure 2 illustrates the result.

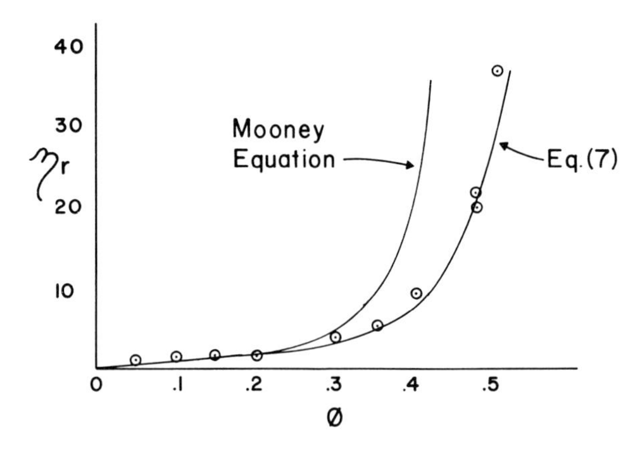

Fig. 2. A comparison of predicted viscosity and the experimental results of Lewis and Nielsen. The data for the curves was obtained using ϕ_M equal to 0.59 which is the value of the sedimentation volume reported by Lewis and Nielsen (3).

The coordination number at the maximum possible concentration can be calculated from Equation [4]. The values for the maximum coordination number varied from 4.5 to 5.1. In experiments with large spheres Bernal and Mason reported a coordination number of 5.5 for loose random packing and 6.7 for close random packing (8).

In a novel set of experiments Barclay, Harrington and Ottewill (10) have measured the pressure required to squeeze the water out of latex. They found that the pressure rose dramatically at ϕ equal to 0.60 in one case and ϕ equal to 0.64 in another case. This is in the range for ϕ_M determined in the present work.

The volume of liquid which may be considered to be occluded when two particles interact in a sheared suspension can be estimated from the values of k in Table 1, the particle diameter and Equation [6] which neglects a number of effects which may be important. Vand (4) has suggested that v_o/v should be of the order of 0.5. Experiments with aggregated suspensions indicate that v_o/o is difficult to measure with precision and appears to range between 0.12 and 0.52, (3). Calculations for the systems in Table 1 yield v_o/v equals 0.28 ± 0.8.

REFERENCES

1. Einstein, A., Ann. Physik. 34, 591 (1911).
2. Einstein, A., Ann. Physik. 19, 289 (1906).
3. Lewis, T. B., and Nielsen, L. E., 12, 421 (1968).
4. Vand, V., J. Phys. and Colloid Chem. 52, 277 (1948).
5. Manley, R. St. J. and Mason, S. G., Canad. J. Chem. 32 (1954).
6. Saunders, F. L., J. Colloid Sci. 16, 13 (1961).
7. Robinson, J. V., Trans. Soc. Rheology, 1, 15 (1957).
8. Bernal, J. D. and Mason, J., Nature, 188, 910 (1960).
9. Scott, G. D., Nature, 188, 908 (1960).
10. Barclay, L., Harrington, A. and Ottewill, R. H., Kolloid Z. u. Z. Polymere, 250, 655 (1972).

HIGH-INTERNAL-PHASE-RATIO EMULSIONS; PUMPING STUDIES

Kenneth J. Lissant
Tretolite Division/Petrolite Corporation

ABSTRACT
The rheology of High-Internal-Phase-Ratio emulsions has received only limited attention. Single point viscometers yield meaningless data. Limited attempts in the use of laboratory rheometers have also produced confusing results. This paper reports the results of pumping High-Internal-Phase Ratio emulsions through a small, pilot scale, pipe loop.

I INTRODUCTION

Previous studies on High-Internal-Phase Ratio emulsions have shown them to have unusual rheology.[1,2] Attempts have been made to measure precisely the pumping properties of these materials, and in several cases, difficulties have been encountered in securing reproducible results.[3,4] This study was undertaken in an attempt to resolve some of the fundamental questions concerning High-Internal-Phase-Ratio emulsion rheology.

II EXPERIMENTAL PROCEDURE

The test were conducted in an experimental pipe loop consisting of an emulsion reservoir, a Viking pump driven by a 10 H.P. electric motor through a hydraulic variable speed drive, and a test loop into which various test sections could be inserted. The test sections each consist of an 8-foot length of pipe with pressure transducers installed in taps 1' in from each end thus making it possible it measure the pressure drop across a 6' section.

Preliminary tests with various size pipe indicated that pipe sizes less than 1" gave results which did not scale up well to larger commercial equipment. Therefore, a nominal 1" pipe was used for the test section to obtain the reported data.

The system had a capacity of approximately 20 gallons. The material could either be recirculated through the test section and back to the reservoir or could be diverted from the reservoir to a 5 gallon pail mounted on a balance. Pumping rates were determined by measuring the time required to pump 10 kilograms of emulsion.

It was found that pressure differentials and pumping rates could be duplicated with reasonable precision.

The emulsion used in these tests was a water-in-oil emulsion stabilized by a proprietary emulsifier supplied by the Tretolite Division of Petrolite Corporation and designated EM-12. The emulsifier is completely soluble in kerosene and the external phase consisted of a 20% solution of EM-12 in deodorized kerosene. The external phase is available from Petrolite Corporation, designated EM-120.

Tests were made on emulsions having internal to external phase ratios of 80/20, 85/15, 90/10, and 95/5.

The raw data consisted of pressure drops across the test section in pounds per square inch and pumping rates expressed as the time, in seconds, required to pump 10 kilograms. From these data the apparent viscosity, in centipoise, and the shear rate were calculated assuming non-turbulent flow.

The required amount of EM-120 was placed in an, open head, 55 gallon drum and recirculated through a Jabsco pump and 1" hose lines. While recirculating the EM-120, the required amount of tap water was slowly added until an emulsion of the desired phase ratio was obtained. A portion of this emulsion was then transferred to the pipe loop and tests conducted.

III EXPERIMENTAL RESULTS

Table 1 and Figures 1,2,3, and 4 illustrate the data obtained working with an 80% internal phase emulsion. A study of Table 1 shows that three cycles of tests were made on this emulsion. Initially, the emulsion was pumped at a shear rate of about 1600 reciprocal seconds for a period of 12 minutes.

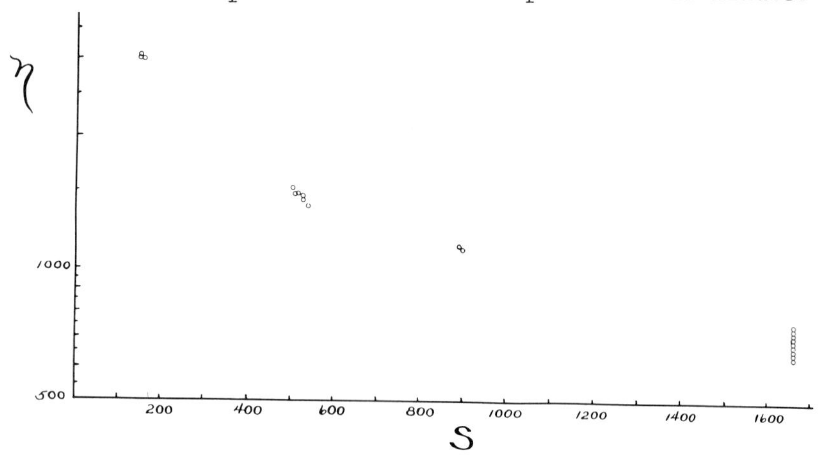

Fig. 1. - 80% Internal Phase, Water-In-Oil Emulsion. First cycle.

TABLE I

80/20 Phase Ratio Water In Oil Pumping Data

η	S	COMMENT	η	S	COMMENT
634	1665	High shear wind up	987	1150	4 min
662	"		991	"	5 min
668	"	1 min	1513	406	
709	"	3 min	1486	411	
721	"	4 min	Let stand overnight		
734	"	5 min	1207	167	Relaxed
736	"	6 min	1216	166	
741	"	7 min	839	483	Winding up
750	"	8 min	881	472	1 min
753	"	9 min	932	462	2 min
756	"	10 min	959	462	
754	"	11 min	992	458	3 min
756	"	12 min	1028	452	5 min
1133	904		1077	434	
1143	891		1094	439	6 min
1474	527		1117	439	
1485	518		1143	433	7 min
1440	527		1175	427	8 min
1383	545		1198	422	
1503	502		1216	419	10 min
1471	510		1230	419	
3050	148		1245	416	12 min
3010	148	Unwinding	1254	416	
2963	149		1247	422	
Let stand 40 min			1308	395	
2323	164		1304	411	17 min
2060	183		5259	37	
1938	193		882	1150	
6609	28		1060	1150	10 min
922	1150	Winding up	968	1150	10 Kg dumped in
949	"	1 min	1062	1154	
968	"	2 min	2634	194	
978	"	3 min			

During this time, the apparent viscosity rose from about 630 to about 750. Then, a series of tests were made covering the shear range from 900 reciprocal seconds down to about 150 reciprocal seconds. At the end of this series, it was found that the emulsion was beginning to relax. These data are shown on Figure 1 and on Figure 4 as the open circles.

The emulsion was then allowed to stand for 40 minutes and some low shear determinations made before winding the emulsion up again at a shear rate of 1150 reciprocal seconds. Two mid-range determinations were then made and the emulsion was allowed to stand overnight. These data are depicted on Figure 2 and on Figure 4 as open squares.

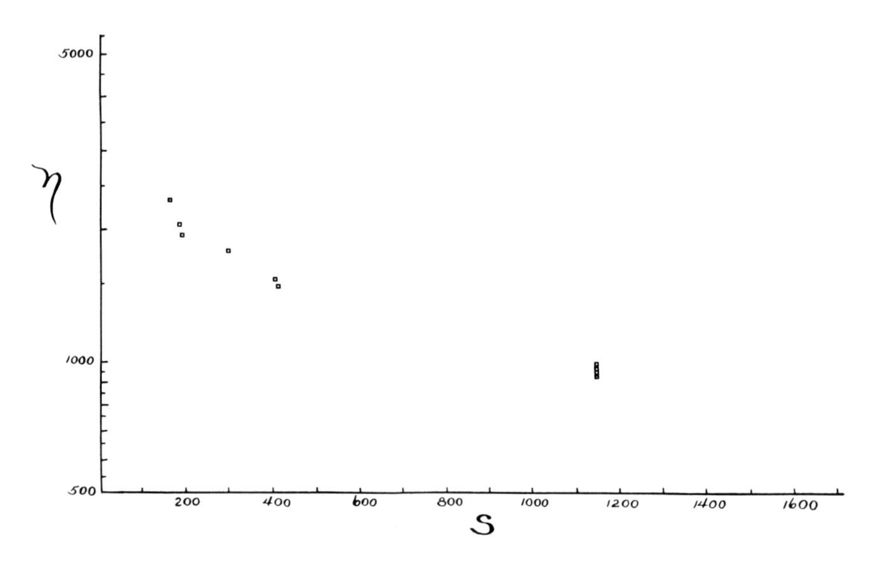

Fig. 2. - 80/20 W/O Emulsion, Second Cycle.

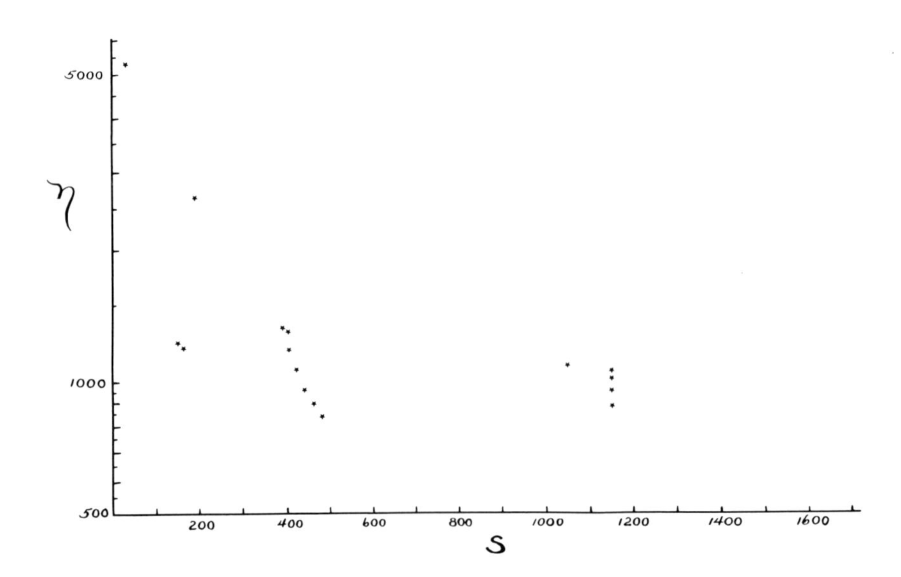

Fig. 3. - 80/20, W/O Emulsion, Third Cycle.

In the morning, two low range relaxed determinations were made and then the emulsion was wound up for 12 minutes at a shear rate of approximately 400 to 450 reciprocal seconds during which time the apparent viscosity rose from a little over 800 to 1300 centipoise. A low shear determination was made and then the emulsion was further wound up by pumping at shear rate of 1150 reciprocal seconds for ten minutes. These data are depicted on Figure 3 and Figure 4 as open stars. Figure 4 combines the data in Figure 1,2, and 3.

Table 2 and Figure 5 depict a similar series of tests conducted on a water-in-oil emulsion containing 85% internal phase. Table 3 and Figure 6 record corresponding data for a 90% internal phase emulsion and Table 4 and Figure 7 display the data for a water-in-oil emulsion containing 95% internal phase.

It can be seen in each case that a smooth curve could be drawn thru the maximum or "wound up" values. This has been done to produce Figure 8. The completely "relaxed" values are difficult to reproduce or obtain since the act of getting a value "winds-up" the emulsion.

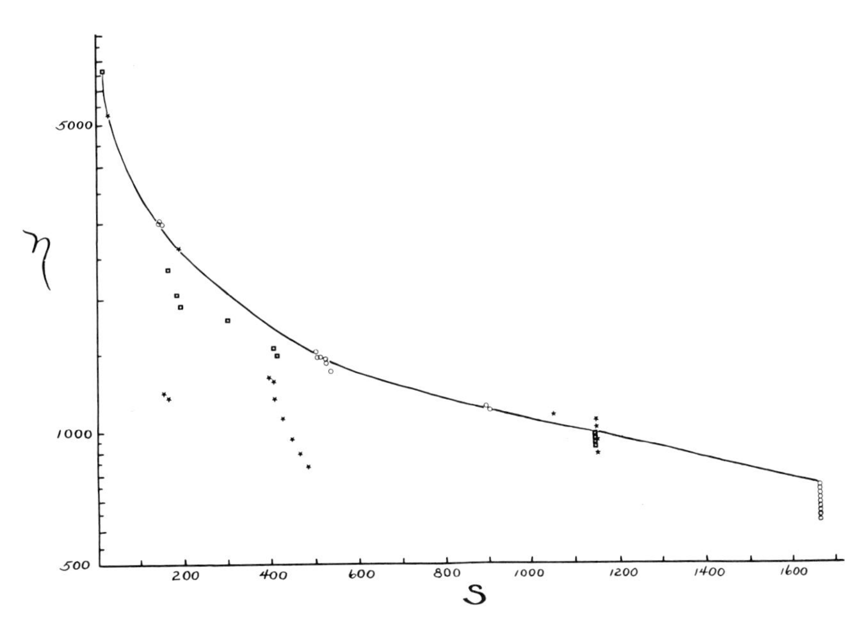

Fig. 4. - 80/20, W/O Emulsion. Apparent Viscosity in Centipoise vs. Shear Rate.

TABLE 2

85/15 W/O Pumping Data

η	S	COMMENTS
3325	63.3	Start low shear
3029	61.7	
3280	62.6	
1171	502.1	Med shear
1236	487	A little wind up
1194	502.1	
1272	479	
1091	973	Winding up
1155	"	1 min
1196	"	2 min high shear
1229	"	3 min
1260	"	4 min
1286	"	5 min
1310	"	6 min
1329	"	7 min
1337	"	8 min
1355	"	9 min
1364	"	10 min
1372	"	11 min
1386	"	12 min
1385	"	
1774	591	
1750	602	
2613	307	
2584	308	
4404	130	
4225	139	
7879	53	
1519	666	
1554	659	
Shut down 48 hours		
1082	114	Relaxed wind up
1518	115	a little
1503	114	
794	652	Moderate shear
939	652	2 min some wind up
978	652	
1048	614	3 min
Shut down one hour		
8165	24	Low shear relaxed
1563	338	
1079	973	High shear windup
1179	973	1 min

η	S	COMMENTS
1216	973	2 min
1237	"	3 min
1288	"	4 min
1320	"	5 min
1344	"	6 min
1366	"	7 min
1384	"	8 min
1393	"	9 min
1406	"	10 min
1610	855	
1996	575	
2035	565	
2701	351	Relaxing slightly
2696	351	
5079	138	
12579	40	
Stands for one week		
3766	51	Relaxed
1583	310	Med shear slight
1622	312	Windup
1092	944	High shear Windup
1166	"	2 min
1208	"	3 min
1247	"	4 min
1267	"	5 min
1292	"	6 min
1309	"	7 min
1327	"	8 min
1348	"	9 min
1356	"	10 min
1604	801	
1629	801	12 min
1662	791	13 min
2408	422	
2374	427	
4069	192	Slight relaxing
4043	192	
7802	75	
7440	78	
11157	44	
10461	46	
15682	25	
Goes to 2.25# on shut down		

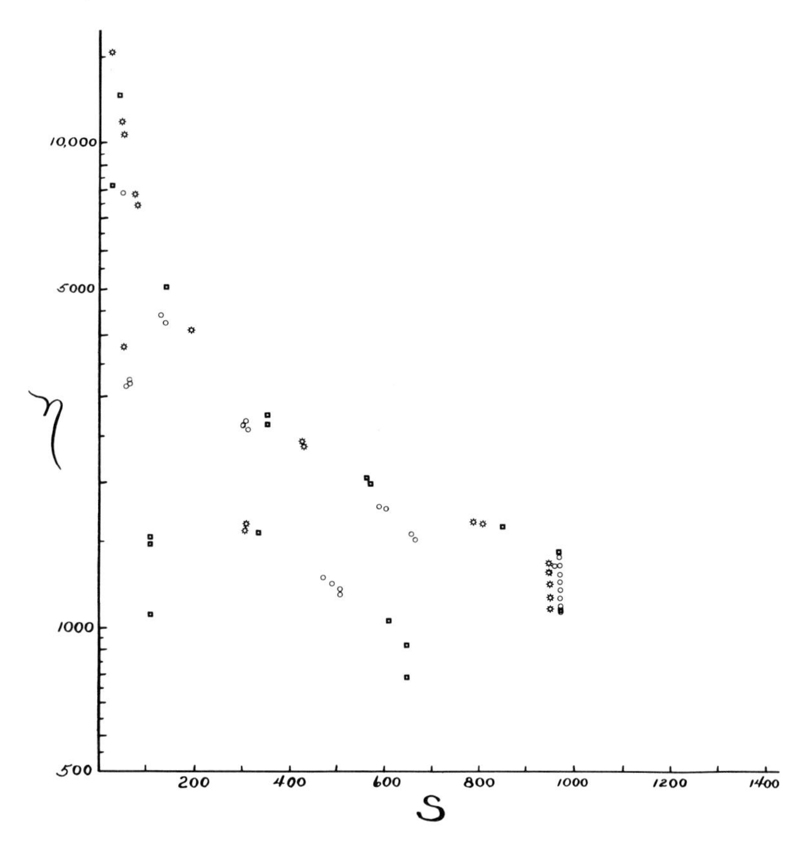

Fig. 5. - 85/15, W/O Emulsion. Apparent Viscosity in Centipoise vs. Shear Rate.

TABLE 3

90/10 W/O Pumping Data

η	S	COMMENT	η	S	COMMENT
14693	28	Start	1774		13 min
1972	494		1784		14 min
1489	879	High shear wind up	1796		15 min
1585		3 min	2347	673	
1615		4 min	2310	688	18 min
1651		5 min	2719	536	
1663		6 min	2588	527	
1685		7 min	3266	416	
1701		8 min	3278	416	
1719		9 min	3940	232	
1735		10 min	3931	323	
1749		11 min	5094	229	
1759		12 min	4999	233	

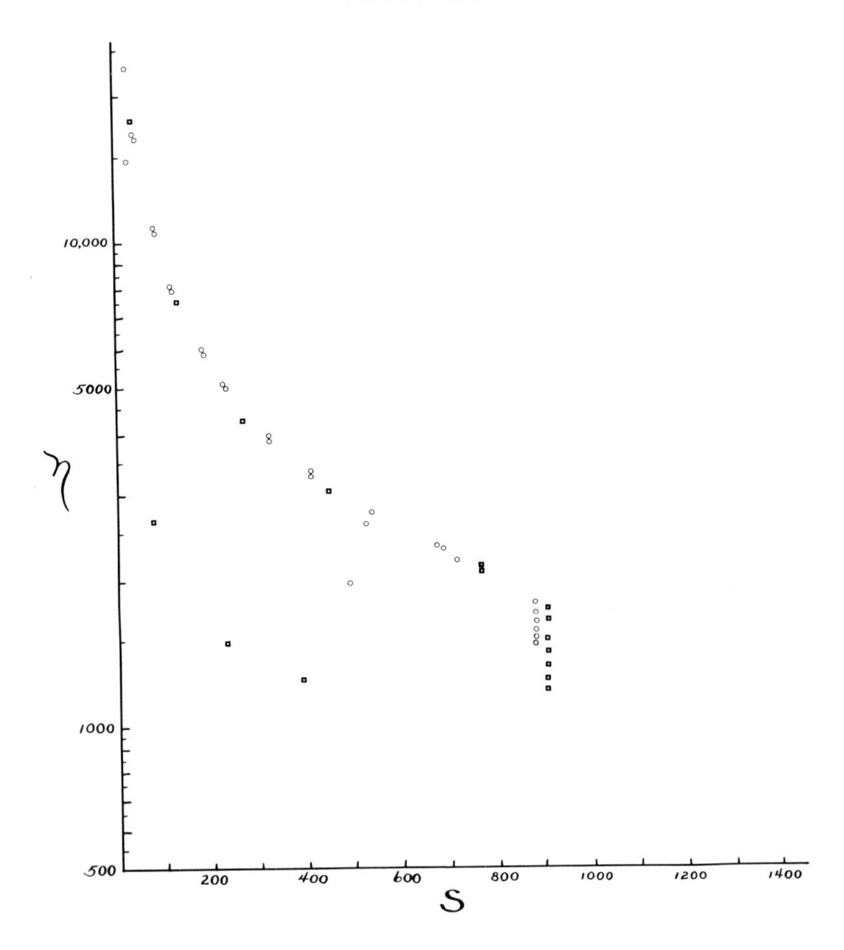

Fig. 6 - 90/10, W/O Emulsion. Apparent Viscosity in Centipoise vs. Shear Rate.

TABLE 3 (con't)

90/10 W/O Pumping Data

η	S	COMMENT	η	S	COMMENT
5929	182		Stand over week nearly zero		P now
5827	185				
8059	121		2681	78	Relaxed
7912	123		1472	236	
10708	83		1258	395	
10610	83		1173	904	High shear wind up
16937	46		1228	"	1 min
16540	47		1347	"	2 min
23781	30		1438	"	3 min
2050	771	On shut down	1498	"	4 min
2212	719	Coasts to 5.17 psi	1550		5 min

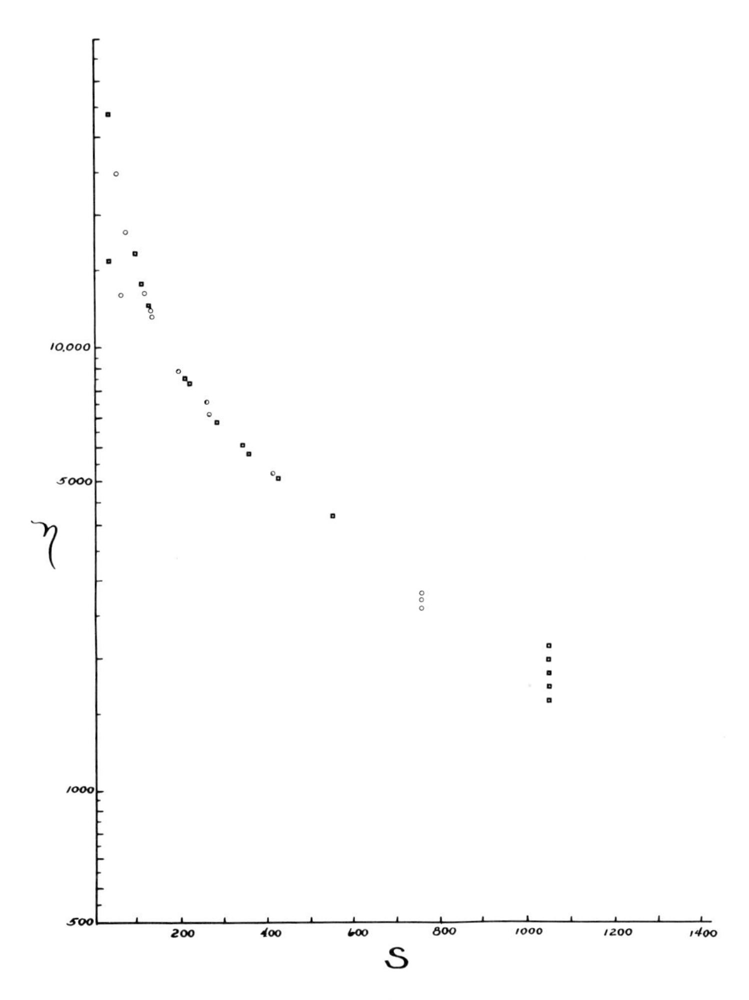

Fig. 7. - 95/5, W/O Emulsion. Apparent Viscosity in Centipoise vs. Shear Rate in Reciprocal Seconds.

TABLE 3 (con't)

90/10 W/O Pumping Data

η	S	COMMENTS	η	S	COMMENTS
1593	"	6 min	3034	452	
1633	"	7 min	4350	272	
1698	"	9 min	7416	132	
1728	"	10 min	17998	42	
1753	"	11 min			
Five minutes shut down					
2063	772				
2120	772				

at 5.81 psi will no longer pump. On shut off coasts down to 4.70 psi.

TABLE 4
95/5 W/O Pumping Data

η	S	COMMENTS	η	S	COMMENTS
		Minimum pumping	10570	33	Relaxed
		Pressure is about	1612	1054	High shear wind up
		5.65 psi	1718	"	
13332	59	Relaxed start	1845	1054	
2510	753	High shear wind up	1989	"	
2693	"	2 min	2081	"	
2741	"	3 min	2119	"	
2760	"	4 min	2146	"	
2776	"	5 min	2161	"	
2782	"	6 min	4201	545	
5215	403		5046	425	
7076	259		5743	357	
7520	243		6047	340	
11912	131		6889	282	
12113	129		8206	226	
24906	52		8610	212	
Pumping stops at about 10.63 psi					
18676	75		12515	132	
12837	122		12241	135	
9233	189		14137	115	
Coasts down to 9.10 on shut off.					
Let stand overnight. Still residual P of 1.15 psi.			16234	96	
In morning minimum pumping pressure is 2.42 psi.			Minimum pumping pressure about 10.50 psi.		
			33954	38	
			Coasts down to 9.06 psi on shut off.		

IV DISCUSSION

An examination of Figures 4 through 7 indicate that satisfactorily reproducible apparent viscosity versus shear curves can be obtained provided the emulsion is maintained at its "wound up" state. It will also be seen that unless one takes into account this peculiar wind-up phenomena, the partially relaxed values appear to show an erratic scatter.

The maximum viscosity curves of the four emulsions are shown on Figure 8. This Figure illustrates the fact that as the internal phase ratio increases the whole apparent viscosity versus shear curve is shifted upwards.

These emulsions have been referred to as "thioxotropic". It is our belief that they are not thixotropic in that they do not show time dependent shear thinning and rather may be said to show time dependent shear thickening. Provided the

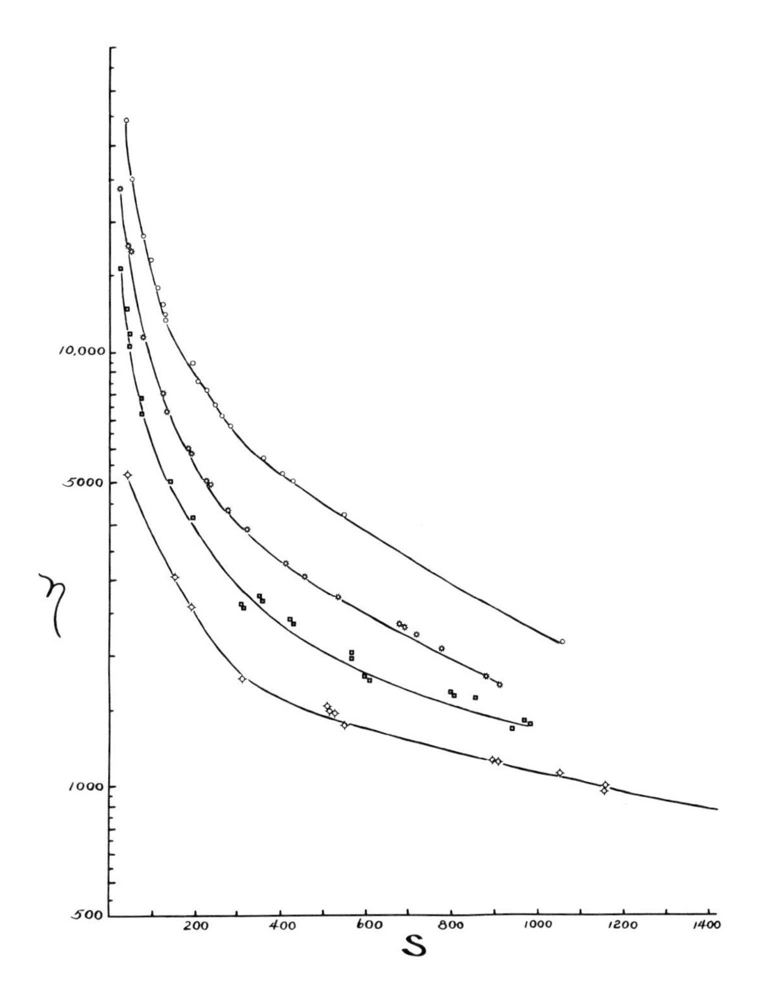

Fig. 8. - Effect of Internal Phase Ratio on Maximum Viscosity Curves.

maximum shear thickening state is obtained, the emulsions appear to follow essentially a pseudo plastic behavior.

We have clearly established that this shear thickening phenomenon does, indeed, occur and is reproducible. A possible explanation may lie in the peculiar structure of these emulsions. The geometry of High-Internal-Phase-Ratio emulsions has been discussed in previous papers(5,6) and some experimental verification of the geometrical configuration has been obtained in certain special cases(7,8). We postulate that the shear thickening action may be explained by assuming that the process of agitation introduces irregularities into the droplet lattice. A two dimensional analogy is depicted in Figure 9, 10, and 11. Figure 9 shows a layer of spheres in their ideal close-packed configuration.

Figure 10 shows the same spheres in an irregular configuration after being disturbed. Figure 11 shows the spheres in a state where they have partially re-arranged themselves in an approach to the ideal configuration. It will

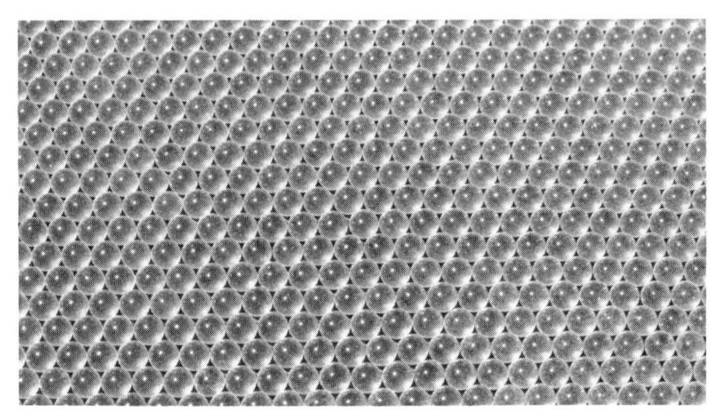

Fig. 9 - Maximum Close Packing

Fig. 10. - Disturbed Packing

be noted that the spheres occupy more space, or pack less economically, when they are in a disturbed condition. We postulate, therefore, that the polyhedral droplets, of which the emulsion has been shown to consist, would prefer to array themselves in one of two ideal configurations, and in this state would occupy the least volume. When agitated, regions of differing configurations are produced, and at the boundaries between these differing configurations, a situation occurs similar to the boundaries in a crystalized solid.

Since these disturbed configurations require more space, the emulsion behaves as though it has a higher internal phase ratio and therefore exhibits a higher apparent viscosity. Since the droplets are liquid and deformable, they can slide over one another and re-arrange themselves into a more

Fig. 11. - Zones of Similar Packing

compact configuration if allows to stand undisturbed. In the more conpact configuration, they would have more room to move and would therefore exhibit a lower apparent viscosity.

This explanation seems to fit the experimental facts, but to date we have found no way of achieving direct verification.

REFERENCES CITED

1. Mannheimer,R.J., J. Colloid Interface Sci., Vol. 40, No. 3, p. 370-382 (1972).
2. Nixon, J. and Beerbower, A., Preprints. Div. Pet. Chem., Am. Chem. Soc. 14 (1) 49-59 (1969).
3. Lissant, K.J., J. Colloid Interface Sci. 22, 462-468 (1966).
4. Lissant, K.J., "Emulsions and Emulsion Technology" Part I Charpter I. pp 61-65, M. Dekker (1975).
5. Lissant, K.J., Hopper, L.R., Harris, J.L., presented at ASME Meeting Feb. 26, 1968.
6. Lissant, K.J., J. Colloid Interface Sci. 22, p. 462 (1966).
7. Lissant, K.J., and Mayhan, K.G., J. Colloid Interface Sci. Vol. 42, p. 201-208 (1973).
8. Lissant, K.J., Mayhan, K.G., et al, J. Colloid Interface Sci., Vol. 47, No. 2, p. 416-423 (1974).

FLOW INSTABILITIES IN COUETTE FLOW IN NEMATIC LIQUID CRYSTALS

P. E. Cladis and S. Torza
Bell Laboratories

Unlike a dilute suspension of ellipsoidal rods, a nematic liquid crystal is characterized by long range orientational order defined by a unit vector, $\underset{\sim}{n}$ (director). In most nematics, the direction of $\underset{\sim}{n}$ with respect to a given velocity field is (almost completely) determined by the ratio of two of the five viscosities of these anisotropic liquids. In some nematics, it is not possible to define such a direction. We show here our experimental results of some of the flow instabilities observed in this latter case. In particular, at the lowest shearing rate, we have observed a solitary wave instability involving a (temporarily) propagating change in the director orientation of $\pi/2$. At somewhat higher shearing rates we have observed a secondary cellular flow only superficially similar to classic Taylor vortices. These latter, we have also observed at shearing rates about a thousand times greater than for the cellular instability.

I. INTRODUCTION

Nematic liquid crystals are anisotropic liquids characterized by long range orientational order. They are generally composed of organic molecules which are longer ($\sim$30Å) than they are wide ($\sim$5Å). The long range orientational order of nematics implies that if at $\underset{\sim}{x}$, the long axis of the nematic is on average aligned parallel to $\underset{\sim}{n}(\underset{\sim}{x})$, $(\|\underset{\sim}{n}^2\|=1)$, then, $\underset{\sim}{n}(\underset{\sim}{x}+\underset{\sim}{\delta}) = \underset{\sim}{n}(\underset{\sim}{x})$ in the absence of external forces and torques. $\underset{\sim}{n}$ is called the director. Deviations from this equality cost energy, called elastic energy, and imply the existence of an elastic restoring force which tries to maintain the equality. The strength of this restoring force depends upon how the nematic is disturbed from its ground state.[1] This can be done in the bulk of the liquid in one or more of the three canonical[1,2] distortions:

- splay with an increase in the energy density of $K_1(\text{div}\underset{\sim}{n})^2$
- twist, resulting in $K_2(\underset{\sim}{n}\cdot\text{curl }\underset{\sim}{n})^2$ increase
- bend, resulting in $K_3(\underset{\sim}{n}\times\text{curl }\underset{\sim}{n})^2$ increase.

Nematics can be oriented by many external agents, such as, surface forces, magnetic and electric fields and by velocity gradients. When two or more external forces or torques compete in the determination of $\underset{\sim}{n}$, then $\underset{\sim}{n}(\underset{\sim}{x}+\underset{\sim}{\delta}) \neq \underset{\sim}{n}(x)$

whenever $\|\delta\|^2$ is smaller than a characteristic length usually determined by the restoring elastic force (K_i) and the energy density of the perturbing force.

A velocity gradient, ∇v, exerts a torque on n. The simplest properly invariant theory to describe the rheology of such liquids was first given by Ericksen.[3] His theory was specialized to the case of nematic liquid crystals by Leslie.[4] Ericksen shows that in simple shear, $v_1(x_3)$, say, there is a torque acting upon n which will always align it in a fixed direction given by $\tan^2\theta = \alpha_3/\alpha_2$ where θ is the angle between $n = (n_1, o, n_3)$ and $v = (v_1(x_3), o, o)$. α_3 and α_2 are two of the six (five of which are linearly independent[5]) Leslie viscosities, α_i.

As is evident, the existence of θ depends crucially upon the fact that α_3 and α_2 are of the same sign (usually both negative) which is not true for some nematics.[6-8] In this latter case ($\alpha_3 > o$ and $\alpha_2 < o$), there is no limit to the viscous torque, Γ_2, and one cannot define a spatially constant stationary value for θ. Γ_2 will act to continually increase θ. In the absence of elastic forces, this leads to a non-stationary θ, i.e. $\partial\theta/\partial t \neq o$ (t = time). But for θ to increase without bound implies increasingly important elastic forces which will act so as to drive $\partial\theta/\partial t \to o$. This is because in nematics, the initial conditions in the bulk of the nematic can be determined by boundary conditions. If n is given at a boundary (x_1) to be $n = (o, o, 1)$, then $n(x_1) \equiv (o, o, 1)$ no matter how large the perturbing force or torque is in the bulk of the liquid ("strong anchoring"[2]). Consequently, if θ increases without bound in the bulk, while constrained to its initial value at the boundaries, the elastic forces of the nematic will increase without bound.[2]

We mention, briefly, at this point, two of the many other liquid crystal phases we are interested in. These are the cholesteric and smectic A phases. The cholesteric phase is similar to a nematic except its ground state is one in which $(n\cdot\text{curl}\ n)^2 = q^2 \neq o$. There is a built in twist to it. A nematic is really a special case of a cholesteric for which $q = o$. A nematic can be transformed into a cholesteric by adding a small quantity of cholesteric (e.g. cholesterol nonanoate) or optically active, but isotropic, material such as Canada Balsam or microscope immersion oil. If the amount of cholesteric material is small enough (usually less than 0.1% by weight), the ground state pitch, $2\pi/q$, can be very large $\gtrsim$ 100 microns.

A smectic liquid crystal is a layered system where the director lies on equidistant surfaces. The smectic A phase is one where there is no additional ordering within a surface and where the director is perpendicular to the surface. Because of the inherent layering of this system, the ground state for this phase is one in which curl $\underset{\sim}{n} \equiv \underset{\sim}{o}$. There is no known case of a temperature driven cholesteric-nematic transition, but there are many such smectic A-nematic transitions occurring at a transition temperature, T_{NS}, the smectic A phase being (usually[9]), the lower temperature phase. Sometimes the smectic A-nematic transition can be very nearly of second order. In this case pretransitional effects due to density fluctuations in the nematic phase owing to the continual production of evanescent, submicroscopic smectic A regions have been predicted and observed. These pretransitional effects consist in a divergance of the elastic constants K_3 and K_2[10] as $T \to T_{NS}$ and also the Leslie viscosities, α_1, α_3 and α_6.[11] It is, thus, in compounds where there is a nearly second order nematic-smectic A transition that we can expect to find the interesting combination $\alpha_3 > o$ and $\alpha_2 < o$, since pretransitional fluctuations in the nematic phase are expected to drive α_3 from its originally negative value far above T_{NS} through zero and towards a large positive value near T_{NS}.

The purpose of this paper is to show some of the acrobatics and contortions the nematic goes through to find its stationary states in the flow regime where $\alpha_3 > o$ and $\alpha_2 < o$. To do this we have studied the liquid crystals in a Couette flow field. In order to observe the director (with a polarizing microscope) we made the inner and outer cylinders of the sample container out of glass. We limit our studies here to only those instabilities which occur in the flow regime $\alpha_3 > o$ and $\alpha_2 < o$. We have observed that for small shears, the first instability is a "tumbling" instability which we show in the next section to be analogous to the appearance of a solitary wave in a long torsion bar to which is attached a dense array of pendulums. That this would be the case was first suggested to us by R. C. Dynes.[12] A second instability is observed at higher (but still small) shear rates which involves the onset of a secondary cellular flow. At still higher shear rates, a dense mass of disclinations forms in the gap. They align with their long axis parallel to the flow field. Finally, at very large shear rates, we have observed the onset of classical Taylor vortices.[13] The secondary flow of a Taylor vortex axially separates the disclination lines into well defined clumps so that each Taylor cell is clearly defined without the aid of "artificial additives".

In the next section we describe briefly the theoretical background specific to our system. In the third section we describe our experimental arrangement and finally in the fourth section we present our observations.

II. THEORETICAL BACKGROUND

When the inner cylinder is at rest, in polar coordinates $(\hat{r}, \hat{\theta}, \hat{z})$, $\underset{\sim}{n} = (1, o, o)$. Once the inner cylinder is set into motion, $\underset{\sim}{n} = (\cos \psi(r), \sin \psi(r), o)$ at small shears. This we will call the "planar couette". Eventually, at a high enough shear rate $\underset{\sim}{n} = [\rho(r,z) \cos \psi(r,z), \rho(r,z) \sin \psi(r,z), (1-\rho^2(r,z))^{1/2}]$, a much more complicated configuration in which the viscous torques and elastic torques must now be balanced in all three directions.

We consider first the "planar couette" for which the velocity field is $\underset{\sim}{v} = (o, r\omega(r), o)$.

A. Planar Couette ($\rho = 1$)

In this case, the only component to the director torque, $\underset{\sim}{n} \times d^2\underset{\sim}{n}/dt^2$, is in the z direction. It can be written as[3,14]

$$I \frac{\partial^2 \psi}{\partial t^2} = - \left(\alpha_3 - \alpha_2 \right) \frac{\partial \psi}{\partial t} + r \frac{d\omega}{dr} \left(-\alpha_2 \cos^2\psi + \alpha_3 \sin^2\psi \right)$$
$$- \frac{1}{2} \left[1 + \left(r \frac{\partial \psi}{\partial r} \right)^2 \right] \left(K_3 - K_1 \right) \sin 2\psi / r^2 \qquad (1)$$
$$+ r \frac{\partial}{\partial r} r \frac{\partial \psi}{\partial r} \left(K_3 \cos^2\psi + K_1 \sin^2\psi \right) / r^2$$

where I is a positive constant. The first term on the R.H.S. represents a frictional torque density, the second term the externally applied viscous torque density due to the rotation of the inner shaft. The last two terms are the elastic torque densities. For small distortions ($|\psi| \leq \pi/2$), both of these terms will oppose the externally applied viscous torque. Once $|\psi|$ exceeds $\pi/2$, the first of these will throw in its lot with the externally applied viscous torque. R. C. Dynes[12] was the first to point out to us the striking similarity between Eq. (1) and that for a long torsion bar to which is attached a dense array of pendulums[15] with position coordinates x and angular coordinates $\phi(x)$. The torsion bar compliance is R, the moment arm density is S(x) and the moment of inertia density is T(x). A viscous drag force U(x) opposes the motion of the pendulums and an external torque of density W(x) is applied along the bar. Newton's law for this array is

$$T(x) \frac{\partial^2 \phi}{\partial t^2} = - U(x) \frac{\partial \phi}{\partial t} + W(x) - S(x) \sin \phi + R \frac{\partial^2 \phi}{\partial x^2} \qquad (2)$$

which is evidently formally very similar to Eq. (1). We know from the case of this mechanical analogue that twisting the pendulum suspended in the middle of the bar through an angle greater than π, results in a change in sign of the gravitational torque for those pendulums which have passed over the bar. Upon continued application of this external torque, the system eventually becomes unstable and two solitary waves (solitons) where ϕ changes by $\pm\pi$, respectively, propagate, one towards each end of the torsion bar. In our case, the term analogous to the moment arm density of the pendulums changes sign at $\psi = \pi/2$. Increasing the applied external torque ($rd\omega/dr$), (which we can do without limit when $\alpha_3 > o$ and $\alpha_2 < o$, but not when they are both negative), we expect to be able to generate, in a manner completely analogous to the torsion-bar pendulum array system, two solitary waves, where ψ changes by $\pm\pi/2$, each propagating towards the walls of the Couette. In section IV we show that this is in fact true and we have called this event "tumbling".

B. Non-Planar Couette

In order to investigate at what point the planar couette becomes unstable, we look at the equation, first suggested by Ericksen(1) but adding to his original expression the elastic terms of a nematic liquid crystal for ρ = constant,

$$\frac{1}{2} \frac{\partial}{\partial t} \rho^2 = - \frac{\rho^2(1-\rho^2)}{(\alpha_3-\alpha_2)r^2} \left[(\alpha_3+\alpha_2) r^3 \frac{d\omega}{dr} \cos\psi \sin\psi + E_c \right]$$

where

$$\begin{aligned} E_c = & \left[1 + \left(r \frac{d\psi}{dr} \right)^2 \right] \left(K_2 \sin^2\psi + K_1 \cos^2\psi \right) \\ & - 2 (K_3 - K_2) \left[\sin\psi + \left(r \frac{d\psi}{dr} \right) \cos\psi \right]^2 \\ & - (K_3 + K_2 - K_1) \cos\psi \sin\psi \, r \frac{d}{dr} r \frac{d\psi}{dr} . \end{aligned} \qquad (3)$$

Clearly, as long as the term in the square brackets remains negative, a small perturbation $\rho \rightarrow 1 - \varepsilon^2$ ($\varepsilon << 1$) results in $\partial\rho/\partial t > 0$ and ρ will be restored to its value $\rho = 1$. Once this term changes sign, then we expect $\rho = 1$ to become unstable at a critical shearing rate given by

$$\left(r^3 \frac{d\omega}{dr} \right) = \frac{-E_c}{(\alpha_3+\alpha_2) \cos \psi \sin \psi} . \qquad (4)$$

In general therefore, at a high enough shearing rate, we can expect the planar Couette to become unstable somewhere in the gap - except if $\alpha_3 \approx -\alpha_2$. In this case Eq. (4) shows $(r\, d\omega/dr)_E = \infty$. Not until $\alpha_3 > -\alpha_2$ (i.e. near to the nematic-smectic A transition) is it once again unstable. Consequently, we chose to study the onset of this cellular flow in a compound in which smectic pretransitional effects are large enough to observe, in a single sample, the onset of a secondary cellular flow regime and its disappearance (even at very large shear rates), merely by lowering the temperature to where $\alpha_3 \approx -\alpha_2$.

III. EXPERIMENT

A. Small Samples

Small glass tubes ($2R_2 \sim 1$mm i.d.) and glass shafts ($2R_1 \sim .5$mm o.d.) are treated with the surfactant of Kahn[15] to ensure that the boundary condition $\underset{\sim}{n}$ is radial at R_2 and R_1. The inner shaft is threaded through the larger tube and held concentric with teflon bearings. The whole assembly is mounted into a microscope hot stage, with index matching oils, so that the long axis of the cylinder is perpendicular to the microscope axis. A motor with a series of gears turns the inner shaft at angular speeds ranging from 10^{-3} to 50 rev/sec.

B. Large Samples

These samples were prepared using the same surface treatment (as in A. above) for the glass cylinders, however $2R_2 \sim 2$cm and $2R_1 \sim 1$cm. Unlike the small samples, these cells were mounted vertically and there was an air liquid crystal meniscus. The length of the column of liquid crystal was about 15cm. Our observations were made approximately at the mid-section of the length using a telescope with polarizer and analyzer.

C. Materials

The two liquid crystals we studied are called HBAB (hexyloxyamino benzonitrile) and CBOOA (cyano benzylidene octyloxy aniline). Both possess the interesting flow regime where $\alpha_3 > o$ but $\alpha_2 < o$. In HBAB, this occurrs for $40°C < T < 91°C$[6] and for CBOOA $83°C \leq T \leq 105°C$.[17]

IV. RESULTS

A. Small Samples

1. *Planar Couette*

When the inner shaft is at rest, the equilibrium configuration for the nematic is one in which the director is radial. This is shown in Fig. 1.1 and in the top figure of Fig. 2 schematically. As the inner shaft starts to turn, the

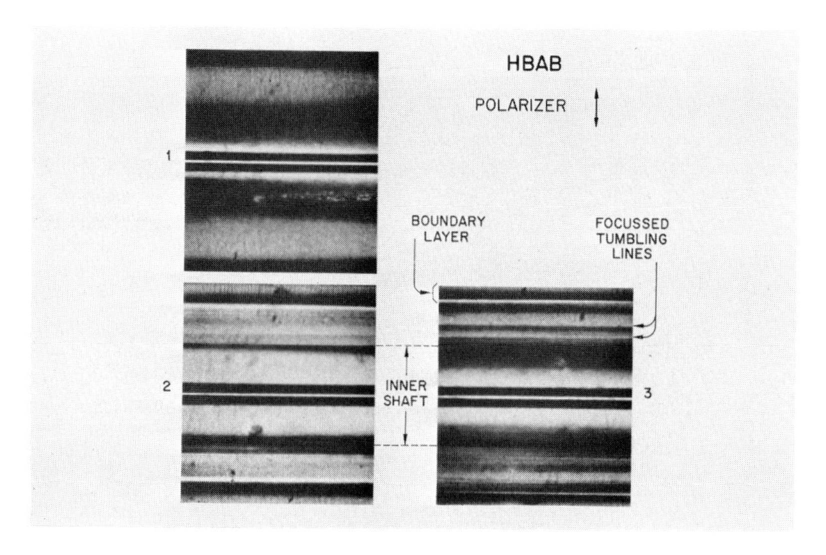

Fig. 1. Tumbling Instability

director experiences a viscous torque turning it into the $\hat{\theta}$ direction. Since the surface treatment requires $\underset{\sim}{n}$ to be radial at the boundary, the director is continuously distorted

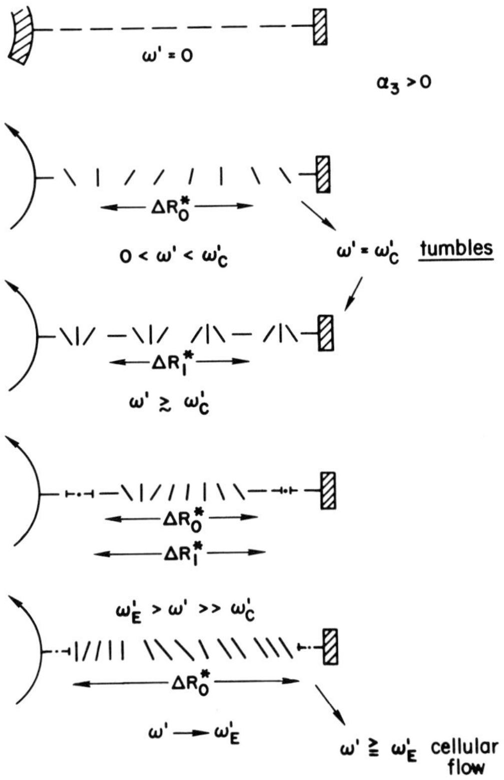

Fig. 2. Schematic of director configuration for tumbling.

from its boundary value to some maximum value in the gap, then back again to its boundary value at the inner shaft. We represent this schematically in the second scheme of Fig. 2. We call this the "planar" couette because the director remains in the $(\hat{r}, \hat{\theta})$ plane during what follows. Eventually at high enough shears, it does develop an axial component. As the rate of angular rotation of the inner shaft, ω' is very slowly increased, the maximum distortion of the director begins to exceed $(-\pi/2)$. As long as ω' is held constant, we observed no changes in the optical pattern in the gap. In this regime the elastic torques are able to balance the small viscous torques and a stationary distribution exists for the director. As soon as ω' increases to ω_C', the maximum distortion for the director suddenly increases by $(-\pi/2)$. We say the director has "tumbled". In the microscope, one observes two focussed lines appearing in the gap region (Fig. 1.3). They are visible if the light is polarized perpendicular to the axis of the lines (extraordinary illumination) but not when the polarizer is parallel to the lines (ordinary illumination). This shows that the director deformation which leads to the appearance of these lines is only in the $(\hat{r}, \hat{\theta})$ plane. The lines owe their occurrence to the radial reorientation of the director in the gap (middle scheme shown in Fig. 2).

Since the tumbling has resulted in a new more energetic configuration for the director in the gap, the elastic torques are once again able to balance the viscous torques and, provide ω' is constant, the director configuration is observed, once again, to be stationary.

Continuing to increase ω', one observes that the two tumbling lines migrate towards the boundaries R_1 and R_2. Eventually, at higher ω', we can expect another tumbling event to occur in the region between these two lines but before this has happened we have observed, instead, the onset of a secondary cellular flow -- the non-planar couette.

Before discussing this new regime, we wish to briefly mention an observation which is at odds with our present understanding. That is, once the director has tumbled, we have observed that even though it remains planar in what we have called the effective gap (the distance between the two boundary layers where $\underset{\sim}{n}$ is mostly radial), two thin shells of axially oriented nematic are observed to appear mysteriously at the interface between the boundary layer and the effective gap.

We also note that if at any stage we turn off the rotation of the inner shaft, we recover the originally uniform radial orientation of the director.

2. *Non-Planar Couette*

Fig. 3.2 shows the appearance of the cellular regime in CBOOA which occurs when $\omega' \geq \omega'_E \geq \omega'_C$. As long as we keep ω' constant. There is no change in this pattern. We call this the non-planar couette because now there is an axial component to $\underset{\sim}{n}$ in the effective gap. It is the appearance of this component which triggers the secondary flow. Increasing $\omega' > \omega'_E$ results in the cells shrinking axially in size (Fig. 3.3) and the eventual nucleation of disclinations ($\omega' \sim 5\omega_E$) (regions where the elastic energy is very large[1]) on or near the inner shaft. The disclinations appear as dark threads and rapidly grow in length until the entire gap is opaque. We note that they lie parallel to the velocity field.

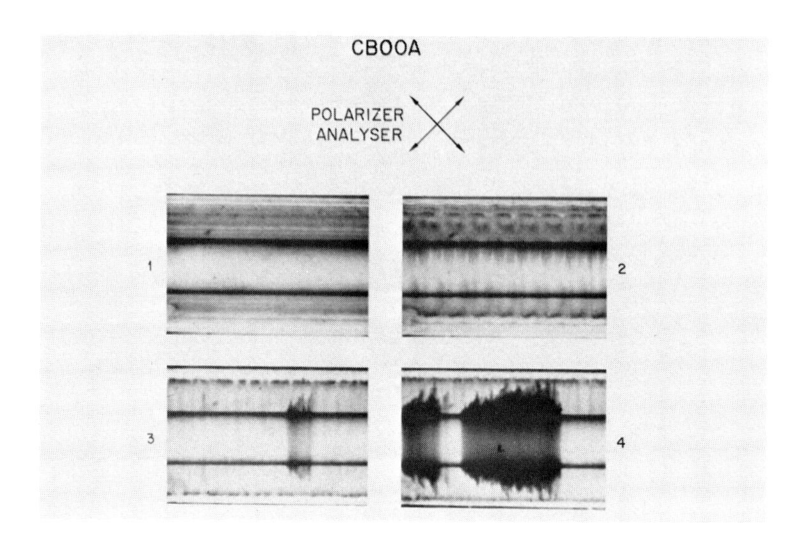

Fig. 3. Cellular Flow in a nematic liquid crystal.

Fig. 4 shows the cellular regime for "cholesterized" CBOOA. It has been cholesterized by adding a small quantity of microscope immersion oil. The cellular regime in this case is similar to the nematic except the pattern appears

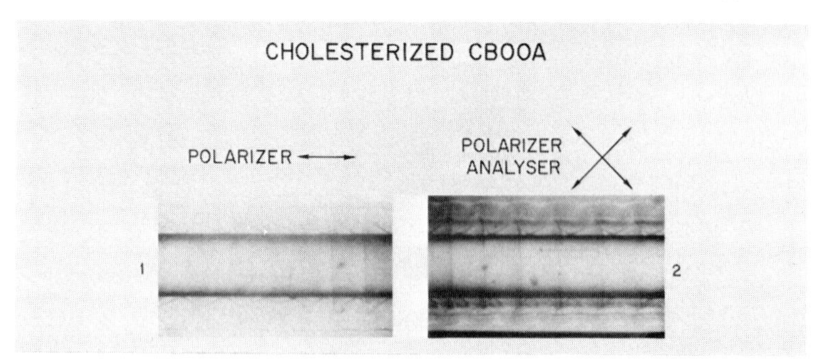

Fig. 4. Cellular Flow in a cholesteric liquid crystal.

to be translating in the $\hat{z}$ direction. Fig. 5 shows our schema for the director configuration for the cellular flows. In the case of the cholesterized samples the cells are actually connected in a helical pattern rather than the nematic tubular one, so that, the whole pattern appears to translate as it rotates just as a threaded shaft appears to translate when it is turned.

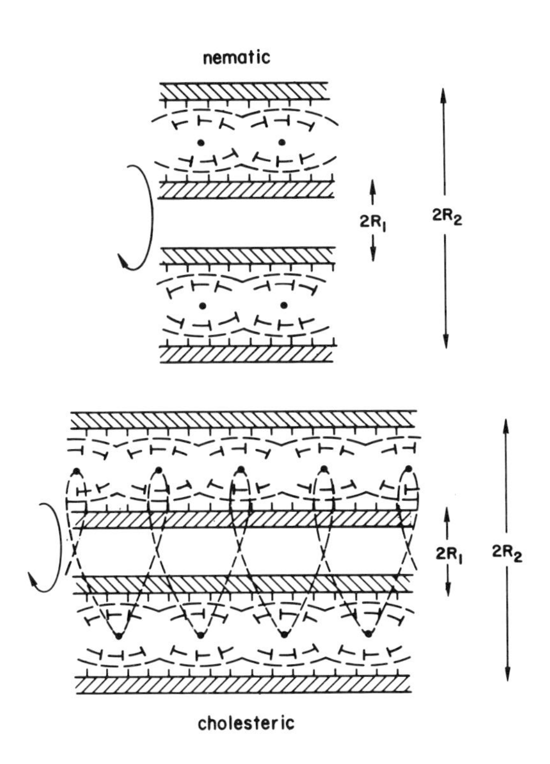

Fig. 5. Schematic of director configuration for cellular flow.

When $T \leq 88°C$, we no longer observe the cellular regime in CBOOA. $\omega_E^{\tilde{i}} = \infty$. From a practical point of view, this is a very convenient regime to rapidly clear the gap of any stray disclinations or defects. Rapid rotation in this temperature range results in a perfectly clear gap very rapidly.

Fig. 6 shows one of the large samples (CBOOA) when $T > 90°C$ in the Taylor vortex regime. The black regions are disclination lines (seen growing in Fig. 3.3 and 3.4). They are segregated by the secondary flow into clumps which neatly indicate the Taylor cell of the secondary flow. The Taylor number for these liquids seems quite high ~ 95.

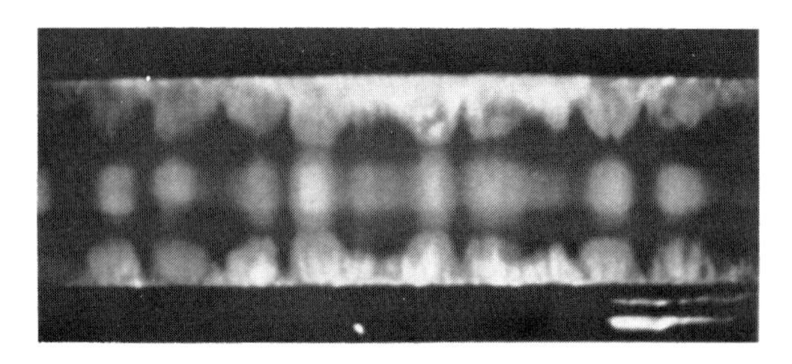

Fig. 6. Disclinations in the Taylor vortex regime.

We have also observed the cells to expand and contract rythmically with a frequency ∿ 1 sec^{-1}. We think the pulsing effect may be due to normal stresses at the liquid-crystal air interface as has been observed and analyzed for other non-newtonian liquids.(18)

CONCLUSIONS

We conclude that there are many interesting flow instabilities in liquid crystals some of which we are only beginning to understand. Another more detailed analysis of this particular flow regime is to be submitted for publication shortly.(19)

ACKNOWLEDGEMENTS

We wish to thank J. Sciortino for the preparation and layout of this manuscript.

REFERENCES

1. Frank, F. C., Discuss. Faraday Soc. 25, 19 (1958)/
2. deGennes, P. G. in "The Physics of Liquid Crystals" (Clarendon Press, Oxford, 1974).
3. Ericksen, J. L., Kolloid-Z. 173, 117 (1960).
4. Leslie, F. M., Q. J. Mech. Appl. Math. 19 Pt.3, 357 (1966).
5. Parodi, O., J. Phys. (Paris) 31, 58 (1970).
6. Gähwiller, Ch., Phys. Rev. Letters 28, 1554 (1972).
7. Pieranski, P. and Guyon, E., Phys. Rev. Letters 32, 924 (1974).

8. Cladis, P. E. and Torza, S., Phys. Rev. Letters 35, 1283 (1975).
9. Cladis, P. E., Phys. Rev. Letters 35, 48 (1975).
10. deGennes, P. G. Solid State Comm. 10, 753 (1972).
11. See for example: McMillan, W. L., Phys. Rev. A9, 1720 (1974); Brochard, F., J. Phys.-(Paris) 34, 28 (1973); Jahnig, F. and Brochard, F., J. Phys.-(Paris) 35, 299 (1974).
12. Dynes, R. C. (private communication).
13. Taylor, G. I., Phil. Trans. Roy. Soc. London A223, 289 (1923).
14. Atkin, R. J. and Leslie, F. M., Quart. Journ. Mech. and Applied Math 23 Pt.2, S4 (1970).
15. Scott, A. C., Am. J. Phys. 37, 52 (1969).
16. Kahn, F. J., Appl. Phys. Letters 22, 386 (1973).
17. Pieranski, P. and Guyon, E. (to be published).
18. Joseph, Daniel D. and Fosdick, Roger L., Arch. Rat. Mech. Anal. 49, 321 (1973).
19. Cladis, P. E. and Torza, S. (to be published).

ANALYSIS OF SEDIMENTATION VELOCITY IN TERMS OF BINARY PARTICLE INTERACTIONS

C. Christopher Reed and John L. Anderson
Cornell University

Abstract. This paper discusses the basic concepts involved in computing sedimentation velocities of spherical particles as functions of concentration when the volume fraction of particles is small. The analysis rests on a consideration of binary interactions between a test particle and each of its neighbors. The procedure described entirely avoids the divergent integrals usually associated with such calculations. Numerical results for one-component systems show that the relation between sedimentation velocity and particle concentration depends on an integral average of the long-range potential energy between two particles. Results for two-component hard spheres are also given.

I. INTRODUCTION

For isolated spherical particles at low Reynolds numbers the sedimentation velocity (U_o) is related to the applied force (F) by Stokes' law, $F = 6\pi\mu a U_o$, where a is the particle radius and μ the fluid viscosity. When the volume fraction of particles is finite, however, the effects of interparticle hydrodynamics must be taken into account. This is done by means of an averaging procedure which expresses the velocity of a test particle as a sum of effects due to binary interactions with each of its neighbors, this sum being then averaged over all configurations of neighbors. In doing so, it is essential to take into account the fact that binary interactions in a bounded fluid are slightly different from those in an unbounded fluid: in particular, the downward flux of particles will be accompanied by a corresponding upward flux of fluid.

Furthermore, the manner in which the interactions are averaged will depend on any long range potentials which may exist between particles.

This problem has been treated before, the most thorough analysis having been given by Batchelor (Ref. 1). In this paper we give a more physical description of his theory and show how the problem of divergent integrals, which Batchelor successfully overcomes, can be avoided altogether. We also extend his theory to include the effects of interparticle potentials, and to sedimentation in multicomponent systems. Our computational procedure is similar to his, differing mainly in the computation of the $\overline{V}_2''$ term (Eq. (2.12)).

II. BASIC CONCEPTS

The system which we envision consists of a dilute dispersion of N identical spherical particles falling through a bounded Newtonian fluid. The macroscopic concentration of particles is uniform throughout the system, and it is assumed that the container dimension $V^{1/3}$ is much larger than the particle radius a, in order to ensure that the sedimentation velocity will be independent of container shape. Furthermore, the particles are assumed small enough so that the creeping flow equations of motion may be used.

In the discussion that follows, we will be looking at how a test particle is affected by its interaction with the N particles of the system. The test particle need not be of the same size or density as the remaining particles, and it is this fact which will allow us to obtain the results for one- and multi-component systems almost simultaneously.

A. Physical Concepts

There are two essential physical concepts which underlie the sedimentation analysis. First is the fact that the system is dilute (i.e., the volume fraction ϕ of particles is much less than 1). This means that to an excellent approximation, only binary interactions between the test particle and each of its neighbors need be considered. In other words, the velocity of the test particle will be the velocity it would have if there were no neighbors present, plus a correction due to its (binary) interaction with neighbor #1, plus a correction due to its interaction with neighbor #2, etc. The sum of these corrections is then averaged over all possible configurations of neighbors in order to obtain an average sedimentation velocity. Since the probability of any configuration is determined by the two-particle potential energy, the average sedimentation velocity is a measure of direct long-range forces between particles.

The second important concept is that of backflow. A particle falling through a fluid drags some fluid along with it, and in a bounded system this downflow of particle plus fluid must be compensated for by a reverse flow of fluid elsewhere in the system, as illustrated in Fig. 1. Furthermore one observes that, to a first approximation, if two

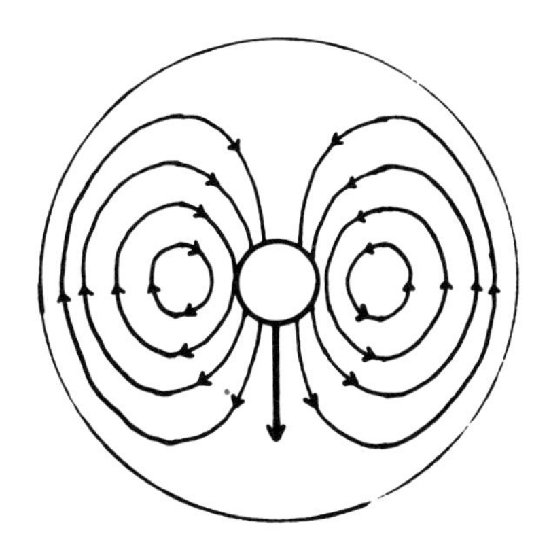

Fig. 1. Fluid streamlines for particle motion in a bounded system. Adapted from an illustration by Williams (Ref. 2).

particles are close to each other, then each will feel the downflow due to the other, and so both particles will "push" each other, resulting in an increase in velocity. On the other hand, if two particles are far apart, then each will feel the backflow generated by the other, and a slowing down will occur.

In terms of binary interactions, then, those neighbors near a test particle will have the effect of increasing its velocity, while those particles far away will tend to slow it down. These two effects will then be weighted together according to the radial distribution function for neighbors in the vicinity of the test particle. In view of the fact that most of the neighbors are far away, one expects that the overall effect will be a slowing down of the test particle.

Of course, if the boundaries of the system are very far away, then the backflow generated by the motion of a single particle will be spread out more or less uniformly throughout the fluid, so that its magnitude at any point is extremely small. Nevertheless the backflow experienced by a test particle is the cumulative backflow generated by all of the remaining particles in the system, a quantity which is not negligible. This fact also underscores the importance of using the velocity fields for particles in a bounded fluid; although the difference between the velocity fields for a particle in bounded and unbounded fluids is extremely small, it is a difference which cannot be neglected.

B. Mathematical Concepts

In order to compute the average sedimentation velocity of a test particle, we would in principle proceed as follows: first, the N neighbors are arranged about the test particle in some configuration C_N; the creeping flow equations are then turned on and the velocity of the test particle, $U_t(C_N)$, computed. Then the neighbors are rearranged into a new configuration, and the test particle velocity is computed for this new configuration. This procedure is repeated until an average over all configurations is obtained.

More precisely, a configuration C_N of N particles means that particle 1 is located at $\mathbf{r}_1$, particle 2 is at $\mathbf{r}_2$, ..., particle N is at $\mathbf{r}_N$. If $P(C_N|\mathbf{r}_t = 0)$ is the conditional probability density for C_N, given the presence of the test particle at the origin ($\mathbf{r}_t = 0$), then the velocity of the test particle, averaged over all configurations, may be expressed as

$$<U_t> = \frac{1}{N!} \int_{C_N} U_t(C_N)\, P(C_N|\mathbf{r}_t)\, dC_N$$

$$= \frac{1}{N!} \int_{\mathbf{r}_1} \cdots \int_{\mathbf{r}_N} U_t(\mathbf{r}_1 \ldots \mathbf{r}_N)$$

$$P(\mathbf{r}_1 \ldots \mathbf{r}_N|\mathbf{r}_t)\, d\mathbf{r}_1 \ldots d\mathbf{r}_N \,, \qquad (2.1)$$

where $1/N!$ is a normalization factor which takes into account the fact that there are $N!$ possible permutations of identical spheres.

In order to convert Eq. (2.1) into a more tractable

form, three steps are involved. First, (2.1) is reduced to an average involving only binary interactions. Mathematically, this is accomplished by rewriting $U_t(C_N)$ in "cluster expansion" form; this is a physically convenient way of adding zero to $U_t(C_N)$:

$$U_t(C_N) = U_t(C_o) + \frac{1}{N}\sum_{i=1}^{N} [U_t(C_1(i) - U_t(C_o)]$$

$$+ \frac{1}{N(N-1)} \sum_{\substack{j=1 \\ j \neq i}}^{N} \sum_{i=1}^{N} [U_t(C_2(i,j)) - U_t(C_1(i))]$$

$$+ \ldots \tag{2.2}$$

$U_t(C_o)$ is the velocity which the test particle would have if there were no neighbors (i.e., $U_t(C_o) = U_o = F/6\pi\mu a$); $U_t(C_1(i))$ is the velocity which the test particle would have if it and a single neighbor (i) were falling together through a bounded fluid; $U_t(C_2(i,j))$ is the corresponding velocity in the presence of two neighbors (i and j), etc.

If Eq. (2.2) is now substituted into the configurational integral (2.1), a straightforward but tedious calculation[1] leads to

$$U_t = U_o + n \int_V [U^{(t)}(0,\mathbf{r}) - U_o]\, g(\mathbf{r})\, d\mathbf{r}$$

$$+ O(n^2) , \tag{2.3}$$

where $U^{(t)}(0,\mathbf{r})$ is the velocity which a test particle located at the origin would have if there were a neighbor located at $\mathbf{r}$ and both particles were falling together through a bounded fluid; $g(\mathbf{r})$ is the radial distribution function for the neighbors about the test particle, given by $g = \exp(-W/kT)$ for dilute dispersions, where $W(\mathbf{r})$ is the interparticle potential, and n is the macroscopic concentration (number

---

[1] Reed, C. C. and Anderson, J. L., article in preparation.

density) of neighbors.

Equation (2.3) merely says that the average velocity of the test particle is the velocity which it would have if there were no neighbors present, plus a correction due to binary interactions. This correction is averaged according to the radial distribution function, and since it is a cumulative effect, the over-all correction is proportional to the concentration of neighbors in the system. If the system is dilute, as we have assumed, then the $O(n^2)$ terms contribute negligibly to the sedimentation velocity.

The second step involves the conversion of the quantity $\mathbf{U}^{(t)}(0,\mathbf{r})$ to a form more amenable to computation. For the sake of notation we will refer to the test particle as being of species 1, and the neighbor as being of species 2. Recall then that $\mathbf{U}^{(1)}(0,\mathbf{r}_2)$ is the velocity which a test particle located at the origin would have if it were falling through a bounded fluid together with a neighbor located at $\mathbf{r}_2$. Again we use the idea of adding zero in a physically meaningful way. Thus

$$\begin{aligned} \mathbf{U}^{(1)}(0,\mathbf{r}_2) &= \mathbf{U}_o^{(1)} + \mathbf{v}^{(2)}(0,\mathbf{r}_2) \\ &+ \frac{1}{6} a_1^2 \left(\nabla_1^2 \mathbf{v}^{(2)}(\mathbf{r}_1,\mathbf{r}_2)\right)_{\mathbf{r}_1=0} + [\mathbf{U}^{(1)}(0,\mathbf{r}_2) - \mathbf{U}_o^{(1)} \\ &- \mathbf{v}^{(2)}(0,\mathbf{r}_2) - \frac{1}{6} a_1^2 \left(\nabla_1^2 \mathbf{v}^{(2)}(\mathbf{r}_1,\mathbf{r}_2)\right)_{\mathbf{r}_1=0}] \, . \end{aligned} \tag{2.4}$$

The physical meaning of the terms is as follows: $\mathbf{U}_o^{(1)}$ is the velocity which the test particle would have in the absence of the neighbor; on the other hand, if the test particle were for the moment absent, then $\mathbf{v}^{(2)}(\mathbf{r}_1,\mathbf{r}_2)$ would be the velocity field at $\mathbf{r}_1$ generated by the neighbor (at $\mathbf{r}_2$) moving with velocity $\mathbf{U}_o^{(2)}$. If the test particle were now placed in this velocity field, (at $\mathbf{r}_1 = 0$) then it would acquire the additional velocity $\mathbf{v}^{(2)}(0,\mathbf{r}_2) + \frac{1}{6} a_1^2 \left(\nabla_1^2 \mathbf{v}^{(2)}(\mathbf{r}_1,\mathbf{r}_2)\right)_{\mathbf{r}_1=0}$ (Faxen's law); here a_1 is the radius of the test particle and ∇_1 indicates differentiation with respect to $\mathbf{r}_1$. Finally, the third term in Eq. (2.4) corrects for the interaction between the two particles (this term will be denoted by $\mathbf{W}^{(1)}(0,\mathbf{r}_2)$).

If Eq. (2.4) for $\mathbf{U}(0,\mathbf{r}_2)$ is now substituted into the integral (2.3), the expression for the test particle velocity becomes

$$\overline{U}^{(1)} = U_o^{(1)} + n_2 \int_V \mathbf{v}^{(2)}(0,\mathbf{r}_2)\, g_{12}(\mathbf{r}_2)\, d\mathbf{r}_2$$

$$+ \frac{1}{6} a_1^2 n_2 \int_V \left(\nabla_1^2 \mathbf{v}^{(2)}(\mathbf{r}_1,\mathbf{r}_2)\right)_{\mathbf{r}_1 = 0} g_{12}(\mathbf{r}_2)\, d\mathbf{r}_2$$

$$+ n_2 \int_V W^{(1)}(0,\mathbf{r}_2)\, g_{12}(\mathbf{r}_2)\, d\mathbf{r}_2 \,, \tag{2.5}$$

and it should be born in mind that $\mathbf{v}$ and W are velocity fields for particles in a bounded fluid.

The third step, finally, is the evaluation of the above integrals in a manner which is independent of container geometry. For ease of reference, we denote these by $\overline{V}_2'$, $\overline{V}_2''$, $\overline{W}_2$, respectively. For the first integral, this is done by making use of the fact that for any system which as a whole is at rest, the configurational average of the velocity at any point in the system must be zero. For a system containing only one particle, this means that

$$\int_V \mathbf{v}^{(2)}(0,\mathbf{r}_2)\, d\mathbf{r}_2 = 0 \,. \tag{2.6}$$

If one now thinks of Eq. (2.6) as being split into two parts, one part of which is the first integral in Eq. (2.5), then

$$\int_V \mathbf{v}^{(2)}(0,\mathbf{r}_2)\, d\mathbf{r}_2 = \int_V \mathbf{v}^{(2)}(0,\mathbf{r}_2)\, g_{12}(\mathbf{r}_2)\, d\mathbf{r}_2$$

$$+ \int_V \mathbf{v}^{(2)}(0,\mathbf{r}_2)\, [1 - g_{12}(\mathbf{r}_2)]\, d\mathbf{r}_2 = 0 \,, \tag{2.7}$$

so that

$$\overline{V}_2' = n_2 \int_V \mathbf{v}^{(2)}(0,\mathbf{r}_2)\, g_{12}(\mathbf{r}_2)\, d\mathbf{r}_2 =$$

$$= n_2 \int_V \mathbf{v}^{(2)}(0,\mathbf{r}_2)\ [g_{12}(\mathbf{r}_2) - 1]\ d\mathbf{r}_2 \ . \tag{2.8}$$

In other words, the top integral is an integral over almost all of V, where $g_{12}(r_2)$ is non-zero, while by virtue of Eq. (2.6) the bottom integral involves the region in the immediate vicinity of the origin, where $g_{12}(\mathbf{r}_2) - 1$ is substantially different from zero.

The advantage of being able to switch integration ranges like this lies in the fact that when the neighbor is near the origin, the velocity field at the origin may be replaced with negligible error by the corresponding velocity field due to a particle moving through an unbounded fluid, since the bounded and unbounded fluid velocity fields in this region are virtually identical. The geometric interpretation of Eq. (2.8) is most apparent when the hard sphere radial distribution function is used (i.e., zero for $r_2 < a_1 + a_2$, and one otherwise).

The second integral in the expression for $\overline{\mathbf{U}}^{(1)}$ is

$$\overline{\mathbf{V}}_2'' = \frac{1}{6}\, a_1{}^2\, n_2 \int_V \left(\nabla_1{}^2\, \mathbf{v}^{(2)}(\mathbf{r}_1,\mathbf{r}_2)\right)_{\mathbf{r}_1 = 0} g_{12}(\mathbf{r}_2)\ d\mathbf{r}_2 \ . \tag{2.9}$$

In order to simplify its evaluation, we first make use of the fact that the container dimension is much larger than the particle diameter, so that we may regard the neighbor as fixed in Eq. (2.9) and allow the position of the test particle to vary. Thus

$$\overline{\mathbf{V}}_2'' = \frac{1}{6}\, a_1{}^2\, n_2 \int_V \nabla_1{}^2\, \mathbf{v}^{(2)}(\mathbf{r}_1,0)\ g_{12}(\mathbf{r}_1)\ d\mathbf{r}_1 \ . \tag{2.10}$$

Since $\mathbf{v}^{(2)}$ is the velocity field that would be generated by a single particle (of species 2) located at the origin, Eq. (2.10) is then just the weighted integral of $\nabla_1{}^2\mathbf{v}^{(2)}$ over the volume of the fluid.

In order to evaluate $\overline{\mathbf{V}}_2''$, we again make use of a physical constraint on the system. It will be shown in a subsequent publication[2] that as the container dimension becomes

---

[2] Ibid.

arbitrarily large (but still finite) in comparison with the particle diameter, then

$$\int_{\text{fluid}} \mu \nabla_1^2 \mathbf{v}^{(2)}(\mathbf{r}_1, 0)\, d\mathbf{r}_1 \to 10\pi\mu a_2 \mathbf{U}_o^{(2)} . \qquad (2.11)$$

Proceeding as before, we obtain

$$\bar{\mathbf{V}}_2'' = \frac{5}{3}\pi a_2 a_1^2 n_2 \mathbf{U}_o^{(2)}$$

$$+ \frac{1}{6} a_1^2 n_2 \int_{\text{fluid}} \nabla_1^2 \mathbf{v}^{(2)}(\mathbf{r}_1, 0)\, [g_{12}(\mathbf{r}_1) - 1]\, d\mathbf{r}_1 . \qquad (2.12)$$

Then, since $g_{12} - 1$ decreases rapidly to zero, the integration may be carried out with negligible error by using the velocity field for a particle (of species 2) in an unbounded fluid.

Finally, the third integral in Eq. (2.5) is evaluated by observing that $W(0, \mathbf{r}_2)$ will be nearly the same in a bounded fluid as in an unbounded fluid, and that the contribution to $W(0, \mathbf{r}_2)$ from a sphere located at $\mathbf{r}_2$ decreases asymptotically as r_2^{-4}. Thus the expression for $W(0, \mathbf{r}_2)$ in a bounded fluid may be replaced by the corresponding expression for an unbounded fluid,[3] and the range of integration extended to infinity with negligible error.

In the integrals for $\bar{V}_2'$ and $\bar{W}_2$ we may regard the position of the neighbor as fixed as we did for $\bar{\mathbf{V}}_2''$. Then our results can be combined to yield

$$\bar{\mathbf{U}}^{(1)} = \mathbf{U}_o^{(1)} + n_2 \int_V \mathbf{v}^{(2)}(\mathbf{r}, 0)\, [g_{12}(\mathbf{r}) - 1]\, d\mathbf{r}$$

$$+ \frac{1}{6} a_1^2 n_2 \int_{r > a_2} \nabla^2 \mathbf{v}^{(2)}(\mathbf{r}, 0)\, [g_{12}(\mathbf{r}) - 1]\, d\mathbf{r} \; +$$

---

[3]The velocity $\mathbf{U}(0, \mathbf{r}_2)$ of a test particle falling through unbounded fluid together with a neighbor **located** at $\mathbf{r}$ may be obtained from the articles by Spielman (Ref. 3) and Davis (Ref. 4).

$$+ \frac{5}{3} \pi a_2^3 \left(\frac{a_1}{a_2}\right)^2 n_2 U_o^{(2)} + n_2 \int_V W^{(1)}(\mathbf{r},0)\, g_{12}(\mathbf{r})\, d\mathbf{r} , \quad (2.13)$$

where all of the velocity fields are for particles in an unbounded fluid. The above expression represents the average sedimentation velocity of a test particle of species 1 when it is in a dispersion made up entirely of particles of species 2. Equation (2.13) may be put in the form $U^{(1)} = U_o^{(1)}(1 - K_2\phi_2)$ if the dimensionless integration variable $\rho = r/2a_2$ is used, since $\phi_2 = 4\pi a_2^3 n_2/3$. Then recalling that for dilute systems the effect of concentration on sedimentation velocity is cumulative (i.e., a sum over binary interactions), it becomes immediately apparent that the sedimentation velocity for particles of component 1 in a mixture of n components is given by

$$\overline{U}^{(1)} = U_o^{(1)}(1 - K_1\phi_1 - K_2\phi_2 - \ldots - K_n\phi_n) , \quad (2.14)$$

where K_n is obtained by replacing 2 by n in all of the appropriate subscripts and superscripts of Eq. (2.13). The value of K_n depends on the radial distribution function $g_{12}(r)$, or equivalently, the potential energy $W(\mathbf{r})$ between two particles.

III. COMPUTATION OF SEDIMENTATION COEFFICIENTS

The coefficient K_n in Eq. (2.14) is a function of a_n/a_1, $(\rho_n - \rho_f)/(\rho_1 - \rho_f)$, and $g_{1n}(r)$, as can be seen from detailed inspection of Eq. (2.13). Here ρ_f is the fluid density. For a single-component system $(n = 1)$, the radial distribution function is given by $g_{11}(r) = \exp(-W_{LR}/kT)$ for $r > 2a$, and zero otherwise, where W_{LR} is the long range inter-particle potential. If $W_{LR} = 0$ (hard spheres) then our calculations yield $K = 5.80$, as compared to the value 6.55 obtained by Batchelor (Ref. 1); the small difference is due to the manner in which $\overline{V}_2''$ is computed here and is discussed in another paper. For sample calculations we have chosen a long range repulsive potential generally used for spherical, charged particles in an electrolyte solution (Debye-Hückel approximation for overlapping double layers): $W_{LR}/kT = A \exp(-m[\rho - 1])/[1 + m/2]^2\rho)$, where $\rho = r/2a$, $A = q^2/(2a\epsilon kT)$, and $m = 2a/L$. q is the charge on the particle, ϵ is the solution dielectric constant, and L is the solution Debye length parameter. Figure 2 shows the results for two aqueous systems. The graph on the left is relevant to proteins (a = 36Å) with charges from 0 to 20 electron units and electrolyte concentrations (monovalent) from 4.5×10^{-4} to 2.8×10^{-2} M at 300°K. The graph on the right could pertain to microcolloids (a = 2000 Å) with charges from 0 to 7.5×10^5 electron units (this latter charge is equivalent to a surface density of 24 μ coulombs per cm^2) and electrolyte concentrations from 5.2×10^{-4} to 4.2×10^{-2}M. These graphs show that repulsive potentials increase K and hence decrease sedimentation rates at finite particle concentrations. For both graphs $K \to 5.80$ as $A \to 0$ or $m \to \infty$.

We know from Stokes' law that $U_o^{(n)}$ (the sedimentation velocity of a type n particle in an unbounded fluid at infinite dilution) is proportional to the buoyant density ($\Delta\rho_n = \rho_n - \rho_f$) and to a_n^2. This dependence is incorporated into the coefficients K_n in Eq. (2.14). The effects of $\Delta\rho_2$ and a_2 on the sedimentation of type 1 particles are shown in Fig. 3. The graphs show that as the neighbors become heavier ($\Delta\rho_2/\Delta\rho_1 > 1$) or larger ($a_2/a_1 > 1$), the coefficient K_2 increases substantially. In these calculations there is no long range interaction between particles ($W_{LR} = 0$). As $a_2/a_1 \to 1$ and $\Delta\rho_2/\Delta\rho_1 \to 1$, then $K_2 \to K_1 = 5.80$.

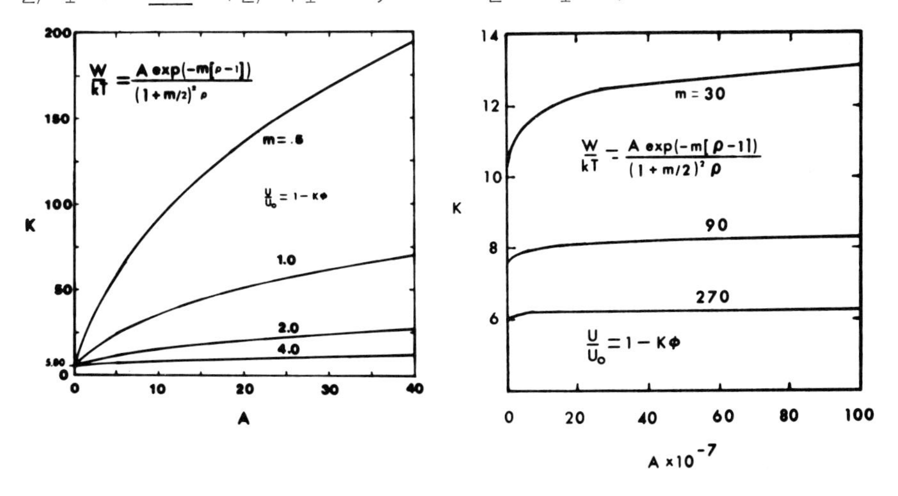

Fig. 2. Sedimentation coefficient (K) vs. interaction

parameters (A, m) for a one-component system. The left graph corresponds to proteins (a = 36 Å), the right one to micro-colloids (a = 2000 Å).

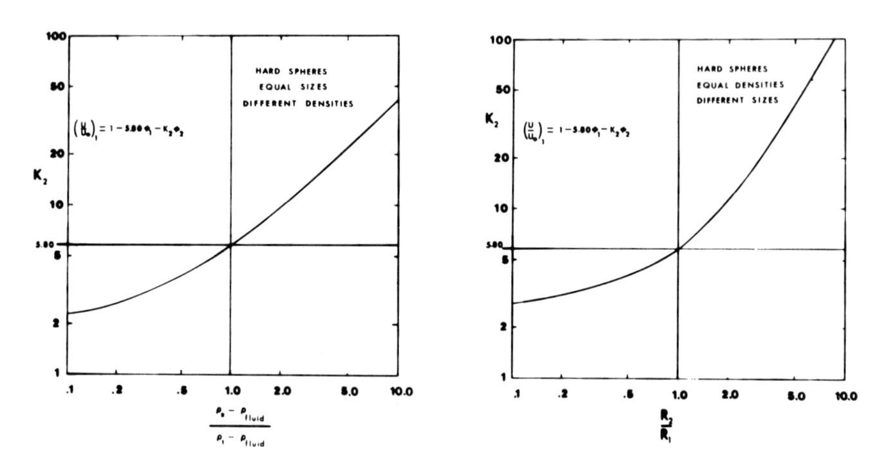

Fig. 3. Sedimentation coefficient (K_2) for the effect of particles of type 2 on the sedimentation of particles of type 1. Long range interactions are absent ($W_{LR} = 0$).

IV. REFERENCES

1. Batchelor, G. K., J. Fluid Mech. 52, 245 (1972).
2. Williams, W. E., Phil. Mag. (6th ser.) 29, 526 (1915).
3. Spielman, L. A., J. Colloid Interface Sci. 33, 562 (1970).
4. Davis, M. H., Chem. Eng. Sci. 24, 1769 (1969).
5. Goldman, A. J., Cox, R. G., and Brenner, H., Chem. Eng. Sci. 21, 1151 (1966).

A SEDIMENTATION STUDY OF POLYMERIC COLLOIDS IN NONAQUEOUS SOLVENTS

Melvin D. Sterman
Eastman Kodak Company

A polymer latex has been prepared in hexane by the precipitation of a vinyl terpolymer which is insoluble in hexane in the presence of a second vinyl terpolymer which is soluble in hexane. It has been demonstrated that (1) the latex particles consist only of the hexane-insoluble polymer, there being no adsorptive interaction between the two polymers, and (2) the effective stabilization mechanism is electrostatic repulsion as a result of surface charges on the latex particles.

I. INTRODUCTION

Although the subject of nonaqueous colloid chemistry is increasingly becoming an area of intense interest from both a theoretical (1,2,3) and a practical point of view, (4) it has not received the attention that aqueous colloids have received. By and large this has been the case since aqueous colloids are easier to treat both experimentally and theoretically, and a variety of methods have been developed for quantitative study of aqueous colloids. Not all of these methods are applicable to the study of nonaqueous colloids.

In nonaqueous colloids the phenomena which are effective in achieving stable dispersions are either electrostatic repulsion or steric stabilization. Electrostatic repulsion requires that the colloidal particle acquire a surface charge. The mechanisms by which this normally occur are either dissociation of ionic surface groups on the colloid particle, or the adsorption of ionic species. Steric stabilization requires a finite polymer adsorption layer which will physically prevent particles from approaching close enough such that van der Waals attraction may cause coagulation. Generally in nonaqueous solvents, a negligible enthalpy change results from the squeezing of the polymer adsorption layers on the approach of the two particles. The steric repulsion barrier is primarily due to a decrease in the configurational entropy of the adsorbed polymer which results in an increase in the free energy of the system when the adsorbed polymer layers of two particles start to interpenetrate.

Recent publications by Fitch and Kamath (5) and Cairns, Ottewill, Osmond and Wagstaff (6) are typical of the procedures reported in the literature for preparation of polymeric colloidal dispersions in nonaqueous media. In both studies the authors describe the preparation of stable polymethyl methacrylate latices in different aliphatic hydrocarbon solvents. The latices were prepared by the polymerization of methyl methacrylate monomer in the hydrocarbon solvent in the presence of a stabilizing polymer which is adsorbed onto the polymer particles, producing sterically stabilized dispersions.

In this report we shall describe a simple method for preparing polymer latices in a nonaqueous solvent which does not require in situ polymerization of the latex polymer. The latex is prepared by mixing together two polymers; one of the polymers must be insoluble in the solvent in which the latex is to be prepared and the second polymer must be soluble in this solvent. We shall demonstrate that the latex particles consist only of the solvent-insoluble polymer, and the effective stabilization mechanism is electrostatic repulsion.

II. EXPERIMENTAL PROCEDURES

Dispersion Preparation

The dispersions were prepared by adding a known volume of a solution of a hexane-soluble polymer (S-polymer) dissolved in xylene to a known volume of a solution of a hexane-insoluble polymer (D-polymer) also dissolved in xylene. Then any additional xylene, if needed, was added. Finally hexane was added at a constant rate, by use of a syringe pump, as the solution was agitated with a magnetic stirrer.

Polymers Used

The composition of the S- and the D-polymers both of which are randomly polymerized terpolymers, are illustrated in Figure 1. The S-polymer contained 2.4 wt% of lithium

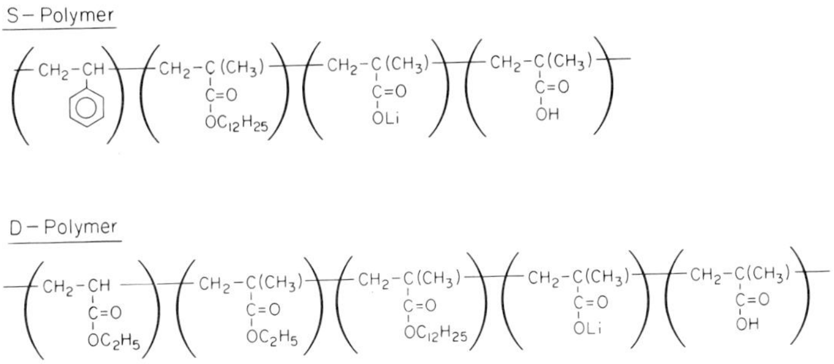

S – Polymer

$$\left(CH_2-CH(C_6H_5)\right)\left(CH_2-C(CH_3)(C{=}O)OC_{12}H_{25}\right)\left(CH_2-C(CH_3)(C{=}O)OLi\right)\left(CH_2-C(CH_3)(C{=}O)OH\right)$$

D – Polymer

$$\left(CH_2-CH(C{=}O)OC_2H_5\right)\left(CH_2-C(CH_3)(C{=}O)OC_2H_5\right)\left(CH_2-C(CH_3)(C{=}O)OC_{12}H_{25}\right)\left(CH_2-C(CH_3)(C{=}O)OLi\right)\left(CH_2-C(CH_3)(C{=}O)OH\right)$$

Fig. 1. The composition of the vinyl terpolymers used in the preparation of the dispersions.

methacrylate and 0.6 wt% methacrylic acid, with the balance almost equally divided between styrene and lauryl methacrylate on a weight basis. This polymer had a GPC $\bar{M}_w$ of 55,000, a $\bar{M}_n$ of 30,000 and an inherent viscosity, $\{\eta\}$, in dioxane of 0.20. The D-polymer contained 4.7 wt% of lithium methacrylate and 1.3 wt% methacrylic acid, 16 wt% lauryl methacrylate with ethyl acrylate and ethyl methacrylate making up the remaining 78 wt%. This polymer had a GPC $\bar{M}_w$ of 58,000, a $\bar{M}_n$ of 11,000 and an $\{\eta\}$ in dioxane of 0.17.

Characterization of the Latex

a. *Particle Size Distribution Measurements*

Transmission electron microscopy is a particularly effective technique for characterizing the shape, size and size distribution of the polymeric particles. Figure 2 is an illustration of a typical electron micrograph of a dispersion prepared using an S-polymer to D-polymer weight ratio of 1/1, having a total polymer content of 5.0 g/liter and in a solvent which contained 12% xylene. The particles are spherical. From the lengths of the shadows it is evident that there was some flattening of the particles during drying. The particle size distribution was determined from these electron micrographs by counting a statistically large particle population. Normally a particle count of between 500 and 1,000 particles is considered to be satisfactory. To ease the tedium of this procedure, it was carried out with a Quantimet 720 Image Analyzing Computer. The particle size distribution of this dispersion is illustrated in

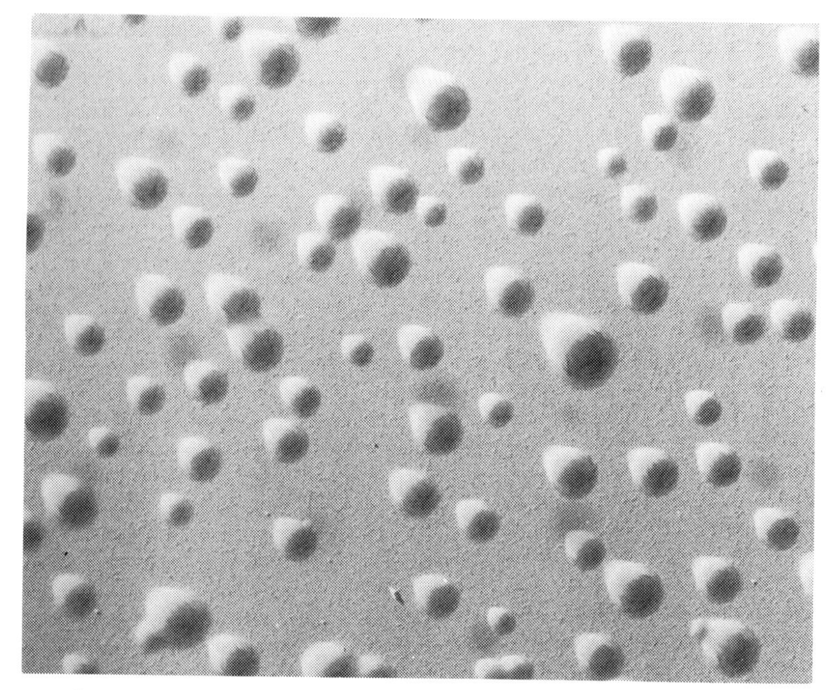

Fig. 2. A transmission electron micrograph of a polymer latex prepared using an S-polymer to D-polymer ratio of 1/1, a total polymer concentration of 5.0 g/l and in a solvent which contained 12% of xylene. The magnification is 20,000X.

Figure 3. The best mathematical fit to the data points was obtained by making the following assumptions: (1) a log normal distribution is applicable, and (2) the particle distribution is bimodal.

The average particle size of this latex was determined by measuring the angular dissymmetry of scattered light (45° and 135° angles) in a BRICE-PHOENIX light scattering photometer. Dissymmetries were measured at three latex concentrations, the original concentration and two dilutions, and extrapolated to zero concentration. The dissymmetry value was corrected for the Fresnel effect, and the average radius was calculated by means of the Rayleigh-Gans theory. This was justified in view of the small relative refractive index, N polymer/N solvent, of less than 1.07. Measurements made at two wavelengths, 436 and 546 nm, were in very good agreement and gave a value of 200 nm as the average particle diameter.

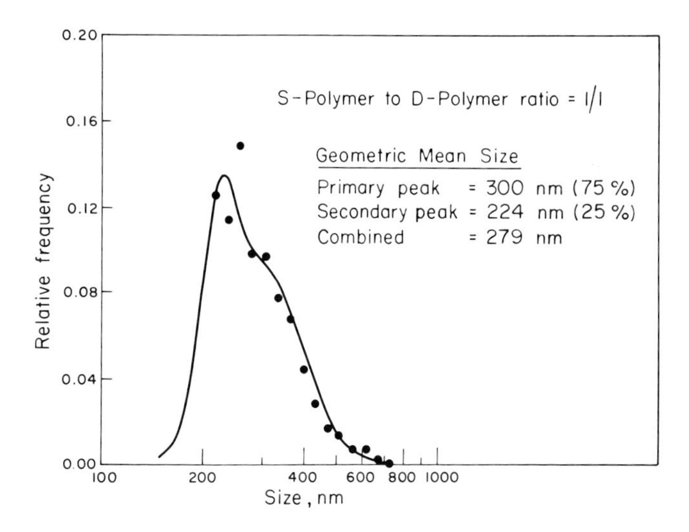

Fig. 3. Particle size distribution of the polymer latex illustrated in Fig. 2.

b. Electrophoretic Mobility Measurements

The electrophoretic mobility of a typical dispersion, which when prepared contained a total polymer content of 5.0 g/liter, an S-polymer to D-polymer weight ratio of 1/1 and the solvent had a xylene content of 12%, was measured by Light Beating Doppler Spectroscopy (7) using a 60 cycle square wave AC field. The electrophoretic mobility as a function of electric field strength is given in Figure 4. The mobility

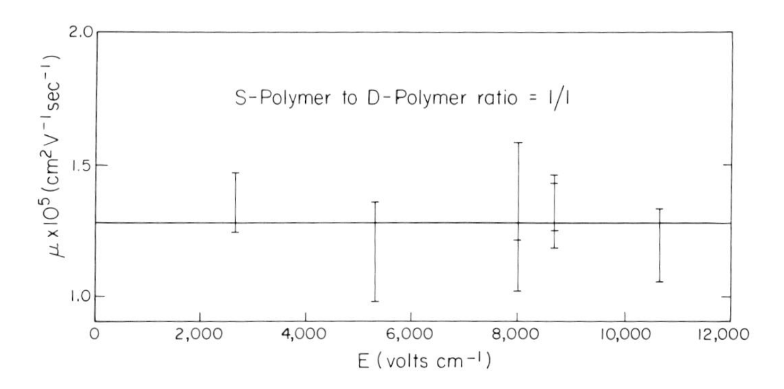

Fig. 4. Electrophoretic mobility of a polymer latex as a function of AC field strength.

Is independent of electric field over the range studied, namely 2,500 to 11,000 Volts cm^{-1}. From a least squares analysis of the data a value of 1.32×10^{-5} cm^2 V^{-1} sec^{-1} was calculated for the electrophoretic mobility. In carrying out this measurement it was necessary to dilute the original dispersion by a factor of 200 using hexane as the diluent. The zeta potential could now be calculated from the electrophoretic mobility data, and a value of 73 millivolts was obtained.

In a separate experiment using a simple electric cell containing two copper electrodes spaced one cm apart, it was demonstrated that the latex particles were positively charged.

Centrifugation Procedure

The centrifugation experiments were carried out in a SORVALL RC-2B refrigerated centrifuge using a type SS-34 rotor. During all runs the rotor compartment was thermostated at 22°C to avoid any problems due to heat buildup during centrifugation.

Ultraviolet Spectrophotometric Analysis

The polymer sediments from each centrifugation was analyzed spectrophotometrically. The sediments were carefully dried, weighed and dissolved in dioxane to yield a polymer concentration in the range from 0.5 to 1.0 g/liter. The UV spectra of these solutions were measured on a PERKIN-ELMER Model 402 Spectrophotometer. The spectra were examined primarily for the styrene absorption in the wavelength region from 240 to 280 nm. This analysis is specific for the S-polymer which contains styrene, whereas the D-polymer does not absorb in this wavelength region. This is demonstrated in Figure 5 in which the UV spectra of individual solutions of the two polymers in dioxane are recorded. In every experiment when the polymer sediment was subjected to this analysis there was no styrene absorption detected in the spectra. We, therefore, concluded that the latex particles consist only of the D-polymer and there is no adsorptive interaction between the two polymers.

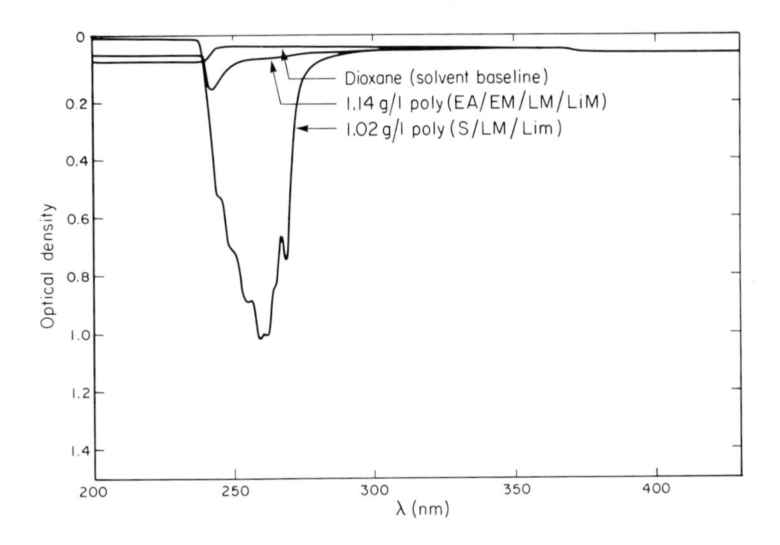

Fig. 5. Ultraviolet spectra of individual dioxane solutions of the S-polymer and the D-polymer.

This is illustrated in Figure 6 which is typical of the results obtained in these experiments. In the two experiments

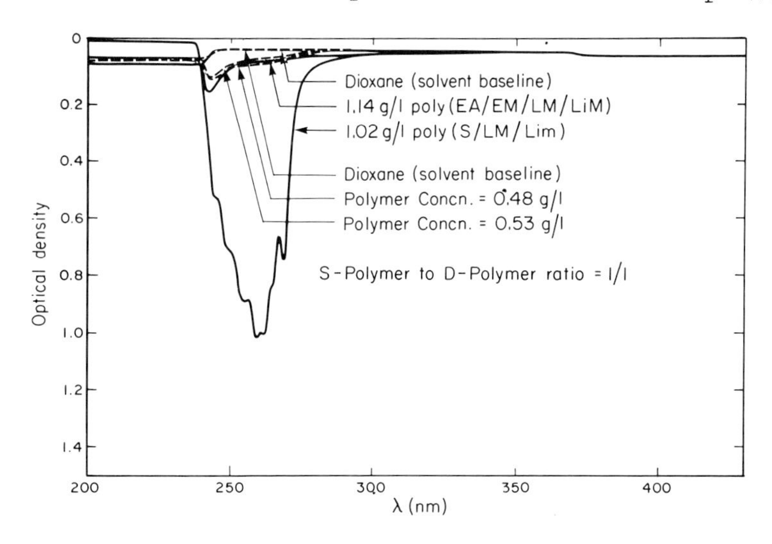

Fig. 6. Ultraviolet spectra of polymer solutions prepared by dissolving the sedimented polymer in dioxane. See text for descriptions of the experiments. For direct comparison the curves from Fig. 5. are included.

illustrated an S-polymer to D-polymer ratio of 1/1, and a total polymer concentration of 5.0 g/l was used in the preparation of these dispersions. One of the dispersions was prepared in a solvent which contained 15% xylene, and the polymer solution, which was prepared from the sedimented polymer for the UV analysis, had a concentration of 0.48 g/l. The solvent for the second dispersion contained 30% xylene and the polymer solution prepared from the sedimented polymer had a concentration of 0.53 g/l. Included in this figure are the reference curves for the two polymers which are found in Figure 5, and which facilitates a direct comparison of these spectral curves.

III. RESULTS OF THE SEDIMENTATION STUDIES

All of our initial experiments were carried out with a dispersion that contained an S-polymer to D-polymer weight ratio of 1 to 1 and a total polymer concentration of 5.0 g/liter. After conducting sedimentation experiments over a range of centrifuge speeds from 1000 rpm to 20,000 rpm, a speed of 3500 rpm was selected as the optimum speed for all further experimentation. In Figure 7 the percentage of dispersed polymer in the supernatant solution after centrifugation is plotted as a function of centrifugation time at a constant speed of 3500 rpm. The solvent for this dispersion contained 12% xylene. The curve shows that the percentage of polymer monotonically decreases with increasing time of centrifugation approaching an asymptotic value of 20% after 360 minutes. It should be pointed out that in calculating the percentage of dispersed polymer which remains in the supernatant solution after centrifugation, a correction has been made for the slight solubility of the D-polymer in the particular solvent mixture being used.

In the preparation of these dispersions it was noted that the stability, i.e., the relative ease with which the latex particles can be sedimented out of the solvent phase, appeared to depend principally on the following factors: (1) the xylene content of the solvent, (2) the total polymer concentration of the dispersion, and (3) the ratio of S-polymer to D-polymer. To study each of these factors, dispersions were prepared and centrifuged at 3,500 rpm for 30 minutes. The supernatants were then analyzed quantitatively for polymer content, and the sedimented polymer was analyzed spectrophotometrically for presence of the styrene containing

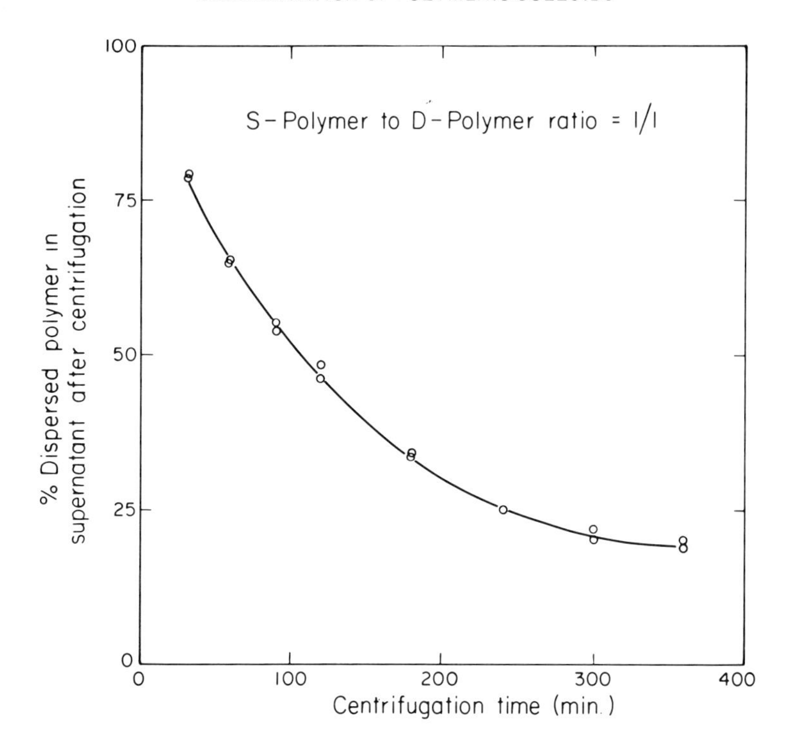

Fig. 7. The effect of centrifugation time upon the percentage of dispersed polymer in the supernatant. The centrifugation speed was 3,500 rpm.

S-polymer. In each of these experiments no S-polymer was found in the polymer sediment. In selected experiments the total weight of the polymer sediment was also determined in order to corroborate the calculated data from the supernatant solution and to demonstrate that our technique yields a material balance.

In Figure 8 the effect of increasing the xylene concentration in the solvent mixture is illustrated for a dispersion containing an S-polymer to D-polymer weight ratio of 1 to 1 and a total polymer concentration of 5.0 g/l. The effect of increasing the concentration of xylene in the solvent phase is to markedly diminish the sedimentation stability of the polymer latex. Increasing the concentration of xylene results in an increase in the soluble concentration of the D-polymer. The remaining insoluble fraction of this polymer will be in a more highly swollen state due to increased solvation. Thus the observed centrifugation behavior

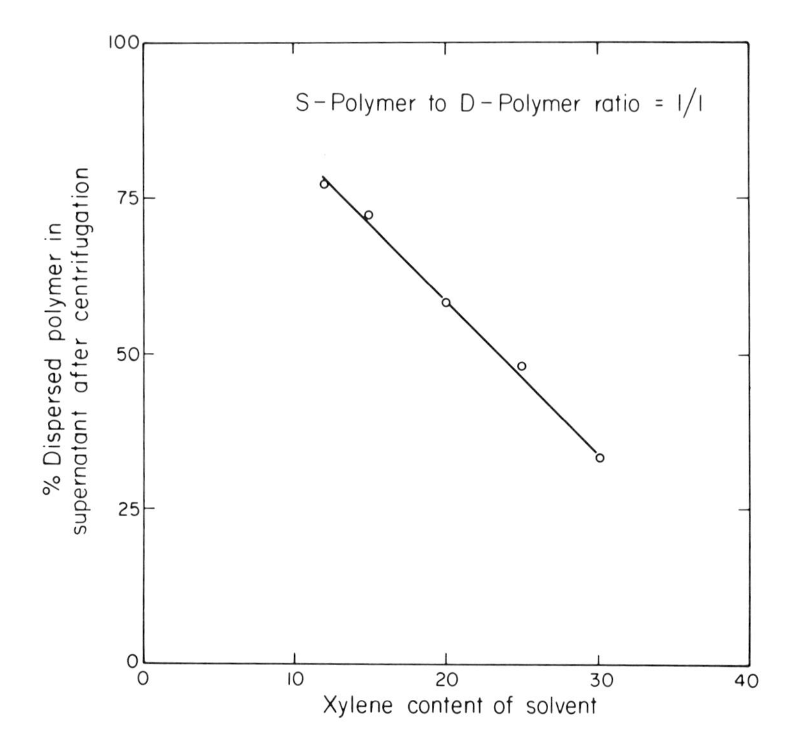

Fig. 8. The effect of solvent composition upon the sedimentation stability of a polymer dispersion in which the total polymer concentration was 5.0 g/l. The centrifugation speed was 3,500 rpm.

should be explainable in terms of a shift in the particle size distribution to much larger particle size with increasing xylene content of the solvent. An attempt was made to verify this hypothesis by measuring the particle size distribution of a latex which was prepared in a solvent that contained 25% xylene. The technique used was the same as previously described and the distribution is illustrated in Figure 9. Comparing these data with the distribution obtained for the reference dispersion given in Figure 3, we observe at best a very small shift in the particle size distribution. The probable reason we do not see a significant shift in the particle size is inherent in the method used for measuring the particle size distribution. In preparing the sample for transmission electron microscopy it is necessary to dry the

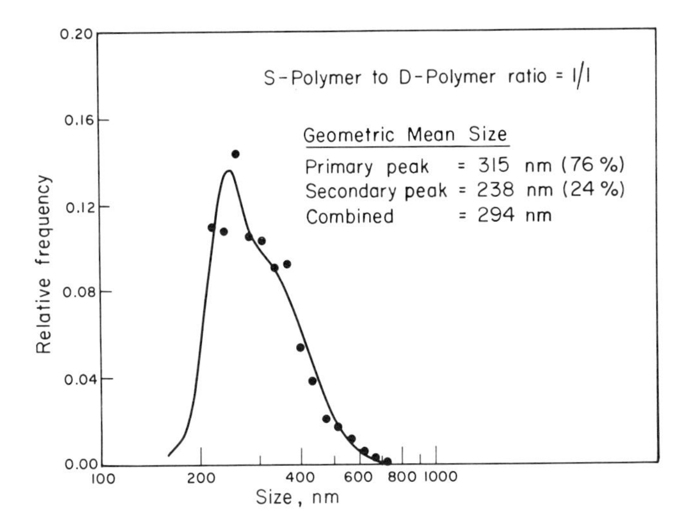

Fig. 9. Particle size distribution of a polymer dispersion in which the total polymer concentration was 5.0.g/l, and was prepared in a solvent which contained 25% xylene.

sample, and any expansion of the polymer particle due to increase solvation and attendent network expansion effects would be lost.

The pronounced effect that the total polymer concentration in the dispersion has upon the sedimentation stability is illustrated in Figure 10 for two sets of dispersions, one prepared in a solvent containing 12% xylene and the second containing 25% xylene. In both examples a dramatic increase in the stability of the dispersions is noted with increasing polymer concentration. It is particularly noteworthy that with increasing polymer concentration the two curves approach one another. This suggests that the increase in stability may be due primarily to a solvency effect of the xylene. At lower polymer concentrations the solvation of the polymer particle may be more pronounced producing a shift in particle size distribution to much larger particle size. With increasing polymer concentration the solvency effect of the xylene becomes less significant. An attempt to verify this hypothesis by measurement of the particle size distribution of a latex, in which the total polymer concentration was 9.0 g/l, and was prepared in a solvent which contained 12% xylene, is illustrated in Figure 11. Comparing this distribution curve with the reference distribution curve in

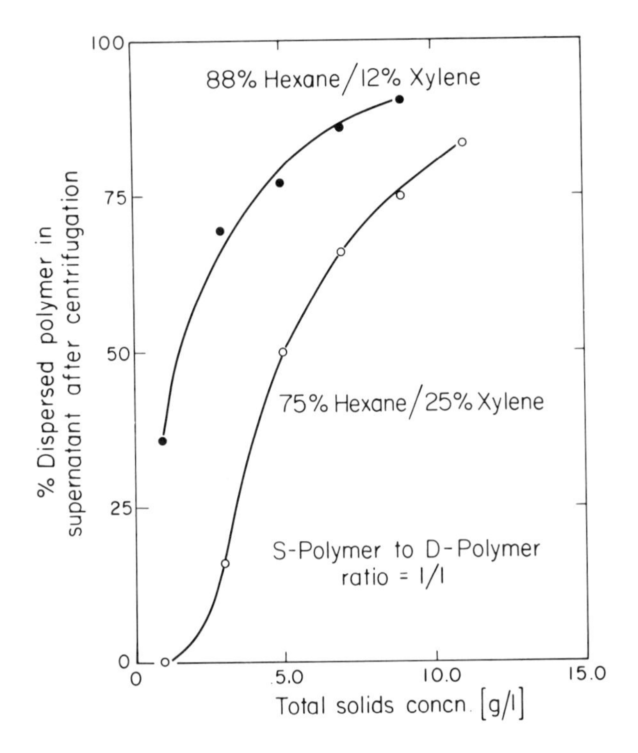

Fig. 10. The effect of polymer concentration upon the sedimentation stability of polymer dispersions. The centrifugation speed was 3,500 rpm.

Figure 3, a slight shift in the distribution curve toward smaller particles is observed. Again the method utilized for measuring the distribution curve, which requires the dispersion to be dried, may very well obscure the solvency effect.

The importance of the concentration of the S-polymer is illustrated in Figure 12. As the S-polymer concentration is increased, keeping the D-polymer concentration constant, the sedimentation stability increases essentially linearly until at an S-polymer to D-polymer ratio of about 1.25 to 1, an equilibrium condition appears to have been reached and no further change takes place.

Lastly, we investigated the importance of the ionic monomer composition in both polymers. If the lithium methacrylate and methacrylic acid are eliminated from the D-polymer, then the latex prepared in conjunction with the

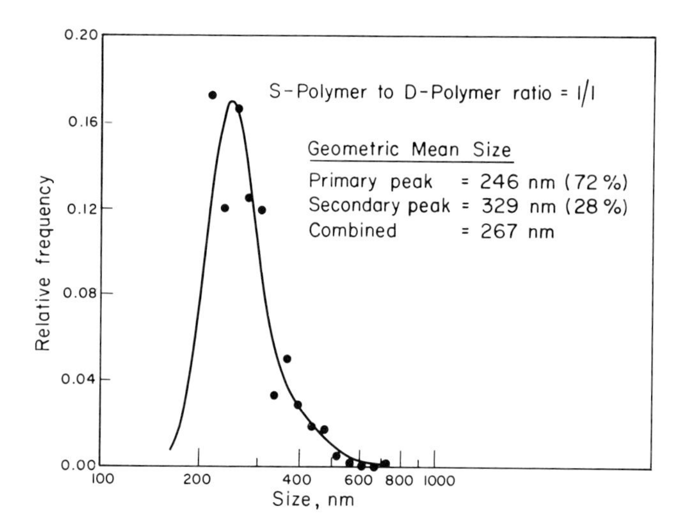

Fig. 11. Particle size distribution of a polymer dispersion prepared using a total polymer concentration of 9.0 g/l, and in a solvent which contained 12% xylene.

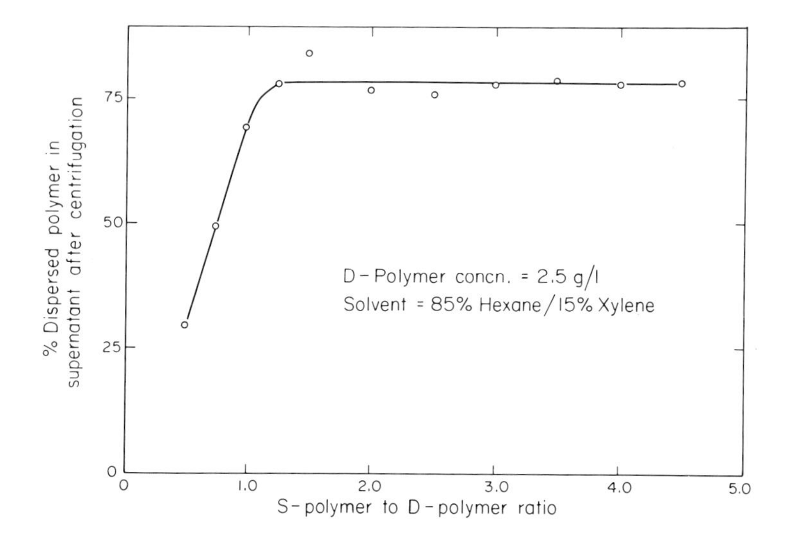

Fig. 12. The effect of S-polymer to D-polymer ratio upon the sedimentation stability of a polymer dispersion. The centrifuged speed was 3,500 rpm.

S-polymer is very unstable, and completely sediments out in a very short period of time on standing. If the lithium methacrylate and methacrylic acid are eliminated from the S-polymer, then upon addition of hexane the D-polymer precipitates out of solution. Conversion of all of the methacrylic acid in the S-polymer to lithium methacrylate does not significantly alter the sedimentation stability of the dispersions prepared with the D-polymer. However, conversion of all of the lithium methacrylate in the S-polymer to methacrylic acid substantially reduces the stability of dispersions prepared with the D-polymer. Substitution of ethyl hexyl methacrylate for the styrene-lauryl methacrylate portion of the polymer does not significantly alter the stability of the dispersions prepared with the D-polymer. These results point to the importance of the lithium cation in the stabilization mechanism, and are suggestive of an interaction between the soluble polymer and the dispersed, insoluble polymer which involves the transport of lithium cations between the ionizable groups of the two polymers.

Summarizing the results of this study, it has been demonstrated that there is no adsorptive interaction between the soluble polymer and the dispersed polymer particles. Rather the data suggest that the effective stabilization mechanism is electrostatic repulsion as a result of surface charges on the latex particle. Support for this conclusion comes from experiments which demonstrated that (1) the latex particles are positively charged, (2) the zeta potential value calculated from the electrophoretic mobility is in accord with this hypothesis, and (3) the lithium cation appears to be involved in the charging mechanism.

IV. ACKNOWLEDGMENTS

The author wishes to gratefully acknowledge the contributions of Dr. G. Buske for his technical assistance, Dr. J. Cheng for the Light Beating Doppler Spectroscopic measurements, Dr. H. Coll for the light scattering measurements, and Mr. B. Wood for the size distribution computations with the Quantimet Image Analyzing Computer. The author also wishes to express his appreciation to Dr. L. Oppenheimer and Dr. P. Bagchi for their helpful discussions.

V. REFERENCES

1. Lyklema, J., Adv. Colloid Interface Sci., 2, 66 (1968).
2. Vincent, B., Adv. Colloid Interface Sci., 4, 193 (1974).
3. Smitham, J. B., Evans, R. and Napper, D. H., J.C.F.T.A.R., 71, 285 (1975).
4. Vijayendran, B., in "Colloid Dispersions and Micellar Behavior" (K. L. Mittal, Ed.), A.C.S. Symposium Series 9, 1975.
5. Fitch, R. M. and Kamath, Y. K., J. Colloid Interface Sci., 54, 6 (1976).
6. Cairns, R.J.R., Ottewill, R. H., Osmond, D.W.J., and Wagstaff, I., J. Colloid Interface Sci., 54, 45 (1976).
7. Ware, B. R., Adv. Colloid Interface Sci., 4, 1 (1974).

THE EFFECT OF THE ADDITION OF A WATER SOLUBLE POLYMER ON THE ELASTICITY OF POLYMER LATEX GELS

J.W. Goodwin and A.M. Khidher

University of Bristol

ABSTRACT

The addition of a soluble polymeric 'thickener' or gelling agent to an aqueous polymer latex produced a marked increase in shear modulus. A model is developed, based on a gel network in which particles are dispersed but increase the number of junctions by adsorption. The shear modulus is calculated from the linear addition of the entropic terms from the network and the electrical terms from the particles. Some preliminary experiments are presented which show that the model results in a satisfactory description of the data.

I. INTRODUCTION

In order to produce the required degree of thixotropy of a concentrated dispersion, a 'thickener' or gelling agent is often added. Soluble polymers can be used for this purpose and the most important rheological properties of the resultant dispersion are the modulus of the gel state, the yield stress, the high shear viscosity and thixotropic recovery rate. (Thixotropic latex paints are, perhaps, a particularly good example). This work attempts to describe the shear modulus of such a gel in terms of a model based on the microstructure. Some preliminary experiments are also reported and the model system chosen for the study was a polystyrene latex with ethylhydroxy ethyl cellulose as the soluble polymeric species. The propagation velocity of shear waves was used to measure the elasticity of these systems. This technique effectively gives the high frequency limiting value of the storage modulus.

Earlier work under continuous shear conditions(1,2) suggested that a suitable model should be based on a polymeric gel network with the particles remaining well dispersed in the network but contributing to it. It has long been known that concentrated latices can form gels on dialyses(3,4,5).

II. THEORETICAL

Polystyrene latices, of small particle diameter, have been shown to exhibit elasticity(6,7,8) at volume fractions of ca. 0.2. These aqueous systems contained only electrolytes with no soluble polymer and were colloidally stable. The value of the shear modulus was found to depend on both the volume fraction (ϕ) and the electrolyte concentration. An expression is developed below relating the shear modulus to the second derivative of the inter-particle potential. In contrast to these dispersions, the elasticity of ethyl hydroxy ethyl cellulose solutions is entropic in origin and can be described in terms of the swollen network theory of rubber elasticity. A model will then be developed in an attempt to describe the behaviour of mixtures of the two.

A. Latex Elasticity

Concentrated monodisperse latices frequently show iridescence when the average separation of the particles in a face-centred cubic array is of the order of wavelength of visible light(9). If two particles are centred at positions O and A in the Cartesian coordinate system shown in Fig.1.

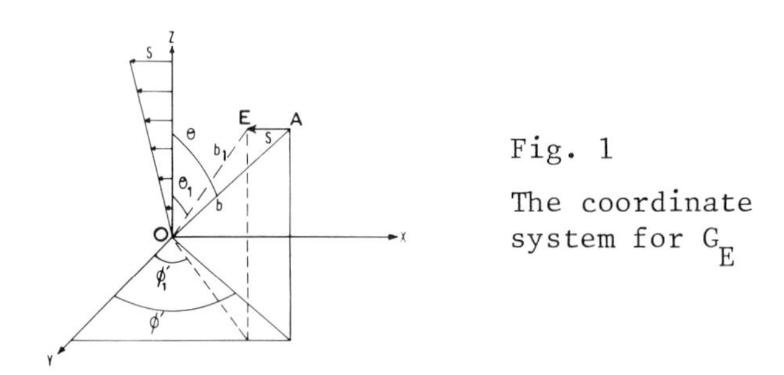

Fig. 1

The coordinate system for G_E

ds/dz is the shear strain and b, θ, ϕ', are the initial coordinates of the centre of the particle in the absence of shear strain and b_1, θ_1, ϕ_1' the new coordinates after a small deformation, s, parallel to the x-y plane. We can define $b = H_o + 2a$, where H_o is the interparticle separation for spheres of diameter 2a and for a f.c.c. array $b = 2a(0.74/\phi)^{1/3}$. After deformation $\Delta H_o = b - b_1$ and it is easily shown that:

$$2b\,\Delta H_o - (\Delta H_o)^2 = 2b\,s \sin\theta \sin\phi' - s^2 \qquad \ldots (1)$$

For small shear strains, both s and ΔH_o are small and using the relation $s = ds/dz \cdot b \cos\theta$:

$$\frac{ds}{dz} = \frac{H_o}{b \cos\theta \sin\theta \sin\phi'} \qquad \ldots (2)$$

Assuming that the force between the particles can be expressed as a Taylor expansion with ΔH_o small, then after deformation the force is increased by an amount F_1 where $F_1 = \partial F/\partial H_o . \Delta H_o$. The restoring force, that is the component along the x-axis, is:

$$F_{1x} = \Delta H_o \sin\theta_1 \sin\phi_1' \frac{\partial F}{\partial H_o} \qquad \ldots (3)$$

that is in terms of θ and ϕ' as ds/dz is small:

$$F_{1x} = b\frac{\partial F}{\partial H_o}\frac{ds}{dz} \sin^2\theta \cos\theta \sin^2\phi' \qquad \ldots (4)$$

In order to calculate the contribution to shear stress, τ_1, due to this pair interaction, it is necessary to assign an area in the x-y plane (A_{xy}) to the interaction. Now the volume, v, occupied by two particles is:

$$v = \frac{2\pi b^3}{0.74 \, . \, 6} \quad \text{and} \quad A_{xy} = \frac{v}{b \cos\theta}$$

Assuming all orientations are equally probable, the average shear stress is found by integration over all angles :

$$\bar{\tau}_1 = \frac{0.7066}{2\pi^2 b}\frac{\partial F}{\partial H_o}\frac{ds}{dz}\int_0^{\pi}\int_0^{2\pi} \sin^2\theta \cos^2\theta \sin^2\phi' d\theta d\phi' \qquad (5)$$

Each particle has twelve nearest neighbours with which it interacts strongly and so, neglecting next nearest neighbours, the total stress τ may be written:

$$\tau = \Sigma_1^{12} \tau_i = 12\,\bar{\tau}_1 \qquad \ldots (6)$$

or

$$\tau = \frac{2.1198}{4b}\frac{\partial F}{\partial H_o}\frac{\partial s}{\partial z} \qquad \ldots (7)$$

The shear modulus for shear between two rotating parallel discs in a latex gel is, for small shear strains :

$$G_E = \frac{3}{2}\,\tau/\frac{ds}{dz} = \frac{0.7949}{b}\frac{\partial F}{\partial H_o} \qquad \ldots (8)$$

The theory of colloid stability of Derjaguin and Landau(10) and Verwey and Overbeek(11) gives the total potential energy of interaction of a pair of particles (V_T) as the sum of the repulsion term arising from the overlap of the electrical double layers around the particles and the London-van-der Waals' attraction. As the first differential of the energy function with respect to distance of separation gives the force-distance relation:

$$\frac{\partial F}{\partial H_o} = \frac{\psi^2 \varepsilon \exp(-\kappa H_o)}{2(x+1)}\left[a\kappa^2 + \frac{\kappa}{(x+1)} + \frac{1}{2a(x+1)^2}\right] - \frac{A}{24a^2}\left[\frac{3(x^2+2x) + 4}{(x^2+2x)^3} + \frac{3}{(x+1)^4} - \frac{6(x^2+2x) + 4}{(x^2+2x)^2(x+1)^2}\right] \quad (9)$$

for $\kappa a < 3$ where κ is the Debye reciprocal length, ε is the relative permittivity, ψ is the surface potential, x is $H_o/2a$ and A is the Hamaker constant. Equation 8 can therefore be solved for G_E as b is known from the volume fraction.

B. The Elasticity of Ethyl Hydroxyl Ethyl Cellulose Solutions

The swollen network theory of rubber elasticity gives the shear modulus of an ideal network as:

$$G = 2A\ n\ kT \quad \ldots (10)$$

(with no volume change on deformation), where n is the number of crosslinks per cm^3 and is proportional to the polymer concentration c g dl^{-1}, kT is the product of the Boltzmann constant and absolute temperature. Much of the recent discussion has centred on the value of the front factor, A. Flory and Wall(12) gave a value of A = 1 whilst James and Guth(13) (also Edwards and Freed(14)) gave a value of A = ½. Recently, Graessley(15,16) has derived the expression A = f-2/f where f is the functionality of the junction, so that for a tetrafunctional junction A = ½ (a case derived by Eichinger(17)) and approaches 1 for N-functional junctions.

Any real network is likely to have the following defects(18):

1) chain entanglements which increase n and therefore G,
2) closed loops which decrease n (and hence G) by removing two sites per loop,
3) free chain ends with unreacted functionalities (or sites) which also decrease n.

The expression for the shear modulus may therefore be written:

$$G = 2A\ n\ kT\ c' \quad \ldots (11)$$

where c' is the correction factor for network defects and c' = f(c).

N_i is defined as the number of crosslinking sites along a polymer chain and so the number of crosslinks per chain is $N_i/2$ provided there are no defects. However, chains which are only bifunctional can only form an infinitely long chain which is uncrosslinked. Therefore the number of crosslinks per chain is N_i-2/2, i.e. the functionality of the chain must be greater than 2 for an elastic network to be formed as the deformation of an infinitely long chain should relax through diffusion. As a result, the number of crosslinks per cm^3 is:

$$n = \frac{N_{Av}\, c}{100\ Mn} \quad \frac{N_i - 2}{2} \quad \text{i.e. } n = \frac{N_{Av}\, c}{200\ Mn}(N_i - 2) \qquad (12)$$

where N_{Av} is the Avogadro number, Mn is the number average molecular weight and c is the polymer concentration in g dl^{-1}.

C. Network Defects

1. Entanglements

The effect of polymer chains being entangled in the network is to increase the effective number of crosslinks and has been shown by Edwards (from Allen et al(19)) to be proportional to the square of the polymer concentration. The resulting equation for G is:

$$G = \frac{2A\ kT\ N_{Av}\ (N_i - 2)\ c}{200\ M_n} + 2A\ R\ c^2 \qquad \ldots (13)$$

where the constant, R, has been found to be in the range of 200 to 300 $cm^5 g^{-1} s^{-1}$ for swollen uncrosslinked rubber (see (19)).

2. Closed Loops

A closed loop results in the loss of two adjacent crosslinking sites on the chain and Walsh et al(20) have estimated this effect. The probability that two adjacent crosslinking sites on a freely-jointed chain is in a volume element dV (21,22):

$$W_L(dV) = \left(\frac{3}{2\ n'\ L^2\ \pi}\right)^{3/2} dV \qquad \ldots (14)$$

where L is the average bond length and n' is the number of bonds between the sites. Restriction of the conformation of the chain due to restricted rotation of bonds and fixed bond angles is given by the Characteristic Ratio(23), c_∞, so that equation (5) becomes :

$$W_L(dV) = \left(\frac{3}{2\ c_\infty\ n'\ L^2\ \pi}\right)^{3/2} dV \qquad \ldots (15)$$

The value of the characteristic ratio can be calculated from experimental data using the following equation(23):

$$c_\infty = \frac{K}{\Phi}^{2/3} \frac{M_b}{L^2} \qquad \ldots (16)$$

where M_b is the mean molecular weight per skeletal bond, the constant $\Phi = 2.6 \times 10^{21}$ at the θ-point and $K_\theta = [\eta]_o/M^{\frac{1}{2}}$ ($[\eta]_o$ is the intrinsic viscosity at the θ-point and M is the molecular weight). Generally $K = [\eta]/M^a$.

The probability that the crosslinking site is occupied by another chain in the volume dV is:

$$W\,(dV) = \frac{N_{Av}\; c\; N_i}{100\; M_n}\, dV\,, \qquad \ldots (17)$$

i.e.,it is equal to the total number of crosslinking sites in volume dV minus N_i (the chain under consideration). Assuming that there are no energetic differences in the formation of a bond formed during crosslinks and loops, the relative probability that a given site is reacted with either of its nearest neighbours is f_E (2 sites have 1 nearest neighbour and $N_i - 2$ sites have 2, therefore the average number of nearest neighbours is:

$$\frac{2 + 2N_i - 4}{N_i} \quad \text{i.e.} \quad 2\left(1 - \frac{1}{N_i}\right) \;).$$

$$f_L = \frac{2\left(1 - \frac{1}{N_i}\right) W(dV)_L}{\left(1 - \frac{1}{N_i}\right) 2W(dV)_L + W(dV)} = 1 \Bigg/ 1 + \frac{N_{Av} N_i c}{200\overline{M}n \left(\frac{3}{2c_\infty n' L^2 \pi}\right)^{3/2} \left(1 - \frac{1}{N_i}\right)} \qquad (18a)$$

$$\text{and } n' = \frac{\overline{M}n}{MbN_i}\,, \text{ so } f_L = 1 \Bigg/ 1 + \frac{N_{Av} \overline{M}n^{\frac{1}{2}} c}{200 \left(\frac{3Mb}{2c_\infty L^2 \pi}\right)^{3/2} N_i^{\frac{1}{2}} \left(1 - \frac{1}{N_i}\right)} \qquad (18b)$$

If there are no unreacted sites the expression for the elastic modulus is:

$$G = 2A\; kT\; \frac{N_{Av}\; (N_i - 2)\; c}{200\; Mn}\; (1 - f_L) + 2A\; R\; c^2 \qquad \ldots (19a)$$

$$G = 2A\; kT\;\; B\; c\;\; \left(1 + \frac{R}{B\; kT}\, c - f_L\right) \qquad \ldots (19b)$$

$$\text{where } B = \frac{N_{Av}\; (N_i - 2)}{200\; Mn}\,.$$

$$\text{and from equation (2)} \quad c' = \left(1 + \frac{R}{B\; kT}\, c - f_L\right)$$

D. Ethyl Hydroxy Ethyl Cellulose - Latex Mixtures

The model chosen for these mixtures is the formation of an infinite polymer network of the polymer chains in the aqueous phase and the effect of the addition of latex particles is: 1) to produce an additional number of crosslinks due to the adsorption of the polymer onto the particle surface (i.e. the original network is not destroyed) and the interconnection of adsorbed molecules via the surface, and 2) to make no modification to the electrical interactions between the particles. Therefore the linear addition of shear moduli can be made :-

$$G_T = G_N + G_E$$

where G_N is the shear modulus of the network (including the contribution from adsorption), and G_E is the modulus produced by electrical interactions in the absence of a network i.e.

$$G_E = \frac{0.7949}{H_o + 2a} \frac{\partial^2 V}{\partial H_o^2}$$

When it is assumed that no bridging flocculation occurs, i.e. each adsorbed molecule is attached to only one particle, the number of crosslinks is increased by the number of adsorption sites per cm^3 of aqueous phase, n_1,:

$$n_1 = \frac{\phi}{1 - \phi} \frac{\rho_1 \Gamma N_{Av}}{Mn}$$

where ϕ is the final volume fraction of the latex, Γ is the weight of soluble polymer adsorbed on one g of latex particles and ρ_1 is the density of the particles. However, each adsorbed chain is linked to each other chain on the same particle through the surface of the particle (i.e. if a particle moves every adsorbed molecule on that particle must move also) and so n' must be multiplied by the number of adsorbed molecules per particle to give the number of additional crosslinks per cm^3 resulting from the adsorption process:

$$n_a = n_1 \frac{\pi d^3 \rho \Gamma N_{Av}}{6 \ Mn}$$

where d is the particle diameter, i.e.

$$n_a = \frac{\phi}{1 - \phi} \frac{\pi d^3 \rho^2 \Gamma^2 N_{Av}^2}{6 \ Mn^2} \qquad \ldots (20)$$

This is equivalent to saying that each adsorbed molecule has an additional number of crosslinking sites which is equal to the number of adsorbed molecules per particle. If n_1 were used without modification, the model would imply that the latex particles were 'broken up' and the fragments dispersed throughout the medium which would be unrealistic.

When Γ is expressed as a function of the area occupied by an adsorbed molecule (A_m) the expression for n_a may be simplified:-

$$n_a = \frac{\phi}{1 - \phi} \frac{b \pi d}{A_m^2} \qquad \ldots (21)$$

In the absence of network defects, the modulus is:

$$G_N = 2A \ kT \left[\frac{N_{Av} \ c}{200 \ Mn} (N_i - 2) + n_a\right] \qquad \ldots (22)$$

Entanglements can only increase the modulus by occurring in the aqueous phase and so:

$$G_N = 2A\ kT\left[\frac{N_{Av}}{200\ Mn}(N_i - 2)\ c + n_a\right] + 2A\ R\ c^2 \quad \ldots (23)$$

The effect of closed loops on the network in the aqueous phase has been given by equation (18) and so with closed loops in the aqueous phase equation (23) becomes :-

$$G_N = 2A\ kT\left[\frac{N_A(N_i - 2)}{200\ Mn}(1 - f_L)c + n_a\right] + 2A\ R\ c^2 \ldots (24)$$

At the surface the situation is not so clear as a closed loop should now be defined as the adsorption of two adjacent cross-linking sites. Treating the surface as a plane (with z as the normal to the plane) a surface loop will occur when a nearest neighbour-site enters the volume element $dz\ \pi\ \langle r_N^2\rangle$ where $\langle r_N^2\rangle^{\frac{1}{2}}$ is the r.m.s. distance between crosslinking sites, i.e. a thin element spread over the surface. The probability of a loop being formed in this volume is (c.f. equation (15)) :

$$W_{SL} = \left(\frac{\beta'}{\pi^{\frac{1}{2}}}\right)^3 \iint \exp|-\beta'^2(x^2 + y^2)|\ dx\ dy\ dz,$$

$$0 < x = y < \langle r_b^2\rangle^{\frac{1}{2}},$$

$$W_{SL} = \left[\beta'/\pi^{\frac{1}{2}} - \frac{\beta'\ e}{\pi^{\frac{1}{2}}}^{\beta'^2\langle r_N^2\rangle}\right] dz$$

$$\text{where}\quad \beta' = \left(\frac{3}{2\ n'\ L^2}\right)^{\frac{1}{2}}$$

It is assumed that when a site is within the distance of one repeat unit from the surface (X = dz) and allowance is made for restricted rotation and fixed bond angles by inclusion of the characteristic ratio, then the probability of a surface loop being formed by either of the two nearest neighbouring crosslinking sites is:

$$2\ W_{SL} = X\left(\frac{6}{c_\infty\ n'\ L^2\ \pi}\right)^{\frac{1}{2}}\left[1 - \exp - \left(\frac{3}{2\ c_\infty\ n'\ L^2}\langle r_N^2\rangle\right)\right] \quad \ldots (25)$$

Thus the probability of finding a surface loop in a volume corresponding to 1 cm^3 of aqueous phase is $2W_{SL}$ multiplied by the number of adsorbed molecules per cm^3. From equation (25) and n_1

$$W'_{SL} = 2\ W_S\ L\ n_1\ X$$

$$W'_{SL} = \frac{\phi}{1-\phi}\frac{X\rho\Gamma N_{Av}}{\overline{M}n}\left(\frac{6}{c_\infty n' L^2\pi}\right)^{\frac{1}{2}}\left[1 - \exp - \left(\frac{3\langle r^2{}_N\rangle}{2c_\infty n' L^2}\right)\right] \qquad (26)$$

Therefore, in unit volume of aqueous phase, the relative probability of a loop, either surface or otherwise, being formed to all the interaction is :

$$f_L = \frac{W'_{SL} + W_L}{W'_{SL} + W_L + Bc\, N_i/(N_i - 2) + n_a} \quad \ldots (27)$$

i.e. the fraction of crosslinks formed is

$$1 - f'_L = 1 - \cfrac{1}{1 + \cfrac{(Bc + n_a)N_i/N_i - 2}{W'_{SL} + W_L}} \quad \ldots (28a)$$

and equation (23) must now be written as

$$G_N = 2A\,kT \left[\frac{N_{Av}(N_i - 2)c}{200\ Mn} + n_a\right](1 - f'_L) + 2A\,Rc^2 \quad \ldots (28b)$$

The complete expression for f'_L is:-

$$f'_L = 1 \Bigg/ 1 + \frac{\dfrac{N_{Av}(N_i)c}{200\ \bar{M}n} + \dfrac{\phi}{1-\phi}\dfrac{\pi d^3\rho^2\Gamma^2 N_{Av}^2}{6\bar{M}n^2}}{\dfrac{\phi}{1-\phi}\dfrac{X\rho\Gamma N_{Av}}{\bar{M}n}\left(\dfrac{6}{c_\infty n' L^2\pi}\right)^{\frac{1}{2}}\left|1-\exp-\left(\dfrac{3\langle r_N^2\rangle}{2c_\infty n' L^2}\right)\right| + 2\left(1 - \dfrac{1}{N_i}\right)\left(\dfrac{3}{2c_\infty n' L^2\pi}\right)^{3/2}} \quad (28c)$$

Second nearest neighbour loops have not been included, and sites at greater distances should not effect the network as the intermediate sites must be linked to other chains. The total shear modulus can be calculated from:

$$G_T = 2AkT\left[\frac{N_{Av}(N_i-2)c}{200\ \bar{M}n} + n_a\right](1 - f'_L) + 2ARc^2 + \frac{0.7949}{H_o+2a}\frac{\partial^2 V_T}{\partial H_o^2} \quad (29)$$

Only N_i and R are left as unknown variables in equation (29) as all the other parameters can be determined from osmometry, viscosity measurements, adsorption isotherms and structural studies. R could be determined from the concentration dependence of the shear modulus in the absence of particles if the effect of closed loops could be minimised (i.e. large values of c). N_i can then be obtained by using equation (19) to give the best fit to the experimental data.

E. Bridging of Particles

Up to this point is has been assumed in the model that none of the adsorbed molecules forms a bridge between two particles. That is the particles are only connected indirectly by the infinite polymer network. Although this may be the case at low volume fractions, i.e. where the average interparticle separation is large, this is unlikely to be the case when the separation is much less than the thickness of the adsorbed layer.

When bridging occurs, the number of crosslinks per unit volume resulting from adsorption must be increased from n_a to $(1+\beta)n_a$, where β represents the probability of a polymer molecule being adsorbed on the surfaces of two particles simultaneously. If δ is the extension of the adsorbed molecule from the surface and H_o is the surface to surface separation of the particles then:

$$\beta = f\left(\frac{\delta}{H_o}\right)$$

and $\quad 0 < \beta \; ; \; \beta < 1$

when $\quad 1 > \frac{\delta}{H_o} \; ; \; \frac{\delta}{H_o} > 1$

Equation (24) may now be written as

$$G_T = 2AkT\left[Bc + (1+\beta)n_a\right](1 - f'_L) + 2ARc^2 + \frac{0.7949}{H_o+2a}\frac{\partial^2 V_T}{\partial H_o^2} \qquad (30a)$$

Although it is clear that, in order to derive an expression for β, a distribution function for the polymer at the interface is required, the upper and lower bounds for the shear modulus can be calculated by taking $\beta = 1$ and 0 respectively in equation (30a). It should be noted that the value of f'_L must also be affected by β so that:

$$f'_L = \frac{W'_{SL}\,(1+\beta) + W_L}{W'_{SL}(1+\beta) + W_L + \frac{BcN_i}{N_i+2} + n_a(1+\beta)} \qquad \ldots(30b)$$

i.e. crosslinking sites can become unavailable for network links by being adsorbed as surface loops on both particles.

As a first approximation to the value of β, it was assumed that β could be described by a step function so that bridging occurs as soon as the surface of an incoming particle approaches a reference particle to a distance which is equal to, or less than, the extension of an adsorbed molecule from the surface. Fig.2 gives the geometry for the model and, for the face centred cubic arrangement, the incoming area

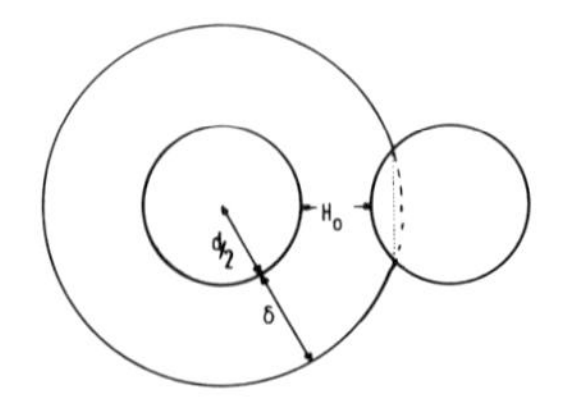

Fig. 2

The geometry for the calculation of β

available for bridging adsorption is given by 12 A_c, where A_c if the area of the cap of the incoming spheres which are within bridging distance. The ratio of the incoming available area to the surface area of the reference sphere with its adsorbed layer (A_s) gives the fraction of molecules on the reference sphere that are bridging i.e. $\beta = 12\ A_c/A_s$.
From Fig.2 it can be seen that :

$$A_c = \frac{\pi\ 2a\ (\delta - H_o)(2a + \delta + H_o)}{2(2a + H_o)}$$

and as $H_o = 2a\left[\left(\frac{0.74}{\phi}\right)^{1/3} - 1\right]$ and assuming that $\delta = \langle r^2\rangle^{\frac{1}{2}}_{25}$, that is the polymer molecules are not significantly perturbed by adsorption:

$$\beta = \frac{6\left[\left(1 + \frac{\langle r^2\rangle^{\frac{1}{2}}_{25}}{2a} - \left(\frac{0.74}{\phi}\right)^{1/3}\right)\left(\frac{\langle r^2\rangle^{\frac{1}{2}}_{25}}{2a} + \left(\frac{0.74}{\phi}\right)^{1/3}\right)\right]}{\left(\frac{0.74}{\phi}\right)^{1/3}\left(1 + \frac{\langle r^2\rangle^{\frac{1}{2}}_{25}}{a}\right)^2} \quad \ldots (31)$$

with the constraint that $0 < \beta < 1$. Equation (30) may now be written as

$$G_T = 2AkT\left[Bc + \frac{\phi}{1-\phi}\Gamma^2 D(1+\beta)\right]\left|1 - \left[1\Big/1 + \frac{\frac{B}{(N_i-2)}cN_i + \frac{\phi}{1-\phi}\Gamma^2 D(1+\beta)}{\frac{\phi}{1-\phi}\Gamma(1+\beta)E + F}\right]\right| + 2ARc^2 + G_E \quad (32)$$

where $B = \frac{N_{Av}(N_i - 2)}{200\ Mn}$, $D = \frac{\pi\ d^3\ \rho^2\ N_{Av}}{6\ Mn^2}$,

$$E = \frac{X\ \rho\ N_{Av}}{Mn}\left(\frac{6}{c_\infty\ n'\ L^2\ \pi}\right)^{\frac{1}{2}}\left[1 - \exp - \left(\frac{3\langle r_N^2\rangle_{25}}{2\ c_\infty\ n'\ L^2}\right)\right],$$

$$F = 2\left(1 - \frac{1}{N_i}\right)\left(\frac{3}{2\ c_\infty\ n'\ L^2\ \pi}\right)^{3/2} \text{ and } n' = \frac{Mn}{N_i\ M_b}.$$

III. EXPERIMENTAL

A. Apparatus

A pulse 'shearometer' was built from the design of H.van Olphen and W.L. Roever(24). The apparatus consisted of two parallel discs mounted in a cylindrical chamber with a 4.7 cm internal diameter . One disc was mounted in the top of the chamber, consisting of a close fitting cylinder to minimise evaporation which could be raised or lowered by means of a screw to vary the gap between the discs. Each disc was 2.5 cm in diameter and was connected to a piezo-electric crystal by a 1.5 mm rod. Silicone rubber sealant was used to isolate the crystals from the dispersions or solutions under investigation. This was essential as the crystals, which were taken from Ronette DSC 400 SCM audio pick-up cartridges, were water soluble. The crystals fitted in to a V-shaped grove at the end of the rod on which the disc was mounted. Intimate contact was maintained between the rod and the crystal by means of the rubber seal. The shearometer cell is shown in Fig.3.

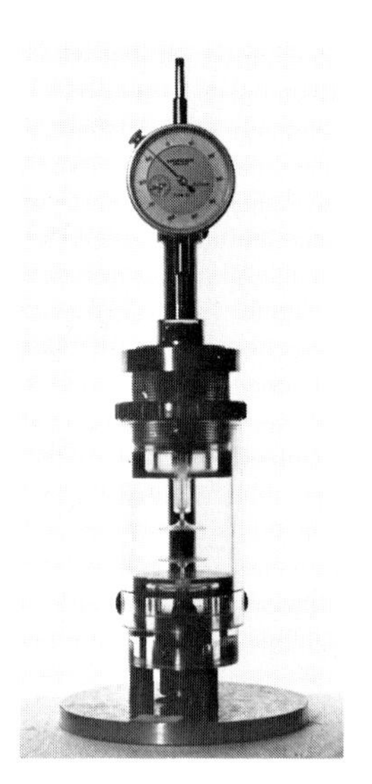

Fig. 3 The pulse shearometer

The lower disc was chosen to be the transmitter and hence the upper disc was the receiver. A pulse generator applied a single electrical pulse to the transmitter disc which produced a rotational displacement of the disc and simultaneously triggered a twin-channel oscilloscope (type SM 113, S.E. Laboratories Ltd., with a maximum sensitivity of 0.002 volts cm^{-1}). The maximum displacement of the disc was 8×10^{-5} rad. A 10^3 Hz oscillator was used to display millisecond timing marks on the second channel of the oscilloscope. The first channel displayed the receiver response, which gave a linear trace with a damped sine wave at the downstream end when the material in the cell showed an elastic response. The trace was photographed and the propagation time of the shear wave was found by measuring the distance from the start of the trace to the first peak of the sine wave using a projection micrometer (Projectorscope, Precision Grinding Ltd.) Calibration of each photograph was carried out by measuring the distance between the timing marks. This procedure was repeated for as wide a range of disc separations as possible for each sample and the velocity of the wave was found from the slope of plots of disc separation against propagation time. This procedure eliminated any uncertainty in determing the position at which the discs first made contact. It was found that the frequency of the output wave was about 300 Hz. For the very small strains the shear modulus (G) can be found from the relation(25):

$$G = \rho_m \nu^2 \qquad \text{... (33)}$$

where ρ_m is the density of the material and ν is the propagation velocity of the wave. A viscous fluid such as water, produced no response of the receiving crystal and a value of $G = 30\ Nm^{-2}$ is the lower limit for the apparatus at present.

B. Materials

The polystyrene latex was prepared by emulsion polymerisation at c.a. 20% solids using potassium persulphate as the initiator and sodium dodecyl sulphate as the emulsifier. A reaction temperature of 70° was used. Ethyl hydroxy ethyl cellulose EX600 (E.H.E.C.) was supplied by Mo Och Domsjö A.B., Sweden and was used as supplied. The sodium chloride, sodium hydroxide, and potassium persulphate were 'Analar' grade materials supplied by British Drug Houses Ltd., but the potassium persulphate was recrystallised before use. The sodium dodecyl sulphate was synthesised in the laboratory from dodecanol (Fluka Ltd.) and was liquid/liquid extracted, using an aqueous ethanol solution and petroleum ether, until no minimum was found in a plot of the surface tension of aqueous solutions as a function of concentration. Except where explicitly stated, the distilled water was distilled twice with the second distillation carried out in an all glass apparatus.

Measurement of Shear Moduli

Polystyrene Latex

After preparation the latex was extensively dialysed (about fifty changes of dialysate) against distilled water, adjusted to pH 9 with sodium hydroxide, to remove electrolyte, surface active agent and any free monomer.

The 'Visking' dialysis tubing was boiled several times before use to remove contaminants. At the end of the dialysis the latex was concentrated by pressure ultra-filtration. The dialysis tube was placed in a flat bottom fused quartz cylinder (5 cm in diameter). A small volume of distilled water (at pH 9) was added and a capped quartz bottle containing 4 kg of mercury was placed on top of the dialysis tube. The initial addition of distilled water prevented any drying out of the latex in the early stages, whilst the 1 to 2 mm gap between the bottle and the cylinder allowed the filtrate to escape. Fused quartz was used to minimise ionic contamination. In principle, this technique is that described by

Hachisu et al(26). The number average particle diameter was found to be 57.1 nm from electron micrographs obtained with an Hitachi HU11B electron microscope and using a Carl Zeiss TGZ3 particle size analyser. A specific surface area of 83.9 $m^2 g^{-1}$ was calculated from the particle size distribution.

The stock latex was in the form of a gel and two samples were taken and sodium chloride solutions were added to give final aqueous phase electrolyte concentrations of 1 x 10^{-3} mol dm^{-3} and 5 x 10^{-3} mol dm^{-3}. More dilute systems were then prepared by dilution with the appropriate sodium chloride solution and the polymer concentration of each solution was determined by drying samples to constant weight in an oven at 80°. The shear moduli of these dispersions were measured using the pulse shearometer. All the latex samples examined were in the volume fraction range of $0.16 < \phi < 0.28$.

Ethyl Hydroxy Ethyl Cellulose

A stock solution E.H.E.C., EX600 at a concentration of 3% w/w was made up by the addition of distilled water to the material at room temperature with shaking. The solution was stored at 4° to complete dissolution. At the same time the moisture content of the E.H.E.C. sample was found by drying a sample to constant weight in an oven at 105°. The shear modulus of the stock solution was measured and that of lower concentrations after dilution of the stock sample. The range of concentrations examined was 2.86% w/w to 1.43% w/w.

Latex-Ethyl Hydroxy Ethyl Cellulose Gels

Latex - E.H.E.C. dispersions were made up by weight at electrolyte concentrations in the aqueous phase of 1 x 10^{-3} mol dm^{-3} and 5 x 10^{-3} mol dm^{-3}. Perks(1) has measured the adsorption isotherm of E.H.E.C. on a similar latex but with a specific surface area of 25.9 $m^2 g^{-1}$. The amount of E.H.E.C. required for each solution was calculated from this adsorption data to give solutions corresponding to the plateau value of the adsorption isotherm and one-third of this coverage. For the full coverage condition, 154.5 mg of E.H.E.C. per g of polystyrene was required at an equilibrium aqueous phase concentration of 0.025 g/100 g and for the one-third coverage solutions 51.5 mg/g at an equilibrium concentration of 1 x 10^{-4} g/100 g. The dependence of the shear moduli of these four systems on their volume fraction was measured. In all cases the volume fractions were in the range $0.09 < \phi < 0.26$.

IV. RESULTS AND DISCUSSION

Latices

The experimental values of the shear modulus of both the polystyrene latices are shown in Fig.4 as a function of volume fraction. Fig.4a gives the results at a sodium chloride concentration of 5 x 10^{-5} mol dm^{-3} and Fig.4b those at 1 10^{-3} mol dm^{-3}. The full curves are the values calculated from equation (8) taking the surface potential as 50 mV. Both the theoretical and experimental curves showed a rapid increase in shear

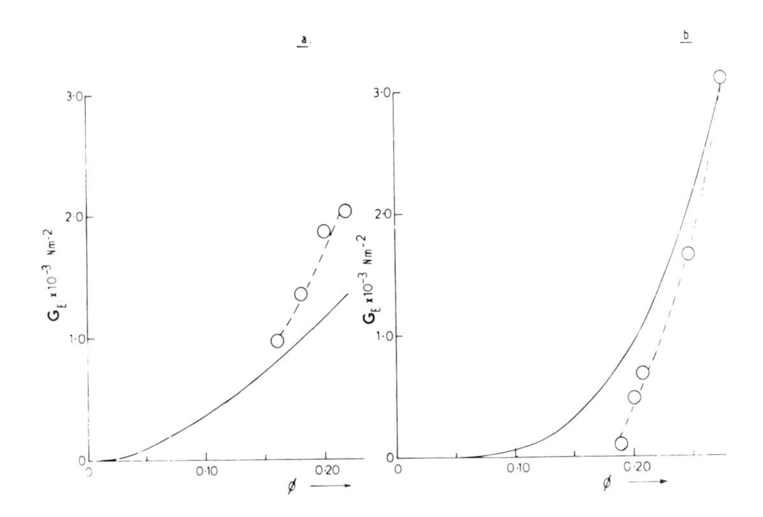

Fig. 4 Experimental data (broken curve) and calculated values of shear modulus as a function of volume fraction for polystyrene latex at (a) 5 x 10^{-5} mol dm^{-3}, and (b) 1 x 10^{-3} mol dm^{-3} sodium chloride concentrations

modulus as the volume fraction increased and the agreement between the two was, at worst, within a factor of 2.

E.H.E.C. Solutions

The variation of the experimental values of the shear modulus of E.H.E.C. solutions with polymer concentration are shown in Fig.5. The best fit to the data, using equation

(19) was given using a value of N_i = 22 cross-linking sites per chain. This is shown plotted through the data in the figure and the network sites were assumed to be tetra-functional, i.e., A = ½. A summary of the characterisation

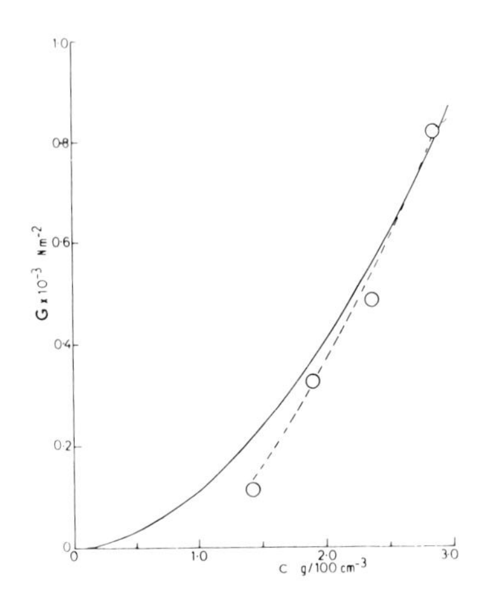

Fig. 5 Shear modulus as a function of polymer concentration for ethyl hydroxy ethyl cellulose. Experimental points and curve calculated from equation (19) with N_i = 22

constants used is given below with the source references:

molecular weight: $\overline{M}v = 1.83 \times 10^5$, $[\eta] = 6.45$ ($\overline{M}n/\overline{M}v = 1$) (1)

mol.wt. of repeat unit: $\overline{M}r = 234$ (degree of polymerisation = 782) - manufacturer

mol.wt. of skeletal bond: $\overline{M}_b = \overline{M}r/3 = 78$ using Benoit's model (12)

mean skeletal bond length: L = 0.187 nm using Benoit's model (12)

repeat unit length: X = 0.505 nm (1)

$\langle r^2\rangle^{\frac{1}{2}}_{OF} = (\overline{M}n)^{\frac{1}{2}}.0.0545$ nm (freely jointed chain) (27)

$\langle r^2\rangle^{\frac{1}{2}}_{\theta} = (\overline{M}n)^{\frac{1}{2}}.0.13$ nm (at the θ-point) (27)

K and a = 3.98×10^{-4} and 0.8 (28)

$\Phi_\theta = 2.6 \times 10^{21}$. (23)

The derived constants are:

$c_\infty = 6.38_3$, $[\eta]_\theta = 2.44$, $\langle r^2\rangle^{\frac{1}{2}}_{25} = 76.86$ nm and

$\langle r^2_N\rangle_{25} = \overline{M}n/N_i \;.\; 0.03228.$

Latex-E.H.E.C. Mixtures

The experimental data for the four latex-EHEC mixtures are plotted in Figs.6 and 7. Using the best fit value of N_i obtained from the data on EHEC solutions, curves were calculated from equation (32) and are plotted as the continuous curves on Figs.6 and 7.

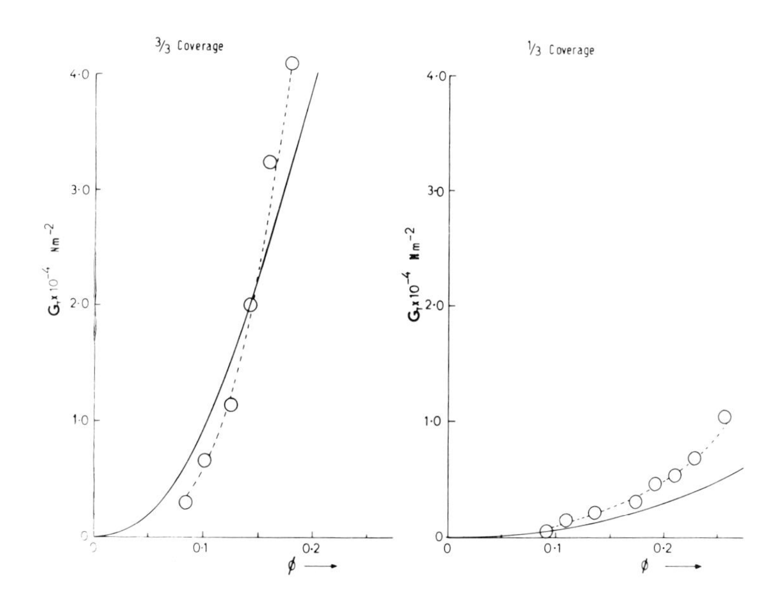

Fig. 6 Latex-EHEC mixtures. G_T as a function of ϕ at 5 x 10^{-5} mol dm^{-3} sodium chloride. Full curves calculated from equation (32); broken curve through experimental points

The value of the front factor, A, was again taken as $\frac{1}{2}$, although as the latex particles provide N-functional cross-linking sites, a value of A = 1 could well be justified. This would of course improve the fit of the calculated curves to the data for the low coverage cases. However, the general trends are correctly predicted by the model as is also the very large increase in the modulus of the mixtures (between one and two orders of magnitude) when compared with that of either of the components. So it is felt that agreement to better than a factor of 2 is at least satisfactory especially in view of the limited amount of experimental data available. An investigation of the effect particle size and molecular weight of the soluble component is currently being carried

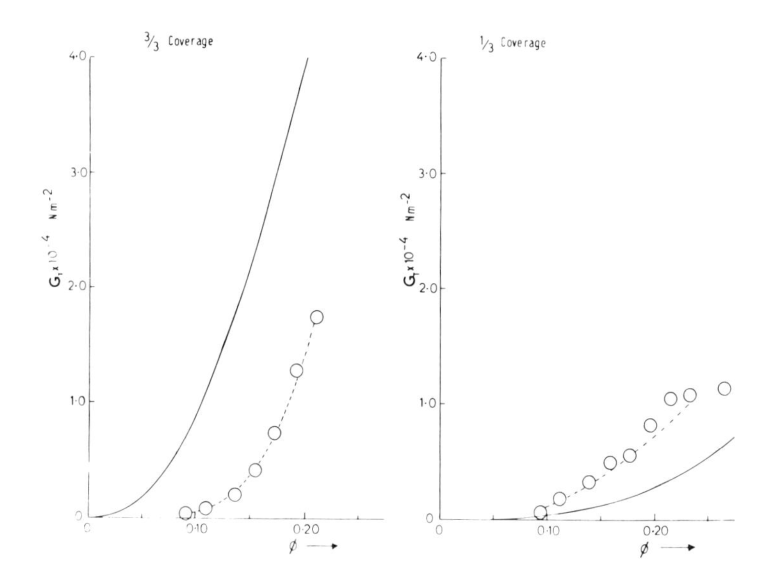

Fig. 7 Latex-EHEC mixtures. G_T as a function of ϕ at 1×10^{-3} mol dm^{-3} sodium chloride. Full curves calculated from equation (32); broken curve through experimental points

out and should provide a more detailed test of the model.

In conclusion, the model described by equation (32) appears to predict the large changes in elastic modulus on the addition of a gelling agent to a particulate dispersion. Also implied in equation (30) is the effect of adding a well dispersed particulate system to a polymer gel and this may have some applications in the study of particular types of composite materials.

REFERENCES

1. Perks, K.B., Ph.D. Thesis, University of Bristol 1971
2. Ho. C.C., Ph.D. Thesis, University of Bristol 1972
3. Fryling, C.F., J.Coll.Sci., 18, 713 (1963)
4. Brodnyan, J.G. and Kelley, E.L., J.Coll.Sci., 19, 488 (1964)
5. Brodnyan, J.G. and Kelley, E.L., J.Coll.Sci., 20, 7 (1965)
6. Smith, R.W., M.Sc. Thesis, University of Bristol 1973
7. Goodwin, J.W. and Smith, R.W., 'Gels and Gelling Processes'

Faraday Disc.Chem.Soc., 57, 126 (1974)
8. Macdonald, R.L.M., M.Sc. Thesis, University of Bristol 1974
9. Hiltner, P.A. and Krieger, I.M., J.Phys.Chem., 73, 2386 (1969)
10. Derjaguin, B.V. and Landau, L., Acta physicochim., 14, 633 (1941)
11. Verwey, E.J.W. and Overbeek, J.Th.G., 'Theory of the Stability of Lyophobic Colloids', Elsevier, Amsterdam, 1948
12. Flory, P.J., 'Principles of Polymer Chemistry', Cornell University Press, Ithaca, N.Y., 1953
13. James, H.M. and Guth, E., J.Chem.Phys., 15, 669 (1947)
14. Edwards, S.F. and Freed, K.F., J.Phys.Chem., 3, 739 (1970)
15. Graessley, W.W., Macromolecules, 8, 186 (1975)
16. Graessley, W.W., Macromolecules, 8, 865 (1975)
17. Eichinger, B.E., Macromolecules, 5, 496 (1972)
18. Dusek, K. and Prins, W., Adv.Polym.Sci., 6, 1 (1969)
19. Allen, G., Holmes, P.A. and Walsh, D.J., 'Gels and Gelling Processes', Faraday Disc.Chem.Soc., 57, 19 (1974)
20. Walsh, D.J., Allen, G. and Ballard, G., Polymer, 15, 366 (1974)
21. Jacobson, H. and Stockmayer, W.H., J.Chem.Phys., 18, 1600 (1930)
22. Kuhn, W., Kolloid Z., 68, 2 (1934)
23. Flory, P.J., 'Statistical Mechanics of Chain Molecules', Interscience, N.Y., 1969
24. van Olphen, H., Clays and Clay Min., 4, 204 (1956)
25. Love, A.E.H., 'Treatise on the Mathematical Theory of Elasticity', Dover, New York, 1944
26. Hachisu,S., Kobayashi, Y. and Kose, A., J.Coll.Int.Sci., 42, 342 (1973)
27. Polymer Handbook, ed. Brandrup, J. and Immergut, E.H., Interscience, N.Y., 1965
28. Manley, R.St.J., Arkiv fir Kemi, 9, 44 (1956)

HYDRODYNAMIC CHROMATOGRAPHY OF LATEX PARTICLES

A. J. McHugh, C. Silebi,
G. W. Poehlein, and J. W. Vanderhoff
Emulsion Polymers Institute
Lehigh University

ABSTRACT

Hydrodynamic chromatography (HDC) is a method developed by Small (1) for the measurement of the particle sizes of colloidal sols. The diluted colloidal sol is injected into an eluant stream, which is pumped through a series of beds packed with solid spheres. The larger colloidal particles pass through the beds faster than the smaller particles, and the separation achieved is detected turbidimetrically.

This paper describes a mathematical model of the flow dynamics of the colloidal particles within the packed beds. This model is based on the analysis of flow of colloidal particles through a collection of equal-size parallel capillaries which are connected in series by mixing regions. The model takes into account the disturbance of the eluant flow profile by the particles and the forces (electrostatic repulsion and London-van der Waals attraction) between the colloidal particles and the bed packing.

The predictions of the model are in excellent agreement with data for monodisperse polystyrene latexes reported by Small (1) and obtained in our laboratory, except for large particles and eluants of low ionic strength. The reasons for these discrepancies are outlined.

I. INTRODUCTION

Recently, Small (1) described a chromatographic method for fractionating a colloidal sol according to the size of its particles and demonstrated the resolution of this method using different-size monodisperse polystyrene latexes. The colloidal sol was diluted and injected into an eluant stream, which was pumped through a series of beds packed with nonporous, uniform-size spheres (e.g., styrene-divinylbenzene copolymer spheres of 20 μm diameter). The rate of transport of the colloidal particles was found to depend upon the sizes of the packing spheres and the colloidal particles. For a given size of packing spheres, the colloidal particles were found to pass through the bed at higher average transport rates than the eluant (the eluant transport rates were determined by adding an ionic marker species to the colloidal sol), with the rates increasing with increasing particle size. When a mixture of two different-size monodisperse latexes was injected into the eluant stream, the larger particles exited first. Since the packing spheres were non-porous, the separation was attributed solely to particle-velocity gradient interactions in the interstices between the packing spheres. Hence, this separation method was referred to as hydrodynamic chromatography or HDC (1).

An earlier theoretical analysis by DiMarzio and Guttman (2) applied the HDC concept to the fractionation of flexible macromolecules. Since this analysis concerned macromolecules in organic solvent solution, ionic effects were not taken into account. In addition, this flow model did not account for disturbances in the eluant flow profile caused by the presence of the polymer molecules --- the so-called "wall effect" (3). Small, however, found that the ionic strength of the eluant had a strong effect on the separation, the resolution increasing with decreasing ionic strength (1).

Recently, we applied the theoretical analysis of DiMarzio and Guttman to HDC (4), to explain quantitatively the experimental separations of mixtures of monodisperse polystyrene latexes observed first by Small (1) and later in this laboratory (4). In addition, the particle-size distributions of the monodisperse latexes determined by electron microscopy were correlated with the breadth of the experimental chromatograms (4). The present paper describes a more complete analysis of the colloidal particle flow dynamics, which explicitly accounts for the ionic effects and presents a mechanism of the flow-separation process; moreover, the relative merits of HDC and liquid exclusion chromatography (LEC) for separation of colloidal particles are discussed.

II. DESCRIPTION OF HDC

The HDC instrument used in our laboratory is essentially the same as that used by Small (1) and is described in detail elsewhere (4, 5). Briefly, the instrument comprises a series of three 9-mm-diameter by 110-cm-long glass columns packed with 20μm-diameter styrene-divinylbenzene copolymer beads. The eluant is pumped through the column continuously, and the latex sample (diluted with eluant to ~0.01% polymer and containing a small amount of sodium dichromate marker) is injected at the top of the first column by a sample injector value. The exit stream from the third column passes through the flow cell of a photometric detector, where its turbidity (254 nm) is measured to detect the latex particles and the marker.

Small experimented with different sizes and types of column packing materials, including styrene-divinylbenzene beads, ion exchange resins, and glass spheres, and studied samples of monodisperse polystyrene latexes of 88 to 1099 nm diameter in eluants of different ionic strength (1). Thus far, the work in this laboratory has used 20μm-diameter styrene-divinylbenzene copolymer beads for the column packing[1] to separate polystyrene, polyvinyl chloride, and other latex types in various eluants. The theoretical analysis to be discussed was applied to the results for monodisperse polystyrene latexes obtained with the styrene-divinylbenzene copolymer bead packing by Small (1) and to similar results obtained in this laboratory.

III. FLOW-SEPARATION TECHNIQUE

Small (1) defined the rate of transport of colloidal particles through the HDC columns in terms of a flow factor R_f:

R_f = rate of particle transport through the bed/rate of eluant transport through the bed (1)

Small also determined the dependence of R_f on the bed packing and the ionic strength of the eluant, which will be discussed later. In the present analysis, it is more convenient to describe the rate of transport of particles through the bed in terms of the elution volume Δv between the peaks on the recorder chart due to marker absorbance and particle turbidity (4):

$$\Delta v = q\Delta\theta \tag{2}$$

where q is the eluant volumetric flow rate, $\Delta\theta = \theta_m - \theta_p$,

---

[1]Kindly supplied by H. Small, The Dow Chemical Company, Midland, Michigan 48640.

and θ_m and θ_p are the mean column residence times for the marker and the particles, respectively, i.e., the time increment between the marker and particle peaks. This Δv can be related to R_f by:

$$\Delta v = V_c \left[1 - (1/R_f)\right] \tag{3}$$

where the effective column volume $V_c = q/\theta_m$. Since θ_m varies with the ionic strength of the eluant (1, 5), the theoretical analysis must include a correction for the effect of ionic strength on the flow of the marker species.

Figures 1 and 2 show the variation of log particle diameter D_p of monodisperse polystyrene latexes with Δv as a function of ionic strength of the eluant from our results and Small's results, respectively. The following features are of particular interest:

(i) The variation of log D_p with Δv is linear, with the slopes varying only slightly with ionic strength of the eluant; the single curved line in Figure 2 coincides with the conditions in Figure 6 of Reference 1 where the R_f-D_p curves

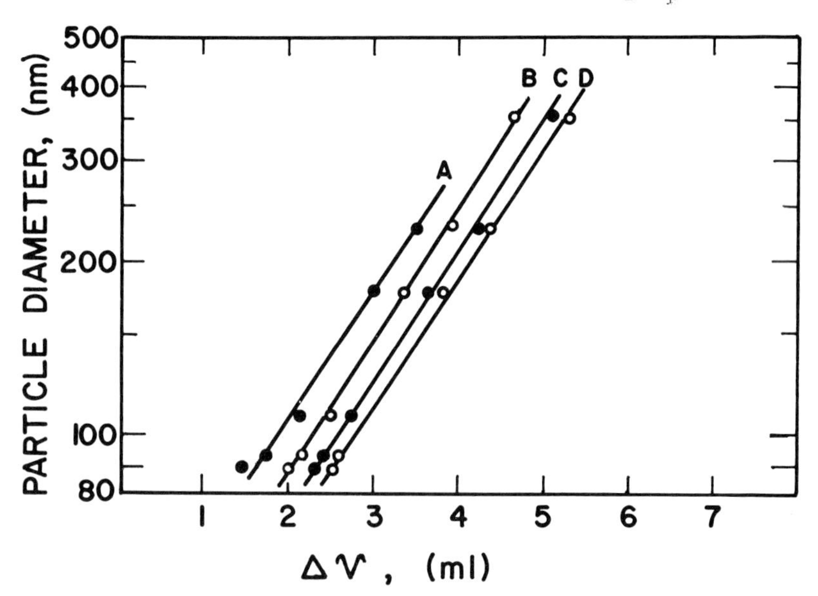

Fig. 1. Particle Diameter - Δv Data from Lehigh Instrument -- Eluant Emulsifier: Aerosol MA; Na_2HPO_4 Normality: A = 2.96×10^{-2}, B = 7.0×10^{-3}, C = 3.0×10^{-3}, D = 8.0×10^{-4}.

begin to show bending; moreover, the linear log $\underline{D}_p$- $\Delta \underline{v}$ relation is similar to the linear relation between the log polymer molecular weight (or intrinsic viscosity times molecular weight) and eluant volume often observed in gel permeation chromatography (GPC) (6);

(ii) At each level of eluant ionic strength, there is a maximum particle size that can pass through the bed, indicated by the last data points of the log $\underline{D}_p$- $\Delta \underline{v}$ plots. Small (1) showed that this maximum particle size also depends upon the size of the packing, i.e., larger packing sizes allowed larger particles to pass through the column; however, in all cases, the maximum particle size that passed through the column was considerably smaller than the minimum pore diameter for the most compact packing arrangement (i.e., 0.155 times the diameter of the packing spheres, or about 3μm for 20μm-diameter packing, for the rhombohedral arrangement), indicating possible adsorption effects (1, 4).

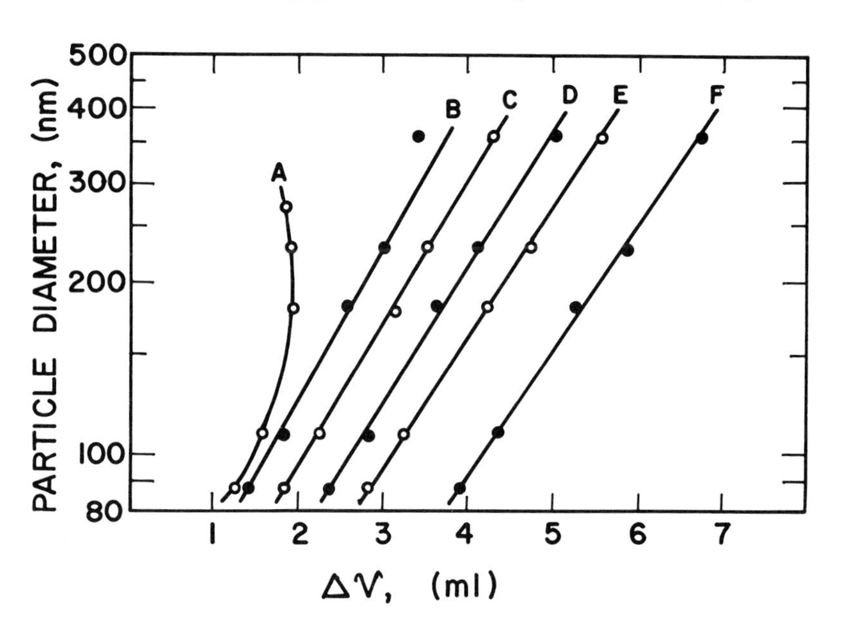

Fig. 2. Particle Diameter - Δv Data from Small's Instrument -- Eluant Emulsifier: SLS; NaCL Normality: $A = 1.76 \times 10^{-1}$, $B = 9 \times 10^{-2}$, $C = 2.96 \times 10^{-2}$, $D = 4.6 \times 10^{-3}$, $E = 1.7 \times 10^{-3}$, $F = 4.25 \times 10^{-4}$.

If one assumes, as a first approximation, that the HDC packed bed can be modeled by a collection of equal-size parallel capillaries connected in series with mixing regions, then a simplified analysis of the hydrodynamic factors is possible (2), and the problem becomes one of calculating the velocity of the colloidal particles in the eluant flowing through such a capillary. Qualitatively, the separation according to particle size is due to two effects:

(i) the void volume available to the smaller particles is greater than that available to the larger particles;

(ii) the larger particles cannot approach the capillary wall as closely as the smaller particles and therefore they sample fluid streamlines of higher velocity, moving at speeds greater than the average eluant velocity.

Let us also assume that the Brownian motion of the particles is sufficient for them to sample all radial positions up to the distance of closest approach to the capillary wall during their axial trajectory through the capillary. Therefore, the analysis can be carried out in terms of average flow rates since the capillary model (2) predicts that:

$$\Delta v/V_c = 1 - (\langle v_m \rangle / \langle v_p \rangle) \quad (4)$$

where $\langle v_p \rangle$ is the average velocity of the particles in the capillary and $\langle v_m \rangle$ the average velocity of the marker. For the volumetric flow rates used in HDC, the tube Reynolds numbers are very much smaller than unity; therefore, the eluant flow profile in the capillaries is laminar. The particle velocity for laminar flow in a tube is given by:

$$v_p = v_o\,[1 - (r^2/R_o^2)] - \gamma\, v_o\, (R_p/R_o)^2 \quad (5)$$

where R_p is the particle radius, R_o the duct radius, and γ a numerical factor to account for the retardation disturbance effect (3). In Equation 5, v_o is twice the average eluant velocity. In general, γ is a function of radial position, with a value of 2/3 along the centerline (7, 8) and increasing to values as high as 10-20 along the wall, depending upon the relative particle and duct radii. The average particle velocity is the result of radial Brownian excursions up to the distance of closest approach to the wall a; therefore:

$$\langle v_p \rangle = \int_0^{R_o - a} v_p r \, dr \Big/ \int_0^{R_o - a} r \, dr \qquad (6)$$

The distance of closest approach normally exceeds the particle radius R_p because of the distortion of the electrical double layer due to flow and the electrostatic double-layer repulsion between the charged latex particles and packing spheres.

The first electroviscous effect introduces a factor k which is related to the intrinsic viscosity by:

$$k = \lim_{\phi \to 0} (\eta_{sp}/\phi)/2.5 \qquad (7)$$

where ϕ is the particle volume fraction and the constant 2.5 is from the Einstein equation for hard spheres. The value of k is unity for hard, non-interacting spheres.

For the second electroviscous effect, let us assume that, on the average, the particles approach the capillary wall only to the point where the double-layer repulsion between the particles and the wall is balanced by the London-van der Waals attraction. Figure 3 shows graphically that the particles will be repelled from the wall at distances smaller than a, but will be attracted to the wall at distances greater than a. (Actually, the particles have sufficient thermal energy to oscillate closer to the wall during Brownian fluctuations; however, on the average, the particle trajectory should behave as if it approaches the capillary wall only up to the distance a). Therefore, at the separation distance a:

$$(\partial\phi_{dl}/\partial a) = (\partial\phi_{vw}/\partial a) \qquad (8)$$

where ϕ_{dl} is the double-layer interaction potential between the tube wall and the colloidal particles and ϕ_{vw} is the London-van der Waals attractive energy. The relative sizes of the latex particles and packing spheres are such that their interaction potentials can be represented by those between a sphere (latex particle) and a plane (packing sphere), which is given by (9):

$$\phi_{dl} = 16\varepsilon R_p (kT/e)^2 \tanh(e\psi_{s1}/4kT) \tanh(e\psi_{s2}/4kT) exp(-\kappa a) \quad (a)$$

$$(9)$$

$$\phi_{vw} = (A/6)[\ln\{(a + 2R_p)/a\} - (2R_p/a)\{(a + R_p)/(a + 2R_p)\}] \quad (b)$$

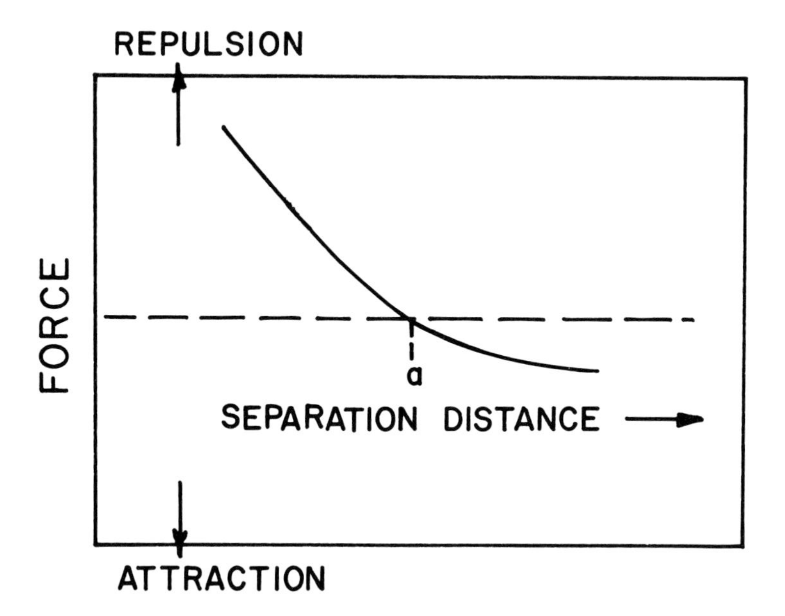

Fig. 3. Generalized force-distance relationship illustrating distance of closest approach concept used in flow calculations

where ε is the dielectric constant of the fluid, e the electronic charge, ψ_{si} the surface electrostatic potential (the subscript i is 1 for the particle and 2 for the packing), κ the double layer thickness, and A the Hamaker constant.

Equation 9b applies when $\kappa a > 2$ and $\kappa R_p >> 1$ (9). Substitution into Equation 6 yields:

$$\frac{\Delta V}{V_c} = \frac{2(R_1 + R_2 - R_3) - (R_1 + R_2)^2 + R_3^2 - 2\bar{\gamma}k^{2/3}R_1^2}{1 + 2(R_1 + R_2) - (R_1 + R_2)^2 - 2\bar{\gamma}k^{2/3}R_1^2} \quad (10)$$

where $R_1 = R_p/R_o$, $R_2 = a/R_o$, and $R_3 = a^*/R_o$. Equation 10 takes into account the fact that, at low eluant ionic strengths, the marker species ($Cr_2O_7^{-2}$) also exhibits a distance of closest approach to the tube wall a^*. The analysis is simplified by using an average value of γ ($\bar{\gamma}$). An examination of Small's data (1) was made using the approach outlined above. The factor k has negligible influence under most conditions and therefore was taken as unity. The appropriate value of R_o was obtained from Small's work by extrapolating the data for the highest eluant ionic strength and smallest particle size on a linear log D_p-Δv plot where the linear slope at $D_p \approx 0$ is proportional to R_o^{-1} (4). Small's data, analyzed in this manner, gives a value of $R_o = 4.05\mu m$. This value is in good agreement with an equivalent packed-bed $R_{o,b}$ calculated from a hydraulic radius formula. The $R_{o,b}$ value for a column packed with 20μm-diameter spheres is computed from the following relation given by Bird et al. (10):

$$R_{o,b} = 2R_h = (D/3)\,[\varepsilon/(1 - \varepsilon)] = 3.72\mu m \qquad (11)$$

where D is packing diameter (20μm) and ε the bed void fraction (0.358). (Similar agreement has been obtained for our columns (4)). Values of a* were obtained empirically at each eluant ionic strength by computing θ_m from equation (10) with $\bar{\gamma} = R_2 = 0$ and comparing these values to Small's values of θ_m, assuming a* = 0 at the highest ionic strength (.176N). Table 1 summarizes these results. Figure 4 compares

TABLE 1

Parameters Used in the Analysis of Small's Data

NaCl Conc. (Normality)	*Range of a (nm)*	*a* (nm)*	$\bar{\gamma}$ *(dimensionless)*
4.25×10^{-4}	135 - 170	56	2.50
1.70×10^{-3}	64 - 72	28	3.50
4.60×10^{-3}	34 - 42	8.5	4.00
2.96×10^{-2}	10 - 17	7.5	4.25
9.00×10^{-2}	4 - 7	4.0	4.50
1.76×10^{-1}	0	0	

Hamaker Constant = *(4 - 8) x 10^{-14}erg*
Surface Potential of Latex Particles = *25 - 90 mv*
Surface Potential of Packing Beads = *10 - 25 mv*

calculated values (solid curves) and experimental points for the Δv-D_p relationship. The agreement for the lower eluant ionic strengths is very good, but could be made more exact by adjusting the values of $\bar{\gamma}$ and a. Even with these approximate values, however, the calculated curves fit the experimental points well.

The values of a in Table 1 are average values for the range of particle diameters used in the calculations. Compu-

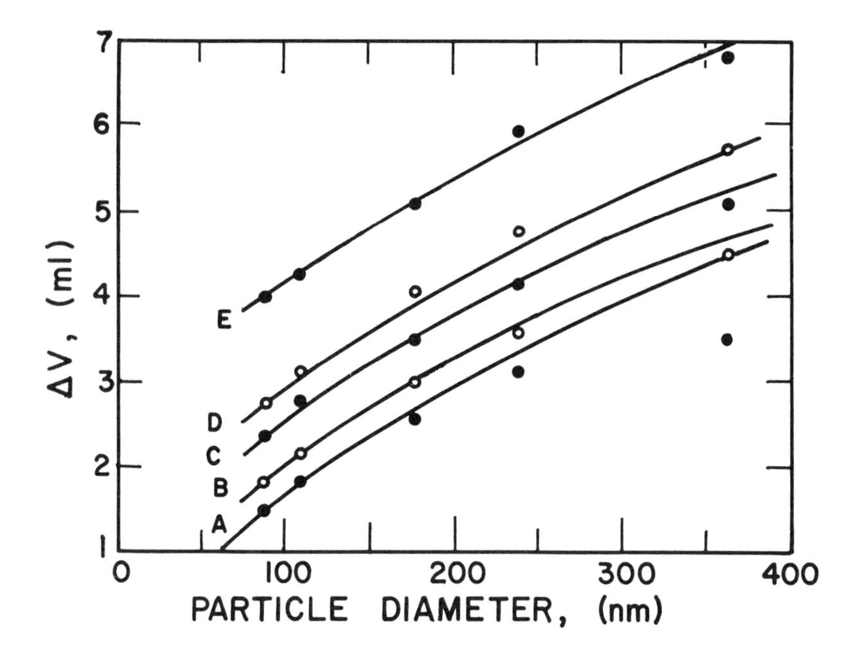

Fig. 4. Linear Plot of Δv *vs* D_p *Data of Small -- Solid Lines Calculated for Theory and Parameters in Table 1; NaCl Normality:* $A = 9 \times 10^{-2}$, $B = 2.96 \times 10^{-2}$, $C = 4.6 \times 10^{-3}$, $D = 1.7 \times 10^{-3}$, $E = 4.25 \times 10^{-4}$

tations carried out over the parameter value ranges given in Table 1 resulted in only small changes in the value of a, with the exception of the lowest eluant ionic strength data, where the double-layer thickness is comparable to the particle size, as indicated in Table 2. Furthermore, the use of smaller values of the surface potential of the packing spheres and colloidal particles (values more closely associated with the von Smoluchowski zeta potential (14)) yields values for the primary minimum in the order of 1 kT for the 357 nm-diameter particles at the highest ionic strength (0.176 N). These small maximum values indicate the possibility of irreversible adsorption in the primary minimum and, possibly, a qualitative explanation of the disappearance of the largest particles in the column at the highest eluant ionic strengths.

It is important to note that the Δv-D_p curves for the lower eluant ionic strengths do not extrapolate to zero since the interaction potential expression given by Equation 9a does not hold rigorously, hence the $-\phi/kT$ values given in Table 2 would be inaccurate. It is expected that a more accurate expression for the double layer potential would give curves that converge on zero. From calculations of small spherical particles, the distance of closest approach is estimated to decrease further from the values computed using Equation 9a. Figure 3 shows that the two highest D_p values at 0.09 N NaCl and the highest D_p value at 0.029 N NaCl deviate significantly from the calculated flow curves. As Table 2 indicates, these points fall in the region of secondary minimum energies close to 6kT, the value generally considered as critical for the onset of a loose, reversible adsorption of the colloidal

TABLE 2

Double Layer Thickness and Secondary Minimum Values Computed from the Average Parameter Values of Table 1.

NaCl Conc. (Normality)	*Double Layer Thickness* (κ^{-1}), *(nm)*	$R_p \times \kappa$ *(dimensionless)*	*Secondary Min.* $(-\phi/kT)$
4.25×10^{-4}	14.95	3.0	.001
"	"	3.7	.002
"	"	6.0	.016
"	"	7.9	.030
"	"	12.9	.074
1.70×10^{-3}	7.47	6.0	.02
"	"	7.5	.03
"	"	12.0	.09
"	"	15.8	.14
"	"	25.8	.31
4.60×10^{-3}	4.54	9.7	.07
"	"	12.1	.11
"	"	19.4	.26
"	"	25.8	.37
"	"	39.0	.72
2.96×10^{-2}	1.81	24.3	.46
"	"	30.4	.65
"	"	48.6	1.25
"	"	64.6	1.82
"	"	97.4	3.05
9.0×10^{-2}	1.03	42.7	1.20
"	"	53.4	1.64
"	"	85.4	2.93
"	"	113.4	4.13
"	"	171.1	6.65
1.76×10^{-1}	0.75	60.5	8.45
"	"	75.6	12.10
"	"	120.8	19.90
"	"	158.7	27.10
"	"	259.0	43.20

particles on the packing spheres (11). We believe for this reason that the flow model breaks down at this point because it does not include the effects of adsorption-desorption on the particle flow. Figure 2 indicates that adsorption increases with increasing particle size, causing a reversal of the particle-residence time behavior. In all cases, adsorption in the primary minimum is unlikely, since our calculations show that the primary maximum exceeds 20kT, the value usually considered to give stability (11).

IV. APPLICATION OF HDC AND LEC TO LATEXES

The foregoing analysis is currently being developed further and applied to HDC data for various latexes. Our calculations for the flow curves in the HDC regime lead to the conclusion that, in the absence of adsorption effects, the separation-flow process can be described accurately by the approach taken in this paper. The dispersion analysis in Reference 4 to compute particle population ratios for bimodal systems shows that the modified Taylor dispersion analysis of DiMarzio and Guttman (2) gives reasonably accurate predictions from first principles. Further work on the measurement of adsorption rates is being carried out to support the conclusions regarding adsorption made on the basis of the flow model. In our opinion, HDC can be modeled to a good first approximation and therefore can be applied as a tool for measuring colloidal particle sizes. The drawback to HDC as a practical tool is exemplified by Figure 5, which shows qualitatively the flow-separation behavior differences between HDC (non-porous packing) and LEC (porous packing). This curve has been sketched qualitatively from data reported by Krebs and Wunderlich (12) and illustrates dramatically the great improvement in resolution which appears to be possible with LEC (resolution is essentially inversely proportional to the slope of the D_p-elution volume flow curve (13)). The dotted lines show qualitatively the improvement in HDC resolution which could be achieved by decreasing the packing size. The data of Krebs and Wunderlich (12) predicts at least a tenfold improvement in resolution in the LEC region, possibly eliminating the need for sophisticated dispersion analysis for size distribution. The drawback of LEC is that it poses a far more complicated hydrodynamics problem, essentially that of GPC, for which no known models based on first principles are available.

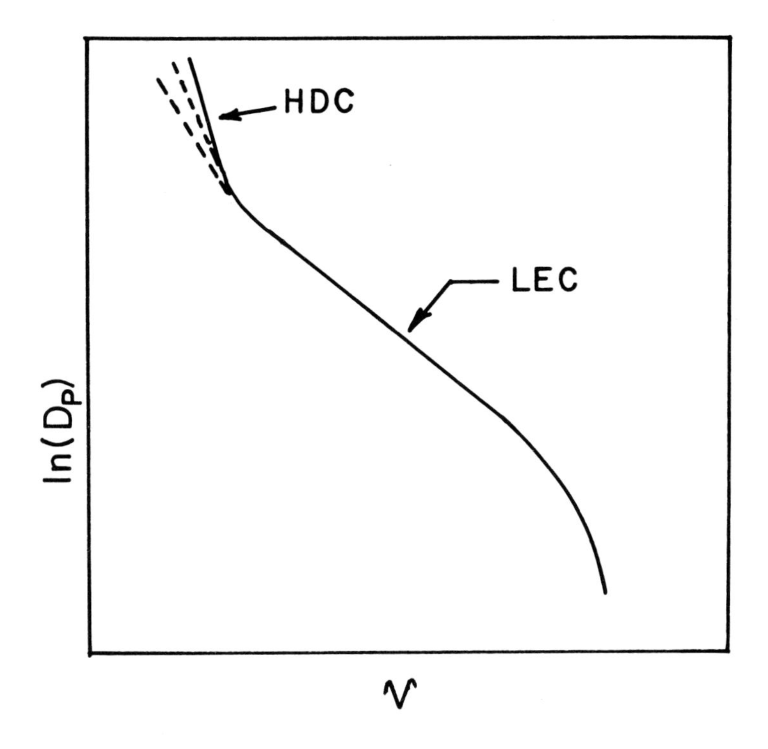

Fig. 5. ln D_p vs elution volume. Generalized plot from this work and reference (12) to show relative ranges of HDC and LEC.

V. ACKNOWLEDGMENTS

The experimental data reported in Figure 1 were obtained by Mr. Joseph Scolere. This work was supported by the International Institute of Synthetic Rubber Producers, Inc. and by the Emulsion Polymers Liaison Program at Lehigh University.

VI. REFERENCES

1. Small, H., J. Colloid Interface Sci. 48, 147 (1974).
2. DiMarzio, E. A., and Guttman, C. M., J. Polymer Sci. Part B 7, 267 (1969); ibid., Macromolecules 3, 131 (1970); Guttman, C. M., and DiMarzio, E. A., ibid. 3, 681 (1970).
3. Brenner, H., Chem. Eng. Sci. 16, 242 (1961).

4. Stoisits, R. F., Poehlein, G. W., and Vanderhoff, J. W., J. Colloid Interface Sci., in press.
5. Stoisits, R. F., M.S. Thesis, Lehigh University, 1975.
6. Altgelt, K. H., Advances in Chromatography 7, 1 (1970).
7. Galman, A. J., Cox, R. G., and Brenner, H., Chem. Eng. Sci. 22, 653 (1967).
8. Goldsmith, H. L., and Mason, S. G., J. Colloid Sci. 17, 448 (1962).
9. Ruckenstein, E., and Prieve, D. C., AIChE J. 22, 276 (1976).
10. Bird, R. B., Stewart, W. E., and Lightfoot, E. N., "Transport Phenomena," John Wiley & Sons, New York, 1960.
11. Verwey, E. J. W., and Overbeek, J. Th. G., "Theory of the Stability of Lyophobic Colloids," Elsevier, Amsterdam, 1948.
12. Krebs, V. K. F., and Wunderlich, W., Angew. Makromol. Chem. 20, 203 (1971).
13. Friis, N., and Hamielic, A. E., Advances in Chromatography 13, 41 (1975).
14. Von Smoluckowski, M., Bull. Acad. Sci., 183 (1903).

COLLOID CHEMICAL ASPECTS OF DRILLING FLUID RHEOLOGY

William C. Browning
SCT Associates

I. ABSTRACT

This paper presents a capsulized account of the microrheology of aqueous oil well drilling fluids from the viewpoint of modern colloid chemistry. The relationship of hydrodynamics and interfacial interactions of concentrated clay suspensions to rheological properties is given. The significance of various colloid stability mechanisms is discussed and related to drilling fluid chemical treatment. The specific importance of viscous layer stabilization as exemplified by lignosulfonates is discussed. Emerging trends in drilling fluid technology are cited.

II. INTRODUCTION

The industrial colloid chemist, in dealing with drilling fluids, must work with impure, heterodisperse, concentrated, disperse systems of various pH values, highly contaminated with electrolytes of various kinds up to and including saturated salt water. In the field these complex systems must be chemically and physically controlled in a practical manner on a day to day basis.

In oil well drilling the success or failure of a multimillion dollar operation may depend upon engineering decisions that involve or are dependent upon colloid chemistry. The prevention of undesirable reactions of drilling fluids upon earth formations exposed in the borehole, the control of dispersion media flow into porous formations, the avoidance of damage to oil producing sands and the control of the rheological properties of the circulating drilling fluid, are all highly dependent upon knowledgeable control of colloidal phenomena. The colloid chemical aspects of drilling fluids offer abundant opportunity for continuing investigation. Fortunately a hundred years of colloid science may be drawn upon to gain perspective and solve problems.

Current drilling fluid engineering is predominately concerned with macrorheology as it applies to drilling hydraulics, particularly as it relates to laminar and turbulent flow in the drill pipe or in annuli.

The fundamentals of macrorheology of Newtonian and non-Newtonian fluids have been reviewed in some detail by Metzner (1) and in the literature of drilling fluid technology.

In recent years, with the growing use of computerized techniques and more sophisticated hydraulic calculations, the multiphase, colloidal drilling fluid disperse system is increasingly regarded as a single phase fluid. Single phase flow concepts facilitate hydraulics calculations but obscure the fact that the rheological properties of the drilling fluid are but the manifestations of the colloid chemistry and interfacial-interactions of a disperse system.

This dichotomy is not a peculiarity of drilling fluid technology. Physical rheologists tend toward slider-spring-dashpot explanations of flow behavior. Chemists concerned with colloid stability infrequently relate the state of disperse systems to rheological properties.

This paper will treat only selected items of colloid interest that are rheology related and will consider only disperse systems in which water is the continuous phase.

The laboratory derived rheological evidence cited in this paper was for the most part obtained by approved procedures of the American Petroleum Institute (2).

Drilling rig instrumentation that provides data regarding pump pressures, drill pipe torque, etc., make the drilling rig a practical rheometer. In many instances the instrumentation and equipment of the rig and the drilling operation itself is the only reliable means of assessing drilling fluid rheology under the temperature and pressure conditions of downhole use.

There is current interest in eliminating the use of clay based drilling fluids. Because of problems incident to the use of bentonite, clay solids are increasingly being removed by mechanical means and rheological control is being secured by the use of processed and synthetic polymers.

However, because clay suspensions are the classic drilling fluids -- the drilling muds -- and will continue to be widely used for some years to come, it is meaningful to direct some attention to clay based fluids. The newer "low solids" and shear thinning polymer fluids can only be briefly referred to because of space limitation. For the same reason significant colloid chemical aspects of disperse system rheology are omitted or drastically reduced in treatment.

Finally, it should be emphasized that to systematize the rheology of suspensions, all suitable experimental techniques should be co-ordinated. Data derived from x-ray diffraction, scanning electron microscope, infra-red, chromatography, etc., are of basic value. In addition to the essential variable shear rheometer -- such as the Fann V-G Meter (3) an ultramicroscope assembly is needed to directly observe the "living" disperse system. It is unfortunate that today the "Colloidiscope" of an older generation of colloid chemists is frequently neglected.

III. HYDRODYNAMICS

The much cited Einstein equation (4) relating the hydrodynamic effect of suspended particles to suspension viscosity is fundamental to drilling fluid rheology. By this equation specific viscosity is shown to increase with increase in effective volume of the suspensoid in the suspension but independently of particle size.

Thus, under the conditions stipulated by Einstein a suspension exhibits Newtonian flow and the slope of the stress-strain curve increases in direct proportion to the volume ratio of the dispersed solids.

In the real world it is well known that particle interaction and adsorption are functions of the particle size of the dispersed phase. The various "assumptions" of ideality made by Einstein in setting up the conditions for his mathematical derivation are, however of direct practical value. By directing attention to the items specified in the "assumptions" the various factors affecting the resistance to shear of the suspension may be individually evaluated, their effect on suspension rheology indicated and the resulting trends in flow properties projected.

Equation correction to real world conditions may be suggested by addition of appropriate terms and more than half a hundred of such "corrections" have been published. Mathematical exercises to formalize these "corrections" or fix precise numerical values are of no practical engineering consequence. The flow property trends projected from consideration of the Einstein equation, however, are of considerable value in analyzing the colloidal aspects of suspension rheology.

The shape and configuration of suspended particles have their well-known effect upon suspension flow properties (5). Rod-like and plate-like particles have a greater effective hydrodynamic value per unit weight of dry substance and also have a greater propensity for steric interference than do spheres. Hence, minerals such as attapulgite, sepiolite or chrysotile are used to build rheological structure in drilling muds, particularly in brine dispersion media.

Organic polymers that swell (imbibe dispersion media) or have a long chain deformable structure when in suspension, also create hydrodynamic and steric resistance to shear far in excess of their true density spherical equivalents. Hence polymers such as starch or sodium carboxymethyl cellulose, used primarily for filtration rate control, may occasionally increase the resistance to shear of a drilling fluid to an undesirable extent.

IV. COLLOID STABILITY EFFECTS

Colloid stability is determined by the interaction between particles during a Brownian motion encounter. When attraction predominates, the particles will adhere and the dispersion will flocculate. Flocculation of the suspended particles is the sticking together of the particles in the form of loose and irregular clusters in which the original particles can still be recognized.

A suspension is considered to be in a state of colloid stability if the particle interaction is one in which repulsion predominates and the dispersed particles remain as discrete entities and in Brownian motion.

Suspensions of finely divided calcium carbonate, zinc oxide, iron oxide or kaolin clay that are in a state of maximum colloid stability, all exhibit Newtonian flow.

Although most of the published work concerning colloid stability has been concerned with dilute suspension phenomena, the relation of the balance of attraction and repulsive forces to rheological properties such as yield value and thixotropy has been noted (6). Drilling fluid engineering, in common with many industrial processes, involves concentrated suspensions. It is therefore of general significance that theories of colloid stability derived originally from experimentation with dilute suspensions may be appropriately applied to concentrated suspensions such as drilling fluids.

A. Attractive Forces

The possibility that the London-van der Waals forces could be responsible for attraction between suspended particles was outlined as a quantitative theory based upon equilibria between London forces of attraction and Coulomb forces of repulsion by Kallmann and Willstatter. Potential energy curves were calculated by Rubin and published in part by Freundlich. Subsequently, the mathematical treatment of the potential energy equilibria was refined by DeBoer and Hamaker. The possibility that the attractive forces might not be strictly additive but modified by passage through a dense medium has been treated by Lifshitz and others. For the purposes of drilling fluid rheology, the refinements of retarded force theory are not of practical consequence (7).

The van der Waals attractive force, although diminishing at an exponential rate with distance has been shown to radiate for distances of 1000 Å or more from a particle surface. This force is electromagnetic in nature and like gravitation is not characterized by "positive" or "negative" as are electrostatic or valence forces. The van der Waals long range attractive force is not specific. All suitably sized (8) dispersed particles will flocculate unless opposed by some repulsive mechanism, regardless of whether the particles are clays, carbon black or plant pollen grains.

It is basic and important to the understanding of disperse system rheology to visualize that the London-van der Waals attractive forces radiate in three dimensions from each suspended particle and that this sphere of electromagnetic attraction can act within or through electrostatic force fields or layers of adsorbed molecules. This concept is implicit in the well known energy potential diagrams which illustrate the co-existence of attractive and repulsive forces.

It is also implicit in the energy potential diagrams that due to the rapid decay of the radiating attractive force field any interface mechanism that interposes a physical barrier resisting the close approach of two colloidal particles will modify the kinetics of flocculation in dilute suspensions and the shear stress---rate of shear relationship in concentrated suspensions.

The algebraic sum of attractive and repulsive forces as given in the energy potential diagrams may be adequate for calculation purposes, but for interpretation and chemical control of drilling fluid rheology the co-existence and simultaneous presence of attractive and repulsive three dimensional force fields must be emphasized.

For example, disperse system colloid stability is achieved if any repulsive barrier at the surface of dispersed particles extends in space to a greater distance than does the radiating attractive force. If the repulsive barrier extends in space to a lesser distance than does the radiating attractive force, then two particles approaching each other in Brownian motion will be attracted to each other but will be prevented from close approach by the presence of the repulsive barrier. Thus varying residual forces of interparticle attraction will be present depending upon the relative spatial extent of the radiating attractive force and the opposing repulsive barrier. If the spatial extent of the repulsive barrier is such that it is nearly that of the radiating attractive force field the attractive force residual will be relatively weak and the suspension will exhibit some appropriate degree of thixotropy. This condition is represented by the secondary minimum of the conventional energy potential diagram.

Flocculation of dispersed particles will always increase the resistance to shear of a given suspension. Reduction of yield value and decreased resistance to shear is the attribute of deflocculation in suspensions of dispersed solids.

The frequently mentioned edge-to-face flocculation of

clays (9) occurs only at pH values less than 7.0. Inasmuch as drilling fluids are universally maintained at pH values of 9.5 or more, the theoretical edge-to-face house-of-cards flocculate structure does not appear to be relevant to field drilling muds.

In this connection it may be noted that most earth minerals have a positive charge in acidic media and a negative charge in alkaline media. The iso-electric point given for quartz is 1.5 pH; for clay, 5.6 pH and for alumina, 9.4 pH.

B. Induced Flocculation

Clay suspensions may be flocculated by treatment with certain long chain synthetic macromolecules by a bridging mechanism in which two or more clay particles are bound together by adsorption to one long chain molecule (10). By this mechanism negatively charged dispersed particles can be flocculated by negatively charged polyelectrolytes. This flocculation process is entirely different from traditional flocculation by simple electrolytes and does not invoke electrostatic repulsion or long range attraction. Flocculation by the bridging mechanism does not invalidate the DLVO theory, it is merely a different and specific flocculation mechanism.

Certain polyelectrolytes such as hydrolyzed polyacryamide are used in drilling technology as bridging flocculants for drilled clays as in "clear water" drilling. In some low solids mud systems vinyl acetate-maleic anhydride copolymers are used to remove drilled clay solids. The bridging mechanism apparently functions more effectively between larger clay aggregates than between colloidal bentonite particles and thus facilitates a kind of particle size fractionation.

C. Repulsive Forces

Of the various mechanisms that stabilize dispersed particles and prevent flocculation, the electric double layer (11) has received the most attention and has dominated colloid stability thinking in industry and academia.

The zeta potential concept and electrokinetic effects applicable to colloidal dispersions have been reviewed by Sennet and Oliver (12).

Traditionally it has been taught in drilling fluid practice that broken valence bonds occuring at the edges of clay platelets are the attractive forces causing flocculation (13). Adsorption of sodium compounds such as polyphosphates is said to satisfy these broken valence bonds and thereby achieves deflocculation. It is still common to refer to "nullifying the electrical charges on clays" when reduction in yield point or apparent viscosity is desired.

Sodium ions being monovalent are thought to be unable to link two clay particles together. The calcium ion, being divalent, is regarded as being able to "bind" two clay particles together and cause flocculation.

The sequestering of calcium ions is sometimes thought to be the mechanism of deflocculant action. This calcium sequestration hypothesis continues to be advanced as the mechanism whereby sodium tannates and lignosulfonates are able to achieve deflocculation in lime or calcium treated muds (14).

Clearly colloid stability concepts involving London-van der Waals attractive forces are not yet universal in drilling fluid field practice.

The general characteristics of the interplay of the electrical double layer and the London-van der Waals forces are the basis of the classical theory of the flocculation of lyophobic dispersions, first presented by Derjaguin and Landau (15) and independently by Verwey and Overbeek (16) --- the "DLVO" theory.

It is well known that double layer stabilization of lyophobic systems is very sensitive to electrolyte contamination. The early drilling fluid use of condensed phosphates as "thinners" (deflocculants) was beset by problems due to contamination by anhydrite, halite and portland cement. In the advent of calcium contamination the obvious remedy was to "treat out" the calcium by sodium carbonate or bicarbonate. These chemicals, if used copiously, could themselves cause adverse effects due to sodium ion contamination in accord with the principles of the DLVO theory.

Deflocculation by sodium hydroxide treated quebracho (sodium tannate) and sodium hydroxide treated lignite (sodium humate) has traditionally been interpreted in terms of double layer effects. These so-called "fresh water" muds were empirically used and treated by the principles of the DLVO theory.

The logical extensions of the DLVO theory provide the general theoretical basis for colloid stabilization by means other than electrical double layer repulsion. Derjaguin and Overbeek have both explicity mentioned other stabilization mechanisms.

Recognized "barrier" mechanisms that act to prevent flocculation include ordered water, steric stabilization, structurized layers and viscous layer stabilization. These barriers do not necessarily involve any active mutual repulsion as does the electrical double layer. The barrier mechanisms simply interpose a physical resistance to the close approach of two particles in Brownian motion. If particles are prevented from approaching within the "critical distance" of van der Waals attraction, flocculation does not occur.

1. Ordered water. The early concept of a non-electrical, thick shell of adsorbed water as a colloid stability mechanism

was discredited by developments in diffuse double layer theory. In recent years evidence has accumulated establishing the presence of water structures as a stabilizing mechanism (17).

Ordered water at the solid-liquid interface increases the effective hydrodynamic volume of the dispersed solid. Therefore, in accord with the Einstein relation, surface hydration will increase the specific viscosity of a suspension over that calculated from true density values of the dispersed phase.

Ordered water structures are dissipated by increasing temperature unless stabilized by adsorbed surfactants. Agents such as ethoxylated nonylphenol apparently act in this manner to decrease high temperature gelation in some drilling muds.

2. Steric stabilization. Various aspects of this stabilization mechanism have been recently reviewed by Napper (18). Steric stabilization is created by long chain hydrophilic-hydrophobic molecules (19). This mechanism, sometimes referred to as "entropy stabilization", is not generally effective for drilling fluids. It finds application, perhaps, in the use of the so-called "mud detergents".

3. Structuromechanical barriers. Establishing maximum colloid stability in concentrated suspensions such as drilling fluids or other industrial applications, requires stabilization mechanisms more rugged and less sensitive than the electrical double layer. Rehbinder has pointed out that such stabilization requires the application of some sort of protective colloid, which is identified as a high molecular weight substance which will form a rather strong spatial structure of the gel type by adsorption on the surface of the dispersed phase (20).

Rehbinder considers materials that establish colloid stability in this manner to have two specific characteristics, viz., adsorbability at various interfaces, and the ability to form spatial structures of the gel type either in the bulk of the dispersion medium or on the surface of the disperse phase. Rehbinder emphasizes that this stabilization ranges from structuralization of the adsorption layer alone to structuralization of the entire volume of the dispersion medium (21).

The adsorbed protective colloid is considered to be so solvated and hydrophilic as to not cause mutual attraction between the protected dispersed particles. In advent the protective colloid is rendered lyophobic by precipitation with electrolytes, colloid stability of the suspension is lost.

4. Viscous multilayer adsorption. As noted by Overbeek viscous layer stabilization has been proposed as a distinct barrier stabilization mechanism by Derjaguin (22). This mechanism is not widely recognized but it is quite important in drilling fluid technology. Treatments with lignites, tannin compounds, bark extracts and lignosulfonates frequently involve multilayer adsorption.

Viscous multilayer colloid stabilization and its

application to oil well drilling fluids was discussed by Browning in 1958 (23).

The physical fact of multilayer adsorption has been subjected to doubt and disbelief. Under the Langmuir influence adsorption was considered to be limited to a monolayer. In 1914 Polanyi (24) proposed the existence of long range attractive potentials that could extend from a solid surface several molecular diameters. This force was conceived to be of sufficient strength to condense adsorbed molecules as a multilayer at the solid interface. Such an idea was heretical to the scientific orthodoxy of the time. It is a footnote to history that professor Polanyi was forbidden to teach his theory and was subjected to censures that adversely affected his whole scientific career.

After many years the fact of multilayer adsorption is now accepted and specific experimental verification has been conducted by Derjaguin and others (25). An interesting review of the differing emphasis placed on short-range forces, long-range forces and statistical mechanics has been presented by Ninham and Israelachvili (26).

In viscous layer stabilization the adsorbed multilayer is presumed to have Newtonian flow characteristics instead of the yield-pseudoplasticity of the structurized adsorption layer postulated by Rehbinder.

Substances capable of imparting viscous multilayer colloid stability have been found to be macromolecules of globular shape and moderate molecular weight that exhibit Newtonian flow in solution concentrations of 25% or more. Such substances have viscosities differing little from that of water in 2% to 5% solutions, but at 30% to 40% concentrations have viscosities of hundreds or even thousands of centipoise.

When such macromolecules are adsorbed on a particle surface the multilayer, due to surface condensation will have a viscosity much greater than that of the dispersion medium. The high viscosity at the interface will decrease sharply with distance from the surface, merging finally with that of the dispersion medium. Therefore, two particles thus stabilized, on approaching each other encounter increasing viscous resistance. If the quality of the viscous layer prevents approach within the critical attraction distance in the time interval of Brownian motion, the dispersed particles will remain as deflocculated entities.

Lignosulfonates have proved to be the most effective deflocculating agents to use with electrolyte media drilling fluids. Lignosulfonates being sulfonated polyelectrolytes are capable of deflocculating kaolin and pigments by electrical double layer effects in accord with the DLVO theory. Typical of this mechanism, only small quantities of lignosulfonate are required for maximum deflocculation in the absence of electrolyte contamination. More than optimum quantities of lignosulfonate will decrease the colloid stability by "overtreatment" causing double layer compression (27).

Addition of sodium chloride or other electrolytes also cause the typical destabilizing effect resulting from double layer compression as indicated by zeta potential values. However, for a given amount of electrolyte contamination, increasing amounts of lignosulfonate will again stabilize the suspension even if the determined zeta potential remains negligible. This is not true of electrokinetic stabilization as achieved with polyphosphates. Neither is it in accord with the DLVO theory. It should be well noted that viscous layer stabilization by its nature requires higher levels of stabilizer treatment than double layer stabilization.

It was proposed in 1956 that deflocculation of clays in solutions of electrolytes by lignosulfonates was accomplished by the formation of viscous condensed adsorption layers to create a barrier which prevented the dispersed particles from flocculating due to long range van der Waals attractive forces (28).

The ability of lignosulfonates to form condensed adsorption layers at oil-water interfaces and the application to oil well drilling fluids has been shown (29).

The solution properties of lignosulfonate may therefore be reviewed to illustrate the requirements of viscous multilayer stabilization.

Lignosulfonates are water soluble, sub-colloidal, spheroidal, sulfonated, macromolecular polyelectrolytes, having no oil soluble groups in their molecular structure but capable of strong adsorption on hydrogen bonding surfaces (30).

Lignosulfonate solutions up to 5% solids have viscosities essentially the same as water. At 30% and 40% concentration the viscosity at 30°C may be 150 centipoise or more. The exponential increase in viscosity begins to develop rapidly at about 20% concentration. Lignosulfonate solutions of concentrations of 40% or more behave as Newtonian fluids (31).

The Newtonian behavior of lignosulfonate solutions is believed due to the mutual electrokinetic repulsion of the globular, non-swelling, polyelectrolyte macromolecules. The dissociation of the sulfonic salt groups at the surface of the macromolecule creates an electrostatic repulsion that exceeds the intermolecular van der Waals and hydrogen bonding attractive forces. The greater viscosity of calcium lignosulfonate solutions and the lesser viscosity of sodium lignosulfonates is in accord with electrokinetic theory. The dissociation potential of sodium sulfonate is greater than that of calcium sulfonate, hence the repulsive force is greater for sodium lignosulfonates. The result is less internal friction for sodium lignosulfonate solutions and greater internal friction for calcium lignosulfonate solutions (31).

Contamination of lignosulfonate solutions with a strongly

dissociating electrolyte such as sodium chloride results in an increase in solution viscosity, a result opposite to that obtained with solutions of expanding coil macromolecular polyelectrolytes. This effect would follow from repression of sulfonic group dissociation resulting in decreased electrokinetic repulsion, thus permitting the intermolecular attractive forces to increase the internal friction of the solution.

Lignosulfonate solutions continue to show a linear relation between shear stress and shear rate until increasing electrolyte contamination renders the lignosulfonate insoluble (salts out). With incipient salting out, lignosulfonate solutions begin to show anomalous flow characteristics. The solutions become "structurized" i.e., show yield values and pseudoplasticity because the repulsive forces are so reduced that the intermolecular attractive forces begin to predominate (32).

Concentrated clay suspensions treated with lignosulfonates to achieve maximum colloidal stability (viscous layer) in the presence of electrolytes exhibit Newtonian flow. Such Newtonian suspensions show increasing resistance to shear with increasing additions of lignosulfonate. This increase in the specific viscosity of the deflocculated suspensions is in accord with the previously cited Einstein equation. The increased effective hydrodynamic volume of the dispersed clay is interpreted to be caused by the condensed multilayers of increasing amounts of adsorbed lignosulfonate (27).

Because the high viscosity exists only in the multilayer at the interface, and the dispersion medium remains low in viscosity, concentrated clay suspensions of low voscosity in the presence of electrolyte contamination may be prepared by means of lignosulfonate viscous layer stabilization. Stabilization with classical protective colloids does not permit preparation of low viscosity concentrated clay suspensions. This illustrates a significant difference between structurized layer and viscous layer stabilization.

It has been previously indicated that a correlation exists between the properties of lignosulfonates in concentrated solutions and in the condensed adsorption layers (31).

The question of why the protective adsorbed viscous layers on dispersed solids do not adhere one to another under the influence of van der Waals forces is answered by the rheology of concentrated lignosulfonate solutions. Obviously lignosulfonate macromolecules do not cohere if they exhibit Newtonian flow in concentrated solutions. Consequently, the lignosulfonate viscous protective "barriers" on the dispersed solids do not adhere one to another.

Sodium lignosulfonates are sufficiently soluble to show Newtonian flow in 6% sodium chloride or 1.0% calcium chloride solutions and will deflocculate kaolin clay in such solutions.

When the lignosulfonate solutions become "structurized" due to salting out (indicating that the macromolecules are becoming mutually attracted), then adsorbed layers at the solid interface also become attractive to each other and a previously stable suspension flocculates. Therefore, structurization *per se* is not essential to colloid stability. The rheology of lignosulfonates and other viscous layer stabilizers in electrolyte solutions is indicative of their ability to effect colloid stability in the presence of contaminating electrolytes. The flow properties of concentrated solutions are more informative in this regard than conventional flocculation tests conducted with dilute suspensions.

Humic acid compounds (lignite) or plant tannins (quebracho, etc.,) may create colloidal stability by DLVO concepts at low treatment levels or by viscous layer stabilization at higher levels of treatment as do the lignosulfonates.

Electrokinetic charge effects and solubilization of the humates is governed by carboxylic group dissociation. In the plant tannins, phenolates are the dissociating groups, whereas sulfonate groups are the primary dissociating structures in lignosulfonates. Sulfonic dissociation is notably stronger than carboxylic or phenolate dissociation, consequently, lignites and quebracho salt out at much lower electrolyte concentrations than do the lignosulfonates.

Both tannins and humates are notably less effective in establishing colloid stability in the presence of electrolytes than lignosulfonates. Higher sodium hydroxide content in dispersion media help maintain the necessary dissociation of phenolates and carboxylates and also act to keep the calcium ion content low to minimize the formation of insoluble phenolate and carboxylate calcium salts. Calcium lignosulfonate, however, is a soluble and ionizing lignosulfonic acid salt.

Increasing the drilling mud pH to 11-12 by addition of sodium hydroxide to increase the efficiency of tannins and lignosulfonates is a common expedient in field practice. In similar electrolyte environments increased sodium hydroxide may make lignosulfonates less effective by decreasing their adsorption at the solid-liquid interface.

The colloid stability of a suspension is fundamentally dependent upon adsorption of the stabilizer by the dispersed phase. Generally, increase in the molecular weight of a polymer increases adsorption. However, the solubility of a polymer in electrolyte solutions also decreases with increasing molecular weight. Hence, increasing molecular weight tends to be self defeating with respect to achieving colloid stability in electrolyte solutions.

Fortunately, the development of supplementary adsorption mechanisms can increase the adsorbability of macromolecules without increasing molecular weight or decreasing solubility in solutions of electrolytes. One such approach is the adaptation of the mordanting technique of dyestuff technology (33) (34). Other means of chemical modification of lignosulfonates have been used (35) (36) (37). The rationale of lignosulfonate technology may be applied to synthetic polymers (38).

D. Kinetic Energy Effect

The dispersive effect of kinetic energy input to drilling fluids must not be overlooked. The van der Waals attractive force between colloidal particles can be overcome by the energy of shear applied to a suspension. Flocculated particles may be dis-aggregated into independent entities solely by the kinetic energy input of high shear rates. This effect was beautifully illustrated recently by cinematography (39).

The less the residual attractive force in instances of partial stabilization, the lower will be the rate of shear (energy input) required to dis-aggregate the floccules. Thus all suspensions having some degree of flocculation exhibit shear thinning properties. In drilling fluid field practice it is conventional to assume that in all muds all floccules are dis-aggregated into original primary particle entities by shear rates of 480 sec^{-1}, an assumption that is not always justified.

If a lyophobic suspension is in a state of complete colloidal stability no floccules will be present, no shear energy will be consumed in causing dis-aggregation of flocculate structures and the system will show a linear relation between shear stress and rate-of-shear i.e., exhibit Newtonian flow.

V. CLAY SUSPENSIONS

To correctly interpret the rheological properties of clay suspensions, hydration effects must be considered. Formation clays incorporated into drilling fluids include illite, calcium montmorillonite and kaolinite.

Non-lattice expanding clays such as kaolin and illite adsorb water in ordered structures on their external surfaces (40). Increasing temperature would be expected to dissipate these water structures, incurring the expected rheological results. These results are observed, for example kaolin suspensions show decreasing specific viscosity with increasing temperature (31).

Kaolin and illite formation clays are subject to delamination, dispersion and structural weakening in massive formations by water uptake and cleavage caused by hydroxyl promoted wetting and penetration reactions (41).

Bentonite (smectite) also has surface hydration but the rheological effect is obscured by swelling phenomena.

In evaluating reports concerning the flocculation and dispersion of bentonite, it should be well noted that bentonite forms single silica-gibsite-silica platelets only with special preparation and near infinite dilution. In ordinary water suspensions it has been shown by x-ray diffraction techniques that bentonite is present in particles composed of stacks or packets of crystallites (42).

Water may enter between the silica-to-silica surfaces of the stacked platelets causing the entire packet to expand in accordion fashion. Thus, in water suspension the bentonite particle swells and the effective hydrodynamic volume of a given weight of dry bentonite can be 20 times or more the volume calculated on the basis of dry density.

This swelling mechanism is in accord with electrokinetic theory (43), sodium bentonite exhibiting greater swelling than calcium bentonite. Obviously, as indicated by the previously cited Einstein hydrodynamic relationship, whether or not the dispersed bentonite particles are swollen or non-swollen has a profound effect on the flow properties of the suspension.

A chrome treated lignosulfonate will deflocculate an untreated bentonite suspension as indicated by its rheogram and ultra-microscopy. A second bentonite suspension first treated with chrome lignosulfonate and then with calcium sulfate will have an almost identical rheogram as the first suspension, indicating deflocculation in the presence of calcium sulfate and no volume change in the bentonite particles. If, however, a third sample is prepared in which the bentonite is first treated with calcium sulfate to convert it to low swelling calcium bentonite and then treated with chrome lignosulfonate, the suspension will have a stress-strain rheogram that is Newtonian and have a resistance to shear much less than the two previous suspensions. This is the expected result of deflocculation of a suspension with a dispersed phase of lower hydrodynamic volume (42). Thus depending upon conditions favoring multilayer adsorption and order of addition, lignosulfonate deflocculated bentonite suspensions may by base exchange have the clay packets in unexpanded calcium bentonite form, or may have sodium bentonite packets stabilized in the swollen state protected from the effect of calcium ions by lignosulfonate condensed multilayers.

This dual property of bentonite, flocculation-deflocculation; swelling-deswelling, in its various combinations, causes considerable confusion in the interpretation and control of drilling fluid rheology.

The fact that bentonite in ordinary field drilling muds is present primarily in the form of stacked packets is substantiated by the investigations of Mering (44), Norrish (45), and Browning (31) (42); the pre-hydration and post-hydration effect shown by Rodgers (46), and the current field practice of pre-hydrating bentonite in freash water to maximize its viscosifying effect in sea water.

Nevertheless, bentonite drilling fluids continue to be interpreted in terms of the bentonite being present as single layer platelets that upon addition of electrolyte flocculate and re-aggregate by re-stacking to re-form the primary packet aggregate (47) (48).

It is noteworthy that the experimental evidence cited for this hypothesis may be explained equally well by volume changes induced in bentonite primary aggregate packets.

The assumption of reforming of stacked primary aggregates by flocculation of single layer platelets without redrying is statistically improbable and is not demonstrated by direct observation or direct measurement.

Work with carefully prepared fractionated monodisperse colloidal bentonite sols has failed to reveal by ultramicroscopic observation any house-of-cards structure or any specific alignment of dispersed particles (49).

Direct observations thus corroborate the hypothesis of spheres of attractive force acting within or through the various possible spatial extensions of physical or electrostatic repulsive barriers to produce and account for all types of suspension flow behavior --- Newtonian, thixotropic, plastic or viscoelastic.

VI. POLYMER FLUIDS

Because of the economics of faster drilling, attention has been directed toward obtaining shear thinning fluids with apparent viscosities of no more than 10 centipoise at shear rates of 100,000 sec^{-1} and approximately 50 to 100 centipoise at 100 sec^{-1} (50). Deflocculated clays tend to have Newtonian flow, consequently, a drilling fluid of good hole cleaning characteristics with an apparent viscosity of 100 centipoise at 100 sec^{-1} will have a very undesirable 100 centipoise at 100,000 sec^{-1} (50).

Hydrated bentonite suspensions of approximately 4% solids, flocculated by bridging polymers are shear thinning and can provide desirable rheology for low weight muds in calcium free environments. However, for various reasons including formation damage, it is desirable to eliminate the use of bentonite.

A bacteria modified carbohydrate has been developed that has demonstrated in the field the practicality of providing desirable shear thinning drilling fluids even in sea water media (51).

This biopolymer uniquely differs from conventional pseudoplastic polymers by developing yield stress characteristics when in solution. Using this biopolymer, weighted non-bentonitic drilling fluids may be prepared having lower viscosity at high shear than weighted clay muds (52). In accord with the Einstein equation the barite may be increased until the barite volume effect causes the resistance to shear at high shear rates to reach an undesirable level.

The use of synthetic polymers, particularly acrylic derived, to replace clays in drilling fluids is only emerging and future improvements may be anticipated as knowledge of the colloid chemistry of these systems increases.

VII. SUMMATION

Bentonite is present in drilling fluids not as single layer mineral sheets or platelets but as aggregates composed of stacks of platelets. These aggregates are capable of expansion because of hydration and electrokinetic repulsion between the adjacent silica surfaces of the stacked platelets. The aggregate packets will swell or deswell depending upon the electrolytes present in the dispersion media. The variable volume of the bentonite packets have a profound effect upon the suspension rheology independent of whether the suspension is flocculated or deflocculated. In colloidally stable suspensions the degree of expansion of the packet will affect specific viscosity in the manner predicted by the Einstein equation regarding the hydrodynamic volume of the suspended particles.

Drilling fluid rheology is also dependent upon the colloid stability of the dispersed phase solids. There is no single or exclusive mechanism of colloidal stabilization, there are several. All of these mechanisms involve the electromagnetic long range attraction of the London-van der Waals forces opposed by the so-called repulsive forces.

The London-van der Waals attraction is regarded as a three dimensional radiating sphere of attractive forces that may operate within or through three dimensional protective zones. The protective zone most considered in fresh water muds is the double layer electrostatic repulsive force field of the DLVO theory.

Physical barriers may be created by interface adsorption structures including ordered water, adsorption of long chain molecules (steric stabilization), adsorption of protective colloids (structuromechanical stabilization) and adsorption of spheroidal macromolecular polyelectrolytes (viscous multilayer stabilization). Any or all of these stabilization mechanisms may be utilized in appropriate situations in the conditioning of oil well drilling fluids.

The physical barrier structures do not necessarily result in active mutual repulsion between dispersed particles. The physical barriers act to prevent dispersed particles from approaching each other within the critical zone of attraction for non-elastic collisions in the time interval of Brownian motion.

The conventional evaluation of colloid stability by flocculation of dilute sols is not particularly meaningful with respect to drilling fluids. The rheology of concentrated suspensions is not only more relevant, it is a more sensitive means of evaluating the colloid state of a suspension.

The concept of the co-existence and independent interaction of three dimensionally radiating long range attractive forces acting through the zones of repulsion that oppose particle-to-particle adhesion, does not only explain but facilitates the chemical control of suspension rheology.

The variation in effective attractive force acting upon dispersed particles depends upon the effective thickness of the protective barrier. The greater the spatial extent of the physical barrier the greater will be the separation of the particles during the Brownian motion collision interval. The greater the separation the less will be the residual attraction of the London-van der Waals force because of the exponential decrease of attractive force with increasing distance. Varying attraction between dispersed particles is thus accounted for without invoking the hypothesis of retarded forces.

The above discussed colloid stability concept, with the interpolation of suspension concentration, hydrodynamic volume, Brownian motion and kinetic effects, provides a comprehensive explanation of flow behavior that accounts for suspension rheology as a continuity varying from plastic flow through thixotropy and viscoelasticity to Newtonian flow.

VIII. REFERENCES

1. Metzner, A.B., "Advances in Chemical Engineering," Vol. 1, pp. 78-153. Academic Press Inc., New York, 1956.
2. API Production Department RP 13B, Standard Procedure for Testing Drilling Fluids (1976).
3. Melrose, J.C., and Lilienthal, W.B., J. Petrol. Technol. T.P. 3061, (1951).
4. Einstein, A., Ann. Physik, Vol. 19, 289 (1906); Vol. 34, 591 (1911).
5. Goldsmith, H.L., and Mason, S.G., in "Rheology" (F.R. Eirich, Ed.), Vol. 4, p. 86, Academic Press, New York, 1967.
6. DeWaele, A., and Mardles, E.W.J., in "Proceedings of the International Congress on Rheology" (Scheveningen Holland 1948), II-166. North-Holland Publishing Co., Amsterdam, 1949.
7. Overbeek, J.Th.G., in "Colloid Science" (H.R. Kruyt, Ed.), Vol. I, p. 264. Elsevier Publishing Co., 1952.
8. Buzagh, A. Von, "Colloid Systems", p. 37. London Technical Press LTD., 1937.

9. Marshall, E.C., "The Physical Chemistry and Mineralogy of Soils," Vol. 1, p.325. John Wiley & Sons, Inc., New York, 1964.
10. La Mer, V.K., Jour. Colloid Science, 19, 291 (1964).
11. Devanathan and Tilak, Chem. Rev., 65, 635 (1965).
12. Sennet, P. and Oliver, J.P., in "Chemistry and Physics of Interfaces," American Chemical Society Publications, Washington, D.C.
13. Rodgers, W.F., "Composition and Properties of Oil Well Drilling Fluids," 3rd ed., p. 369. Gulf Publishing Co., Houston, Texas, 1963.
14. van Olphen, H., "An Introduction to Clay Colloid Chemistry," p. 164. Interscience Publishers, New York, 1963.
15. Derajaguin, B.V. and Landau, L.D., Acta Physiocochim. USSR, 14, 633 (1941).
16. Verwey, E.J.W. and Overbeek, J.Th.G., "Theory of the Stability of Lyophobic Colloids," Elsevier, Amsterdam, 1948.
17. Eagland, D. in "Water," (F. Franks, Ed.), Vol. 5, p. 1. Plenum Press, New York, 1975.
18. Napper, D.H., "Steric Stabilization," paper presented at the 50th Colloid & Surface Sci. Symp., June 21-25, 1976. San Juan, Puerto Rico.
19. Pugh, R.L. and Heller, W., Jour. Polymer Science, 47, 219 (1960).
20. Rehbinder, P.A., Colloid Jour. USSR, 20 (1958).
21. Rehbinder, P.A., Colloid Jour. USSR, 23 (1961).
22. Overbeek, J.Th.G., "Recent Developments in the Understanding of Colloid Stability," paper presented at the 50th Colloid & Surface Sci. Symp., June 21-25, 1976. San Juan, Puerto Rico.
23. Browning, W.C., Can. Mining & Metallurgical Bull., November, 709 (1958).
24. Polanyi, M., Science, 141, 1010 (1963).
25. Patat, F. et al., Makromol. Chem., 49, 200 (1961).
26. Ninham, B.W. and Israelachvili, J., "Intermolecular Forces," paper presented at the 50th Colloid & Surface Sci. Symp., June 21-25, 1976. San Juan, Puerto, Rico.
27. Browning, W.C., paper 557-G (1955), 30th Ann. Fall Meeting, Petroleum Branch, AIME, New Orleans, La.
28. Browning, W.C., "The Deflocculation of Clays in Solution of Electrolytes," paper presented at the 5th National Clay Conference, October 8-10, 1956. University of Illinois, Urbana, Illinois.
29. Browning, W.C., Jour. Petrol. Tech., 7, 9 (1955).
30. Browning, W.C., in "Chemistry, Physics and Applications of Surface Active Substances," (F. Asinger, Ed.), Vol.1 Proc. IV Int. Congr. Surf. Act. Subs., Brussels, 7-12 September, 1964. Gordon & Beach Science Publishers, London, pp. 141-154.
31. Browning, W.C., Ind. Eng. Chem., 49, 1401 (1957).

32. Browning, W.C., "Flow Properties of Lignosulfonates from Hardwood and Softwood," paper presented at the Lignin Symposium, 130th National ACS Meeting, September 16-21, 1956. Atlantic City, N.J.
33. Browning, W.C., paper SPE 432 (1962). 37th Ann. Falling of SPE of AIME, Los Angeles, Calif.
34. Browning, W.C., Grt. Brit. Pat. 1,068,686 (1967).
35. Browning, W.C., "Applied Polymer Symposia 28," Proc. 8th Cellulose Conf., SUNY. Vol. 1, 109(1975). John Wiley & Sons, Inc.
36. Browning, W.C. and Chesser, B.G., U.S. Pat. 3,697,498 (1972).
37. King, E.G. and Adolphson, C., U.S. Pats. 2,935,473 (1960) 2,935,504 (1960)
38. Perricone, A.C., et al., U.S. Pat. 3,843,524 (1974).
39. Mason, S.G., "Orthokinetic Phenomena in Disperse Systems," paper presented at 50th Colloid & Surface Sci. Symp., June 21-25, 1976. San Juan, Puerto Rico.
40. Grim, R.E., "Clay Mineralogy," Chap. 8. McGraw-Hill Book Co., New York, 1968.
41. Browning, W.C., Trans. AIME, 231, 1177 (1964).
42. Browning, W.C. and Perricone, A.C., paper SPE 540 (1963). 1st U Texas Conf. on Drilling & Rock Mechan., Austin, T.
43. Foster, M.D., in "Clays and Clay Minerals," (W.O. Milligan, Ed.) p. 205, Publ. 395, Natl. Acad. Sci.-- Natl. Research Council, Washington, D.C., 1955.
44. Mering, J., Trans. Faraday Soc., 42B, 205 (1946).
45. Norrish, K., Trans. Faraday Soc., 18, 120 (1954).
46. Rodgers, W.F., "Composition and Properties of Oil Well Drilling Fluids," 3rd ed., p.356. Gulf Publishing Co., Houston, Texas, 1973.
47. van Olphen, H., "An Introduction to Clay Colloid Chemistry," p. 104. Interscience Publishers, New York, 1973.
48. Rodgers, W.F., "Composition and Properties of Oil Well Drilling Fluids," 3rd ed., p. 361. Gulf Publishing Co., Houston, Texas, 1973.
49. Hauser, E.A. and LeBeau, D.S., Jour. Physical Chem., 42, 961 (1938).
50. Eckel, J.R., Oil & Gas Jour., 69 (June 17, 1968).
51. Deily, F.H. et al., Oil & Gas Jour., 62 (June 26, 1967).
52. Browning, W.C. and Chesser, B.G., Jour. Petrol. Technol., 24, 1255 (1972).

RHEOLOGY OF DILUTE POLYMER SOLUTIONS IN POROUS MEDIA

Richard J. Mannheimer
Southwest Research Institute

ABSTRACT

Dilute solutions of high molecular weight linear polymers are shown to exhibit Newtonian behavior in capillary tube flow but to be highly non-Newtonian in that they offer a very high resistance to flow through different types of porous media. Measurements with solutions in different viscosity solvents showed that the minimum flow rate at which the observations departed significantly from Newtonian behavior was inversely proportional to the product of the solvent viscosity and the intrinsic viscosity. These results are in agreement with recent microrheological considerations that predict that the minimum elongation rate required to produce appreciable uncoiling of flexible macromolecules is also inversely proportional to the solvent viscosity and intrinsic viscosity.

PARTICLE INTERACTION IN COLLOIDAL SOL FLOW

Robert J. Hunter
University of Sydney

ABSTRACT

The behaviour of sols of spherical particles undergoing minimal interaction is well understood. When the interaction is repulsive and double layer effects are important the first and second electroviscous effects must be taken into account. Some limitations of the description of the second effect will be discussed.

When attractive forces exist between the particles at all separations the colloidal forces dominate over the hydrodynamic forces between individual particles and the basic flow unit is a floccule. A satisfactory description of the behaviour of such systems can be developed by considering the energy dissipation processes occurring during flow. Apart from the viscous energy involved in the flow of the liquid around the flocs, the predominant energy dissipation process occurs within the flocs as the particle-particle links are stretched (but not broken) by the hydrodynamic force as it is transmitted to the contact area between two colliding flocs. Rupture of inter-floc bonds involves slightly more energy but since there are far fewer of them, the energy involved is significantly smaller than that dissipated within the floc.

This model satisfactorily describes the dependences of the flow characteristics (Bingham Yield Value, critical shear rate and plastic viscosity) on the colloidal parameters (particle size, interaction energy and solids concentration).

SHEAR THICKENING IN DILUTE SOLUTIONS OF POLYMERS AND COLLOÏDAL DISPERSIONS

Claude Wolff
Université de Bretagne Occidentale

ABSTRACT

A short review of the different types of shear thickening behaviours observed in colloïdal dispersions and polymers solutions is presented, with special emphasis to the case of dilute solutions of high molecular weight flexible coil macromolecules. For the latter, the shear thickening may be explained by a permanent deformation of the spherical coil into an ellipsoïd (coil → stretch transition) above a critical Weissenberg number. Assuming that the effect is maximum at a concentration such as the free volume allowed to the deformed coils rotating randomly around their center is approximatively equal to the volume of the solution, the axial ratio of the ellipsoïd can be calculated. Moreover, calculations on the hydrodynamically dissipated energy show that aggregation ↔ desaggregation processes may also be involved in this phenomenon. The comparison made with colloïdal dispersions shows that, despite the differences between the materials, there is some similarity with the dilute solution case and in the way in which both types of shear thickening may be analysed.

A 6
B 7
C 8
D 9
E 0
F 1
G 2
H 3
I 4
J 5

rned

S